Atmosphere, Weather and Climate

Atmosphere, Weather and Climate

FIFTH EDITION

ROGER G. BARRY AND RICHARD J. CHORLEY

METHUEN
LONDON AND NEW YORK

First published in 1968 by
Methuen & Co. Ltd
11 New Fetter Lane, London EC4P 4EE
Second edition 1971
Third edition 1976
Fourth edition 1982
Fifth edition 1987

Published in the USA by
Methuen & Co.
in association with Methuen, Inc.
29 West 35th Street, New York, NY 10001

Typeset by
Scarborough Typesetting Services
and printed in Great Britain by
Richard Clay (The Chaucer Press) Ltd.,
Bungay, Suffolk.

British Library Cataloguing in Publication Data
Barry, R. G.
Atmosphere, weather and climate. –
5th ed.
1. Meteorology
I. Title II. Chorley, Richard J.
551.5 QC861.2

ISBN 0–416–07142–2
ISBN 0–416–07152–X (Pbk)
(University paperback 208)

Library of Congress Cataloging in Publication Data
Barry, Roger Graham.
Atmosphere, weather, and climate.

Bibliography: p.
Includes index.
1. Meteorology. 2. Atmospheric physics.
3. Climatology. I. Chorley, Richard J.
II. Title.
QC861.2.B36 1987 551.5 87–12337

ISBN 0–416–07142–2
ISBN 0–416–07152–X (pbk.)

Contents

Acknowledgements

This book developed from an original manuscript by R. J. Chorley and A. J. Dunn, and the present authors wish to record their appreciation of Mr Dunn's important contribution to the earlier draft.

The authors are also very much indebted to Dr F. Kenneth Hare of Birkbeck College, London, now at the University of Toronto, Ontario, for his thorough and authoritative criticism of the preliminary text and his valuable suggestions for its improvement; also to Mr Alan Johnson of Barton Peveril School, Eastleigh, Hampshire, for helpful comments on chs. 1–3; and to Dr C. Desmond Walshaw, formerly of the Cavendish Laboratory, Cambridge, and Mr R. H. A. Stewart of the Nautical College, Pangbourne, for offering valuable criticisms and suggestions at an early stage in the preparation of the original manuscript. Gratitude is also expressed to the following persons for their helpful comments with respect to the fourth edition: Dr Brian Knapp of Leighton Park School, Reading; Dr L. F. Musk of the University of Manchester; Dr A. H. Perry of University College, Swansea; Dr R. Reynolds of the University of Reading; and Dr P. Smithson of the University of Sheffield. Dr C. Ramage, of the University of Hawaii, made numerous helpful suggestions on the revision of ch. 6 for the fifth edition. Thanks are also due to Professor R. A. McCance, of Cambridge University, for his daily interest in the problems treated here. The authors accept complete responsibility for any remaining textual errors.

The figures were prepared by the cartographic and photographic staffs in the Geography Departments at Cambridge University (Mr R. Blackmore, Mr R. Coe, Mr I. Gulley, Mrs S. Gutteridge, Miss R. King, Mr C. Lewis, Mrs P. Lucas, Miss G. Seymour, Mr A. Shelley and Mr M. Young); at Southampton University (Mr A. C. Clarke, Miss B. Manning and Mr R. Smith); and at the University of Colorado, Boulder (Mr T. Wiselogel).

Our grateful thanks go to our wives for their constant encouragement and forbearance.

The authors would like to thank the following learned societies, editors, publishers, organizations and individuals for permission to reproduce figures, tables and plates.

Learned societies

American Geographical Society for fig. 1.37 from the *Geographical Review*.
American Geophysical Union for figs. 1.18B, 4.11 and 4.12 from the *Review of Geophysics and Space Physics*; for figs. 1.28A–C and 1.32 from the *Journal of Geophysical Research*; and for fig. 7.6 from the *Transactions*.
American Meteorological Society for fig. 4.21 from the *Bulletin*; for figs. 3.29, 4.8, 7.1 and 8.6 from the *Journal of Applied Meteorology*; for figs. 4.2B and 4.4B from the *Meteorological Monographs*; for figs. 4.33 and 6.35 from the *Journal of Atmospheric Sciences*; and for figs. 2.24, 3.11, 3.24, 3.36, 3.37, 3.39, 4.6B, 4.23, 6.2B, 6.3, 6.12, 6.13, 6.28 and 6.32A from the *Monthly Weather Review*.
American Planning Association for fig. 7.27 from the *Journal*.
American Society for Testing and Materials for fig. 1.1.
Association of American Geographers for fig. 2.33 from the *Annals*; and for fig. 4.38 from *Resource Paper 11*.
European Space Agency, Darmstadt, for pls. 1 and 26.
Geographical Association for fig. 2.2 from *Geography*.
Institute of British Geographers for figs. 2.26, 2.27B, 7.20 and 7.28 from the *Transactions*; and for figs. 2.34 and 8.7 from the *Atlas of Drought in Britain 1975–76* by J. C. Doornkamp and K. J. Gregory (eds.).
Institution of Civil Engineers for fig. 2.27A from the *Proceedings*.
Marine Technology Society, Washington, DC, for fig. 6.34C and D from the *Journal*.
National Geographic Society for pl. 22 from the *National Geographic Picture Atlas of Our Fifty States*.
Royal Meteorological Society for figs. 2.20, 4.10, 5.6, 5.7 and 6.3 from the *Quarterly Journal*; for fig. 8.3 from *World Climate* 8000–0 BC; for figs. 2.28, 5.8, 6.30 and 8.8 from the *Journal of Climatology*; and for figs. 1.16, 2.5, 2.29, 4.29, 5.4, 5.11, 5.26 and 7.21, and for pls. 14, 19 and 20 from *Weather*.
Royal Society of London for fig. 4.26 and pl. 15 from the *Proceedings, Section A*.

Editors

Endeavour for fig. 2.22.
Erdkunde for figs. App. 1.1B and App. 1.2.
Geographical Magazine for fig. 6.29A.
Geographical Reports of Tokyo Metropolitan University for fig. 6.27.
Meteorological Magazine for figs. 3.34 and 7.2
Meteorological Monographs for figs. 4.2B and 4.4B.
Meteorologische Rundschau for figs. 5.27 and 7.9.
New Scientist for figs. 4.24, 4.27 and 4.31A.
Progress in Physical Geography for figs. 7.24 and 8.9.
Science for figs. 7.23C and 8.5.

Tellus for figs. 5.9, 5.10, 6.17 and 6.24.
Zeitschrift für Geomorphologie for fig. 7.5 from *Supplement 21.*

Publishers

Academic Press, New York, for figs. 4.12, 4.30, 6.11 and 6.13 from *Advances in Geophysics*; for fig. 6.16 from *Monsoon Meteorology* by C. S. Ramage; and for fig. 6.29B from *Quaternary Research.*

Allen and Unwin, London, for figs. 1.19 and 1.21B from *Oceanography for Meteorologists* by H. V. Sverdrup.

Cambridge University Press for fig. 2.17 from *Clouds, Rain and Rain-making* by B. J. Mason; for fig. 3.22 from *World Weather and Climate* by D. Riley and L. Spalton; for fig. 6.36 from the *Warm Desert Environment* by A. Goudie and J. Wilkinson; and for fig. 7.18 from *The Tropical Rain Forest* by P. W. Richards.

Cleaver-Hume Press, London, for fig. 3.14 from *Realms of Water* by Ph. H. Kuenen.

The Controller, Her Majesty's Stationery Office (Crown Copyright Reserved) for fig. 2.8 from *Geophysical Memoir 102* by J. K. Bannon and L. P. Steele; for fig. 1.20 from *Meteorological Office Scientific Paper 6, M.O.685* by F. E. Lumb; for fig. 2.6 from *Ministry of Agriculture Technical Bulletin 4* by R. T. Pearl *et al.*; for figs. 3.34, 4.14 and 7.2, and for pl. 23 from the *Meteorological Magazine*; for figs. 4.9 and 4.13 from *A Course in Elementary Meteorology* by D. E. Pedgley; for fig. 4.15 from *British Weather in Maps* by J. A. Taylor and R. A Yates (Macmillan, London); for fig. 4.28 from *Geophysical Memoir 106* by D. E. Pedgley; for figs. 5.24 and 5.25 from *Weather in the Mediterranean 1*, 2nd ed. (1962); and for the tephigram base of fig. 2.10 from *RAF Form 2810.*

J. M. Dent, London, for fig. 5.21 from *Canadian Region* by D. F. Putnam (ed.).

Elsevier, Amsterdam, for fig. 6.39 from *Climates of Australia and New Zealand* by J. Gentilli (ed.); and for fig. 6.28 from *Palaeogeography, Palaeoclimatology, Palaeoecology.*

Harvard University Press, Cambridge, Mass., for figs. 1.21A, 1.26, 7.12, 7.13B and 7.14A from *The Climate near the Ground* (2nd ed.) by R. Geiger.

Houghton Mifflin Company, Boston, for pls. 7.12 and 18 from *A Field Guide to the Atmosphere* by V. T. Schaefer and J. A. Day.

Hutchinson, London, for figs. 7.21A and 7.25 from the *Climate of London* by T. J. Chandler.

Justus Perthes, Gotha, for fig. 2.32 from *Petermanns Geographische Mitteilungen, Jahrgang 95.*

Macmillan, London, for fig. 4.15 from *British Weather in Maps* by J. A. Taylor and R. A. Yates.

McGraw-Hill Book Company, New York, for fig. 2.25 from *Handbook of Meteorology* by F. A. Berry, E. Bollay and N. R. Beers (eds.); for fig. 3.35

from *Dynamical and Physical Meteorology* by G. J. Haltiner and F. L. Martin; for figs. 7.13A and 7.14B from *Forest Influences* by J. Kittredge; for figs. 2.9, 2.21 and 3.23 from *Introduction to Meteorology* by S. Petterssen; for figs. 6.5 and 6.6 from *Tropical Meteorology* by H. Riehl; for figs. 5.17 and 6.21 from *The Earth's Problem Climates* by G. T. Trewartha; and for fig. 1.36 from *Handbook of Geophysics and Space Environments* by Shea L. Valley (ed.).

Methuen, London, for figs. 1.25, 2.31 and 6.31 from *Mountain Weather and Climate* by R. G. Barry; and for figs. 2.1, 3.31 and 3.33 from *Models in Geography* by R. J. Chorley and P. Haggett (eds.).

North-Holland Publishing Company, Amsterdam, for fig. 2.30 from the *Journal of Hydrology.*

Oliver and Boyd, Edinburgh, for fig. 7.11 from *Fundamentals of Forest Biogeocoenology* by V. Sukachev and N. Dylis.

Pitman, London, for fig. 3.17 from *Tropical and Equatorial Meteorology* by M. A. Garbel.

Princeton University Press for figs. 5.21 and 5.22 from *The Moisture Balance* by C. W. Thornthwaite and J. R. Mather; and for fig. App. 1.5 from *Design with Climate* by V. Olgyay.

D. Reidel, Dordrecht, for figs. 3.25 and 8.4 from *The Climate of Europe: Past, Present and Future* by H. Flohn and R. Fantechi (eds.); for figs. 6.37 and 8.10 from *Climatic Change*; and for fig. 7.24 from *Interactions of Energy and Climate* by W. Bach, J. Pankrath and J. Williams (eds.).

Scientific American Inc., New York, for fig. 1.2 by G. N. Plass; for fig. 1.30 by R. E. Newell; for figs. 1.4 and 1.5 by M. R. Rapino and S. Self; and for fig. 4.31B by J. Snow.

Springer-Verlag, Vienna and New York, for fig. 1.38 from the *Meteorologische Rundschau*; and for figs. 2.24 and 3.10 from *Archiv für Meteorologie, Geophysik und Bioklimatologie.*

Time-Life Inc., Amsterdam, for pl. 5 from *The Grand Canyon* by R. Wallace.

University of California Press, Berkeley, for fig. 6.8 and pl. 31 from *Cloud Structure and Distributions over the Tropical Pacific Ocean* by J. S. Malkus and H. Riehl.

University of Chicago Press for figs. 1.8, 1.12, 1.26, 1.33, 2.4, 7.7, 7.8 and 7.10 from *Physical Climatology* by W. D. Sellers.

University of Wisconsin Press, Madison, for fig. 6.27 from *The Earth's Problem Climates* by G. T. Trewartha.

Van Nostrand Reinhold Company, New York, for fig. 6.33 from *Encyclopedia of Atmospheric Sciences and Astrogeology* by R. W. Fairbridge (ed.)

Walter De Gruyter, Berlin, for fig. 5.1 from *Allgemeine Klimageographie* by J. Blüthgen.

Weidenfeld and Nicolson, London, for fig. 4.20 from *Climate and Weather* by H. Flohn.

Westview Press, Boulder, for fig. 1.3 from *Climate Change and Society by W. W. Kellogg and R. Schware.*

John Wiley, Chichester, for figs. 2.28, 5.8, 6.30 and 8.8 from the *Journal of Climatology*.

John Wiley, New York, for fig. 1.22A from *Physical Geography* (2nd ed.) by A. N. Strahler; for figs. 8.2, App. 1.3, App. 1.4 and Table App. 1.1 from *Physical Geography* (3rd ed.) by A. N. Strahler; for figs. 1.10E, 1.11 and 2.19 from *Introduction to Physical Geography* by A. N. Strahler; for fig. 1.13 from *Meteorology, Theoretical and Applied* by E. W. Hewson and R. W. Longley; for fig. 4.25 from *Weather and Climate Modification* by W. N. Hess (ed.); and for fig. 4.35 from *Paleoclimate: Models and Analysis* by A. D. Hecht.

V. H. Winston and Sons, Silver Spring, Maryland, for fig. 5.2 from *Soviet Geography*.

Organizations

Department of Electrical Engineering and Electronics, University of Dundee, for pl. 23.

Deutscher Wetterdienst, Zentralamt, Offenbach am Main, for fig. 6.26.

Environmental Science Services Administration (ESSA) for pls. 10 and 20.

Geographical Branch, Department of Energy, Mines and Resources, Ottawa, for fig. 5.13 from *Geographical Bulletin*.

National Aeronautics and Space Administration (NASA) for pls. 16, 28, 29, 33, 35 and 36.

National Geophysical Data Center, Boulder, for fig. 1.9.

National Hurricane Center, Miami, for pl. 34.

National Meteorological Center, Washington, DC, for fig. 4.34.

National Oceanic and Atmospheric Administration (NOAA), United States Department of Commerce, Washington, DC, for figs. 4.36, 4.37 and 4.39, and for pls. 2 and 3 from *Technical Memo. NESS 95*; and for pl. 30

National Snow and Ice Data Center, Boulder, for pls. 11, 24, and 32.

Naval Weather Surface Command, Washington, DC, for figs. 3.18 and 3.26.

New Zealand Meteorological Service, Wellington, New Zealand, for figs. 6.25 and 6.32 from the *Proceedings of the Symposium on Tropical Meteorology* by J. W. Hutchings (ed.).

Press Association-Reuters Ltd, London, for pl. 6.

Quartermaster Research and Engineering Command, Natick, Mass., for fig. 5.16 by J. N. Rayner.

United Nations Food and Agriculture Organization, Rome, for fig. 7.17B from *Forest Influences*.

United States Department of Agriculture, Washington, DC, for figs. 7.16B and 7.17A from *Climate and Man*.

United States Department of Energy, Washington, DC, for fig. 1.18A.

United States Geological Survey, Washington, DC, for fig. 6.38 from *Professional Paper 1052*.

United States National Air Pollution Administration, Washington, DC, for figs. 7.19 and 7.22 from *Public Health Service Publication No. AP-63*.

United States Weather Bureau for figs. 2.24, 3.11, 3.24, 3.36, 3.37, 3.39, 4.6B, 4.23, 6.2B, 6.3, 6.12, 6.28 and 6.32A from the *Monthly Weather Review*; and for fig. 4.18 from *Research Paper 40*.

World Meteorological Organization for fig. 1.21, and pls. 4, 17 and 27 from *Technical Note 124*; and for fig. 6.34A and B from *The Global Climate System 1982–84*.

Individuals

Arizona State Climatologist, Phoenix, Arizona, for fig. 5.28.

Dr C. F. Armstrong and Dr. C. K. Stidd, of the Desert Research Institute, University of Nevada, for fig. 2.30.

Dr August H. Auer, Jr., of the University of Wyoming, for pl. 37.

Mr P. E. Baylis, of the University of Dundee, and Dr R Reynolds, of the University of Reading, for pl. 21.

Dr R. P. Beckinsale, of Oxford University, for suggested modification to fig. 4.7.

Mr. R. Bumpas, of the National Center for Atmospheric Research, Boulder, for pl. 8.

Dr G. C. Evans, of the University of Cambridge, for fig. 7.18A.

Dr H. Flohn, of the University of Bonn, for figs. 3.28 and 6.15.

Dr S. Gregory, of the University of Sheffield, for fig. 6.14.

Mr Ernst Haas for pl. 5.

Dr S. L. Hastenrath, of the University of Wisconsin, for figs. 1.37 and 2.30.

Dr L. H. Horn and Dr R. A. Bryson, of the University of Wisconsin, for fig. 5.14.

Dr R. A. Hooze Jr., of the University of Washington, for fig. 4.30.

Dr Y. Kurihara, of Princeton University, for fig. 6.11.

Mr E. Lantz for pl. 25.

Dr F. H. Ludlam, of Imperial College, London, for pls. 14 and 19.

Dr Kiuo Maejima, of Tokyo Metropolitan University, for fig. 6.27.

Dr J. Maley, of The Université des Sciences et des Techniques du Languedoc, for fig. 6.29B.

Dr Brooks Martner, of the University of Wyoming, for pl. 13.

Dr T. R. Oke, of the University of British Columbia, for figs. 3.13A and C, 4.25, 7.3A and B, 7.6, 7.15, 7.19C and D, 7.22, 7.23B and C and 7.26.

Mr D. A. Richter, of Analysis and Forecast Division, National Meteorological Center, Washington, DC, for fig. 4.23.

Dr J. C. Sadler, of the University of Hawaii, for fig. 6.19.

Dr R. S. Scorer, of Imperial College, London, and Mrs Robert F. Symons, for pl. 9.

Dr A. N. Strahler, of Santa Barbara, California, for figs. 1.10E, 1.11, 1.22A, 2.19, 8.2, App. 1.3 and App. 1.4; and for Table App. 1.1.

Preface to the first edition

The rapid advances over the past 10 to 15 years in our understanding of atmospheric processes and global climates make a continual reappraisal of teaching methods and the content of textbooks essential. The traditional view of climatology as mere 'book-keeping' has at long last been abandoned by the majority of those interested in investigating the basic mechanisms of climatic differentiation, but the approaches of synoptic and dynamic climatology have in general not yet found their way beyond the scientific papers into elementary textbooks.

The authors' aim is to help to fill this gap, particularly for those studying weather and climate in introductory courses of college or university geography departments and in the sixth form. At the same time, students in related disciplines such as agriculture, ecology and hydrology, and indeed all who are interested in the atmosphere and its weather, should find a basic introduction to modern ideas in this field in the present book. Some of the concepts which are introduced undoubtedly go rather beyond the general scope of courses at the levels mentioned, so that the book should also serve as a foundation for more advanced study. A guide to further reading is provided by the bibliography. No attempt is made to present a comprehensive coverage of regional climates, but, by an examination of the weather and climate of the mid-latitudes of the northern hemisphere and the tropics in terms of a variety of themes, it is hoped to give the reader sufficient appreciation of climatic controls to apply these ideas elsewhere himself.

The first three chapters deal with the nature of the atmosphere – its energy budget, moisture balance and motion. The fourth chapter discusses air masses and the processes which lead to the development of frontal and other depressions. These basic concepts, together with such additional ones as are required, are then used to examine the climatic characteristics of mid-latitudes and the tropics. The book concludes with a brief consideration of the modifications of climate produced by the urban and forest environments and of the inherent variability of climate with time. A brief summary of the major schemes of climatic classification is given for reference purposes in App. 1. It is worth emphasizing that the distinction between weather and climate is arbitrary. Average climatic conditions can be specified for particular places and time-periods, but every individual element of climate varies continuously in space and time. This fundamental point underlies the philosophy of the book: climate can only be understood through a knowledge of the workings of the atmosphere.

R. G. BARRY
Department of Geography, University of Southampton

R. J. CHORLEY
Sidney Sussex College, University of Cambridge

Preface to the fifth edition

When the first edition of this book appeared in 1968, it was greeted as being 'remarkably up to date' (*Meteorological Magazine*). Since that time several new editions have extended and sharpened its description and analysis of atmospheric processes and global climates. Indeed, succeeding Prefaces provide a virtual commentary on recent advances in meteorology and climatology of relevance to students in these fields and to scholars in related disciplines.

In 1971 major additions were made regarding the energy budget of the earth, the spatial pattern of the heat budget components, atmospheric stability, orographic precipitation, oceanic circulation and associated climatic effects, vorticity, mesoscale systems in middle latitudes, rainfall variability and aspects of the climate of the sub-Arctic, the Mediterranean and eastern Asia.

The third edition of 1976 involved the substantial recasting of chs. 2, 4 and 7 and the addition of a new Appendix on synoptic weather maps. New material was introduced on atmospheric composition and its variations with time, the radiation budget, adiabatic temperature changes, the effects of topographic barriers on winds, land and sea breezes, the southern hemisphere circulation and air masses, depression structure and the spatial distribution of precipitation, long-range forecasting, the climate of the Mediterranean, the inter-tropical confluence, tropical disturbances and subsynoptic systems, and urban climates.

In 1982 chs. 7 and 8 were substantially rewritten and the units were standardized throughout. Major additions and revisions involved the material on solar radiation, thunderstorm mechanisms, drought, mesoscale rainfall systems, tornado structure, disturbances within subtropical high-pressure belts, the energy balance of vegetated surfaces, urban climatology, atmospheric pollution and the nature and causes of climatic change. This fourth edition contained over 100 new or revised figures and many new plates, compared with the first edition.

This process of updating has continued in the fifth edition, although the highly successful overall structure of the book remains unchanged. New sections have been included on Modelling the Atmospheric Circulation and Climate, Meteorological Forecasts (including data sources, 'nowcasting' and long-range forecasting). Chapter 6 has been substantially augmented; the Bibliography has been revised and updated; a new Appendix added on Data Sources, including satellite data; and some 70 new or revised figures and plates introduced. In addition, ch. 1 has been augmented with material on variations in greenhouse gases and volcanic dust, solar energy and sun-spot activity, surface albedo and satellite measurements of the planetary energy budget. Chapter 2 has expanded treatments of adiabatic diagrams and atmospheric stability, condensation nuclei and cloud processes, and mechanisms of orographic precipitation. Chapter 3 elaborates the concept of convergence and divergence, the behaviour of subtropical high-pressure cells and mean hemispheric circulation patterns and concludes with a new section on modelling studies. Chapter 4 gives additional material on southern hemisphere air masses, frontal structures, the conveyor belt concepts and rain bands, and mesoscale convective systems including the super-cell thunderstorm and tornadoes. Chapter 5 describes British circulation types in more detail, blocking conditions, orographic precipitation in England and Wales, the regional winds of Iberia and rainstorms in the south-west USA. The revised Chapter 6 has expanded treatments of the Inter-tropical Convergence, tropical weather systems, cloud clusters, the Asian monsoon circulation and the climate of the Sahara, as well as new sections on the African monsoon and ENSO events. Chapter 8 has further material on climatic change in the Sahel Zone and Europe, the climatic effect of vegetation changes and the use of mathematical models for climatic fore-casting.

Wherever possible, the criticisms and suggestions of colleagues and reviewers have been taken into account in preparing this latest edition.

R. G. BARRY
Cooperative Institute for Research in Environmental Sciences and the Department of Geography, University of Colorado, Boulder

R. J. CHORLEY
Department of Geography, and Sidney Sussex College, University of Cambridge

Introduction

In this book we aim to provide a non-technical account of how the atmosphere works, thereby developing an understanding of weather phenomena and of global climates. The atmosphere, which is vital to terrestrial life, is a shallow envelope equivalent in thickness to less than 1% of the earth's radius. Most weather systems form and decay within its lowest 10 km. It is thought that the earth's atmosphere evolved to its present form and composition at least 400 million years ago when extensive vegetation developed on land. Its presence provides an indispensable shield from harmful radiation from the sun and its gaseous contents sustain the plant and animal biosphere on which human life depends.

Over much of the globe, the state of the atmosphere is far from constant in response to varying *weather* processes. Weather extremes – gales, blizzards, tornadoes, floods – drastically affect human activities and frequently result in loss of life, even when anticipated. Hence, by seeking to understand atmospheric phenomena, we can hope to forecast their vagaries and in some instances control or modify them in a beneficial way. This broad endeavour constitutes the field of the atmospheric sciences. *Meteorology* is specifically concerned with the physics of weather processes. *Weather systems* – which produce the variety of instantaneous states of the atmosphere – differ in their size and life span. Four scales are commonly recognized: *mesoscale* systems, such as thunderstorms, extend some 10 km horizontally with a life time of a few hours; *synoptic-scale* systems, like mid-latitude cyclones and tropical storms, have a diameter of a few thousand kilometres and a life time of about 5 days; *planetary-scale* waves in the atmospheric circulation span 5000–10,000 km and usually persist for several weeks. In addition, small-scale wind eddies near the earth's surface and processes operating within vegetation canopies are the concern of *micrometeorology*.

Climate introduces the longer time scales operating in the atmosphere. It is sometimes loosely regarded as 'average weather', but it is more meaningful to define climate as the long-term state of the atmosphere encompassing the aggregate effect of weather phenomena – the extremes as well as the mean values. It is also usual to distinguish regional and global *macroclimate*, on the one hand, from *local* or *topo-climates* related to terrain features (valleys, hill slopes), on the other hand.

The structure of the book represents this viewpoint. We will look first at the composition and structure of the atmosphere and its role in the global exchange of energy, the moisture balance and wind systems. Then weather and climate in middle and low latitudes are discussed and, finally, small-scale climates and climatic change. The key to atmospheric processes is the radiant energy which the earth and its atmosphere receive from the sun. In order to study the receipt of this energy we need to begin by considering the nature of the atmosphere – its composition and basic properties.

1

Atmospheric composition and energy

A Composition of the atmosphere

1 Total atmosphere

Air is a mechanical mixture of gases, not a chemical compound. Table 1.1, illustrating the average composition of dry air, shows that four gases – nitrogen, oxygen, argon and carbon dioxide – account for 99·98% of the air by volume. Moreover, rocket observations show that these gases are mixed in remarkably constant proportions up to about 80 km (50 miles).

In addition to these gases, water vapour, which is much more variable in its occurrence in time and space, is a vital atmospheric constituent. This will be discussed more fully below. There are also significant quantities of *aerosols* in the atmosphere. These are suspended particles of sea salt, dust (particularly silicates), organic matter and smoke. They come from both natural sources and anthropogenic ones related to human activities.

Having made these generalizations about the atmosphere, we now examine the variations which occur in composition with height, latitude and time.

Table 1.1 Average composition of the dry atmosphere below 25 km

Component	*Symbol*	*Volume %* *(dry air)*	*Molecular weight*
Nitrogen	N_2	78·08	28·02
Oxygen	O_2	20·94	32·00
*‡Argon	Ar	0·93	39·88
Carbon dioxide	CO_2	0·03 (variable)	44·00
‡Neon	Ne	0·0018	20·18
*‡Helium	He	0·0005	4·00
†Ozone	O_3	0·00006	48·00
Hydrogen	H	0·00005	2·02
‡Krypton	Kr	Trace	
‡Xenon	Xe	Trace	
Methane	CH_4	Trace	

* Decay products of potassium and uranium.
† Recombination of oxygen.
‡ Inert gases.

2 Variations with height

The light gases (hydrogen and helium especially) might be expected to become more abundant in the upper atmosphere, but large-scale turbulent mixing of the atmosphere prevents such diffusive separation even at heights of many tens of kilometres above the surface. The height variations which do occur are related to the source-locations of the two major non-permanent gases – water vapour and ozone. Since both absorb some solar and terrestrial radiation the heat budget and vertical temperature structure of the atmosphere are considerably affected by the distribution of these two gases.

Water vapour comprises up to 4% of the atmosphere by volume (about 3% by weight) near the surface, but is almost absent above 10 to 12 km. It is supplied to the atmosphere by evaporation from surface water or by transpiration from plants and is transferred upwards by atmospheric turbulence. Turbulence is most effective below about 10 km and as the maximum possible water vapour density of cold air is anyway very low (see ch. 1, B.2), there is little water vapour in the upper layers of the atmosphere.

Ozone (O_3) is concentrated mainly between 15 to 35 km. The upper layers of the atmosphere are irradiated by ultraviolet radiation from the sun which causes the break-up of oxygen molecules at altitudes above 30 km (i.e. $O_2 \rightarrow O + O$). These separated atoms ($O + O$) may then individually combine with other oxygen molecules to create ozone, as illustrated by the simple photochemical scheme:

$$O_2 + O + M \rightarrow O_3 + M$$

where M represents the energy and momentum balance provided by collision with a third atom or molecule. Such three-body collisions are rare at 80 to 100 km because of the very low density of the atmosphere, while below about 35 km most of the incoming ultraviolet radiation has already been absorbed at higher levels. Therefore ozone is mainly formed between 30 and 60 km where collisions between O and O_2 are more likely. Ozone itself is unstable; collisions with monatomic oxygen may re-create oxygen (i.e. $O_3 + O \rightarrow O_2 + O_2$), but ozone is mainly destroyed through other photochemical cycles involving catalytic reactions with oxygen compounds, particularly nitrogen oxides (NO_x).

The constant metamorphosis of oxygen to ozone and from ozone back to oxygen by photochemical processes maintains an approximate equilibrium above about 40 km, but the ozone mixing ratio is a maximum at about 35 km, whereas maximum ozone concentration (see note 1) occurs lower down between 20 and 25 km. This is the result of some circulation mechanism transporting ozone downwards to levels where its destruction is less likely, allowing an accumulation of the gas to occur. Despite the importance of the ozone layer, it is essential to realize that if the atmosphere were compressed to sea-level (at normal sea-level temperature and pressure) ozone would contribute only about 3 mm to the total atmospheric thickness of 8 km (fig. 1.1).

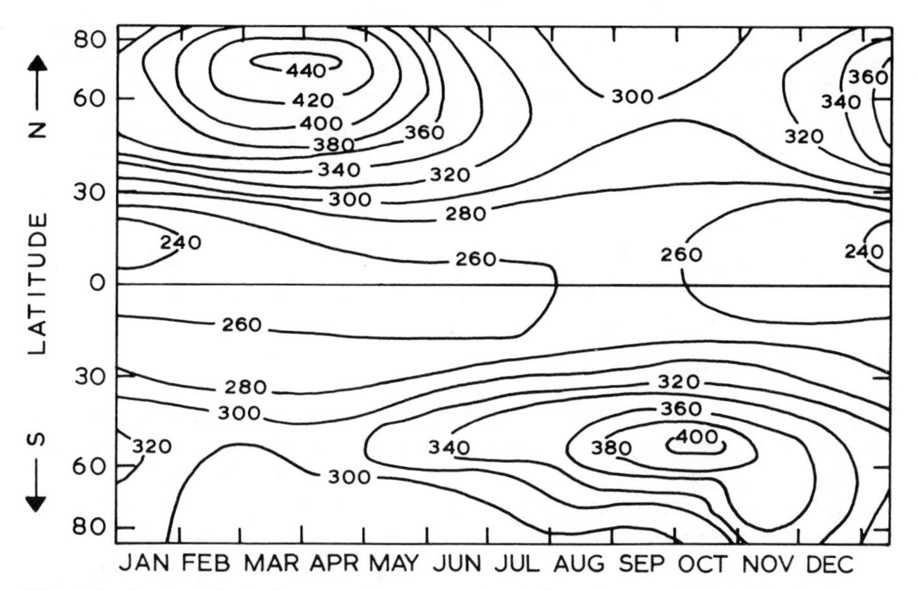

Fig. 1.1. Average latitude–season cross-section of total ozone (1957–75). Values are in Dobson units (10^{-3} cm, at standard atmospheric temperature and pressure) (*from London and Angell 1982; Copyright American Society for Testing and Materials. Reprinted with permission*).

Aerosols enter the atmosphere by urban and industrial pollution and by agricultural practices as well as through forest fires, sea spray, volcanic activity and wind-raised dust. Large particles rapidly sink back to the surface or are washed out (scavenged) by rain after a few days, but fine particles (10^{-5} cm radius) from volcanic eruptions may reside in the stratosphere above the level of weather processes for 1–3 years.

3 Variations with latitude and season

Variations of atmospheric composition with latitude and season are particularly important in the case of water vapour and ozone.

Ozone content is low over the equator and high in subpolar latitudes in spring (fig. 1.1). If the distribution were solely the result of photochemical processes the maximum would occur in June near the equator and the anomalous pattern must be due to a poleward transport of ozone. The movement is apparently from higher levels (30–40 km) in low latitudes towards lower levels (20–25 km) in high latitudes during winter months. Here the ozone is stored during the *polar night*, giving rise to an ozone-rich layer in early spring. The type of circulation responsible for this transfer is not yet known with certainty, although it does not seem to be a simple direct one.

The water-vapour content of the atmosphere is closely related to air temperature (see chs. 1, B.2 and 2, A and B) and is therefore greatest in summer and in low latitudes. There are, however, obvious exceptions to this generalization, such as the tropical desert areas of the world.

The carbon dioxide content of the air (averaging about 345 parts per million (ppm)) has a large seasonal range in higher latitudes in the northern hemisphere associated with photosynthesis and decay of the biosphere. At 50°N the concentration ranges from 338 ppm in late summer to 352 ppm in spring. The low summer values are related to the assimilation of CO_2 by the cold polar seas. Over the year a small net transfer of CO_2 from low to high altitudes takes place to maintain an equilibrium content in the air.

4 Variations with time

The quantities of carbon dioxide, ozone and particles in the atmosphere may be subject to variations over a long time-period and these are of special significance because of their possible effect on the radiation budget.

Carbon dioxide (CO_2) enters the atmosphere mainly by the action of living organisms on land and in the ocean. The decay of organic elements in the soil and the burning of fossil fuels are additional minor sources (fig. 1.2). It is obvious that if this production were not countered in some way the total quantity of carbon would steadily increase. A balance, or dynamic equilibrium, is maintained primarily by photosynthesis which removes approximately 3% of the world's total carbon dioxide annually. In the oceans the carbon dioxide ultimately goes to produce carbonate of lime, partly in the form of shells and the skeletons of marine creatures. On land the dead matter becomes humus which may subsequently form a fossil fuel. These transfers within the oceans and lithosphere involve very long time scales compared with exchanges involving the atmosphere. As fig. 1.2 showed, the exchanges between the atmosphere and the other reservoirs are more or less balanced.

Yet this balance is not an absolute one, for between 1870 and 1986 the total quantity of atmospheric CO_2 is estimated to have increased by 19% (from 290 to 345 ppm) due to the burning of fossil fuels (fig. 1.3). The actual use of fossil fuels should have produced an increase of about 30% but apparently the excess is taken up by the land biosphere and the oceans.

Carbon dioxide has a significant impact on global temperature by its absorption and re-emission of radiation from the earth and atmosphere (see fig. 1.8 and ch. 1, E). Calculations suggest that the increase to 370 ppm expected by AD 2000 could raise the mean air temperature near the surface by 0·5°C compared with the 1960s (in the absence of other factors).

Recent studies show that the atmospheric concentrations of several radiatively important trace gases are also increasing. These include methane and chlorofluorocarbons which, like CO_2 and water vapour, are so-called 'greenhouse gases'. Methane (CH_4), which is produced mainly by the biosphere (the digestive processes of cows, termites and swamps), absorbs terrestrial (infrared) radiation (see ch. 1.E). Records extracted from Antarctic ice cores show that methane increased substantially between 1760 and 1960 from about 0.8 to 1.3 ppm. Chlorofluorocarbons, $CFCl_3$ (F-11) and CF_2Cl_2

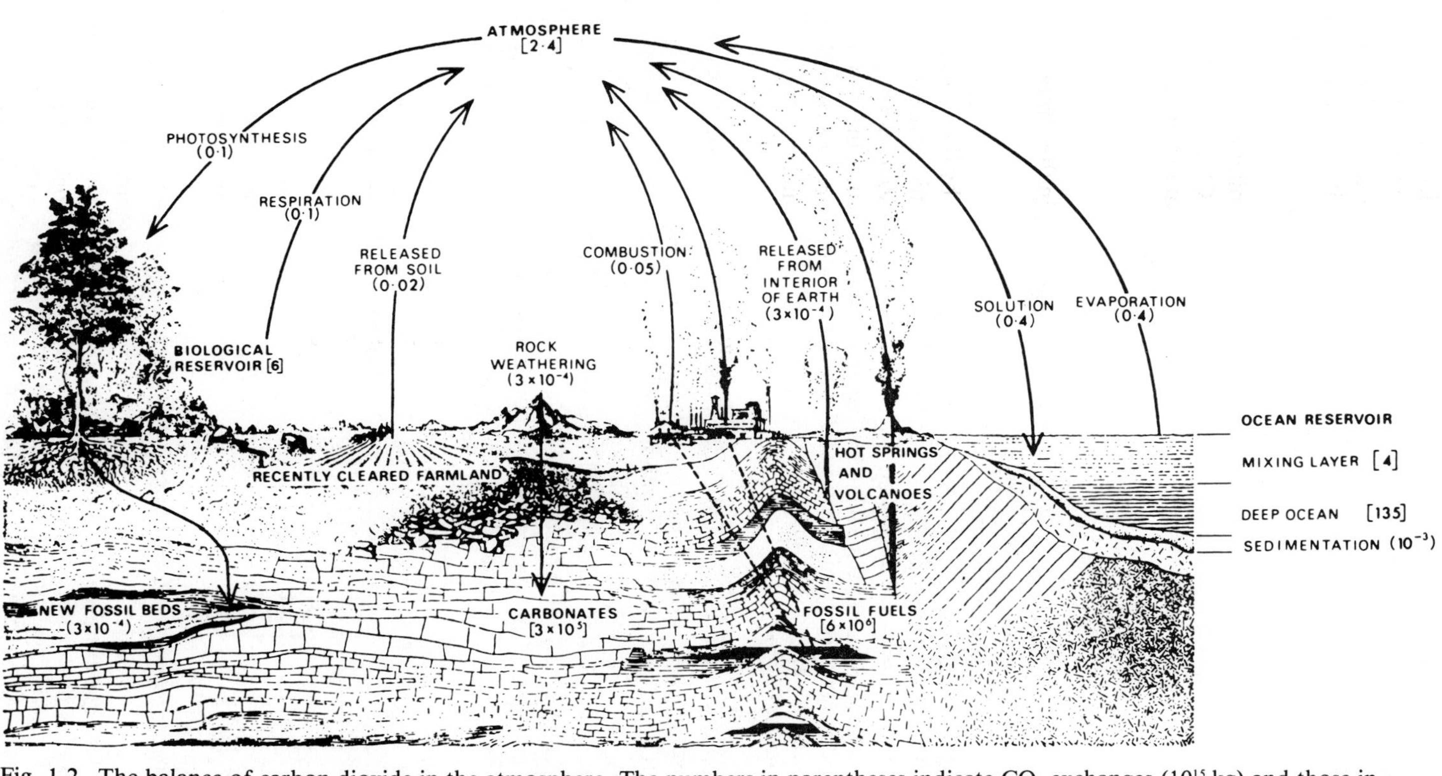

Fig. 1.2. The balance of carbon dioxide in the atmosphere. The numbers in parentheses indicate CO_2 exchanges (10^{15} kg) and those in square brackets storage of CO_2(10^{15} kg) (*partly from Plass 1959*).

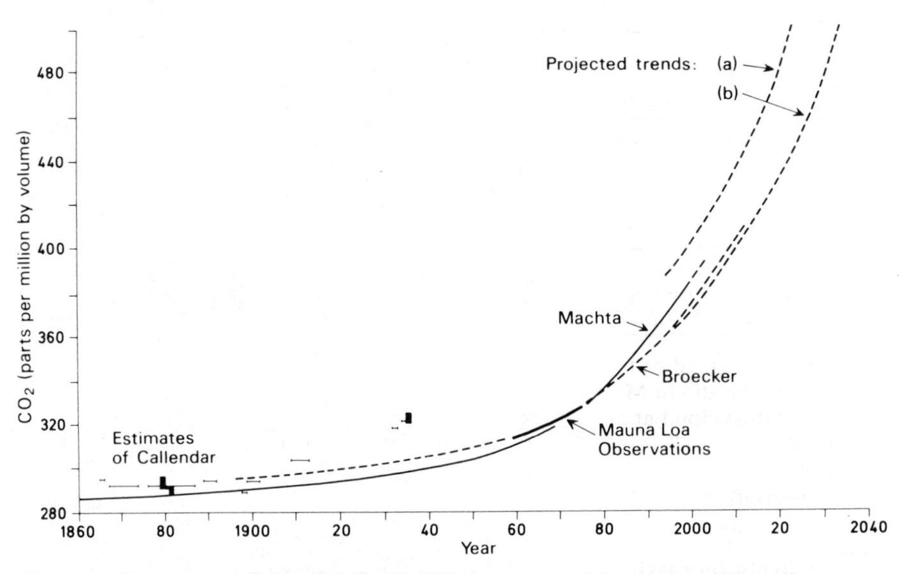

Fig. 1.3. Observations of atmospheric CO_2 increase at Mauna Loa, Hawaii (1957–75), estimates from 1860–1960 based on early measurements, and projected trends into the twenty-first century (*after Keeling, Callendar, Machta, Broecker and others*). (a) and (b) indicate different scenarios of global fossil fuel use (*from Kellogg and Schware 1981*).

(F-12) are produced by industrial processes, the release of propellants in aerosol sprays, and from coolants (such as 'freon'). These gases have catalytic effects on stratospheric and tropospheric ozone. Overall, trace gases could amplify by a factor of almost two the global temperature increase of 3–4°C anticipated as a result of a doubling of atmospheric CO_2 concentrations by late in the twenty-first century.

Changes in atmospheric particle concentration derived from volcanic dust are extremely irregular (fig. 1.4) but individual volcanic emissions are rapidly diffused geographically (fig. 1.5). On the other hand, the contribution of man-made particles (particularly of sulphates and soil) has been

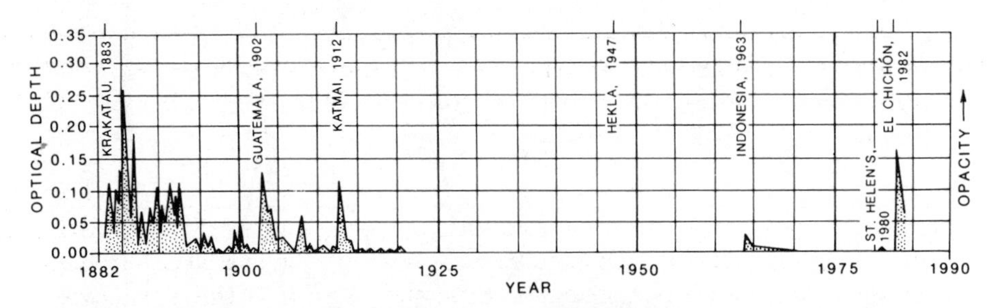

Fig. 1.4. The average opacity of the stratosphere of the northern hemisphere during the past century measured from the intensity of sunlight and starlight. Some general correlation with major volcanic eruptions is indicated (*from The atmospheric effects of El Chicón by M. R. Rampino and S. Self. Copyright © 1984 by Scientific American Inc. All rights reserved*).

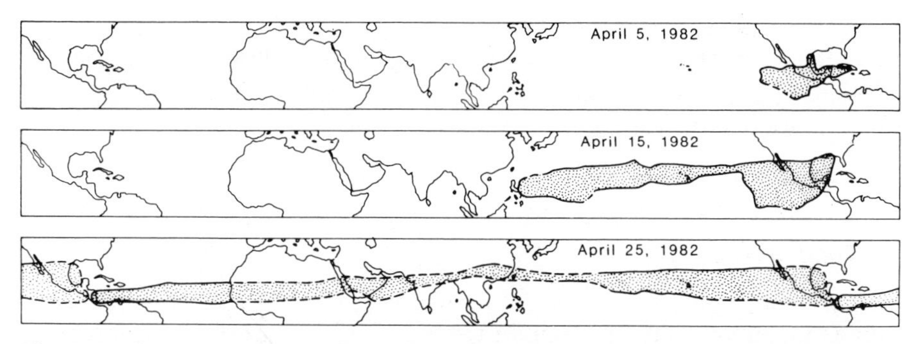

Fig. 1.5. The spread of the volcanic dust cloud following the main eruption of the El Chichón volcano in Mexico on 3 April 1982. A strong zonal wind circulation carried the dust cloud at an average speed of 20 m s^{-1} (45 mph) so that it encircled the globe in less than three weeks (*from The atmospheric effects of El Chicón by M. R. Rampino and S. Self. Copyright © 1984 by Scientific American Inc. All rights reserved*).

progressively increasing, at present accounting for about 30% of the total and this figure could double by AD 2000. The overall effect on the lower atmosphere is now thought to be one of warming, whereas volcanic dust has the opposite effect.

Variations in stratospheric ozone may occur as a result of fluctuations in solar ultraviolet radiation. This has been proposed as a mechanism for climatic change (see ch. 8, C), since ozone absorbs solar and terrestrial radiation, but the hypothesis is largely speculative at present.

B Mass of the atmosphere

It is necessary to examine some of the mechanical laws that the atmospheric gases obey. Two simple laws specify the main factors governing changes in pressure. The first, Boyle's Law, states that, at a constant temperature, the volume (V) of a mass of gas varies inversely as its pressure (P), i.e.

$$P = \frac{k_1}{V}$$

(k_1 is a constant); and the second, Charles's Law, that, at a constant pressure, volume varies directly with absolute temperature (T) measured in degrees Kelvin (see note 2):

$$V = k_2 T$$

These laws imply that the three qualities of pressure, temperature and volume are completely interdependent, such that any change in one of them will cause a compensating change to occur in one, or both, of the remainder. The gas laws may be combined to give the following relationship:

$$PV = RmT$$

where m = mass of air

R = a gas constant for dry air (287 J kg^{-1} K^{-1}) (see note 3).

If m and T are held fixed, we obtain Boyle's Law; if m and P are held fixed, we obtain Charles's Law. Since it is convenient to use density, ϱ (= mass/volume) rather than volume when studying the atmosphere, we can rewrite the equation in the form known as the equation of state

$$P = R\varrho T$$

1 Total pressure

Air is highly compressible, such that its lower layers are much more dense than those above. Fifty per cent of the total mass of air is found below 5 km (fig. 1.6) and the average density decreases from about $1\cdot2\,\mathrm{kg\,m^{-3}}$ at the surface to $0\cdot7\,\mathrm{kg\,m^{-3}}$ at 5000 m (approximately 16,000 ft) close to the extreme limit of human habitation.

Pressure is measured as a force per unit area. The units used by meteorologists are called millibars (mb), 1 mb being equal to a force of 100 newtons acting on 1 $\mathrm{m^2}$ (see App. 2). Pressure readings are made with a mercury barometer which in effect measures the weight of the column of mercury that the atmosphere is able to support in a vertical glass tube. The closed upper end of the tube has a vacuum space and its open lower end is immersed in a cistern of mercury. By exerting pressure downwards on the surface of mercury in the cistern, the atmosphere is able to support a mercury column in the tube of about 760 mm (29·9 in or approximately 1013 mb).

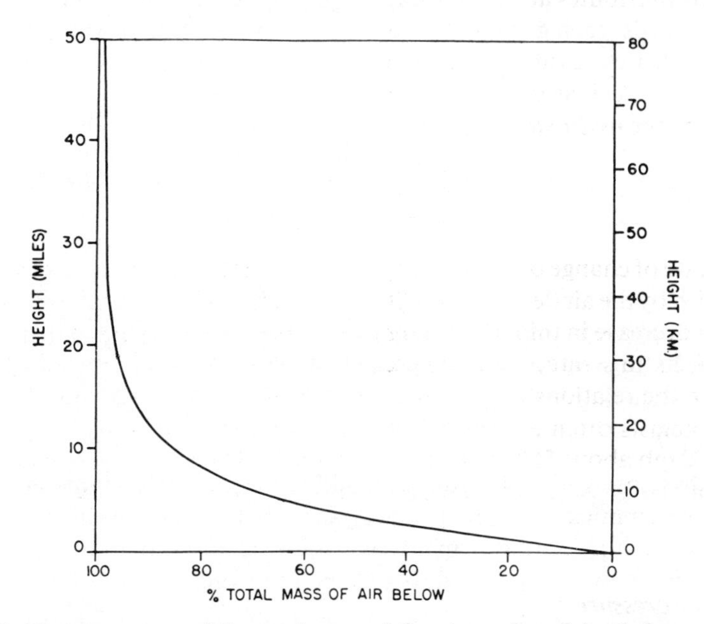

Fig. 1.6. The percentage of the total mass of the atmosphere lying below elevations up to 80 km (50 miles). This illustrates the shallow character of the earth's atmosphere.

Pressure data are standardized in three ways. The readings from a mercury barometer are adjusted to correspond to those for a standard temperature of 0°C (to allow for the thermal expansion of mercury); they are referred to a standard gravity value of 9·81 ms^{-2} at 45° latitude (to allow for the slight latitudinal variation in g from 9·78 ms^{-2} at the equator to 9·83 ms^{-2} at the poles); and they are calculated for mean sea-level to eliminate the effect of station elevation. This third correction is the most significant because near sea level pressure decreases with height about 1 mb per 8 m. A fictitious temperature between the station and sea level has to be assumed and in mountain areas this commonly causes bias in the calculated mean sea-level pressure. (The method is summarized in note 4.)

The mean sea level (p_0) can be estimated by taking account of the total mass of the atmosphere (M), the mean acceleration due to gravity (g_0) and the mean earth radius (R_E):

$$P_0 = g_0\,(M/4\pi R_E^2)$$

where the denominator is the surface area of a spherical earth. Substituting appropriate values into this expression: $M = 5{\cdot}14 \times 10^{18}$ kg, $g_0 = 9{\cdot}8$ $m\,s^{-2}$, $R_E = 6{\cdot}37 \times 10^6$ m, we find $p_0 \simeq 10^5$ $kgm^{-1}s^{-2} = 10^5 Nm^{-2}$, or 10^5 pascals (Pa is the SI unit of pressure). Meteorologists still use the millibar (mb) unit; 1 millibar = 10^6 dyne cm^{-2} = 10^2 Pa.

Hence the mean sea-level pressure is approximately 10^5 Pa or 1000 mb. The global mean value is 1013·25 mb (equivalent to 14·7 lb/in^2). On average, nitrogen contributes about 760 mb, oxygen 240 mb and water vapour 10 mb. In other words, each gas exerts a partial pressure independent of the others.

Atmospheric pressure, depending as it does on the weight of the overlying atmosphere, decreases logarithmically with height. This relationship is expressed by the *hydrostatic equation*:

$$\frac{\partial p}{\partial z} = -g\varrho$$

i.e. the rate of change of pressure (p) with height (z) is dependent on gravity (g) multiplied by the air density (ϱ). With increasing height, the drop in air density causes a decrease in this rate of pressure decrease. The temperature of the air also affects this rate, which is greater for cold dense air (see ch. 3, C.1), although the relationship between pressure and height is so significant that meteorologists often express elevations in millibars: 1000 mb represents sea level, 500 mb about 5500 m and 300 mb about 9000 m. A conversion nomogram for an idealized (standard) atmosphere is given in App. 2.

2 Vapour pressure

At any given temperature there is a limit to the density of water vapour in the air, with a consequent upper limit to the vapour pressure. This is termed the

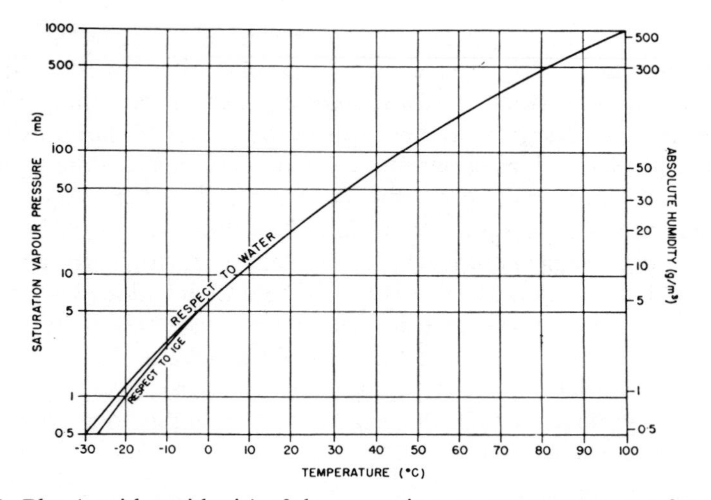

Fig. 1.7. Plot (semi-logarithmic) of the saturation vapour pressure as a function of temperature (i.e. the dew-point curve). Below 0°C the atmospheric saturation vapour pressure is less with respect to an ice surface than with respect to a water drop. Thus, condensation may take place on an ice crystal at lower air humidity than is necessary for the growth of water drops.

saturation vapour pressure (e_s) and fig. 1.7 illustrates how it increases with temperature, reaching a maximum of 1013 mb (1 atmosphere) at boiling point. Attempts to introduce more vapour into the air when the vapour pressure is at saturation produce condensation of an equivalent amount of vapour. Figure 1.7 shows that whereas the saturation vapour pressure has a single value at any temperature above freezing point, below 0°C the saturation vapour pressure above an ice surface is lower than that above a supercooled water surface. The significance of this will be discussed in ch. 2, G.1.

Vapour pressure (e) varies with latitude and season from about 0·2 mb over northern Siberia in January to over 30 mb in the tropics in July, but this is not reflected in the pattern of surface pressure. Pressure decreases at the surface when some of the overlying air is displaced horizontally, and in fact the air in high-pressure areas is generally dry owing to dynamic factors, particularly vertical air motion (ch. 3, D.1), whereas air in low-pressure areas is usually moist.

C Solar radiation

The prime source of the energy injected into our atmosphere is the sun, which is continually shedding part of its mass by radiating waves of electromagnetic energy and high energy particles into space. This constant emission is important because it represents in the long run almost all the energy available to the earth (except for a small amount emanating from the radioactive decay of earth minerals). The amount of energy received by the earth, assuming for the moment that there is no interference from the atmosphere,

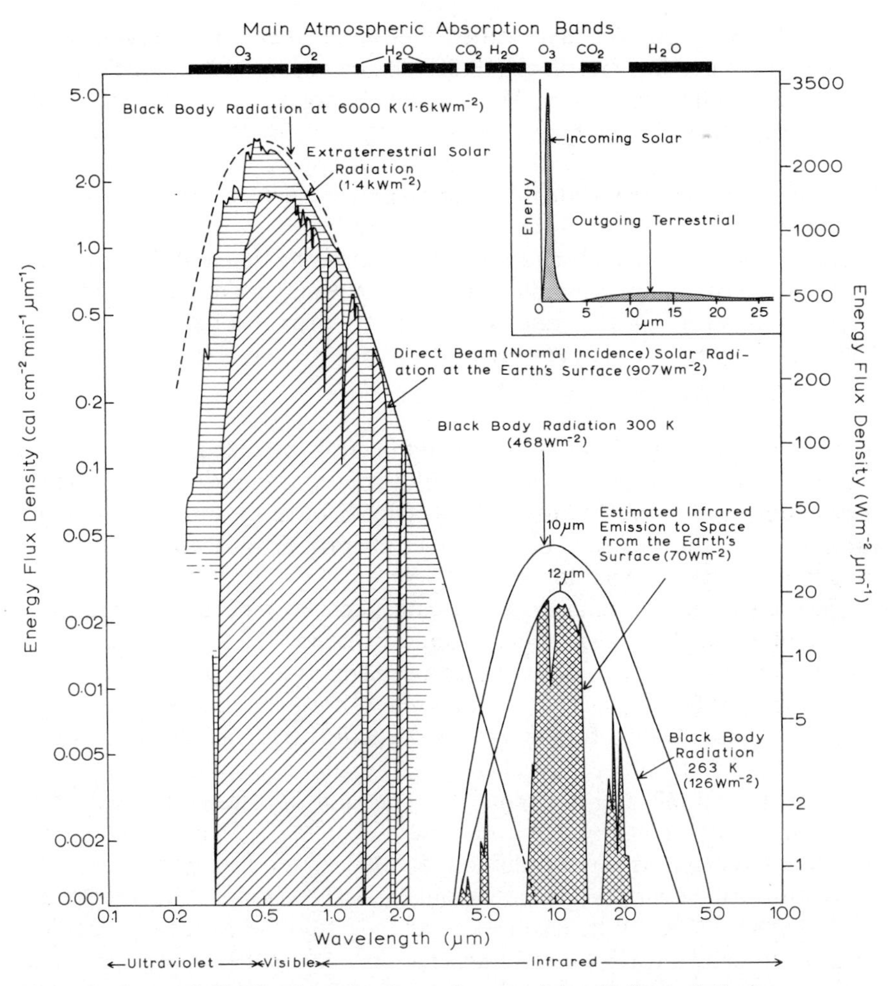

Fig. 1.8. Spectral distribution of solar and terrestrial radiation, plotted logarithmically, together with the main atmospheric absorption bands. The cross-hatched areas in the infrared spectrum indicate the 'atmospheric windows' where radiation escapes to space. The black-body radiation at 6000 K is that proportion of the flux which would be incident on the top of the atmosphere. The inset shows the same curves for incoming and outgoing radiation with the wavelength plotted arithmetically on an arbitrary vertical scale (*mostly after Sellers 1965*).

is affected by four factors: solar output, the sun–earth distance, altitude of the sun, and day length.

1 Solar output

Solar energy, which originates from nuclear reactions within the sun's hot core (16×10^6K), is transmitted to the sun's surface by radiation and hydrogen convection. Visible solar radiation (light) comes from a 'cool'

($\sim$6000 K) outer surface layer called the *photosphere*. Temperatures rise again in the outer chromosphere (10,000 K) and corona (10^6K) which is continually expanding into space. The outflowing hot gases (plasma) from the sun, referred to as the *solar wind* (with a speed of $1\cdot5 \times 10^6$ km hr^{-1}), interact with the earth's magnetic field and upper atmosphere. The earth intercepts both the normal electromagnetic radiation and energetic particles emitted by the sun during solar flares.

The sun behaves virtually as a *black body*, meaning that it both absorbs all energy received and in turn radiates energy at the maximum rate possible for a given temperature. This rate (F) is directly proportional to the fourth power of the absolute temperature of the body:

$$F = \sigma T^4 \text{ (Stefan's Law)}$$

where $\sigma = 5\cdot67 \times 10^{-8}$ W m^{-2} K^{-4} (the Stefan–Boltzmann constant).

Hence, the total solar output to space, assuming a temperature of 6000 K for the sun, is $73\cdot5 \times 10^6$ W m^{-2}. Only 0·0005% of this is intercepted by the earth, because the energy received is inversely proportional to the square of the solar distance (150 million km).

The energy received at the top of the atmosphere on a surface perpendicular to the solar beam for mean solar distance is termed the *solar constant* (see note 5). The most recent satellite measurements indicate a value of about 1370 W m^{-2}, or 1·96 cal cm^{-2} min^{-1}. Figure 1.8 shows the wavelength range of solar (short-wave) radiation and the long-wave (infrared) radiation emitted by the earth and atmosphere. For solar radiation, about 9% is ultraviolet, 45% visible light, and 46% infrared radiation. The figure illustrates the black body radiation curves for 6000 K at the top of the atmosphere (which slightly exceeds the observed extraterrestrial radiation) for 300 K, and for 263 K. The mean temperature of the earth's surface is about 288 K (15°C) and of the atmosphere about 250 K (−23°C). Whereas most solids and liquids behave as black bodies, gases do not, and fig. 1.8 shows the absorption bands in the atmosphere which cause its emission to be much less than that from an equivalent black body. The wavelength of maximum emission (λ_{max}) varies inversely with the absolute temperature of the radiating body:

$$\lambda_{max} = \frac{2897}{T} \times 10^{-6} \text{ m (Wien's Law).}$$

Thus solar radiation is very intense and is mainly short wave between about 0·2 and 4·0 μm, with a maximum (per unit wavelength) at 0·5 μm, whereas the much weaker terrestrial radiation has a peak intensity at about 10 μm and a range of about 4 to 100 μm (1 μm = 1 micrometre = 10^{-6} m).

There have been suggestions that the solar constant undergoes small periodic variations of <0·5%, perhaps related to sunspot activity, but since determinations of the solar constant are subject to errors of similar magnitude, the reality of such fluctuations is in doubt. *Sunspots* are dark (i.e. cooler) areas visible on the sun's surface. Their number and positions change in a regular manner known as the sunspot cycle ($\sim$11 years). Figure 1.9

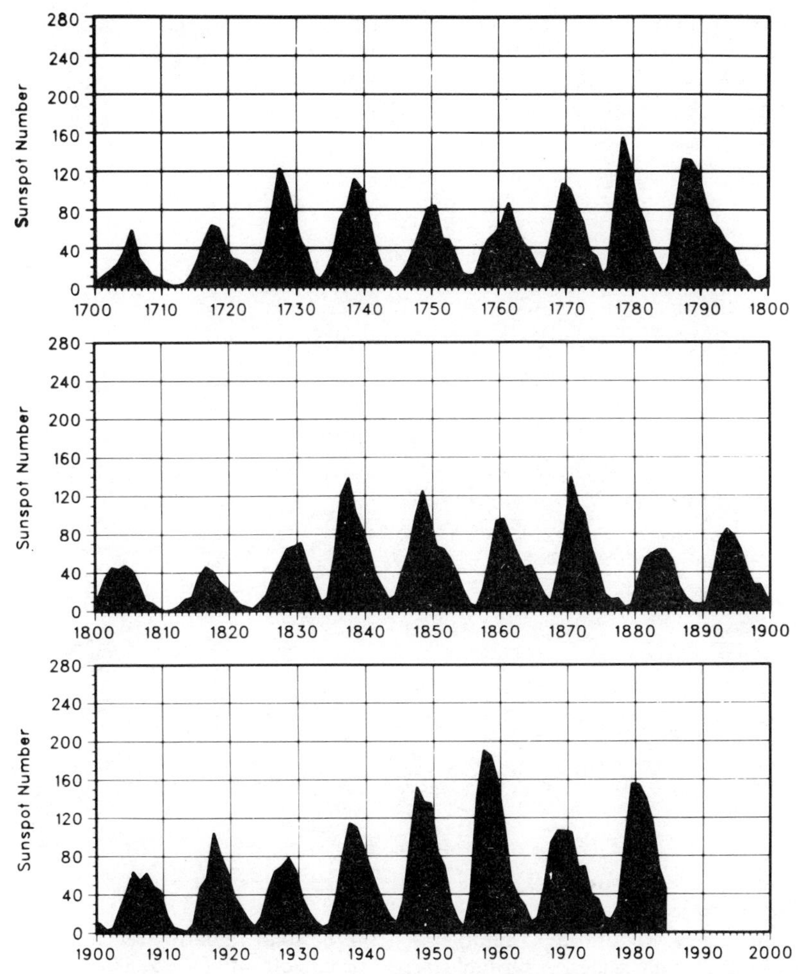

Fig. 1.9. Yearly mean sunspot numbers for the period 1700–1984. (*Courtesy of the National Geophysical Data Center, Boulder, Colorado.*)

shows the variation of sunspot activity over the last few centuries. Output within the ultraviolet part of the spectrum shows considerable variability, with up to twenty times more ultraviolet radiation emitted at certain wavelengths during a sunspot maximum than during a sunspot minimum. However, no clear link between the approximately 11-year sunspot cycle and weather variations has yet been demonstrated, in spite of many attempts to discover such a relationship. In the long term, assuming that the earth behaves as a black body, a long-continued difference of 2% in the solar constant could change the effective mean temperature of the earth's surface by as much as 1·2°C (2·2°F), and a 10% change might alter this temperature by as much as 6°C (10·7°F). The drop in surface temperature often experienced on a sunny day when a cloud temporarily cuts off the direct solar radiation illustrates our reliance upon the sun's radiant energy.

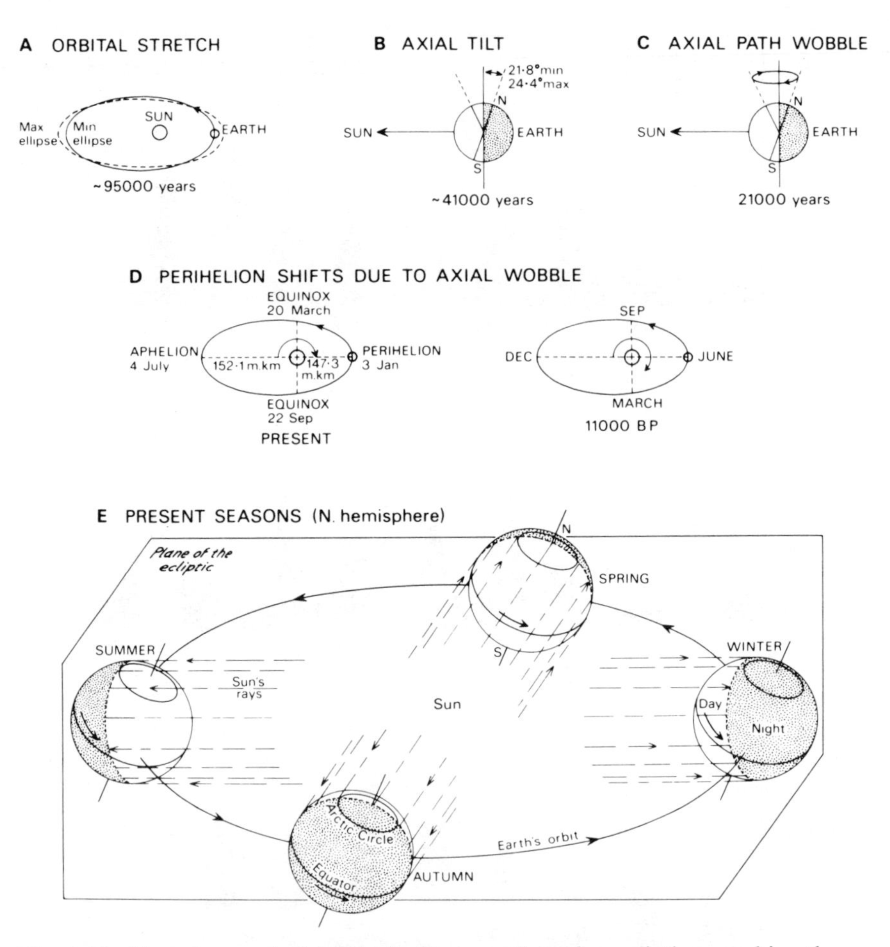

Fig. 1.10. The astronomical (orbital) effects on the solar radiation reaching the earth and their time scales. A Orbital stretch, or eccentricity (≈ 95,000-year period); B Axial tilt (41,000-year); C Wobble of the axial path (21,000-year) which causes a shift in the timing of perihelion (D). E illustrates the geometry of the present seasons (E *after Strahler 1965*).

2 Distance from the sun

The annually changing distance of the earth from the sun produces seasonal variations in our receipt of solar energy. Owing to the eccentricity of the earth's orbit round the sun, the receipt of solar energy on a surface normal to the beam is 7% more on 3 January at the perihelion than on 4 July at the aphelion (fig. 1.10). In theory (that is, discounting the interposition of the atmosphere and the difference in degree of conductivity between large land and sea masses) this difference should produce an increase in the effective January world surface temperatures of about 4°C (7°F), over those of July. It should also make northern hemisphere winters warmer than those in the southern, and southern hemisphere summers warmer than those in the

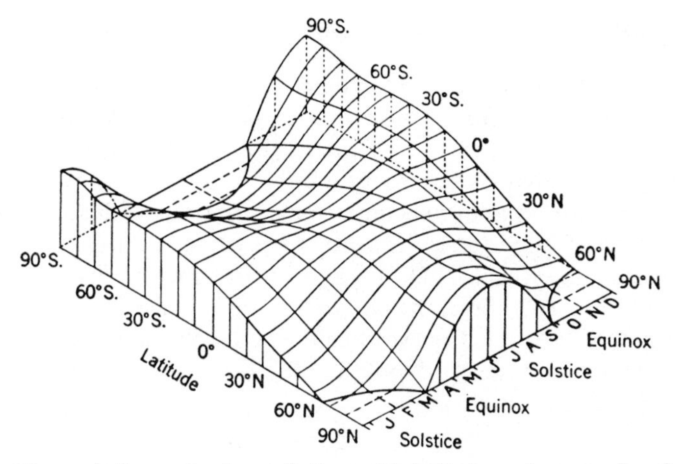

Fig. 1.11. The variations of solar radiation with latitude and season for the whole globe, assuming no atmosphere. This assumption explains the abnormally high amounts of radiation received at the poles in summer, when daylight lasts for 24 hours each day (*after W. M. Davis; from Strahler 1965*).

northern. In practice, atmospheric heat circulation and the effects of continentality substantially mask this global tendency, and the actual seasonal contrast between the hemispheres is reversed. Moreover, the northern summer half-year (21 March–22 September) is 5 days longer than the southern hemisphere summer (22 September–21 March). This difference slowly changes; about 10,000 years ago the aphelion occurred in the northern hemisphere winter and northern summers received 3–4% more radiation than today. This same pattern will return about 10,000 years from now (see fig. 1.10D).

Figure 1.11 graphically illustrates the seasonal variations of energy receipt, with latitude. Actual amounts of radiation received on a horizontal surface outside the atmosphere are given in table 1.2. The intensity on a horizontal surface (I_h) is determined from

$$I_h = I_0 \sin d,$$

where I_0 = the solar constant and d = the angle between the surface and the solar beam.

Table 1.2 Daily solar radiation on a horizontal surface outside the atmosphere: W m^{-2} (*after K. Ya. Kondratiev*)

Date	*90°N*	*70*	*50*	*30*	*0*	*30*	*50*	*70*	*90°S*
Dec 22	0	0	88	233	421	520	528	540	574
Feb 4	0	12	144	284	438	486	454	392	404
Mar 21	0	153	287	387	447	387	287	153	0
May 6	386	350	433	464	418	271	138	12	0
June 22	538	505	494	487	394	218	82	0	0

3 Altitude of the sun

The altitude of the sun (i.e. the angle between its rays and a tangent to the earth's surface at the point of observation) also affects the amount of solar radiation received at the surface of the earth. The greater the sun's altitude the more concentrated is the radiation intensity per unit area at the earth's surface. There are, in addition, important variations with solar altitude of the proportion of radiation reflected by the surface, particularly in the case of a water surface (see ch. 1, D.5). The principal factors that determine the sun's altitude are, of course, the latitude of the site, the time of day and the season (fig. 1.10). At the June solstice the sun's altitude is a constant $23\frac{1}{2}°$ throughout the day at the north pole and the sun is directly overhead at noon at the Tropic of Cancer ($23\frac{1}{2}°$N).

4 Length of day

The length of daylight also affects the amount of radiation that is received. Obviously the longer the time during which the sun shines the greater is the quantity of radiation which a given portion of the earth will be able to receive. At the equator, for example, the daylength is close to 12 hours in all months, whereas at the poles it varies between 0 and 24 hours from winter to summer (see fig. 1.10).

The combination of all these factors produces the pattern of receipt of solar energy at the top of the atmosphere shown by fig. 1.11. The polar regions receive their maximum amounts of solar radiation during their summer solstices, which is the period of continuous day. The amount received during the December solstice in the southern hemisphere is theoretically greater than that received by the northern hemisphere during the June solstice, due to the previously mentioned elliptical path of the earth round the sun (table 1.2). The equator has two radiation maxima at the equinoxes and two minima at the solstices, due to the apparent passage of the sun during its double annual movement between the northern and southern hemispheres.

D Surface receipt of solar radiation and its effects

1 Energy transfer within the earth-atmosphere system

So far we have described the distribution of solar radiation as if it were all available at the earth's surface. This is of course an unreal view because of the effect of the atmosphere on energy transfer. Heat energy can be transferred by the three following mechanisms:

(i) Radiation: Electromagnetic waves transfer energy (both heat and light) between two bodies without the necessary aid of an intervening material medium at a speed of 300×10^6 m s^{-1} (i.e. the speed of light). This is so with solar energy through space, whereas the earth's atmosphere only allows the passage of radiation at certain wavelengths and restricts that at others.

(ii) Conduction: Under this mechanism the heat passes through a substance from point to point by means of the transfer of adjacent molecular motions. Since air is a poor conductor this type of heat transfer can be virtually neglected in the atmosphere, but it is important in the ground.

(iii) Convection: This occurs in fluids (including gases) which are able to circulate internally and distribute heated parts of the mass. The low viscosity of air and its consequent ease of motion makes this the chief method of atmospheric heat transfer. It should be noted that *forced convection* (mechanical turbulence) occurs due to the development of eddies as air flows over uneven surfaces, even when there is no surface heating to set up *free* (thermal) *convection*.

Convection transfers energy in two forms. The first is the *sensible heat* content of the air (called enthalpy by physicists) which is transferred directly by the rising and mixing of warmed air. It is defined as c_pT where T is the temperature and c_p ($= 1004\ \mathrm{J\ kg^{-1}\ K^{-1}}$) is the specific heat at constant pressure (the heat absorbed by unit mass for unit temperature increase). Sensible heat is also transferred by conduction. The second form of energy transfer by convection is indirect, involving *latent heat*. Here, there is no temperature change. Whenever water is converted into water vapour by evaporation (or boiling) heat is required. This is referred to as the latent heat of vaporization (L). At 0°C, L is $2{\cdot}50 \times 10^6\ \mathrm{J\ kg^{-1}}$ of water, or $597\ \mathrm{cal\ g^{-1}}$. More generally,

$$L\ (10^6\ \mathrm{J\ kg^{-1}}) \approx (2{\cdot}5 - 0{\cdot}00235T)$$

where T is in °C. When water condenses in the atmosphere (see ch. 2, C) the same amount of latent heat is given off as is used for evaporation *at the same temperature*. Similarly, for melting ice at 0°C, the latent heat of melting is required which is $0{\cdot}335 \times 10^6\ \mathrm{J\ kg^{-1}}$ ($80\ \mathrm{cal\ g^{-1}}$). If ice evaporates, without melting, the latent heat of this sublimation process is $2{\cdot}83 \times 10^6\ \mathrm{J\ kg^{-1}}$ at 0°C ($676\ \mathrm{cal\ g^{-1}}$) (i.e. the sum of the latent heats of melting and vaporization). In all of these phase changes of water there is an energy transfer. We shall return to other aspects of these processes in ch. 2.

2 *Effect of the atmosphere*

Solar radiation is virtually all in the short-wavelength range, less than 4 μm (fig. 1.8). About 18% of the incoming energy is absorbed directly by ozone and water vapour. Ozone absorbs all ultraviolet radiation below $0{\cdot}29\ \mu$m (2900 Å) and water vapour absorbs to a lesser extent in several narrow bands between about $0{\cdot}9\ \mu$m to $2{\cdot}1\ \mu$m (see fig. 1.8). About 30% is immediately reflected back into space from the atmosphere, clouds and the earth's surface, leaving approximately 70% to heat the earth and its atmosphere. Of this, the greater part eventually heats the atmosphere, but much of this heat

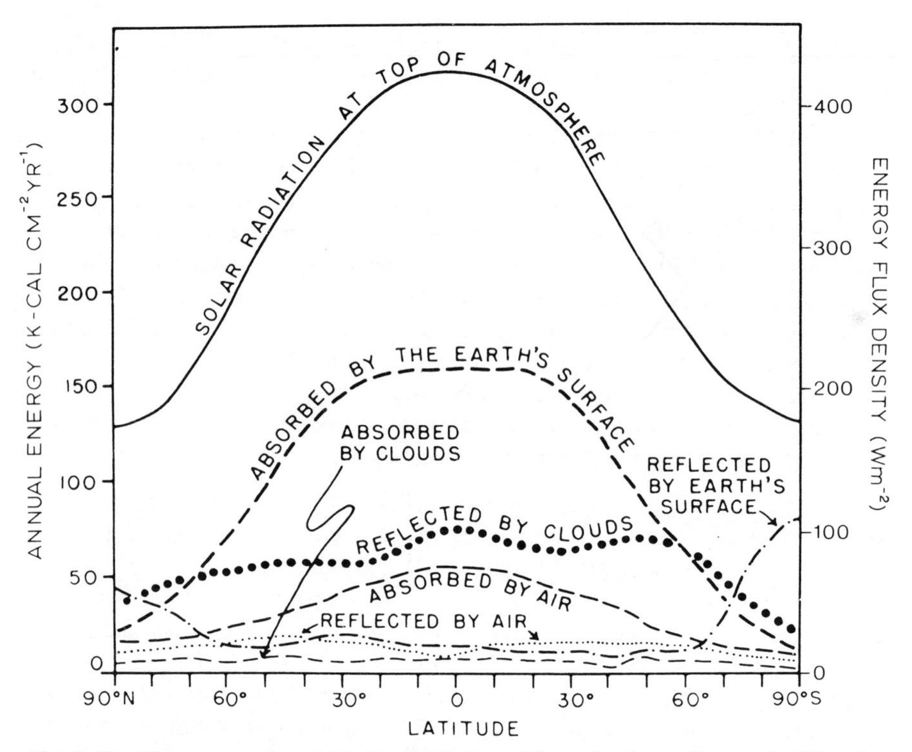

Fig. 1.12. The average annual latitudinal disposition of solar radiation (in W m^{-2} and kcal cm^{-2} yr^{-1}). Of 100% radiation entering the top of the atmosphere, 21% is reflected back to space by clouds, 6% by air (plus dust and water vapour), and 4% by the earth's surface. 3% is absorbed by clouds, 18% by the air, and 48% by the earth (*after Sellers 1965*).

is received secondhand by the atmosphere via the earth's surface. The ultimate retention of this energy by the atmosphere is of prime importance, because if it did not occur the average temperature of the earth's surface would fall by some 40°C (approximately 70°F), making most life obviously impossible. The surface absorbs 48% of the incoming energy available at the top of the atmosphere and re-radiates it outwards as long (infrared) waves of greater than 3 μm (fig. 1.8). Much of this re-radiated long-wave energy can be absorbed by the water vapour, carbon dioxide and ozone in the atmosphere, the rest escaping through atmospheric *windows* back into outer space, principally between 8 and 13 μm (see fig. 1.8). Figure 1.12 illustrates the relative roles of the atmosphere, clouds and the earth's surface in reflecting and absorbing solar radiation at different latitudes. (A more complete analysis of the total heat budget of the earth-atmosphere system is given in ch. 1, F.)

3 *Effect of cloud cover*

Cloud cover can, if it is thick and complete enough, form a significant barrier to the penetration of radiation. How much radiation is actually

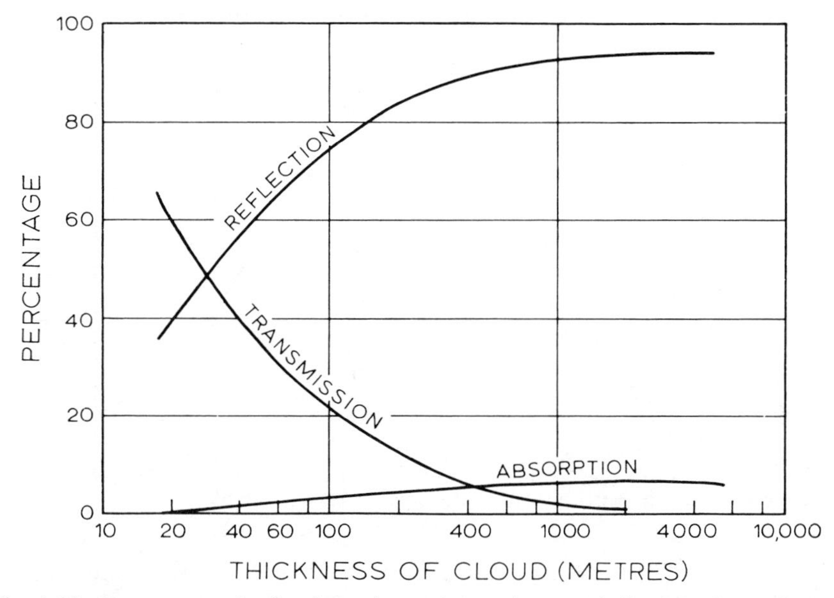

Fig. 1.13. Percentage of reflection, absorption and transmission of solar radiation by cloud layers of different thickness (*from Hewson and Longley 1944*).

reflected depends on the amount of cloud cover and thickness (fig. 1.13). The proportion of incident radiation that is reflected is termed the *albedo*, or reflection coefficient (expressed as a fraction or percentage). Cloud type affects the albedo. Aircraft measurements show that the albedo of a complete overcast ranges from 44 to 50% for cirrostratus to 90% for cumulonimbus. Average albedos, as determined by satellites, aircraft and surface measurements, are summarized in table 1.3 (see note 6).

Table 1.3 The average (integrated) albedo of various surfaces (0·3–4·0 μm)

Planet earth	0·31
Global surface	0·14–0·16
Global cloud	0·23
Cumulonimbus	0·9
Stratocumulus	0·6
Cirrus	0·4–0·5
Fresh snow	0·8–0·9
Melting snow	0·4–0·6
Sand	0·30–0·35
Grass, cereal crops	0·18–0·25
Deciduous forest	0·15–0·18
Coniferous forest	0·09–0·15
Tropical rainforest	0·07–0·15
Water bodies*	0·06–0·10

* Increases sharply at low solar angles.

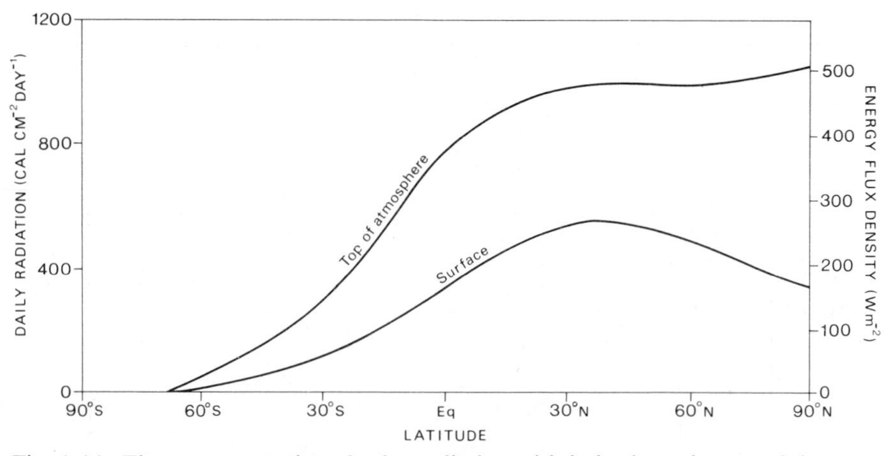

Fig. 1.14. The average receipt of solar radiation with latitude at the top of the atmosphere and at the earth's surface during the June solstice.

The total (or global) solar radiation (direct, Q, and diffuse, q) received at the surface on cloudy days is

$$Q + q = (Q + q)_0[b + (1 - b)(1 - c)]$$

where $(Q + q)_0$ = global solar radiation for clear skies;
c = cloudiness (fraction of sky covered);
b = a coefficient depending on cloud type and thickness; and the depth of atmosphere through which the radiation must pass. For mean monthly values for the United States $b \approx 0 \cdot 35$, so that

$$(Q + q) \approx (Q + q)_0[1 - 0 \cdot 65c].$$

The effect of a cloud cover also operates in reverse, since it serves to retain much of the heat that would otherwise be lost from the earth by radiation throughout the day and night. This largely negative role of clouds means that their presence appreciably lessens the daily temperature range by preventing high maxima by day and low minima by night. As well as interfering with the transmission of radiation, clouds act as temporary thermal reservoirs for they absorb a certain proportion of the energy which they intercept. The effect of this absorption of solar radiation is illustrated in figs. 1.13 and 1.14.

4 *Effect of latitude*

As fig. 1.11 has already shown, different parts of the earth's surface receive different amounts of solar radiation. The time of the year is one factor controlling this, more radiation being received in summer than in winter because of the higher altitude of the sun and the longer days. Latitude is a very important control because the geographical situation of a region will determine both the duration of daylight and the distance travelled through the atmosphere by the oblique rays from the sun. However, actual calculations

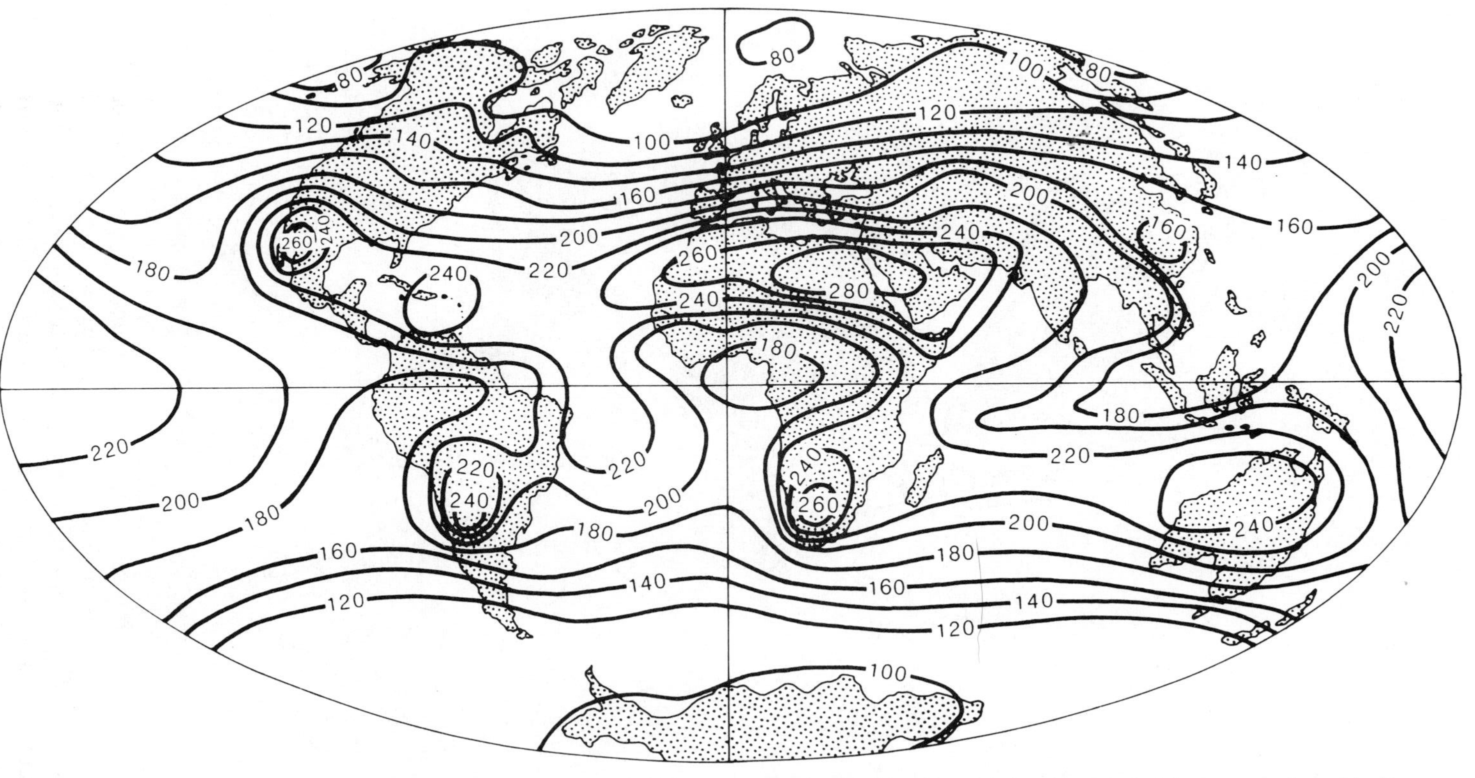

Fig. 1.15. The average annual solar radiation on a horizontal surface at ground level in W m^{-2} (*after Budyko*). Maxima are found in the world's hot deserts, where as much as 80% of the solar radiation annually incident on the top of the unusually clear atmosphere reaches the ground.

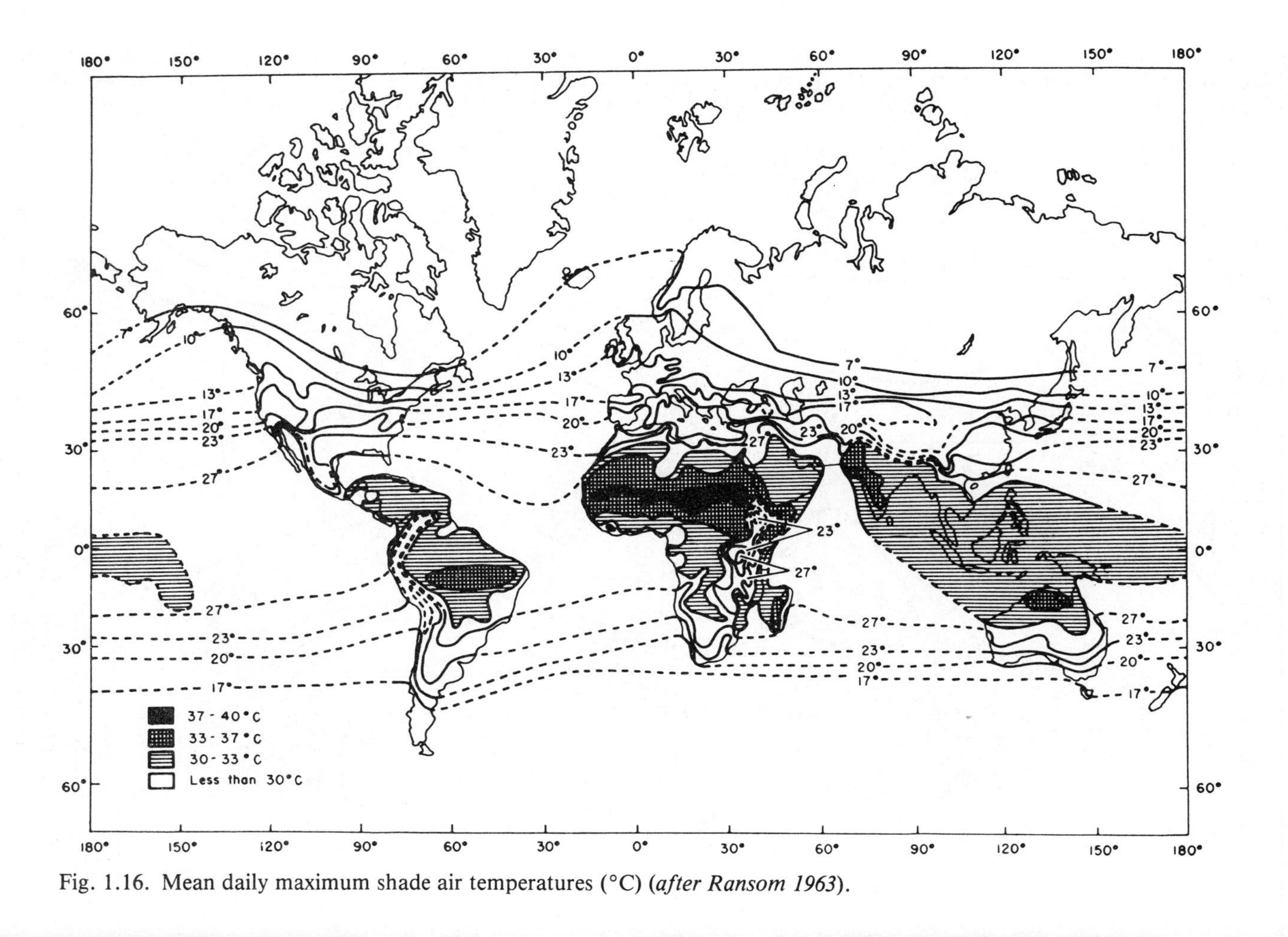

Fig. 1.16. Mean daily maximum shade air temperatures (°C) (*after Ransom 1963*).

show the effect of the latter to be negligible in the Arctic, apparently due to the low vapour content of the air limiting the tropospheric absorption. Figure 1.14 shows that in the upper atmosphere over the north pole there is a marked maximum of solar radiation at the June solstice yet only about 30% is absorbed at the surface. This may be compared with the global average of 45% of solar radiation being absorbed at the surface. The explanation lies in the high average cloudiness over the Arctic in summer and also in the high reflectivity of the snow and ice surfaces. This example illustrates the complexity of the radiation budget and the need to take into account the interaction of several factors.

A special feature of the latitudinal receipt of radiation is that the maximum temperatures experienced at the earth's surface do not occur at the equator, as one might expect, but at the tropics. A number of factors need to be taken into account. The apparent migration of the vertical sun is relatively rapid during its passage over the equator but its rate slows down as it reaches the tropics. Between 6°N and 6°S the sun's rays remain almost vertically overhead for only 30 days during each of the spring and autumn equinoxes, allowing little time for any large build-up of surface heat and high temperatures. On the other hand, between 17·5° and 23·5° latitude the sun's rays shine down almost vertically for 86 consecutive days during the period of the solstice. This longer sustained period, combined with the fact that the tropics experience longer days than at the equator, makes the maximum zones of heating occur nearer the tropics than the equator. In the northern hemisphere this poleward displacement of the zone of maximum heating is emphasized by the effect of *continentality* (see ch. 1, D.5), while low cloudiness associated with the subtropical high-pressure belts is an additional factor. The clear skies are particularly effective in allowing large annual receipts of solar radiation in these areas. The net result of these influences is shown in fig. 1.15 in terms of the average annual solar radiation on a horizontal surface at ground level, and by fig. 1.16 in terms of the average daily maximum shade temperatures. Over the continents the highest values occur at about 23°N and 10°–15°S. In consequence the mean annual *thermal equator* (i.e. the zone of maximum temperature) is located at about 5°N. Nevertheless, the mean air temperatures, reduced to mean sea level, are very broadly related to latitude (figs. 1.17A and B).

5 *Effect of land and sea*

Another important control on the effect of incoming solar radiation stems from the different ways in which land and sea are able to profit from it. Whereas water has a tendency to store the heat it receives, land, in contrast, quickly returns it to the atmosphere. There are several reasons for this.

A large proportion of the incoming solar radiation is reflected back into the atmosphere without heating the earth's surface at all. The proportion depends upon the type of surface (table 1.3). A sea surface reflects very little unless the angle of incidence of the sun's rays is large. The albedo for a calm

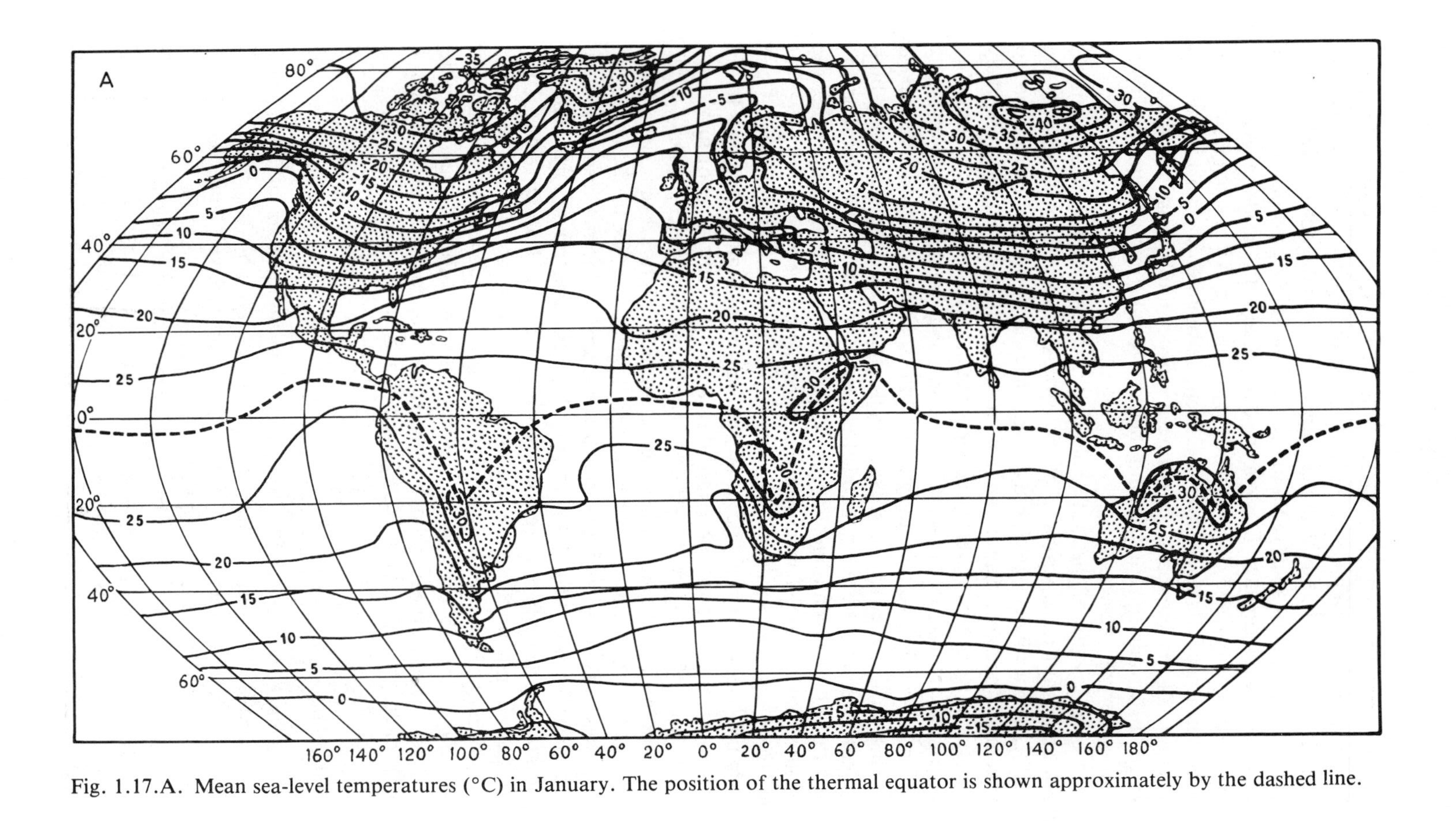

Fig. 1.17.A. Mean sea-level temperatures (°C) in January. The position of the thermal equator is shown approximately by the dashed line.

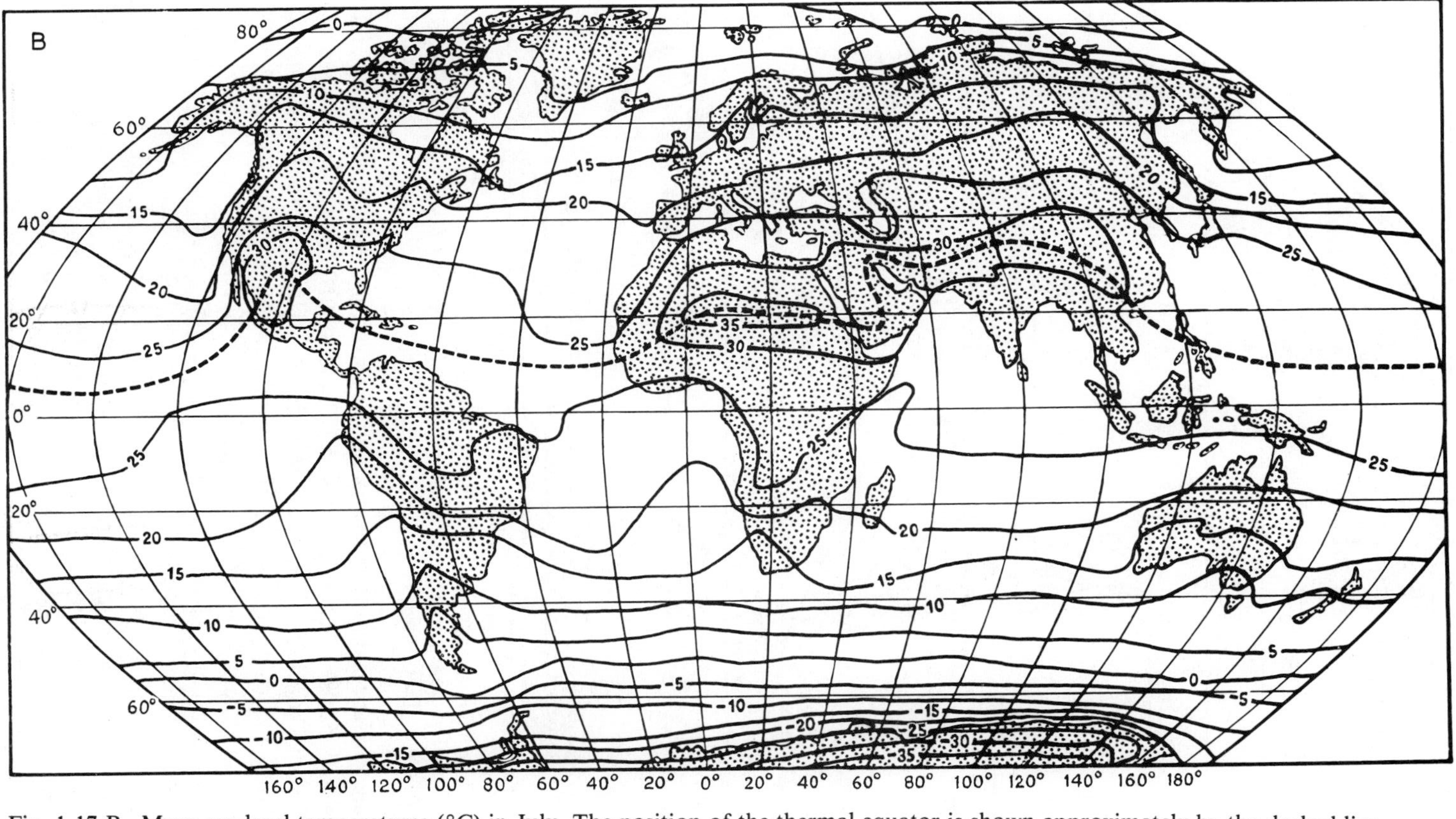

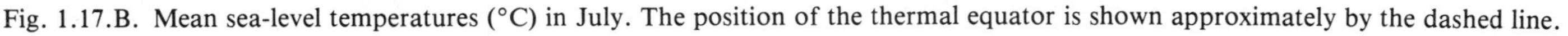
Fig. 1.17.B. Mean sea-level temperatures (°C) in July. The position of the thermal equator is shown approximately by the dashed line.

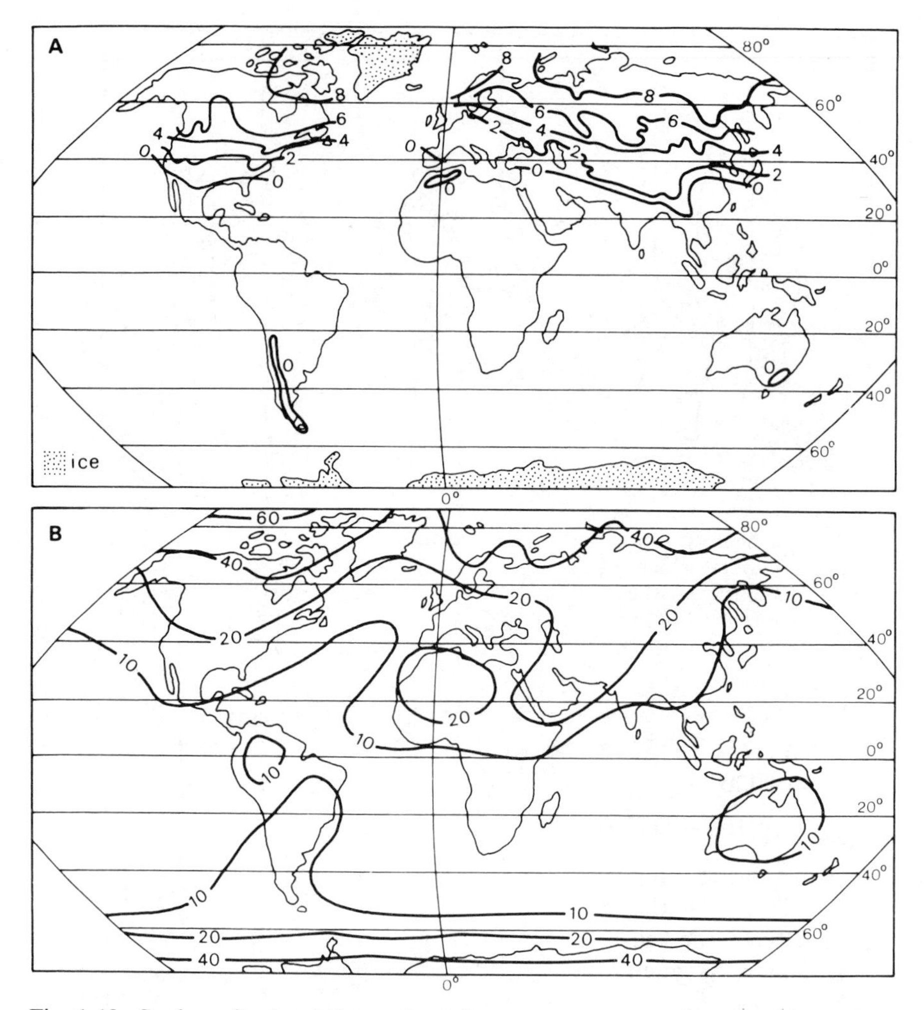

Fig. 1.18. Surface albedo of the earth. A Average snow-cover duration in months. B Annual average surface albedo (%) (*after Hummel and Reck; from Henderson-Sellers and Wilson 1983*).

water surface is only 2 to 3% for a solar elevation angle exceeding 60°, but is more than 50% when the angle is 15°. For land surfaces the albedo is generally between 8 and 40% of the incoming radiation. The figure for forests is about 9 to 18% according to the type of tree and density of foliage (see ch. 7,B), for grass approximately 25%, for cities 14 to 18%, and desert sand 30%. Fresh snow may reflect as much as 85% of solar radiation, but snow cover on vegetated, especially forested, surfaces is much less reflective (30–50%). The long duration of snow cover on the northern continents (fig. 1.18A) causes much of the incoming radiation to be reflected in winter, although global distribution of annual average surface albedo (fig. 1.18B) shows mainly the influence of the snow-covered Arctic sea ice and Antarctic ice sheet (see fig. 1.28A for planetary albedo).

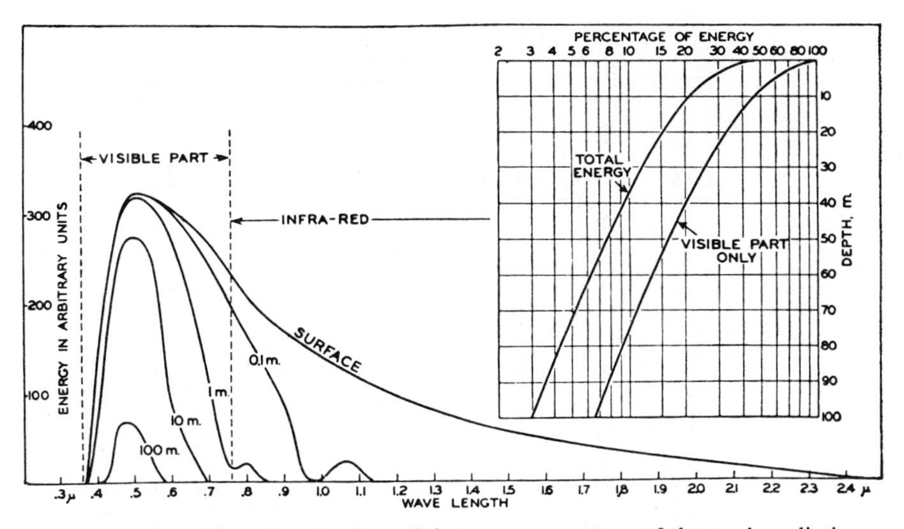

Fig. 1.19. Schematic representation of the energy spectrum of the sun's radiation (in arbitrary units) which penetrates the sea surface to depths of 0·1, 1, 10 and 100 m. This illustrates the absorption of infrared radiation by water, and also shows the depths to which visible (light) radiation penetrates (*from Sverdrup 1945*).

The global solar radiation absorbed at the surface is determined from measurements of radiation incident on the surface and its albedo (a). It may be expressed as

$$(Q + q)(100 - a)$$

where the albedo is a percentage. A snow cover will absorb only about 15% of the incident radiation, whereas for the sea the figure generally exceeds 90%. The ability of the sea to absorb the heat received also depends upon its transparency. As much as 20% of the radiation penetrates as far down as 9 m (30 ft). Figure 1.19 provides some indication of how much energy is absorbed by the sea at different depths. However, the heat absorbed by the sea is carried down to considerable depths by the turbulent mixing of water masses by the action of waves and currents. Figure 1.20, for example, illustrates the warming of the North Sea down to about 40 m in summer. In completely *still* water the annual heat penetration would only be apparent down to about 3–4 m.

A measure of the difference between the subsurfaces of land and sea is given in fig. 1.21, which shows ground temperatures at Kaliningrad (Königsberg) and sea temperature deviations from the annual mean at various depths in the Bay of Biscay. Heat transmission in the soil is carried out almost wholly by conduction and the degree of conductivity varies with the moisture content and porosity of each particular soil.

Air is an extremely poor conductor and for this reason a loose, sandy soil surface heats up rapidly by day, as the heat is not conducted away. Increased soil moisture tends to raise the conductivity by filling the soil pores, but too

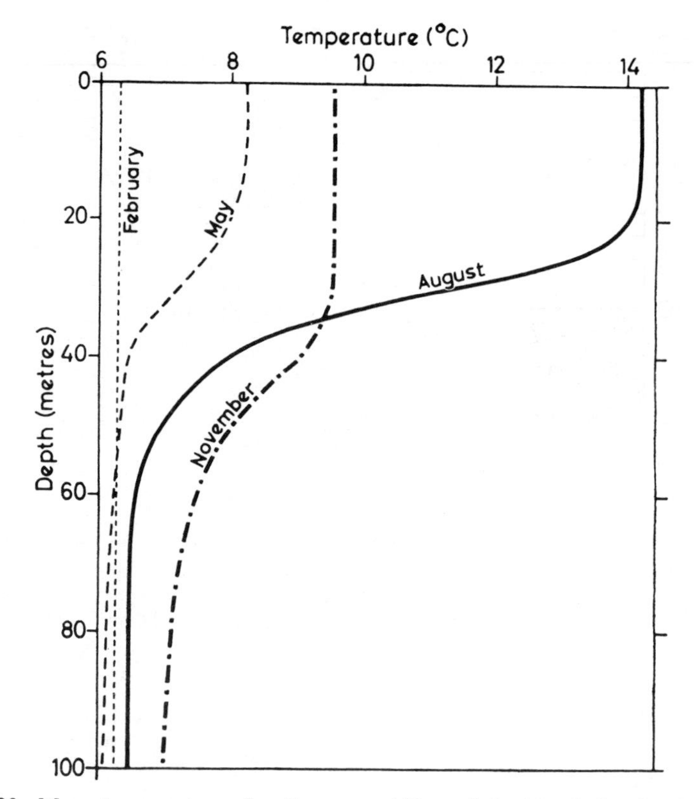

Fig. 1.20. Mean temperatures for the upper 100 m of the North Sea for February, May, August and November (*from Lumb 1961; Crown Copyright Reserved*).

much moisture increases the soil's heat capacity, thereby reducing the temperature response. The relative depths over which the annual and diurnal temperature variations are effective in wet and dry soils are roughly as follows:

	Diurnal variation	*Annual variation*
Wet soil	0·5 m	9 m
Dry sand	0·2 m	3 m

However, the *actual* temperature change is greater in dry soils. For example, the following values of diurnal temperature range have been observed during clear summer days at Sapporo, Japan:

	Sand	*Loam*	*Peat*	*Clay*
Surface	40°C	33°C	23°C	21°C
5 cm	20	19	14	14
15 cm	7	6	2	4

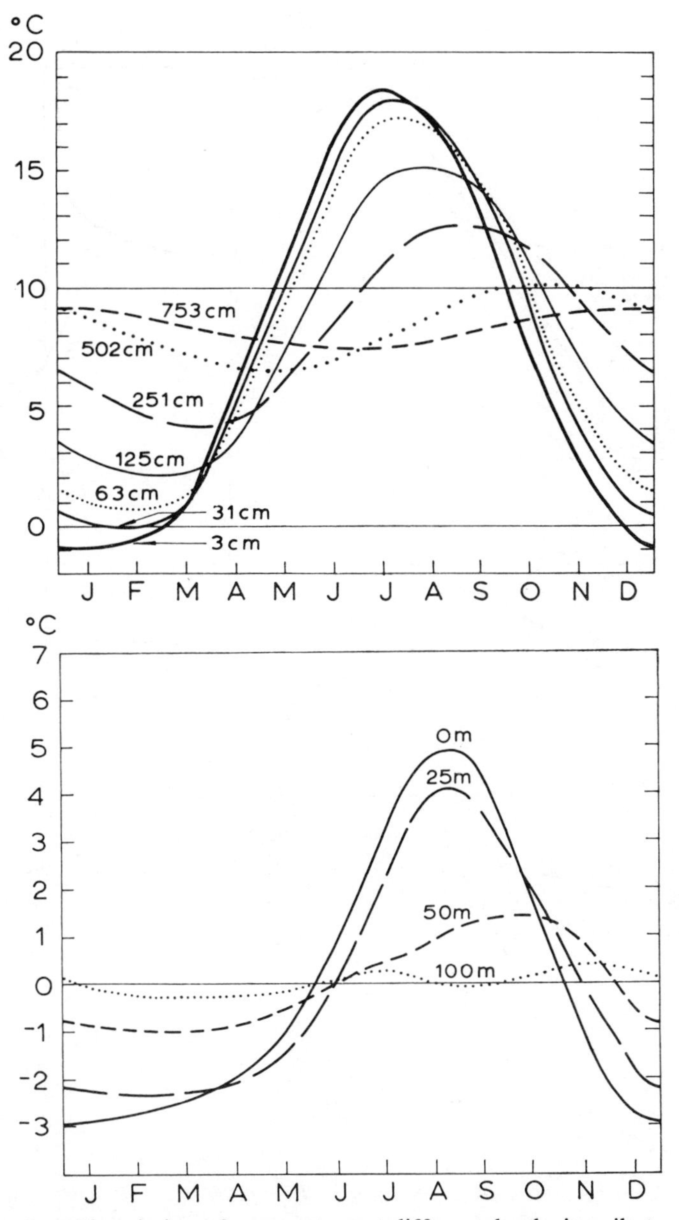

Fig. 1.21. Annual variation of temperature at different depths in soil at Kaliningrad (*above*) and in the water of the Bay Biscay (at approximately 47°N, 12°W) (*below*), illustrating the relatively deep penetration of solar energy into the oceans as distinct from that into land surfaces. The bottom figure shows the temperature deviations from the annual mean for each depth (*from Geiger 1965 and Sverdrup 1945*).

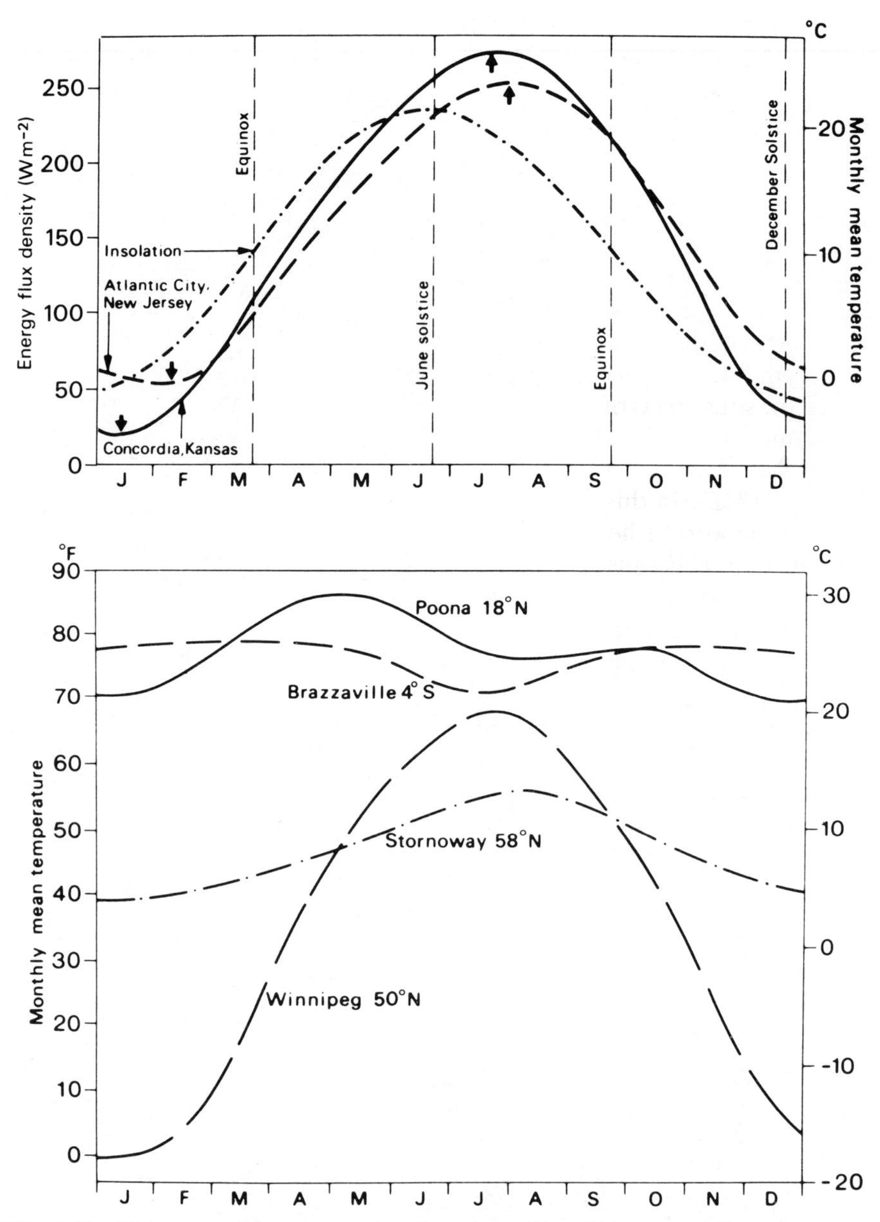

Fig. 1.22. Mean annual temperature regimes in various climates and the relationships with solar radiation. *Above* Temperatures at maritime (Atlantic City) and continental (Concordia, Kansas) locations in the middle latitudes. A curve of representative solar radiation is also given. Maximum and minimum points are indicated on the temperature curves, illustrating the respective time lags behind the radiation curve (*data from Trewartha; after Strahler 1951*). *Below* Mean annual temperature regimes for Poona (Monsoon), Brazzaville (Equatorial), Stornoway (Temperate maritime) and Winnipeg (Temperate continental).

The different heating qualities of land and water are also partly accounted for by their different *specific heats*. The specific heat (c) of a substance can be represented by the number of thermal units required to raise a unit mass of it through 1°C. In cgs units the specific heat of water is conveniently $1 \cdot 0$ cal g^{-1} deg^{-1} (4184 J kg^{-1} K^{-1}). The specific heat of water is much greater than for most other common substances, and water must absorb five times as much heat energy to raise its temperature by the same amount as a comparable mass of dry soil. Thus for dry sand $c = 840$ J kg^{-1} K^{-1} ($0 \cdot 2$ cal g^{-1} deg^{-1}).

If unit volumes of water and soil are considered the heat capacity, ϱc, of the water, where ϱ = density ($\varrho c = 4 \cdot 18 \times 10^6$ J m^{-3} K^{-1}, or $1 \cdot 0$ cal cm^{-3} deg^{-1}), exceeds that of the sand approximately threefold ($\varrho c = 1 \cdot 3 \times 10^6$ J m^{-3} K^{-1}, or $0 \cdot 3$ cal cm^{-3} deg^{-1}) if the sand is dry and twofold if it is wet. When this water is cooled the situation is reversed, for then a large quantity of heat is released. A metre-thick layer of sea water being cooled by as little as 0·1°C will release enough heat to raise the temperature of approximately a 30-m thick air layer by 10°C (18°F). In this way the oceans act as a very effective reservoir for much of the world's heat. Similarly evaporation of sea water causes a large heat expenditure because a great amount of energy is needed to evaporate even a small quantity of water (see ch. 2, A).

These differences between land and sea help to produce what is termed *continentality*. Continentality implies, firstly, that a land surface heats and cools much quicker than that of an ocean. Over the land the lag between maximum and minimum periods of radiation and the maximum and minimum surface temperatures is only 1 month, but over the ocean and at coastal stations the lag is as much as 2 months (fig. 1.22). Secondly, the annual and diurnal ranges of temperature are greater in continental than in coastal locations. Figure 1.22 illustrates the annual variation of temperature at Winnipeg and Stornoway, while fig. 1.29C shows the diurnal ranges experienced in continental and maritime areas. This is described more fully below. The third effect of continentality results from the global distribution of the land masses. The small sea area of the northern hemisphere causes the northern hemisphere summer to be warmer but its winters colder on the average than those of the southern hemisphere (summer, 22·4°C (72·3°F) versus 17·1°C (62·7°F); winter, 8·1°C (46·5°F) versus 9·7°C (49·5°F)). Heat storage in the oceans causes them to be warmer in winter and cooler in summer than land in the same latitude, although ocean currents give rise to some local departures from this rule. The distribution of temperature anomalies for the latitude in January and July (fig. 1.23) illustrates the significance of continentality and also the influence of the warm drift currents in the North Atlantic and the North Pacific in winter (compare fig. 3.38).

Sea temperatures can now be estimated by the use of infrared satellite imagery (see ch. 1, E). Plate 4 is an infrared photograph taken at night of the south-east coast of the United States in which sea surface temperatures appear in various shades of grey with the darkest areas representing the relatively warm, meandering Gulf Stream. From such images, maps of sea-surface temperatures are now routinely constructed, as illustrated in fig. 1.24.

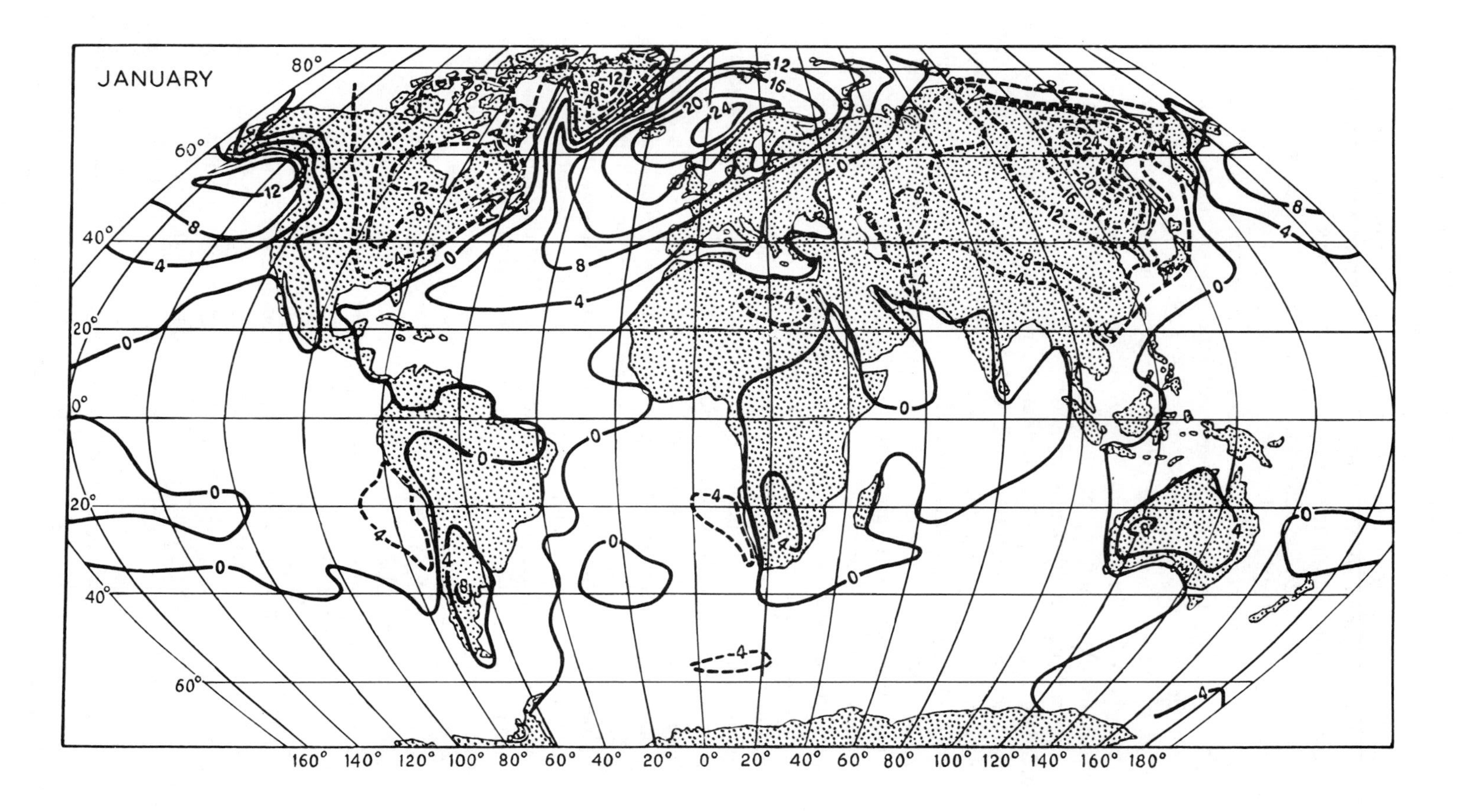
JANUARY

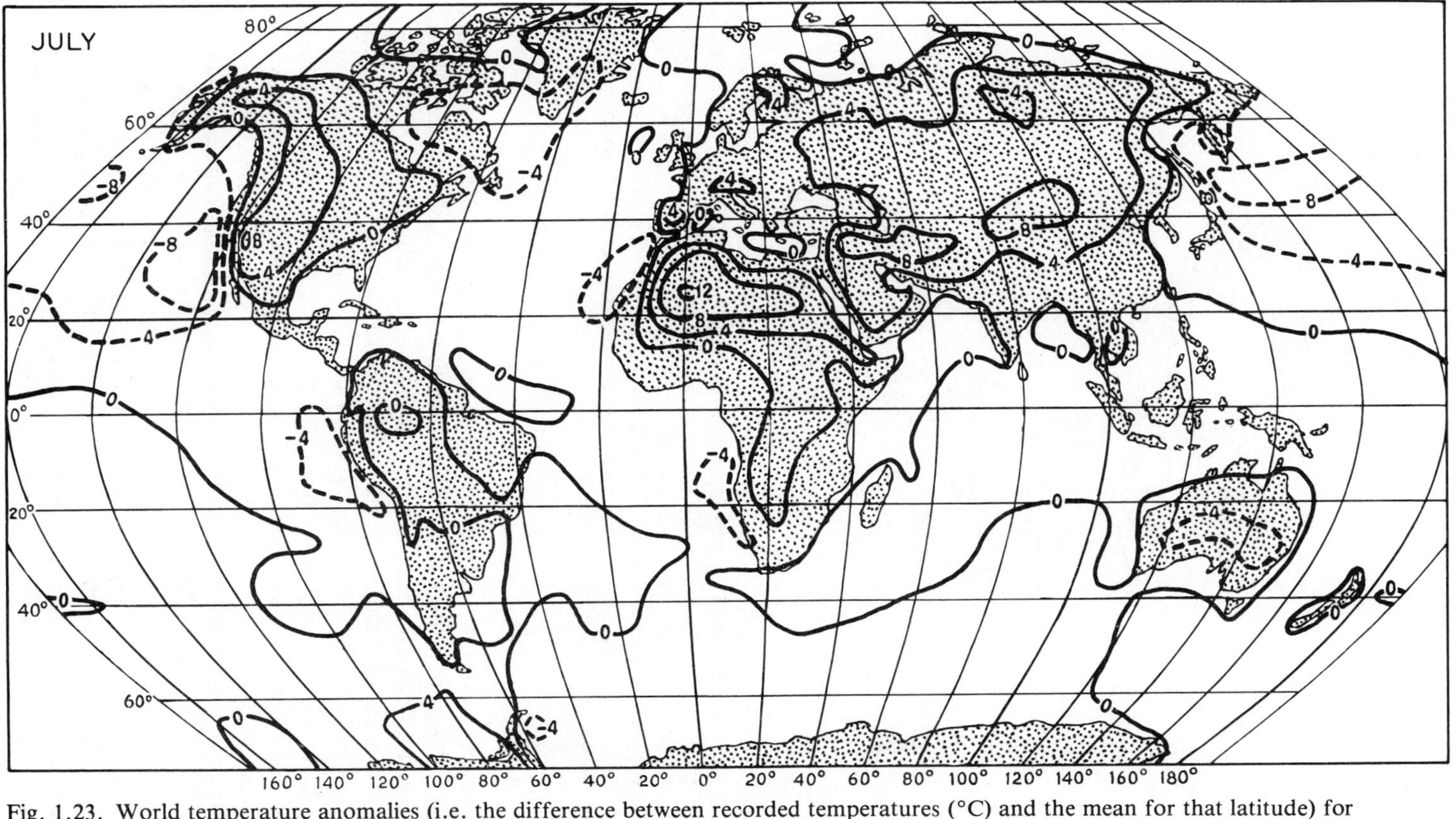

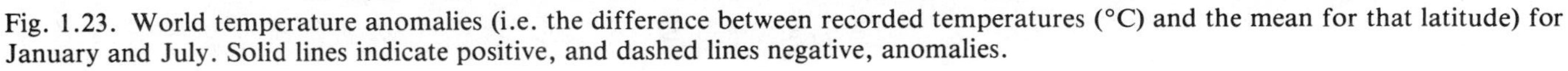
Fig. 1.23. World temperature anomalies (i.e. the difference between recorded temperatures (°C) and the mean for that latitude) for January and July. Solid lines indicate positive, and dashed lines negative, anomalies.

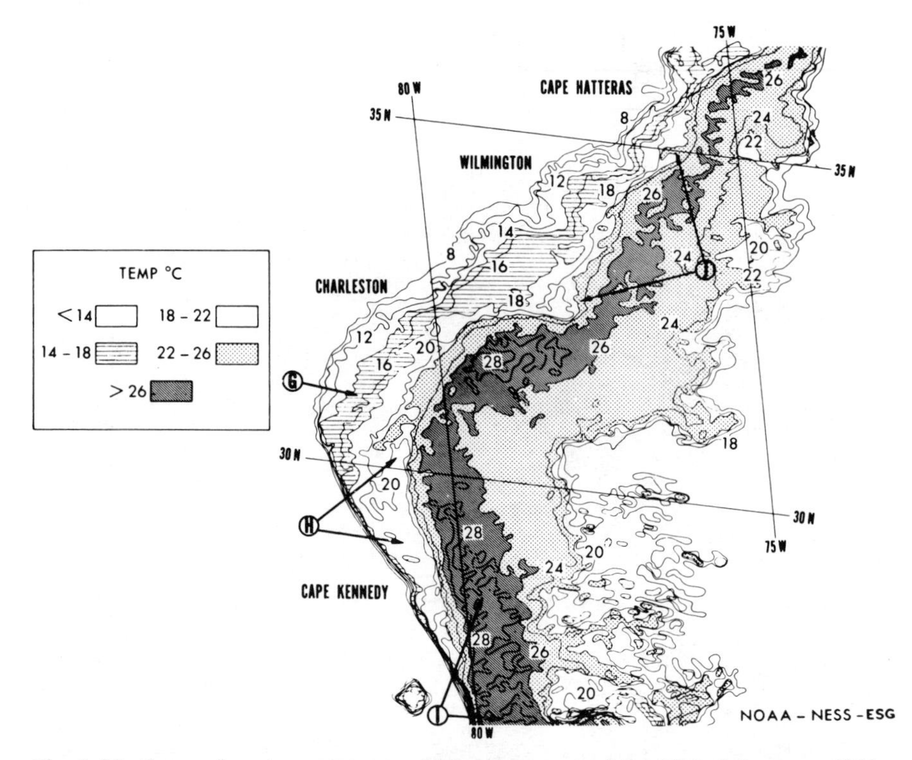

Fig. 1.24. Sea-surface temperatures off the east coast of the United States at 0900 GMT on 15 February 1971, estimated from infrared imagery (see pl. 4). Numbers show the point temperatures which were measured by scanning radiometer. The coldest shelf waters (8°–14°C) are indicated at G; the intermediate slope waters (H) have surface temperatures of 14°–22°C; the Gulf Stream surface (I) is at 26°–28°C and shows steep temperature gradients along certain of its margins (J) (*after Rao et al.; from WMO 1973*).

6 Effect of elevation and aspect

When we come down to the local scale, even differences in the elevation of the land and its *aspect* (that is the direction which the surface faces) will strikingly control the amount of solar radiation received.

Obviously some slopes are more exposed to the sun than others, whereas really high elevations that have a much smaller mass of air above them (see fig. 1.6) receive considerably more direct solar radiation under clear skies than locations near sea-level, particularly below 2000–3000 m due to the concentration of water vapour in the lower troposphere (fig. 1.25). On the average in middle latitudes the intensity of incident solar radiation increases by 5–15% for each 1000 m increase in elevation in the lower troposphere. The difference between sites at 200 and 3000 m in the Alps, for instance, can amount to 70 W m^{-2} on cloudless summer days. However, there is also a correspondingly greater net loss of terrestrial radiation at higher elevations because the low density of the overlying air results in a smaller fraction of the outgoing radiation being absorbed. The overall effect is invariably

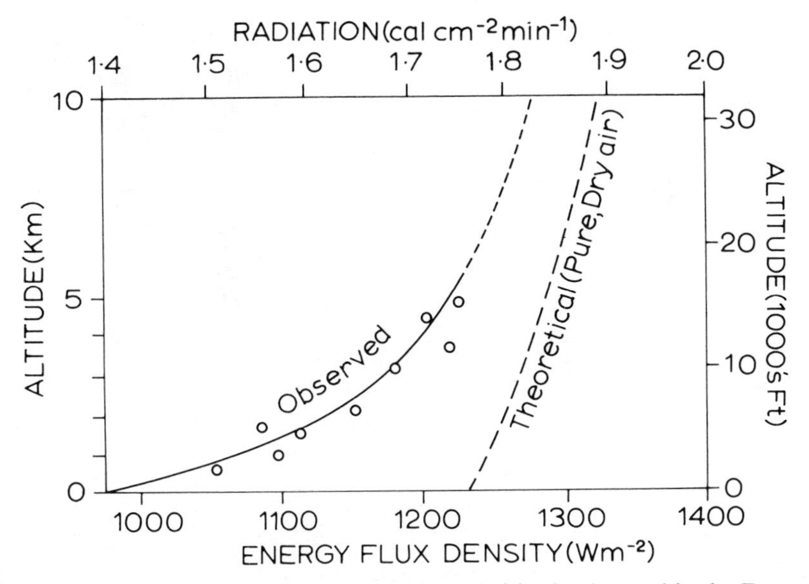

Fig. 1.25. Direct solar radiation as a function of altitude observed in the European Alps. The absorbing effects of water vapour and dust, particularly below about 3000 m, are shown by comparison with a theoretical curve for an ideal atmosphere (*after Albetti, Kastrov, Kimball and Pope; from Barry 1981*).

complicated by the greater cloudiness associated with most mountain ranges and it is therefore impossible to generalize from the limited data at present available.

Figure 1.26 illustrates the effect of aspect and slope angle on theoretical maximum solar radiation receipts at two locations in the northern hemisphere. The general effect of latitude on insolation amounts is clearly shown, but it is also apparent that increasing latitude causes a relatively greater radiation loss for north-facing slopes, as distinct from south-facing ones. The radiation intensity on a sloping surface (I_s) is

$$I_s = I_0 \cos i$$

where i = the angle between the solar beam and a beam normal to the sloping surface. Relief may also affect the quantity of insolation and the duration of direct sunlight when a mountain barrier screens the sun from valley floors and sides at certain times a day. In many alpine valleys settlement and cultivation are noticeably concentrated on southward-facing slopes (the adret or sunny side), whereas northward slopes (ubac or shaded side) remain forested.

E Infrared radiation from the earth

Radiation from the sun is predominantly short wave, whereas that leaving the earth is long wave, or infrared, radiation (see fig. 1.8). The infrared emission from the surface is slightly less than that from a black body at the

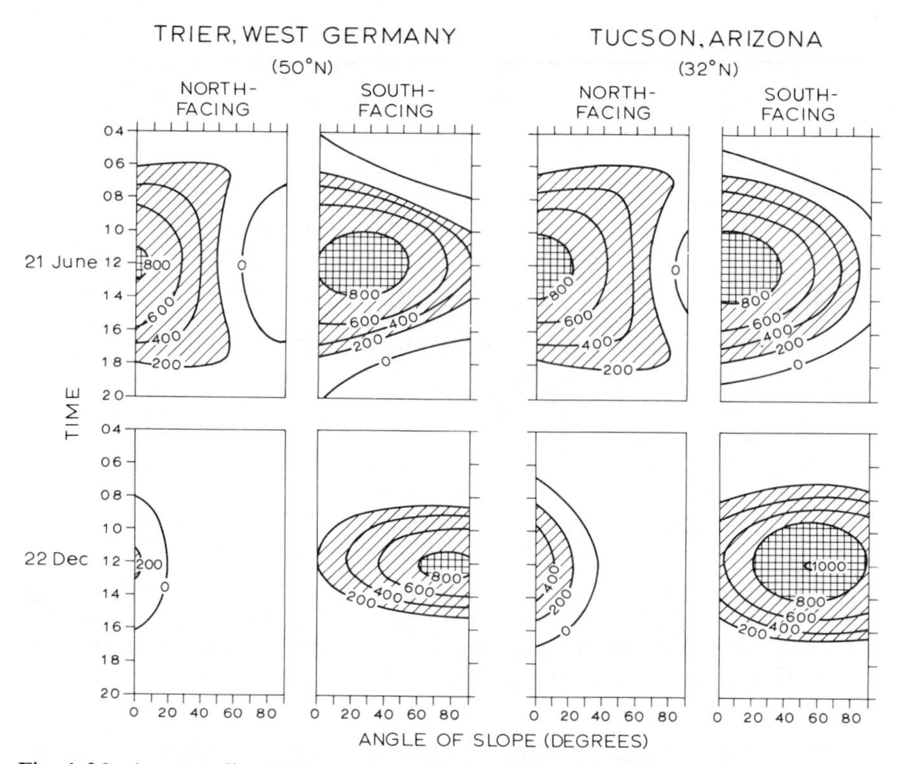

Fig. 1.26. Average direct beam solar radiation (W m^{-2}) incident at the surface under cloudless skies at Trier, West Germany, and Tucson, Arizona, as a function of slope, aspect, time of day and season of year (*after Geiger 1965 and Sellers 1965*).

same temperature and, accordingly, Stefan's equation (p. 12) is modified by an emissivity coefficient (ε) which is generally between 0·90 and 0·95, i.e. $F = \varepsilon\sigma T^4$. Figure 1.8 shows that the atmosphere is highly absorbent to infrared radiation (due to the effects of water vapour, carbon dioxide and ozone), except between about 8·5 and 13·0 μm – the 'atmospheric window'. The opaqueness of the atmosphere to infrared radiation, relative to its transparency to short-wave radiation, is commonly referred to as the 'greenhouse effect'. However, in the case of a greenhouse the effect of the glass is probably as significant in reducing cooling by restricting the turbulent heat loss as it is in retaining the infrared radiation.

The net warming contribution of the greenhouse gases to the mean 'effective' planetary temperature of 254 K (corresponding to the emitted infrared radiation) is approximately 33 K; of this, water vapour accounts for 21 K, carbon dioxide 7 K and ozone 2 K. The present global mean surface temperature is about 287 K, but the surface was considerably warmer during the early evolution of the earth when the atmosphere contained large quantities of methane, water vapour and ammonia.

It is worth emphasizing that long-wave radiation is not merely terrestrial in the narrow sense. The atmosphere radiates to space and clouds are particularly effective since they act as black bodies. For this reason cloudiness and cloud top temperature can be mapped from satellites by day and night using infrared sensors (see pls. 2 and 21). Radiative cooling of cloud layers averages about 1·5°C per day.

F Heat budget of the earth

We can now summarize the net effect of the transfers of energy in the earth–atmosphere system averaged over the globe and over an annual period.

The incident solar radiation averaged over the globe is

$$\text{Solar constant} \times \pi r^2/4\pi r^2$$

where r = radius of the earth and $4\pi r^2$ is the surface area of a sphere. This figure is approximately 342 W m^{-2}, or 11×10^9 J m^{-2} yr^{-1} (10^9 J = GJ); for convenience we will regard it as 100 units. Referring to fig. 1.27, incoming radiation is absorbed in the stratosphere (3 units) by ozone mainly, and 18 units are absorbed in the troposphere by carbon dioxide (1), water vapour (12), dust (2) and water droplets in clouds (3). Twenty-one units are reflected back to space from clouds which cover about 55% of the earth's surface, on average. A further 4 units are similarly reflected from the surface and 6 units are returned by atmospheric scattering. The total reflected radiation is the *planetary albedo* (31% or 0·31). The remaining 48 units reach the earth either directly (27) ór as diffuse radiation (21) transmitted via clouds or by downward scattering. The scattering effect of air molecules and dust particles on the visible wavelengths of radiation (blue light = 0·4 μm, red = 0·7 μm) is greatest at short wavelengths, and hence the sky light appears blue in colour.

The pattern of outgoing terrestrial radiation is quite different (fig. 1.27). The black-body radiation, assuming a mean surface temperature of 287 K is equivalent to 113 units of infrared (long-wave) radiation. This is possible because most of the outgoing radiation is re-absorbed by the atmosphere, as described above. Also, the infrared exchanges involve the whole globe, whereas solar radiation affects only the sunlit hemisphere. Only about 6 units escape through the atmospheric window directly from the surface, but the atmosphere radiates 63 units to space (37 from the emission by water vapour and CO_2 in the atmosphere and 26 from cloud emission), giving a total of 69 units, as well as re-radiating 97 units back to the surface (L_d); L_u + L_d = L_n is negative.

These radiation transfers can be expressed symbolically:

$$R_n = (Q + q)(1 - a) + L_n$$

where R_n = net radiation, $(Q + q)$ = global solar radiation, a = albedo and L_n = net long-wave radiation. At the surface R_n = 32 units. This surplus is

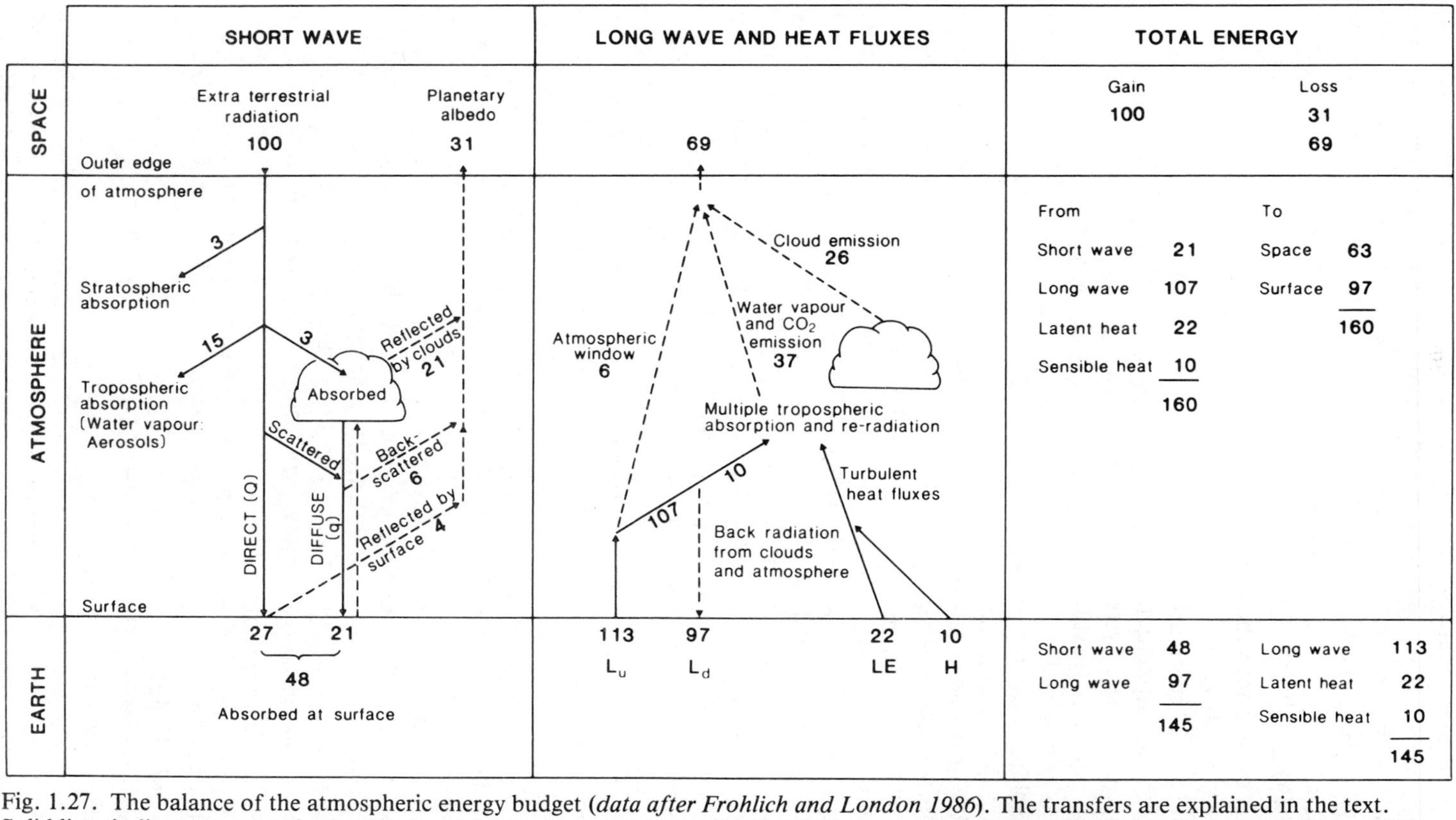

Fig. 1.27. The balance of the atmospheric energy budget (*data after Frohlich and London 1986*). The transfers are explained in the text. Solid lines indicate energy gains by the atmosphere and surface in the left-hand diagram and the troposphere in the right-hand diagram. The exchanges are referred to 100 units of incoming solar radiation at the top of the atmosphere (equal to 342 W m^{-2} or 0·5 cal cm^{-2} min^{-1}).

conveyed to the atmosphere by the turbulent transfer of sensible heat, or enthalpy (10 units) and latent heat (22 units).

$$R_n = LE + H$$

where H = sensible heat transfer and LE = latent heat transfer. There is also a flux of heat into the ground (ch. 1, D.5), but for annual averages this is approximately zero.

Figure 1.27 summarizes the total balances at the surface (± 145 units) and for the atmosphere (± 160 units). The energy balance for the entire earth-atmosphere system is estimated to be ± 7 GJ m^{-2} yr^{-1} (± 66 units). These estimates are still rather crude, however, and various uncertainties are still to be resolved.

Satellite measurements now provide global views of the energy balance at the top of the atmosphere. The incident solar radiation is almost symmetrical about the equator in the annual mean (cf. table 1.2) The mean annual totals on a horizontal surface at the top of the atmosphere are approximately 420 W m^{-2} at the equator and 180 W m^{-2} at the poles. The distribution of the planetary albedo (fig. 1.28A) shows the lowest values over the low-latitude oceans compared with the more persistent areas of cloud cover over the continents. The highest values are over the polar icecaps. Correspondingly, the net (outgoing) long-wave radiation (fig. 1.28B) shows the smallest losses where the temperatures are lowest and largest losses over the largely clear skies of the Saharan desert surface and over low-latitude oceans. The net radiation (fig. 1.28C) shows that the earth-atmosphere system achieves balance about latitude 30°N. The consequences of a low-latitude energy surplus and a high-latitude deficit are examined in the following section (1.G).

The annual and diurnal variations of temperature are directly related to the local energy budget. Under clear skies, in middle and lower latitudes, the diurnal regime of radiative exchanges generally shows a midday maximum of absorbed solar radiation (fig. 1.29A). A maximum of infrared (long-wave) radiation (see fig. 1.8) is also emitted by the heated ground surface at midday when it is warmest. The atmosphere re-radiates infrared radiation downward but there is a net loss at the surface (L_n). The difference between the absorbed solar radiation and L_n is the net radiation, R_n; this is generally positive between about an hour after sunrise and an hour or so before sunset with a midday maximum. The delay in the occurrence of the maximum air temperature until about 1400 hours local time (fig. 1.29B) is caused by the gradual heating of the air by convective transfer from the ground. Minimum R_n occurs in the early evening when the ground is still warm; there is a slight increase thereafter. The temperature decrease after midday is slowed by heat supplied from the ground. Minimum air temperature occurs shortly after sunrise due to the lag in the transfer of heat from the surface to the air. The annual pattern of the net radiation budget and temperature regime is closely analogous to the diurnal one.

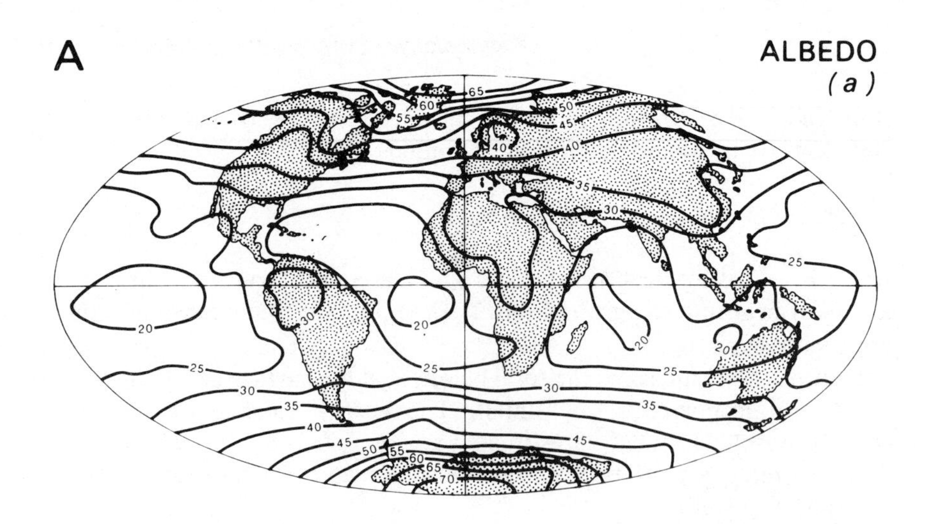

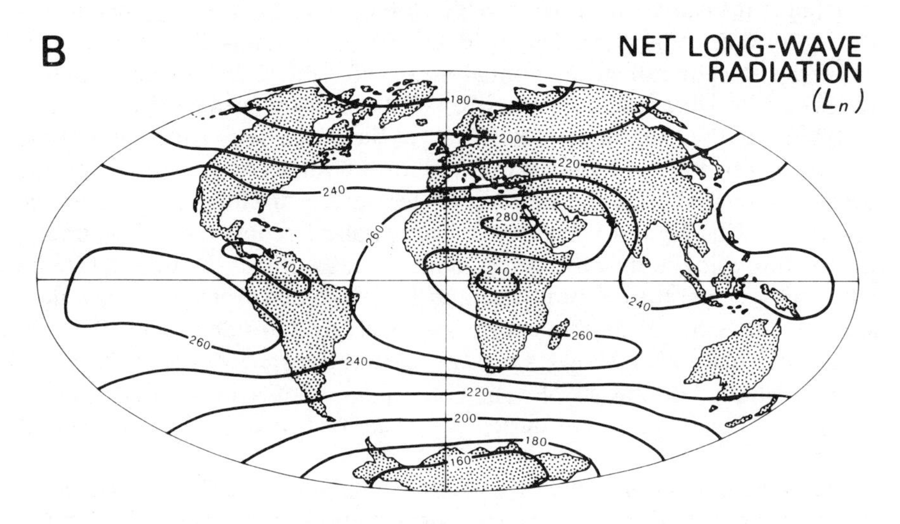

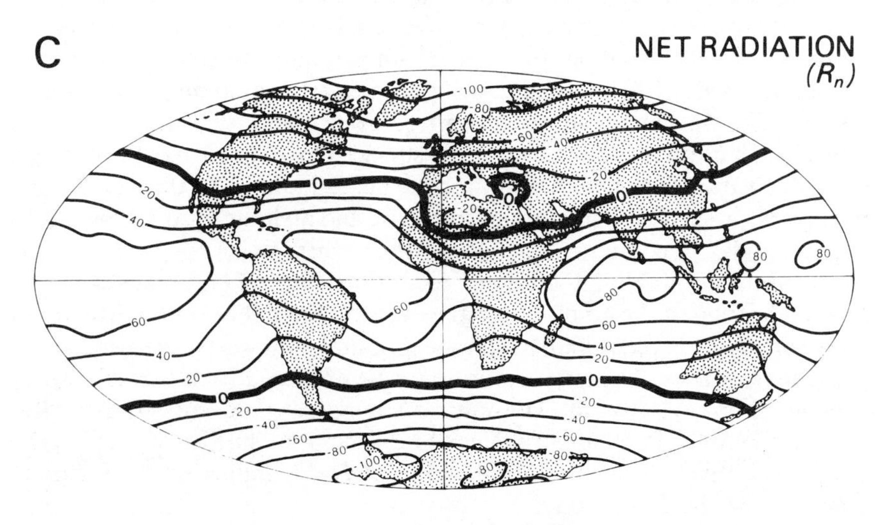

Fig. 1.28. Average annual planetary albedo (%) (A), net long-wave radiation (W m^{-2}) (B) and net radiation (W m^{-2}) (C) on a horizontal surface at the top of the atmosphere (*from satellite data analysed by Stephens et al. 1981*).

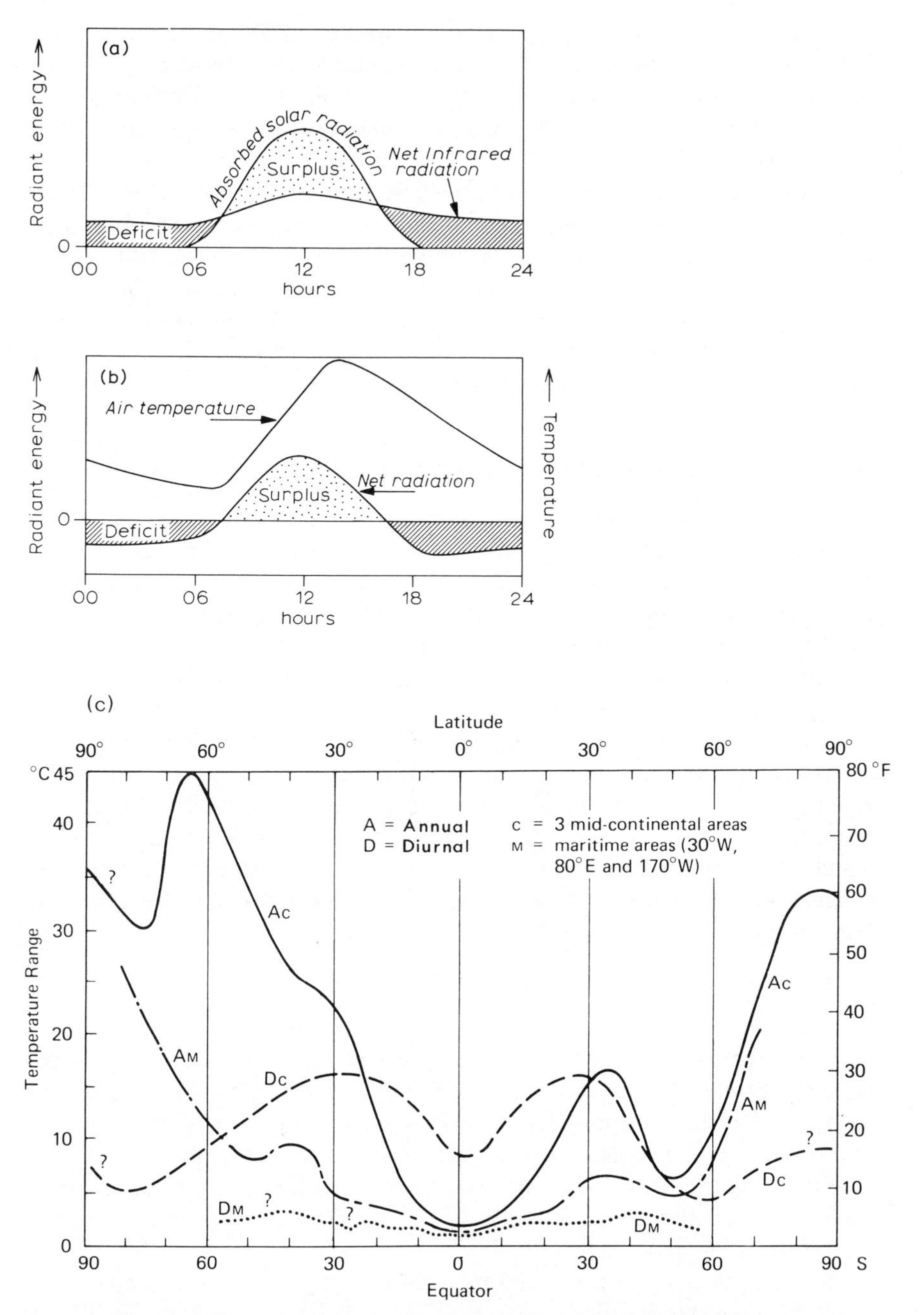

Fig. 1.29. Curves showing diurnal and annual variations of radiant energy and temperature. A Diurnal variations in absorbed solar radiation and infrared radiation in the middle and low latitudes. B Diurnal variations in net radiation and air temperature in the middle and low latitudes. C Annual and diurnal temperature ranges as a function of latitude and of continental or maritime location (*from Paffen 1967*).

There are marked latitudinal variations in the diurnal and annual ranges of temperature. Broadly, the annual range is a maximum in higher latitudes, with extreme values about 65°N related to the effects of continentality in Asia and North America. The diurnal range reaches a maximum at the tropics over land areas, but it is in the equatorial zone that the diurnal variation of heating and cooling exceeds the annual one (fig. 1.29C). This is of course related to the small seasonal change in solar elevation angle at the equator. From the point of view of the total energy budget, the atmosphere, oceans and upper crustal skin of the earth form a complex system of storages and transfers. The global climate as a whole is dominated by the relatively small energy storages provided by the crustal skin and the atmosphere, compared with that of the oceans (see ch. 3).

G Atmospheric energy and horizontal heat transport

So far, we have described the gases and other constituents that make up our atmosphere, and have given some account of the earth's heat budget. We have already referred to two forms of energy: internal (or heat) energy due to the motion of individual air molecules and latent energy which is released by condensation of water vapour. Two other forms of energy are important: geopotential energy due to gravity and height above the surface and kinetic energy associated with air motion.

Geopotential and internal energy are interrelated, since the addition of heat to an air column not only increases its internal energy, but adds to its geopotential as a result of the vertical expansion of the air column. In a column extending to the top of the atmosphere the geopotential is approximately 40% of the internal energy. These two are therefore usually considered together and termed the total potential energy (*PE*). For the whole atmosphere

potential energy $\approx 10^{24}$ J ($23 \cdot 4 \times 10^{22}$ cal)
kinetic energy $\approx 10^{20}$ J

In a later section (ch. 3, F) we shall see how energy is transferred from one form to another, but here we need only be concerned with heat energy. It is apparent that the receipt of heat energy is very unequal geographically and that this must lead to great lateral transfers of energy across the surface of the earth. Much present-day meteorological research is focused on these transfers, since undoubtedly they give rise, at least indirectly, to the observed patterns of global weather and climate.

The amounts of energy received at different latitudes vary substantially, the equator on the average receiving 2·5 times as much annual solar energy as the poles. Clearly if this process were not modified in some way the variations in receipt would cause a massive accumulation of heat within the tropics (associated with gradual increases of temperature) and a corresponding deficiency at the poles. Yet this does not seem to happen, and the earth as a whole is roughly in a state of thermal equilibrium in so far as no one region is

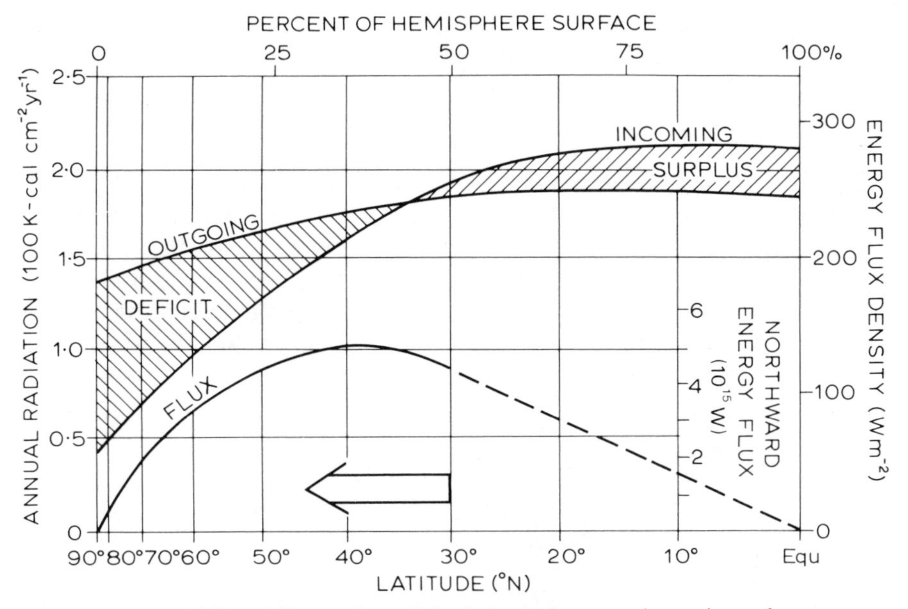

Fig. 1.30. A meridional illustration of the balance between incoming solar radiation and outgoing radiation from the earth and atmosphere (*data from Houghton; after Newell 1964*), in which the zones of permanent surplus and deficit are maintained in equilibrium by a poleward energy transfer (*after Gabites*).

obviously gaining heat at the expense of another. Some authors believe the Ice Ages to have been an exception to this rule. One explanation of this equilibrium could be that for each region of the world there is an equalization between the amount of incoming and outgoing radiation. Observation shows that this is not so (fig. 1.30), however, for, whereas incoming radiation varies appreciably with changes in latitude, being highest at the equator and declining to a minimum at the poles, outgoing radiation has a more even latitudinal distribution owing to the rather small variations in atmospheric temperature. Some other explanation therefore becomes necessary.

1 The horizontal transport of heat

If the net radiation for the whole earth-atmosphere system is calculated, it is found that there is a positive budget between 35°S and 40°N as shown in figs. 1.28c and 1.31. The latitudinal belts in each hemisphere separating the zones of positive and negative net radiation budgets oscillate dramatically with season (fig. 1.32). As the tropics do not get progressively hotter or the high latitudes colder, a redistribution of world heat energy must constantly occur, taking the form of a continuous movement of energy from the tropics to the poles. In this way the tropics shed their excess heat and the poles are not allowed to reach extremes of cold. If there were no meridional interchange of heat, a radiation balance at each latitude would only be achieved if the equator were 14°C warmer and the north pole 25°C colder than now. This

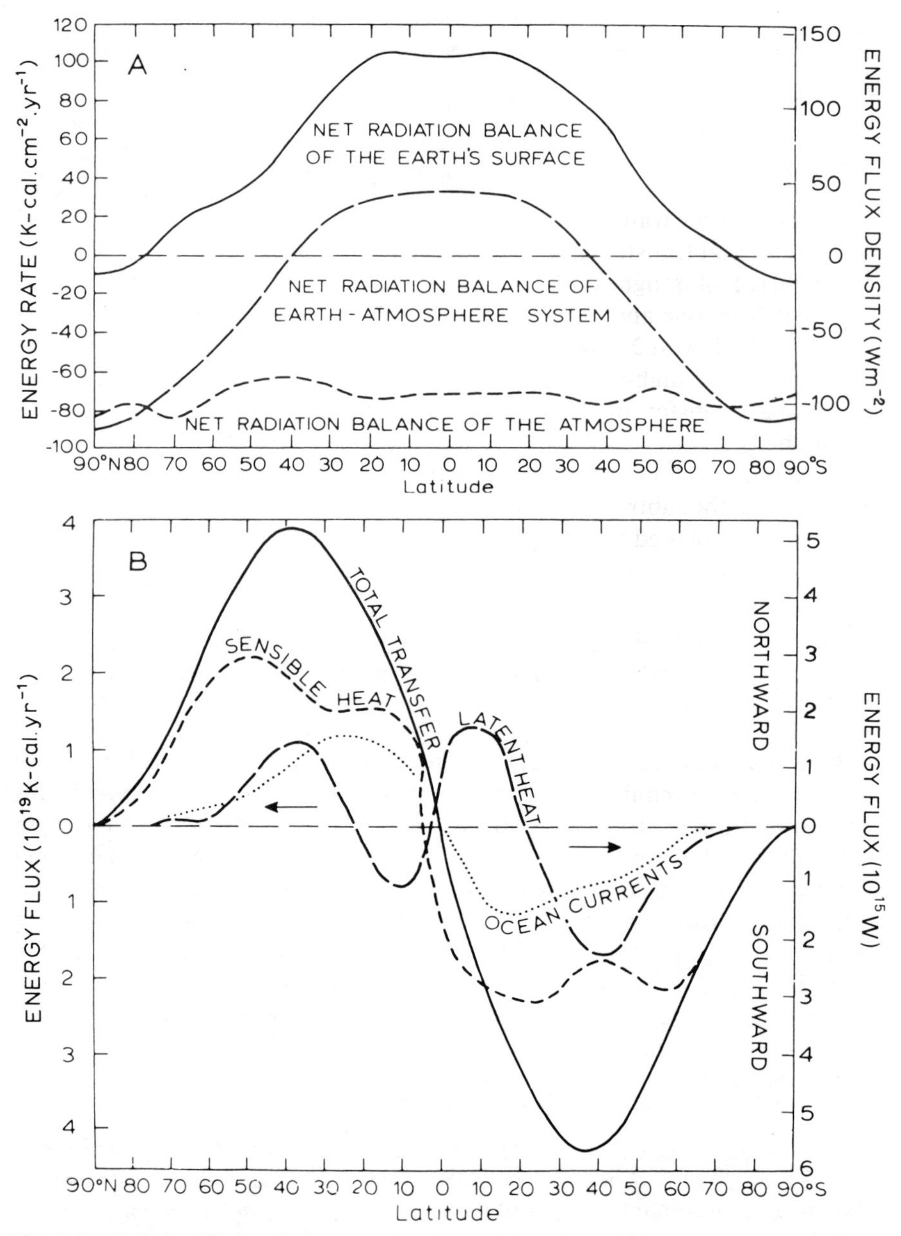

Fig. 1.31 A Net radiation balance for the earth's surface of 101 W m^{-2} or 76 kcal cm^{-2} yr^{-1} (incoming solar radiation of 156 W m^{-2}, minus outgoing long-wave energy to the atmosphere of 55 W m^{-2}); for the atmosphere of -101 W m^{-2} (incoming solar radiation of 84 W m^{-2}, minus outgoing long-wave energy to space of 185 W m^{-2}); and for the whole earth-atmosphere system of zero (*from Sellers 1965*). B The average annual latitudinal distribution of the components of the poleward energy transfer (in 10^{15} W, 10^{19} kcal yr^{-1}) in the earth-atmosphere system (*from Sellers 1965*).

poleward heat transport takes place within the atmosphere and oceans, and it is estimated that the former accounts for approximately two-thirds of the required total. The horizontal transport (*advection* of heat) occurs in the form of both latent heat (that is water vapour which subsequently condenses) and sensible heat (that is warm air masses) (fig. 1.31B). It varies in intensity according to the latitude and the season. Figure 1.31B shows the mean annual pattern of energy transfer by the three mechanisms. The latitudinal zone of maximum total transfer rate is found between latitudes 35° and 45° in both hemispheres, although the patterns for the individual components are quite different from one another. The latent heat transport, which occurs almost wholly in the lowest 2 or 3 km, reflects the global wind belts on either side of the subtropical high-pressure zones (see ch. 3, E). The more important meridional transfer of sensible heat has a double maximum not only latitudinally but also in the vertical plane, where there are maxima near the surface and at about 200 mb. The high-level transport is particularly significant over the subtropics, whereas the primary latitudinal maximum about 50° to 60°N is related to the travelling low-pressure systems of the westerlies.

The intensity of the poleward energy flow is closely related to the meridional (that is, north–south) temperature gradient. In winter this temperature gradient is at a maximum and in consequence the hemispheric air circulation is most intense. The nature of the complex transport mechanisms will be discussed in ch. 3, F.

As shown in fig. 1.31B, ocean currents account for a significant proportion of the poleward heat transfer in low latitudes. Indeed, recent satellite estimates of the required total poleward energy transport indicate that the previous figures are too low. The ocean transport may be 47% of the total at 30°–35°N and as much as 74% at 20°N; the Gulf Stream and Kuro Shio currents are particularly important. As a result of this factor, the energy budget equation for an ocean area must be expressed as

$$R_n = LE + H + G + \Delta A,$$

where ΔA = horizontal advection of heat by currents and G = the heat transferred into or out of storage in the water. The latter is more or less zero for annual averages.

2 *Spatial pattern of the heat budget components*

The mean latitudinal values of the heat budget components discussed above conceal great spatial variations. Figure 1.33 shows the global distribution of the annual net radiation at the surface. Broadly, its magnitude decreases poleward from about 25° latitude, although as a result of the considerable absorption of solar radiation by the sea, the net radiation is greater over the oceans – exceeding 160 W m^{-2} (or 120 kcal cm^{-2} yr^{-1}) in latitudes 15°–20° – than over land areas, where it is about 80–105 W m^{-2} (60–80 kcal cm^{-2} yr^{-1}) in the same latitudes. Net radiation is also rather lower in arid continental areas than in humid ones, because in spite of the increased insolation receipts

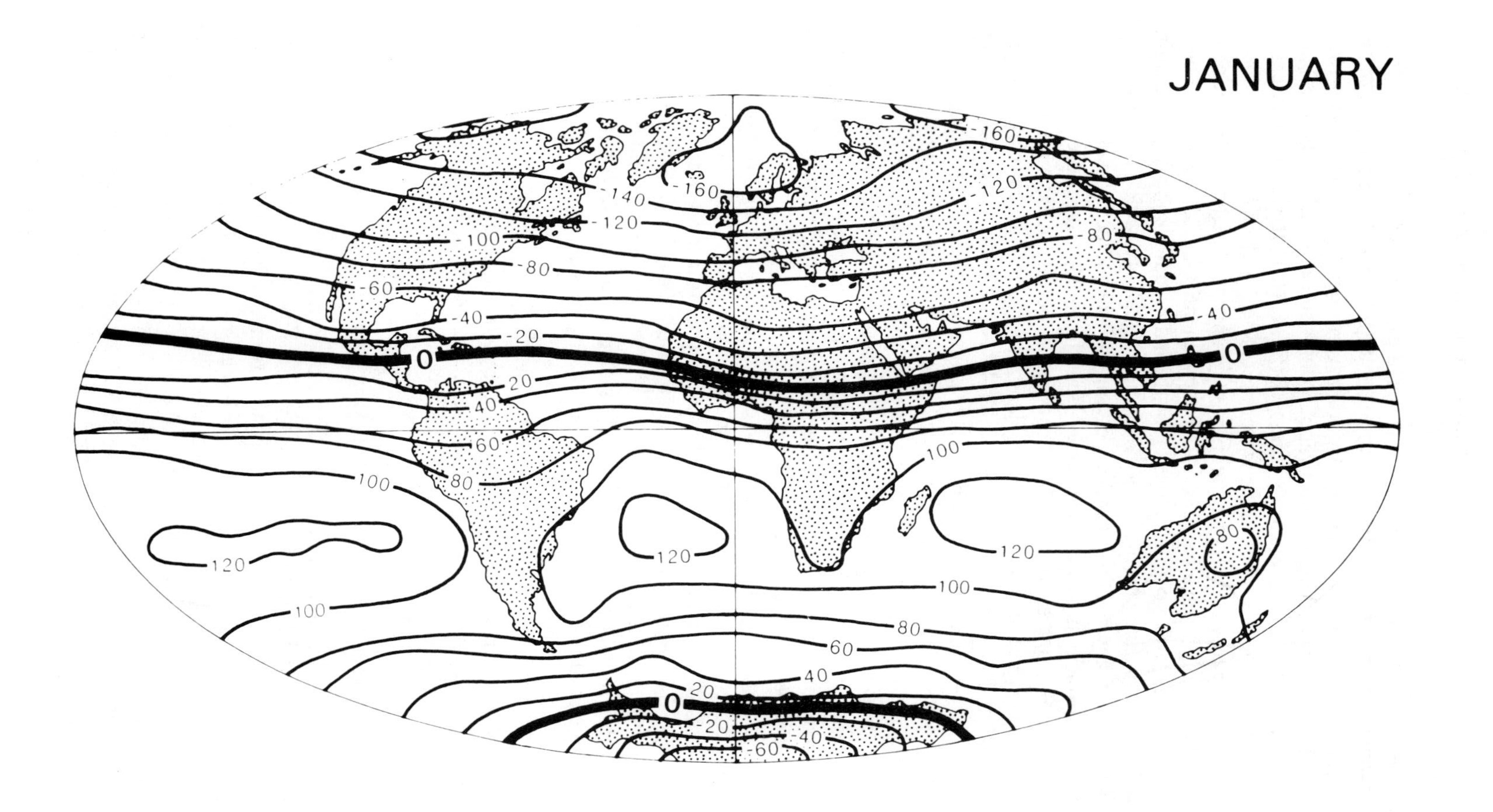
JANUARY
-160
-140
-120
-100
-80
-60
-40
-20
0
20
40
60
80
100
120
-20
-40
-60

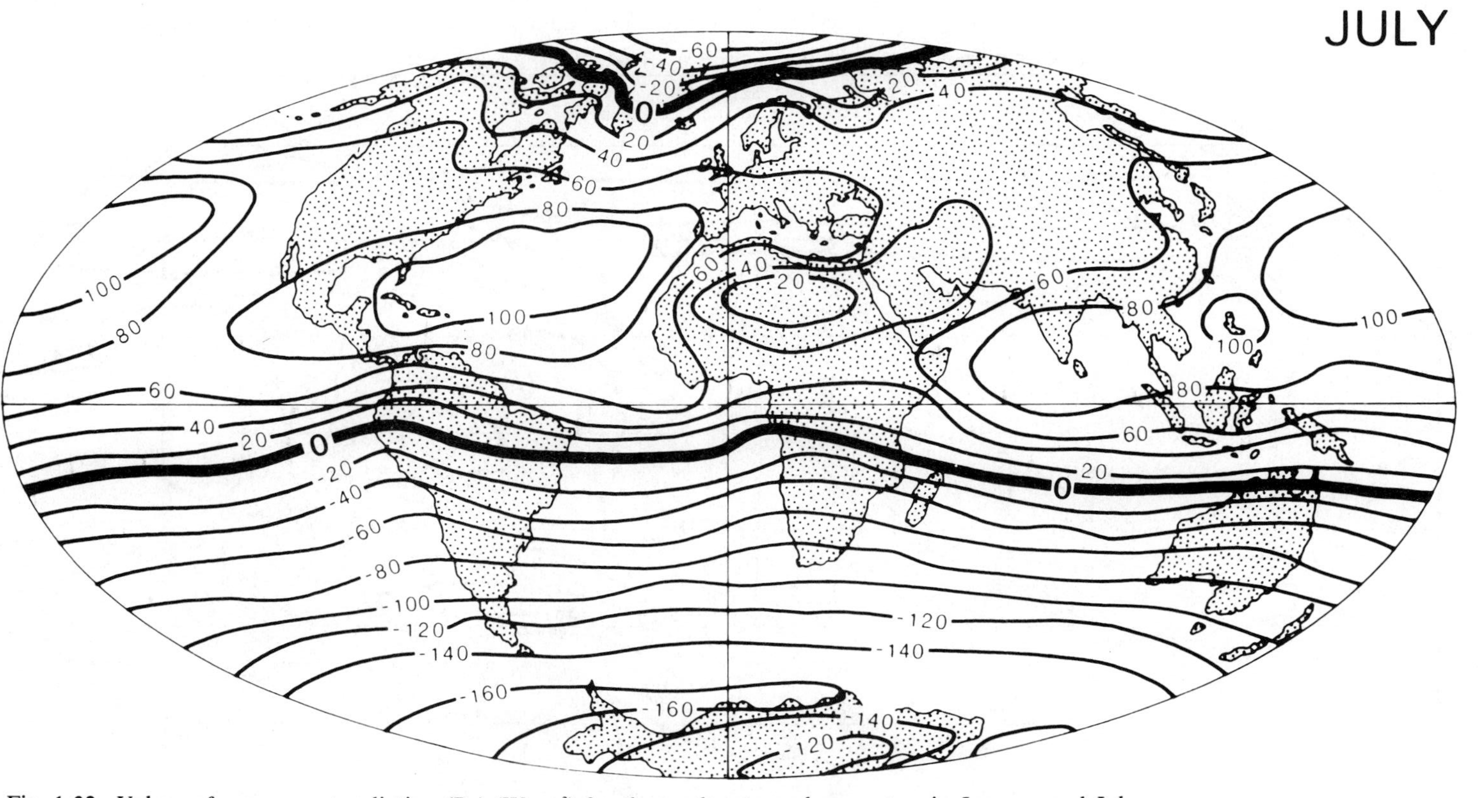

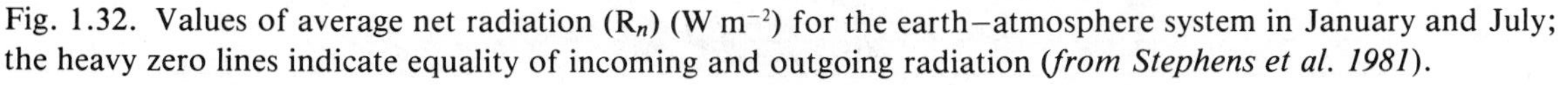
Fig. 1.32. Values of average net radiation (R_n) (W m^{-2}) for the earth–atmosphere system in January and July; the heavy zero lines indicate equality of incoming and outgoing radiation (*from Stephens et al. 1981*).

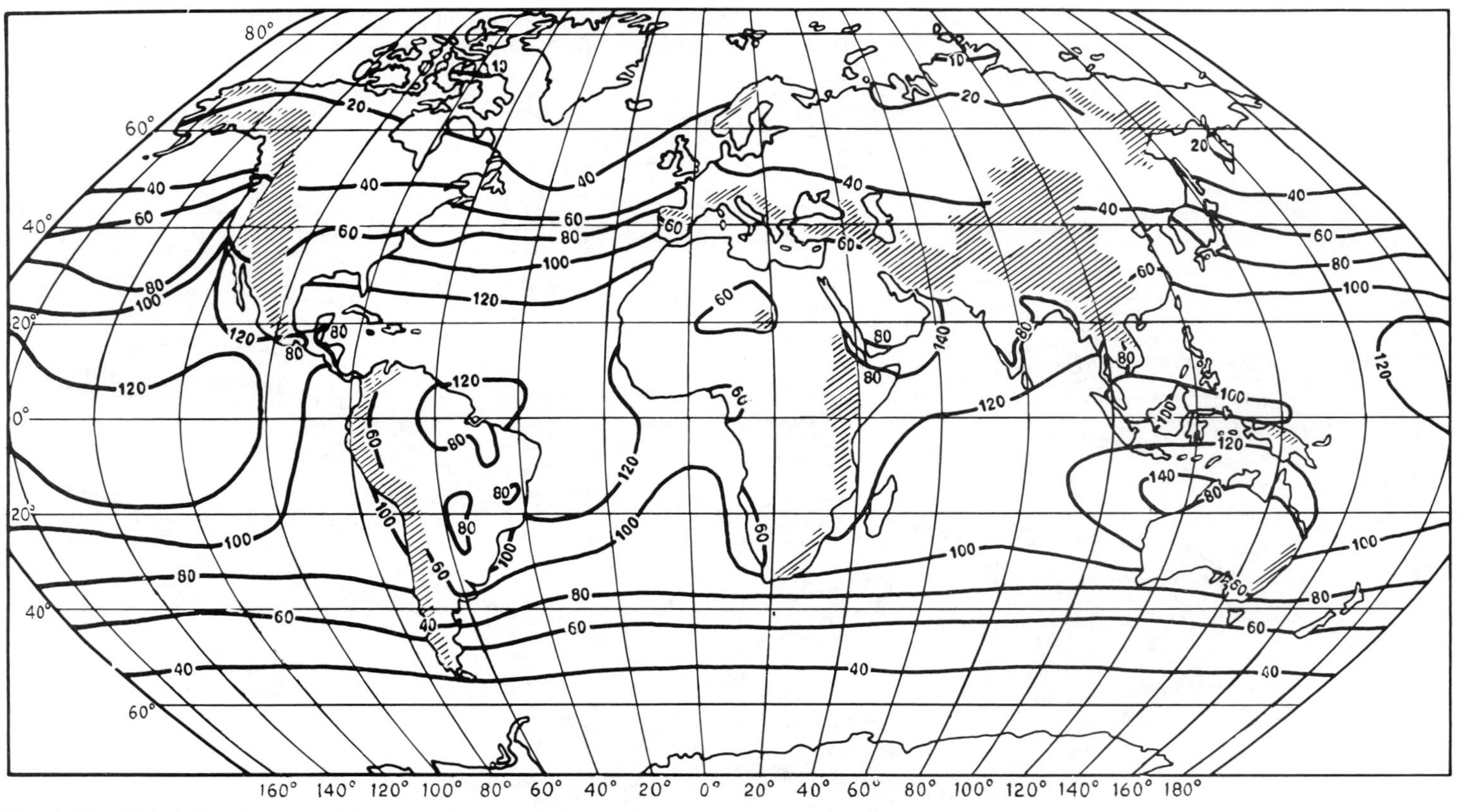

Fig. 1.33. Global distribution of the annual net radiation at the surface, in kcal cm^{-2} (*after Budyko et al. 1962*) (100 kcal cm^{-2} yr^{-1} = 133 W m^{-2}).

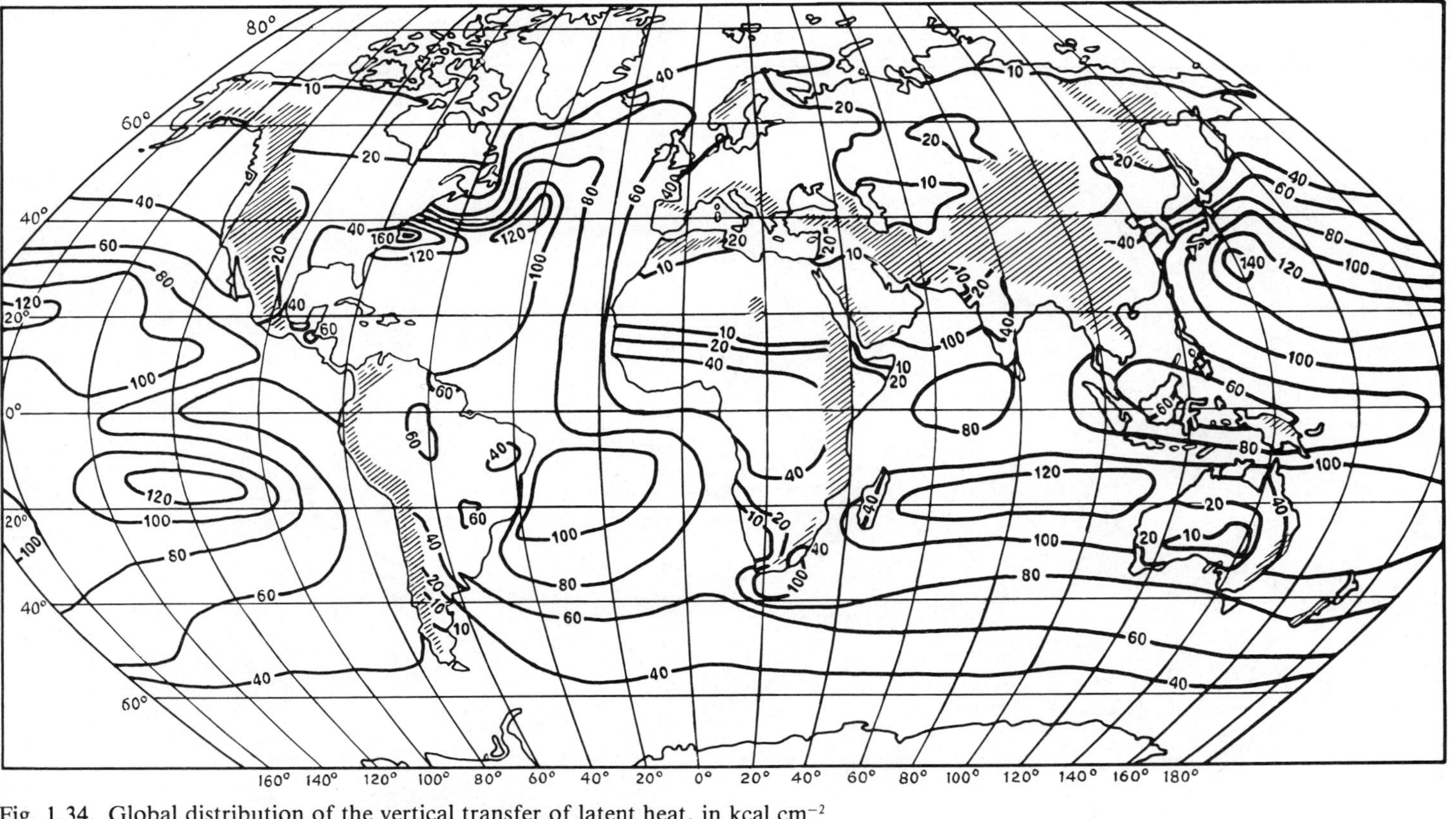

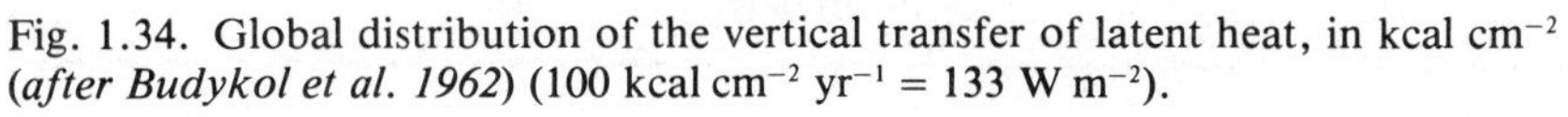
Fig. 1.34. Global distribution of the vertical transfer of latent heat, in kcal cm^{-2} (*after Budykol et al. 1962*) (100 kcal cm^{-2} yr^{-1} = 133 W m^{-2}).

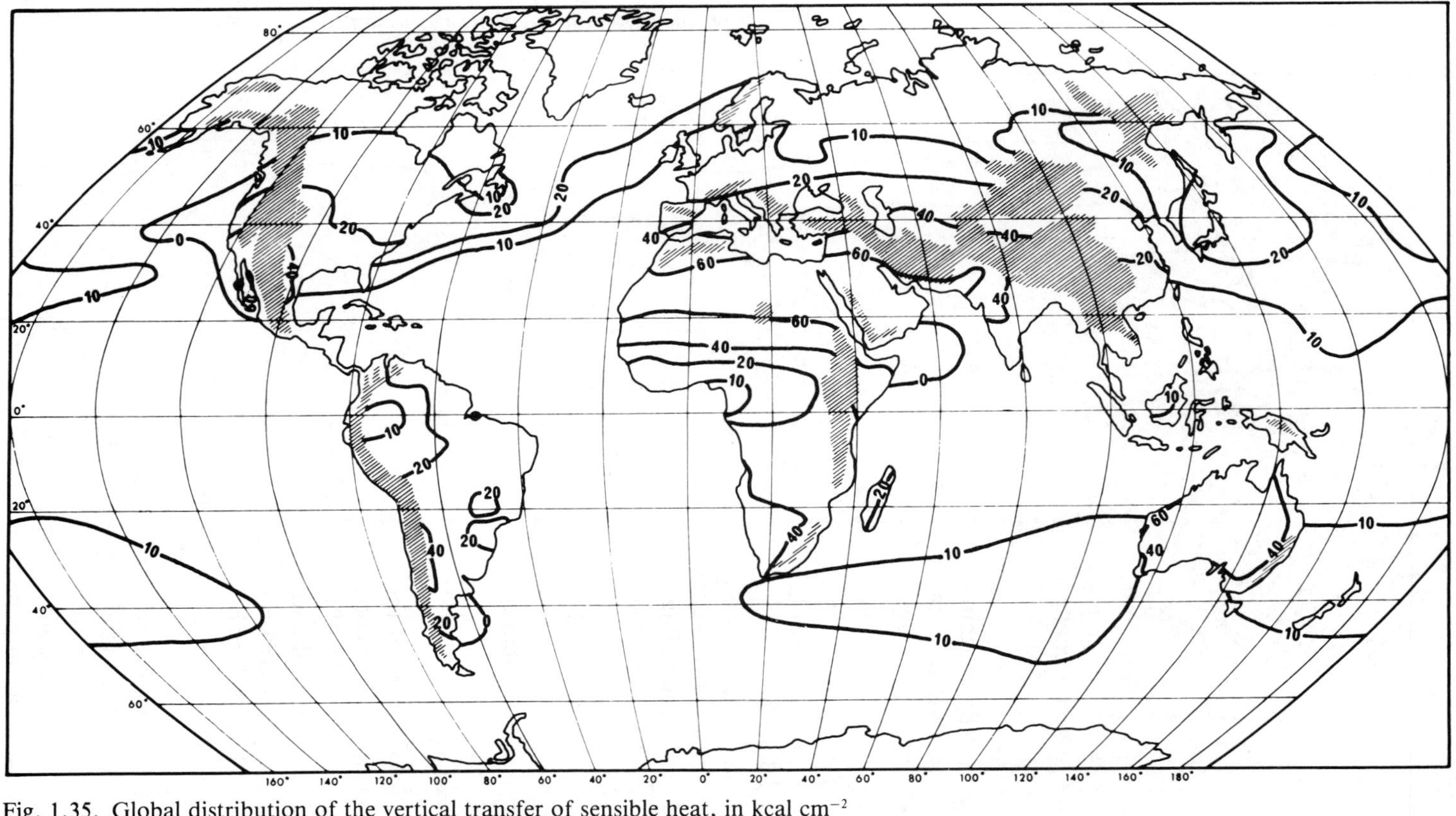

Fig. 1.35. Global distribution of the vertical transfer of sensible heat, in kcal cm^{-2} (*after Budyko et al. 1962*) (10 kcal cm^{-2} yr^{-1} = 13 W m^{-2}).

under clear skies there is at the same time greater net loss of terrestrial radiation.

Figures 1.34 and 1.35 show the annual vertical transfers of latent and sensible heat to the atmosphere. Both maps show that the fluxes are distributed very differently over land and sea. Heat expenditure for evaporation is at a maximum in tropical and subtropical ocean areas, where it exceeds 160 W m^{-2} (120 kcal cm^{-2} yr^{-1}). It is less near the equator where wind speeds are somewhat lower and the air has a vapour pressure close to the saturation value (see ch. 2, A). It is clear from fig. 1.34 that the major warm currents considerably augment the evaporation rate. On land the latent heat transfer is greatest in hot, humid regions. It is least in arid areas due to the low precipitation and in high latitudes where there is little available energy.

The largest exchange of sensible heat occurs in the tropical deserts where more than 80 W m^{-2} (60 kcal cm^{-2} yr^{-1}) is transferred to the atmosphere. In contrast to latent heat, the sensible heat flux is generally small over the oceans, only reaching 25–40 W m^{-2} (20–30 kcal cm^{-2}) in areas of warm currents. Indeed, negative values occur (transfer *to* the ocean) where warm continental air masses move offshore over cold currents.

H The layering of the atmosphere

The atmosphere can be divided conveniently into a number of rather well-marked horizontal layers, mainly on the basis of temperature. The evidence for this structure comes from regular RAWINSONDE (radar wind-sounding) balloons, radio-wave investigations, and, more recently, from rocket flights and satellite sounding systems (pls. 2 and 3). Broadly, the pattern (fig. 1.36) consists of three relatively warm layers (near the surface; between 50 and 60 km; and above about 120 km) separated by two relatively cold layers (between 10 and 30 km; and about 80 km). Mean January and July temperature sections illustrate the considerable latitudinal variations and seasonal trends that complicate the scheme (fig. 1.37).

1 Troposphere

The lowest layer of the atmosphere is called the troposphere. It is the zone where weather phenomena and atmospheric turbulence are most marked, and contains 75% of the total molecular or gaseous mass of the atmosphere and virtually all the water vapour and aerosols. Throughout this layer there is a general decrease of temperature with height at a mean rate of about 6¼°C/km (or 3·6°F/1000 ft), and the whole zone is capped in most places by a temperature inversion level (i.e. a layer of relatively warm air above a colder one) and in others by a zone that is isothermal with height. The troposphere thus remains to a large extent self-contained because the inversion acts as a 'lid' which effectively limits convection (see ch. 2, E). This inversion level or weather ceiling is called the *tropopause* (see note 7). Its height is not constant, either in space or time. It seems that the height of the tropopause at any point is correlated with sea-level temperature and

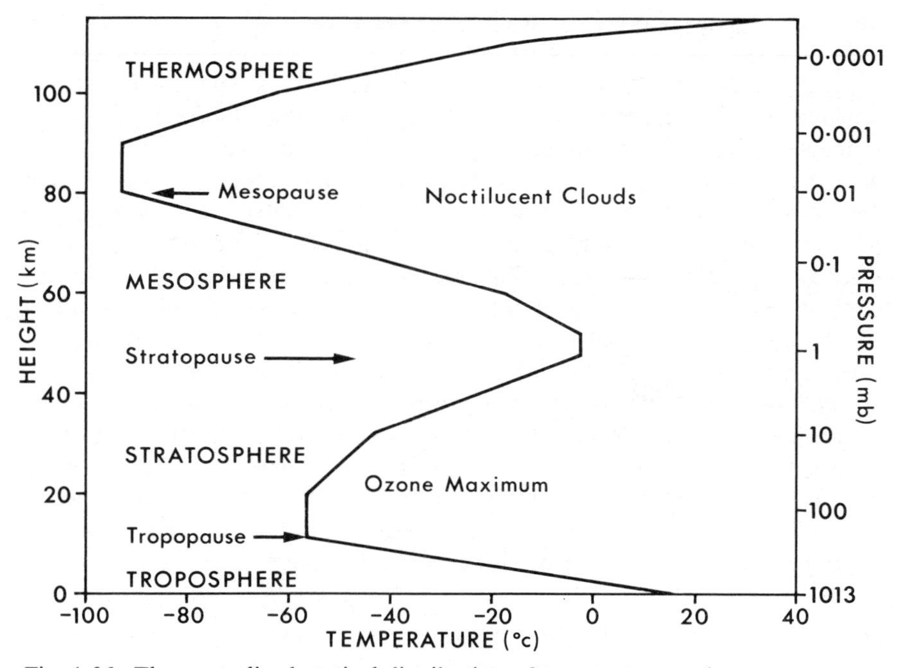

Fig. 1.36. The generalized vertical distribution of temperature and pressure up to about 110 km. Note particularly the tropopause and the zone of maximum ozone concentration with the warm layer above it (*based on data in Valley 1965*).

pressure, which are in turn related to the factors of latitude, season and daily changes in surface pressure. There are marked variations in the altitude of the tropopause with latitude (fig. 1.37), from about 16 km (10 miles) at the equator where there is great heating and vertical convective turbulence to only 8 km (5 miles) at the poles.

The meridional temperature gradients in the troposphere in summer and winter are roughly parallel, as are the tropopauses (fig. 1.37), and the strong lower mid-latitude temperature gradient in the troposphere is reflected in the tropopause breaks (see also fig. 3.21). In these zones important interchanges can occur between the troposphere and stratosphere, or vice versa. Traces of water vapour probably penetrate into the stratosphere by this means, while dry, ozone-rich stratosphere air may be brought down into the mid-latitude troposphere. For example, above-average concentrations of ozone are observed in the rear of mid-latitude low-pressure systems where the tropopause elevation tends to be low. Both facts are probably the result of stratospheric subsidence, which warms the lower stratosphere and causes downward transfer of the ozone.

2 *Stratosphere*

The second major atmospheric layer is the stratosphere which extends upwards from the tropopause to about 50 km (30 miles). Although the

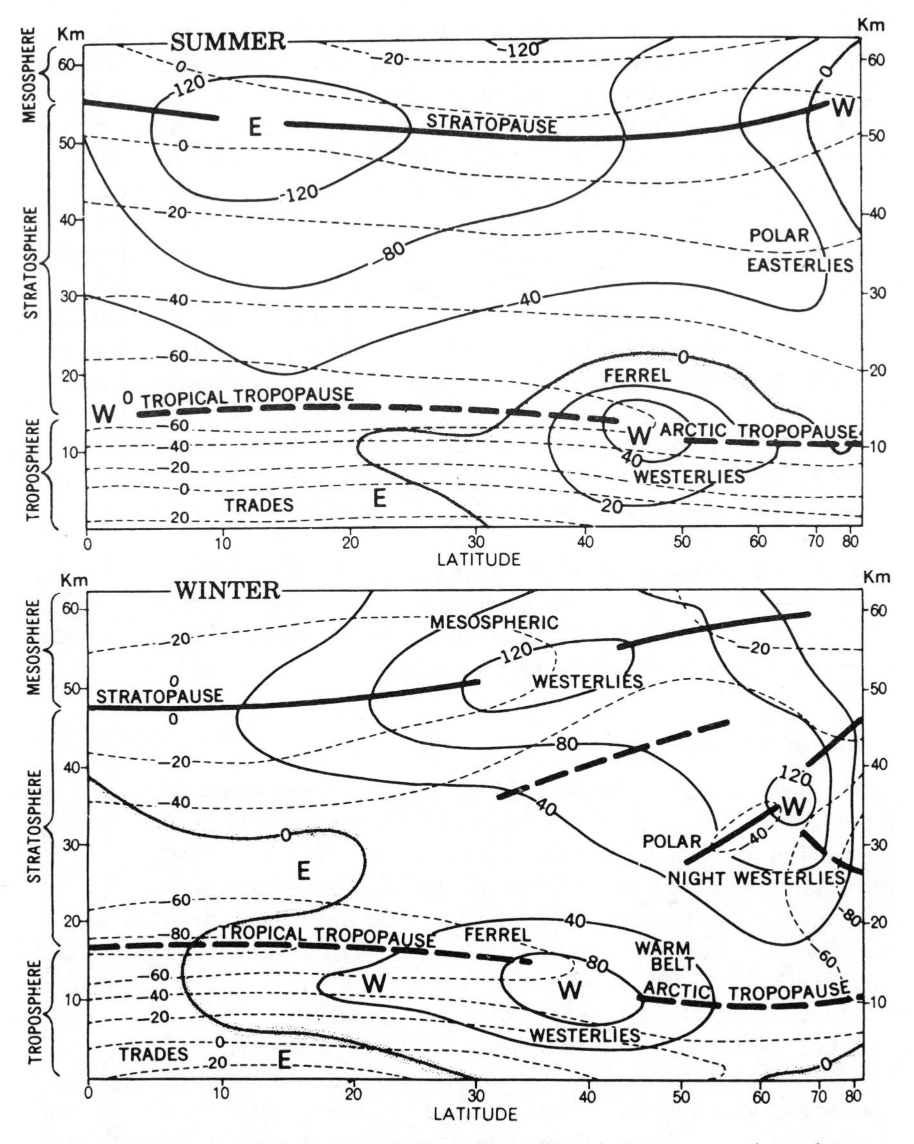

Fig. 1.37. Mean zonal (westerly) winds (solid isolines, in knots; negative values from the east) and temperatures (in °C, dashed isolines), showing the broken tropopause near the mean Ferrel jet stream (*after Boville; from Hare 1962*). The term 'Ferrel Westerlies' was proposed by F. K. Hare in honour of W. Ferrel (see p. 147). The heavy black lines denote reversals of the vertical temperature gradient of the tropopause and stratopause. Summer and winter refer to the northern hemisphere.

stratosphere contains much of the total atmospheric ozone (it reaches a peak density at approximately 22 km), the maximum temperatures associated with the absorption of the sun's ultraviolet radiation by ozone occur at the *stratopause*, where temperatures may exceed 0°C (fig. 1.37). The air density is much less here so that even limited absorption produces a large temperature increase. Temperatures increase fairly generally with height in summer, with the coldest air at the equatorial tropopause. In winter the structure is more complex with very low temperatures, averaging −80°C, at the equatorial tropopause which is highest at this season. Similar low temperatures are found in the middle stratosphere at high latitudes, whereas over 50°–60°N there is a marked warm region with nearly isothermal conditions at about −45°C to −50°C. Marked season changes of temperature affect the stratosphere. The cold 'polar night' winter stratosphere often undergoes dramatic *sudden warmings* associated with subsidence due to circulation changes in late winter or early spring when temperatures at about 25 km may jump from −80°C to −40°C over a 2-day period. The autumn cooling is a more gradual process. In the tropical stratosphere there is a quasi-biennial (26-month) wind regime, with easterlies in the layer 18 to 30 km for 12 to 13 months, followed by westerlies for a similar period. The reversal begins first at high levels and takes approximately 12 months to descend from 30 to 18 km (10 to 60 mb).

How far these events in the stratosphere are linked with temperature and circulation changes in the troposphere remains a topic of meteorological research. Any interactions that do exist, however, are likely to be complex, otherwise they would already have become evident.

3 The upper atmosphere

a Mesosphere Above the stratopause average temperatures decrease to a minimum of about −90°C (183 K) around 80 km. This layer is commonly termed the mesosphere, although it must be noted that as yet there is no universal acceptance of terminology for the upper atmospheric layers. Indeed, some authors refer to the layer between 20 and 80 km as the mesosphere. Above 80 km temperatures again begin rising with height and this inversion is referred to as the 'mesopause'. It is in this region that 'noctilucent clouds' are observed over high latitudes in summer. Their presence appears to be due to meteoric dust particles which act as nuclei for ice crystals when traces of water vapour are carried upwards by high-level convection caused by the vertical decrease of temperature in the mesosphere.

Pressure is very low in the mesosphere, decreasing from about 1 mb at 50 km to 0·01 mb at 90 km.

b Thermosphere Above the mesopause atmospheric densities are extremely low, although the tenuous atmosphere still effects drag on space vehicles above 250 km. The lower portion of the thermosphere is composed mainly of nitrogen (N_2) and oxygen in molecular (O_2) and atomic (O) forms, whereas above 200 km atomic oxygen predominates over nitrogen (N_2 and N). Temperatures rise with height, owing to the absorption of extreme ultraviolet radiation ($<0{\cdot}2\,\mu m$) by molecular and atomic oxygen, probably approaching 1200 K at 350 km, but these temperatures are essentially theoretical. For example, artificial satellites do not acquire such temperatures because of the rarefied air. 'Temperatures' in the upper thermosphere and exosphere undergo wide diurnal and seasonal variations. They are higher by day and are also higher during a sunspot maximum, although the changes are only represented in varying velocities of the sparse air molecules.

Above 100 km the atmosphere is increasingly affected by cosmic radiation, solar X-rays and ultraviolet radiation which cause *ionization*, or electrical charging, by separating negatively charged electrons from oxygen atoms and nitrogen molecules, leaving the atom or molecule with a net positive charge (an *ion*). The Aurora Borealis and Aurora Australis are produced by the penetration of ionizing particles through the atmosphere from about 300 to 80 km, particularly in zones about 20°–25° latitude from the earth's magnetic poles. On occasion, however, the aurorae may appear at heights up to 1000 km, demonstrating the immense extension of a rarefied atmosphere. The term *ionosphere* is commonly applied to the layers above 80 km, although sometimes it is used only for the region of high electron density between about 100 and 300 km. In view of these different designations it seems preferable to avoid confusion by using the terminology adopted here.

c Exosphere and magnetosphere The base of the exosphere is between about 500 and 750 km. Here atoms of oxygen, hydrogen and helium (about 1% of which are ionized) form the tenuous atmosphere and the gas laws (see ch. 1, B) cease to be valid. Neutral helium and hydrogen atoms, which have low atomic weights, can escape into space since the chance of molecular collisions deflecting them downwards becomes less with increasing height. Hydrogen is replaced by the breakdown of water vapour and methane (CH_4) near the mesopause, while helium is produced by the action of cosmic radiation on nitrogen and from the slow, but steady, breakdown of radioactive elements in the earth's crust.

Ionized particles increase in frequency through the exosphere and, beyond about 200 km, in the magnetosphere there are only electrons (negative) and protons (positive) derived from the solar wind. These charged particles are concentrated in two bands at about 3000 and 16,000 km above the surface (Van Allen 'radiation' belts), apparently as a result of trapping by the earth's magnetic field. The magnetosphere has an extended tail on the side of the earth away from the sun, but on the side towards the sun it is compressed by the solar wind to about ten earth radii (57,000 km). Detailed investigation of these regions, made possible since 1958 by satellites, involves the study of *magnetohydrodynamics*. Nevertheless disturbances of these upper regions by solar flares may eventually prove to have meteorological significance at lower levels. At a height of 80,000 km or so the earth's atmosphere probably merges into that of the sun, but even the appropriate definitions of atmosphere, wind and temperature are uncertain in these regions.

I Variation of temperature with height

The last section described the gross characteristics of the vertical temperature profile in the atmosphere. Now we examine in more detail some of the features of the temperature gradient at low levels.

Vertical temperature gradients are determined in part by energy transfers and in part by vertical motion of the air. The various factors interact in a highly complex manner. The energy terms are the release of latent heat by condensation (ch. 1, D), radiational cooling of the air and sensible heat transfer from the ground. Horizontal temperature advection may also be important. Vertical motion is dependent on the type of pressure system. High-pressure areas are generally associated with descent and warming of deep layers of air, hence decreasing the temperature gradient and frequently causing temperature inversions in the lower troposphere. In contrast, low-pressure systems are associated with rising air which cools upon expansion and increases the vertical temperature gradient. This is only part of the story since moisture is an additional complicating factor (see ch. 2, E). It remains true, however, that the middle and upper troposphere is relatively cold above a surface low-pressure area, leading to a steeper temperature gradient.

The overall vertical decrease of temperature, or *lapse rate*, in the troposphere is, as has been stated, about 6·5°C/km. However, this is by no means constant with height, season or location. Average global values calculated by C. E. P. Brooks for July show increasing lapse rate with height: 5°C/km in the lowest 2 km, 6°C/km between 4 and 6 km, and 7°C/km between 6 and 8 km. Winter values are generally smaller and in continental areas, such as central Canada or eastern Siberia, may even be negative (i.e. temperatures increase with height in the lower layer) as a result of excessive radiational cooling over a snow surface (fig. 1.38). A similar situation occurs when dense, cold air accumulates in mountain basins on calm, clear nights. On such occasions the mountain tops may be many degrees warmer than the valley floor below (see ch. 3, C.1). For this reason the adjustment of average temperatures of upland stations to mean sea-level may produce misleading results. Observations in Colorado at Pike's Peak

Table 1.4 Temperature lapse rates in the lowest 1000–1500 metres (*after Lautensach and Bögel*)

Climate	*Season of maximum*	*Rate °C/km*	*Season of minimum*	*Rate °C/km*
Tropical rainy	Dry season	>5	Rainy season	>4·5
Tropical and subtropical deserts	Summer	>8	Winter	>5
Mediterranean	Winter	>5	Summer	<5
Mid-latitudes (cold winter)	Summer	>6	Winter	0–5
Boreal continental	Summer	>5	Winter	<0
Arctic	Summer	⩽0	Winter	<0

(4301 m or 14,111 ft) and Colorado Springs (1859 m or 6098 ft) show the mean lapse rate to be 4·1°C/km (2·°F/1000 ft) in winter and 6·2°C/km (3·4°F/1000 ft) in summer! It should be noted that such topographic lapse rates may bear little relation to free-air lapse rates in nocturnal radiation conditions, and the two must be carefully distinguished.

Table 1.4 summarizes the seasonal characteristics of lapse rates in six major climatic zones and examples of five of these are illustrated in fig. 1.33. The seasonal regime is very pronounced in continental areas with cold

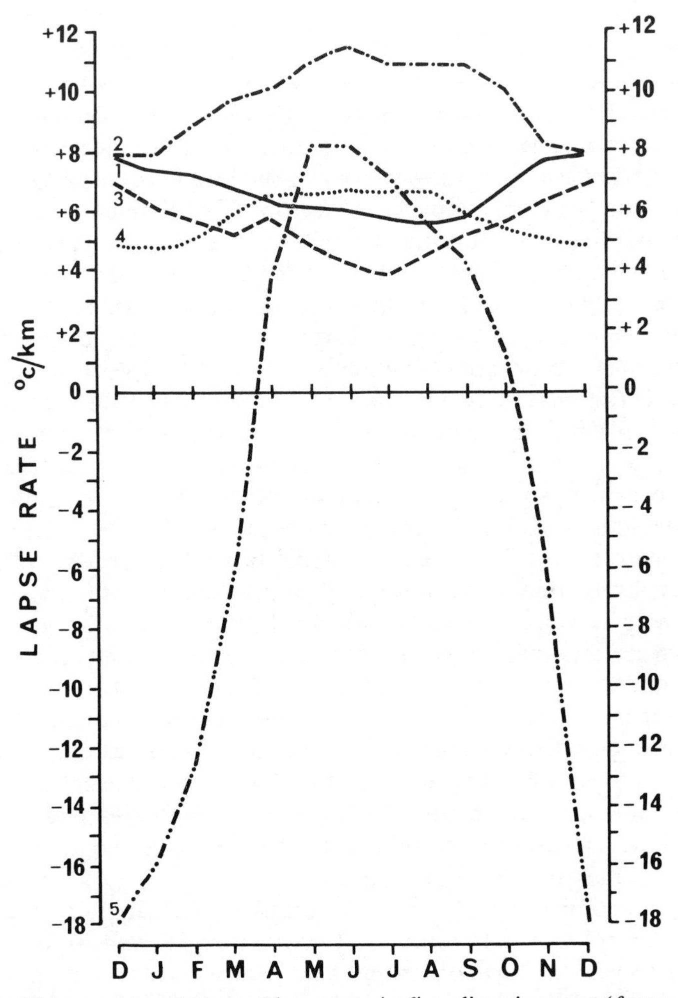

Fig. 1.38. The annual variation of lapse rate in five climatic zones (*from Hastenrath 1968*).
1 Tropical rainy climate (Togo).
2 Tropical desert (Arizona).
3 Mediterranean (Sicily).
4 Mid-latitude, cold winter climate (North Germany).
5 Boreal continental (Eastern Siberia).

winters whereas inversions persist for much of the year in the Arctic. In winter the Arctic inversion is due to intense radiational cooling but in summer it is the result of the surface cooling of advected warmer air. The winter lapse rate is only greater than the summer one in Mediterranean climates. In these regions there is more likelihood of rising air associated with low-pressure areas in winter. In contrast, subsidence is predominant in the desert zones in winter. The tropical and subtropic deserts have very steep lapse rates in summer when there is considerable heat transfer from the surface and generally ascending motion.

Summary

The atmosphere is a mixture of gases with constant proportions up to 80 km or more. The exceptions are ozone, which is concentrated in the lower stratosphere, and water vapour in the lower troposphere. Carbon dioxide is the principal atmospheric gas to vary with time – increasing in this century due to the burning of fossil fuels. Air is highly compressible, so that half of its mass occurs in the lowest 5 km and pressure decreases logarithmically with height from an average sea-level value of 1013 mb.

Almost all energy affecting the earth is derived from solar radiation, which is short wavelength ($<4\mu$m) due to the high temperature of the sun ($\sim$6000 K) (i.e. Wien's Law). The solar constant has a value of approximately 1370W m^{-2}. The sun and the earth radiate almost as black bodies (Stefan's law, $F = \sigma T^4$) whereas the atmospheric gases do not. Terrestrial radiation, from an equivalent black body, amounts only to about 130 W m^{-2} due to its low radiating temperature (263 K) and it is infrared (long-wave) radiation between 4 and 100 μm. Water vapour and carbon dioxide are the major absorbing gases for infrared radiation, whereas the atmosphere is largely transparent to solar radiation (the greenhouse effect). Solar radiation is lost by reflection, mainly from clouds, and by absorption (largely by water vapour). The planetary albedo is 31%; 48% of the extraterrestrial radiation reaches the surface. The atmosphere is heated primarily from the surface by the absorption of terrestrial infrared radiation and by turbulent heat transfer. Temperature usually decreases with height at an average rate of about 6·5°C/km in the troposphere. In the stratosphere and thermosphere it increases with height due to the presence of radiation absorbing gases.

The excess of net radiation in lower latitudes leads to a poleward energy transport from tropical latitudes by ocean currents and by the atmosphere. This is in the form of sensible heat (warm air masses/ocean water) and latent heat (atmospheric water vapour). Air temperature at any point is affected by the incoming solar radiation and other vertical energy exchanges, surface properties (slope, albedo, heat capacity), land and sea distribution and elevation, and also by horizontal advection due to air mass movements and ocean currents.

Notes

1 Mixing ratio = ratio of number of molecules of ozone to molecules of air (parts per million by volume, ppm(v)). Concentration = mass per unit volume of air (molecules/cubic metre).

2 K = degrees Kelvin (or Absolute). The degree symbol is omitted.
°C = degrees Celsius
°C = °K − 273
Conversions for °C and °F are given in App. 2.

3 Joule = 0.2388 cal. The units of the International Metric System are given in App. 2. At present the data in many references are still in calories; a calorie is the heat required to raise the temperature of 1 g of water from 14·5°C to 15·5°C. In the United States, another unit in common use is the Langley (ly) (ly min^{-1} = 1 cal cm^{-2} min^{-1}).

4 The equation for the so-called 'reduction' (actually the adjusted value is normally greater!) of station pressure (p_h) to sea level pressure (p_0) is written:

$$p_0 = p_h \exp\left[\frac{g_0}{R_d \overline{T}_v} Z_h\right]$$

where R_d = gas content for dry air; g_0 = global average of gravitational acceleration (9·8 ms^{-2}); Z_h = geopotential height of the station ($\simeq$ geometric height in the lowest kilometre or so); $\overline{T}_v$ = mean virtual temperature. This is a fictitious temperature used in the ideal gas equation to compensate for the fact that the gas constant of moist air exceeds that of dry air. Even for hot moist air, $\overline{T}_v$ is only a few degrees greater than the air temperature.

5 This can be calculated from (solar output × $4\pi R_s^2$) $(0 \cdot 25\, \pi D^2)$, where the sun's radius, $R_s = 7 \times 10^5$ km and the solar distance, $D = 1 \cdot 5 \times 10^8$ km. A sphere of radius r has a surface area of $4\,\pi r^2$.

6 The albedos refer to the solar radiation received on each given surface; thus the incident radiation is different for planet earth, the global surface and global cloud cover, as well as between any of these and the individual cloud types or surfaces.

7 The official definition is the lowest level at which the lapse rate decreases to less than, or equal to, 2°C/km (provided that the average lapse rate of the 2-km layer does not exceed 2°C/km).

2

Atmospheric moisture

Terrestrial moisture is in a constant state of transformation, termed the *hydrologic cycle*, in which the three most important stages are evaporation, condensation and precipitation. Figure 2.1 indicates the relative average annual amounts of water involved in each phase of the cycle. It shows that the atmosphere holds only a very small amount of water although the exchanges with the land and oceans are very considerable. This is further emphasized by table 2.1. The average storage of water vapour in the atmosphere (about 2·5 cm, or 1 in) is only sufficient for some 10 days' supply of

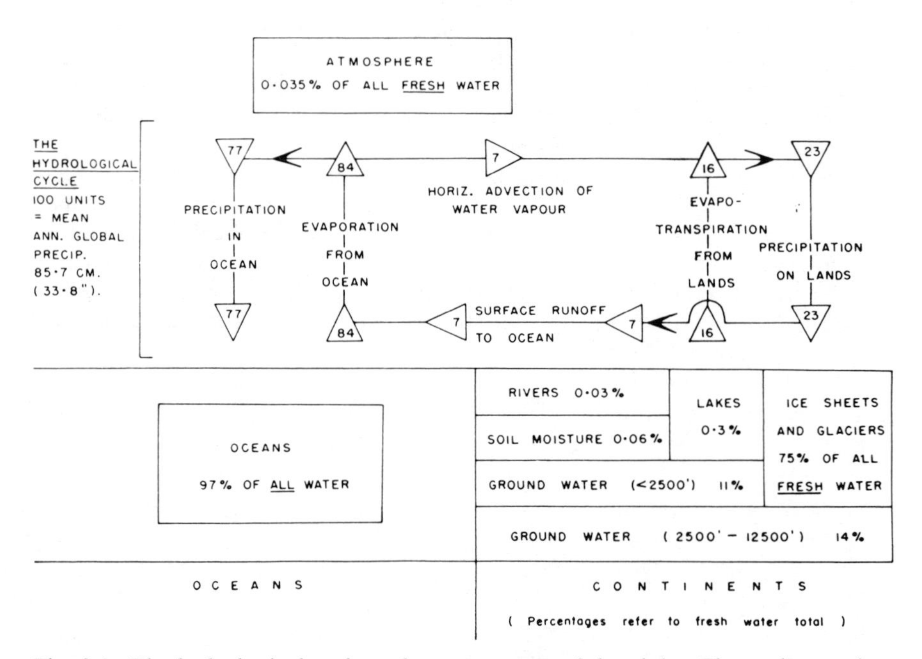

Fig. 2.1. The hydrological cycle and water storage of the globe. The exchanges in the cycle are referred to 100 units which equal the mean annual global precipitation of 85·7 cm (33·8 in). The percentage storage figures for atmospheric and continental water are percentages of all *fresh* water. The saline ocean waters make up 97% of *all* water (*from More 1967*). The horizontal advection of water vapour indicates the *net* transfer.

Table 2.1 Mean water content of the atmosphere (in cm of rainfall equivalent) (*after Sutcliffe 1956*)

	Northern hemisphere	*Southern hemisphere*	*World*
January	1·9 (0·8 in)	2·5 (1·0 in)	2·2 (0·9 in)
July	3·4 (1·3 in)	2·0 (0·8 in)	2·7 (1·1 in)

rainfall over the earth as a whole. However, intense (horizontal) influx of moisture into the air over a given region makes possible short-term rainfall totals greatly in excess of 3 cm. The phenomenal record total of 187 cm (73·6 in) fell on the island of Réunion, off Madagascar, during 24 hours in March 1952, and much greater intensities have been observed over shorter periods (see ch. 2, I.1).

The atmosphere acquires moisture by evaporation from oceans, lakes, rivers and damp soil or from moisture transpired from plants. Taken together, these are often referred to as *evapotranspiration* and the mechanisms involved will now be discussed in detail.

A Evaporation

Evaporation occurs whenever energy is transported to an evaporating surface if the vapour pressure in the air is below the saturated value (e_s). As illustrated in fig. 1.7, the saturation vapour pressure increases with temperature. The change in state from liquid to vapour requires energy to be expended in overcoming the intermolecular attractions of the water particles. This energy is generally provided by the removal of heat from the immediate surroundings causing an apparent heat loss (*latent heat*), as discussed on p. 17, and a consequent drop in temperature. The latent heat of vaporization to evaporate 1 kg of water at 0°C is $2 \cdot 5 \times 10^6$ J (or 600 cal g^{-1}). Conversely, condensation releases this heat, and the temperature of an air mass in which condensation is occurring is increased as the water vapour reverts to the liquid state. The diurnal range of temperature is often moderated by damp air conditions, when evaporation takes place during the day and condensation at night.

Viewed another way, evaporation implies an addition of kinetic energy to individual water molecules and, as their velocity increases, so the chance of individual surface molecules escaping into the atmosphere becomes greater. As the faster molecules will generally be the first to escape, so the average energy (and therefore temperature) of those composing the remaining liquid will decrease and the quantities of energy required for their continued release become correspondingly greater. In this way evaporation decreases the temperature of the remaining liquid by an amount proportional to the latent heat of vaporization.

The rate of evaporation depends on a number of factors. The two most important are the difference between the saturation vapour pressure at the water surface and the vapour pressure of the air, and the existence of a continual supply of energy to the surface. Wind velocity also affects the evaporation rate because the wind is generally associated with the importation of fresh, unsaturated air which will absorb the available moisture.

Water loss from plant surfaces, chiefly leaves, is a complex process termed *transpiration*. It occurs when the vapour pressure in the leaf cells is greater than the atmospheric vapour pressure, and is vital as a life function in that it causes a rise of plant nutrients from the soil and cools the leaves. The cells of the plant roots can exert an osmotic tension of up to about 15 atmospheres upon the water films between the adjacent soil particles. As these soil water films shrink, however, the tension within them increases. If the tension of the soil films exceeds the osmotic root tension the continuity of the plant's water supply is broken and wilting occurs. Transpiration is controlled by the atmospheric factors which determine evaporation as well as by plant factors such as the stage of plant growth, leaf area and leaf temperature, and also by the amount of soil moisture (see ch. 7, C). It occurs mainly during the day when the *stomata* (i.e. small pores in the leaves) through which transpiration takes place are open. This opening is determined primarily by light intensity. Transpiration naturally varies greatly with season, and during the winter months in mid-latitudes conifers lose only 10–18% of their total annual transpiration losses and deciduous trees less than 4%.

In practice it is difficult to separate water evaporated from the soil, *intercepted moisture* remaining on vegetation surfaces after precipitation and subsequently evaporated, and transpiration. For this reason evaporation is sometimes applied as a general term for all these, or, more correctly, the composite term *evapotranspiration* may be used.

Evapotranspiration losses from natural surfaces cannot be measured directly. There are, however, various indirect methods of assessment, as well as theoretical formulae. One approximate means of indirect measurement is based on the moisture balance equation:

$$\text{precipitation} = \text{runoff} + \text{evapotranspiration} + \text{soil moisture storage change}$$

Essentially the method is to measure the percolation through an enclosed block of soil with a vegetation cover (usually grass) and to record the rainfall upon it. The block, termed a *lysimeter*, is weighed regularly so that weight changes unaccounted for by rainfall or runoff can be ascribed to evapotranspiration losses, provided the grass is kept short! The technique allows the determination of daily evapotranspiration amounts.

If the soil block is regularly 'irrigated' so that the vegetation cover is always yielding the maximum possible evapotranspiration the water loss is called the *potential evapotranspiration* (or PE). More generally, PE can be defined as the water loss corresponding to the available energy. Assuming a constant soil moisture storage, potential evapotranspiration is calculated as the difference between precipitation and percolation. A simple evapotranspirometer installation is shown in fig. 2.2; the double-tank installation

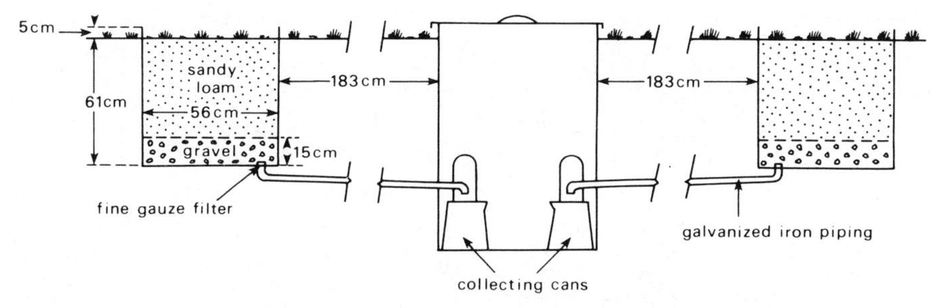

Fig. 2.2. An evapotranspirometer installation for calculating potential evapotranspiration losses. The double installation allows an average of the two results to be determined, giving a more reliable estimate (*from Ward 1963*).

ensures that representative readings are obtained. Potential evapotranspiration forms the basis for one system of climate classification developed by C. W. Thornthwaite (see App. 1).

In regions where snow cover is long lasting, evaporation/sublimation from the snowpack can be estimated by lysimeters sunk into the snow that are regularly weighed. Except in dry, sunny and windy environments, however, evaporation from snow is of minor importance compared to melt.

A meteorological solution to the calculation of evaporation uses sensitive instruments to measure the net effect of eddies of air transporting moisture upward and downward near the surface. In this 'eddy correlation' technique the vertical component of wind and the atmospheric moisture content are measured simultaneously at the same level (say, 1·5 m) every few seconds. The product of each pair of measurements is then averaged over some time interval to determine the evaporation (or condensation). This method requires delicate rapid-response instruments and so it cannot be used in very windy conditions.

Theoretical methods for determining evaporation rates have followed two lines of approach. The first relates average monthly evaporation (E) from large water bodies to the mean wind speed (u) and the mean vapour pressure difference between the water surface and the air ($e_w - e_d$) in the form:

$$E = Ku(e_w - e_d)$$

where K is an empirical constant. This is termed the aerodynamic approach because it takes account of the factors responsible for removing vapour from the water surface. The second method is based on the energy budget. The *net balance* of solar and terrestrial radiation at the surface (R_n) is used for evaporation (E) and the transfer of heat to the atmosphere (H). A small proportion also heats the soil by day, but since nearly all of this is lost at night it can be disregarded. Thus:

$$R_n = LE + H,$$

where L is the latent heat of evaporation ($2{\cdot}5 \times 10^6$ J kg^{-1}). R_n can be measured with a net radiometer and the ratio $H/LE = \beta$, referred to as

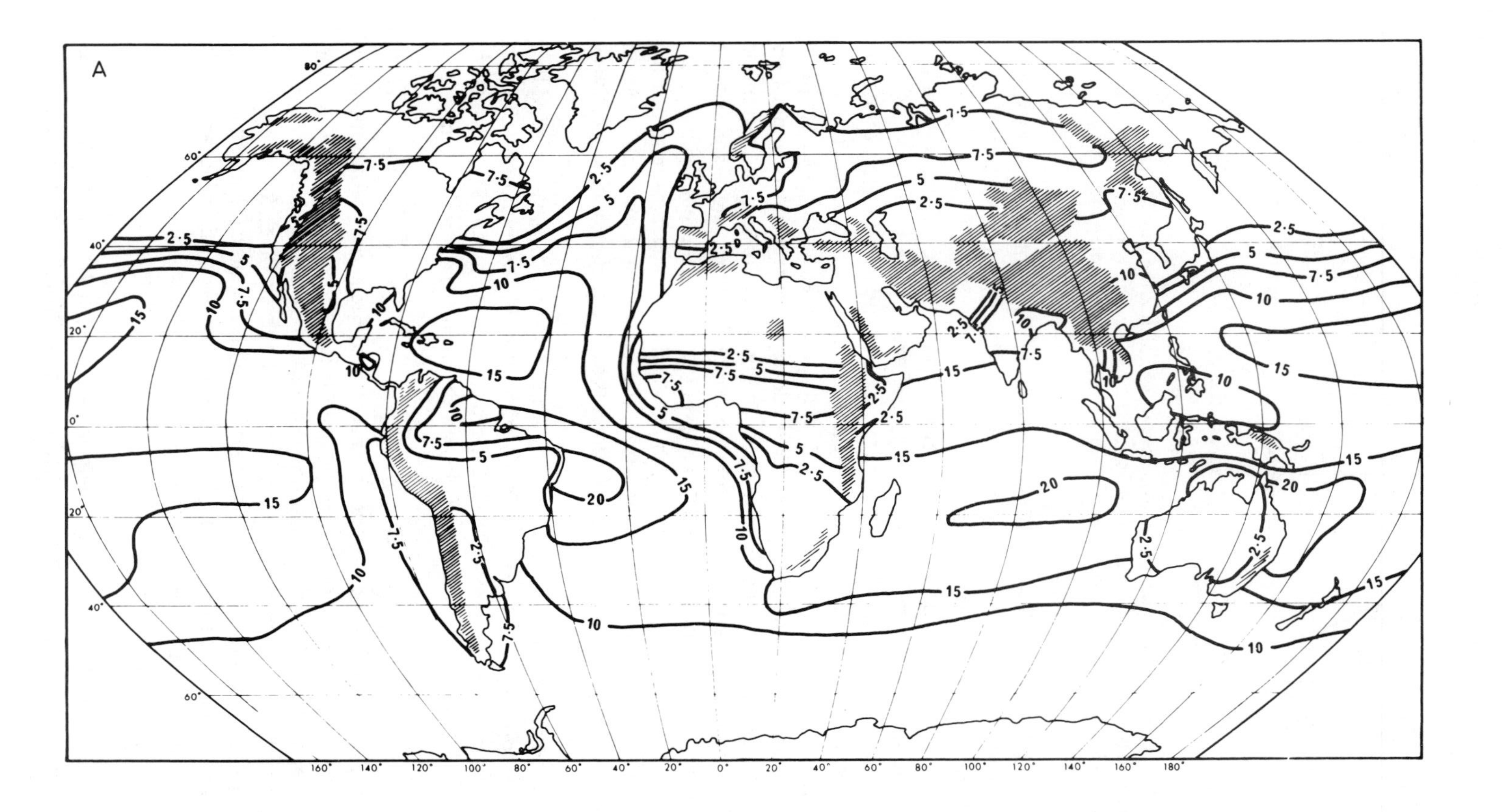
A

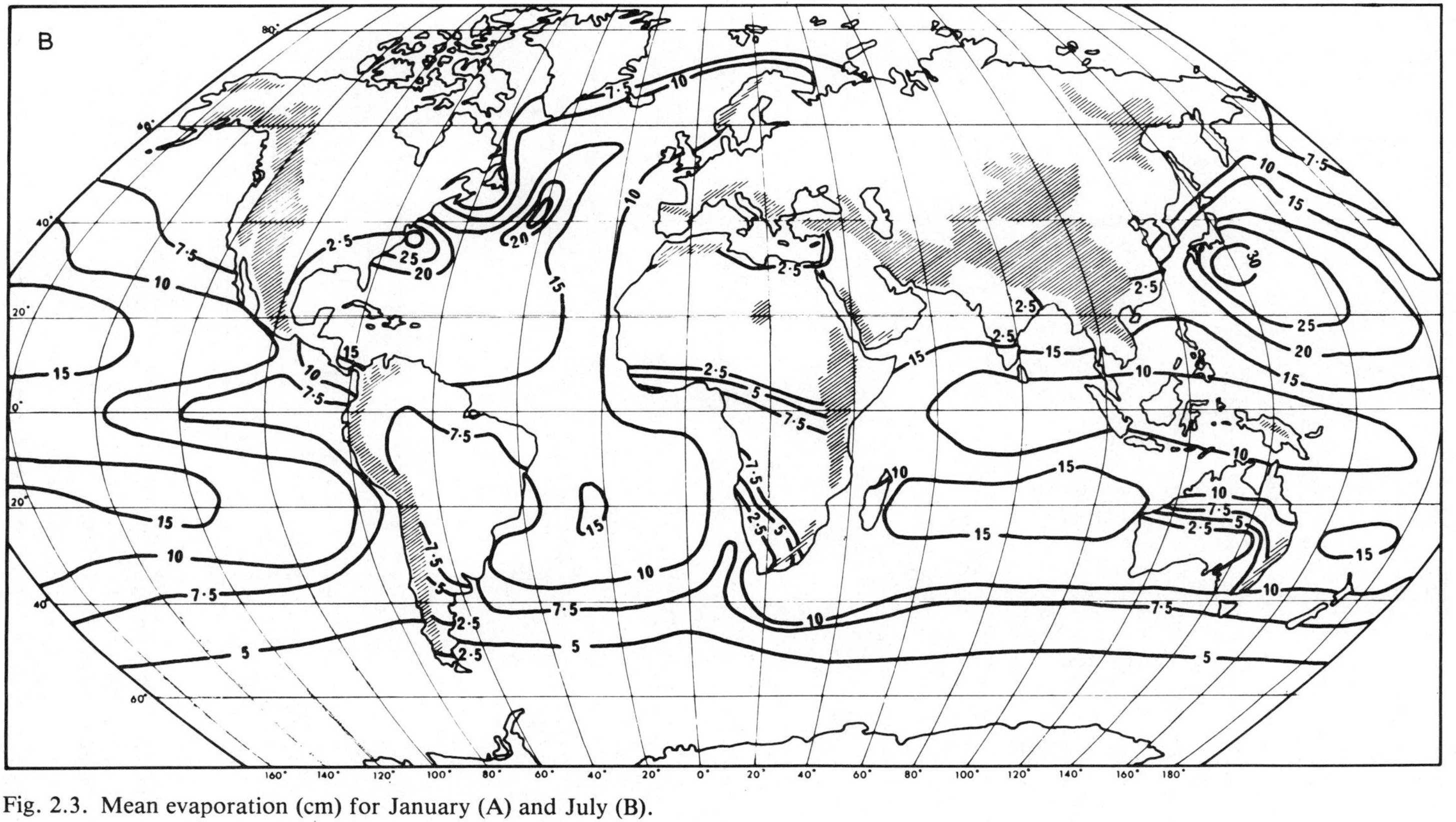

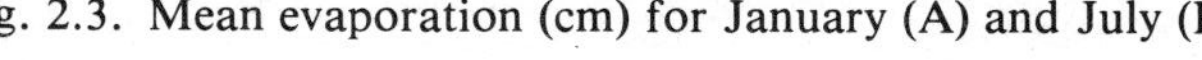
Fig. 2.3. Mean evaporation (cm) for January (A) and July (B).

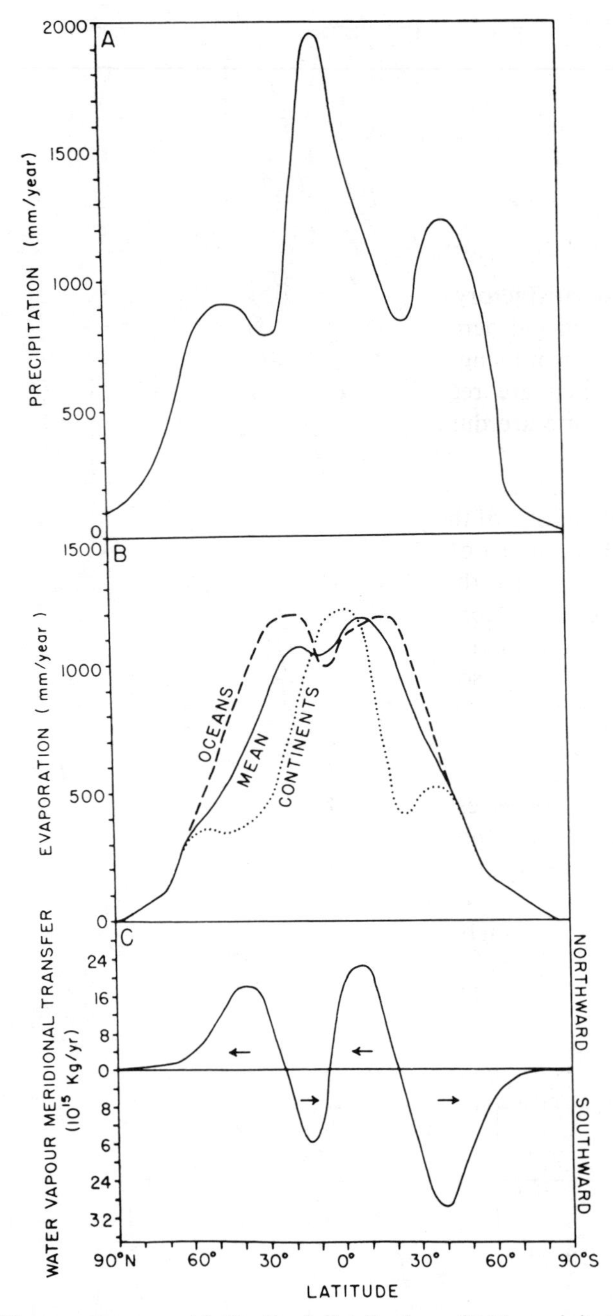

Fig. 2.4. The average annual latitudinal distribution of (A) precipitation (in mm); (B) evaporation (in mm); and (C) meridional transfer of water vapour (in 10^{15} kg) (*mostly from Sellers 1965*).

Bowen's ratio, can be estimated from measurements of temperature and vapour content at two levels near the surface. β ranges from $<0 \cdot 1$ for water to $\geqslant 10$ for a desert surface. The use of this ratio assumes that the vertical transfers of heat and water vapour by turbulence takes place with equal efficiency. Evaporation is then determined from an expression of the form:

$$E = \frac{R_n}{L(1 + \beta)}$$

The most satisfactory climatological method so far devised combines the energy budget and aerodynamic approaches. In this way H. L. Penman succeeded in expressing evaporation losses in terms of four meteorological elements which are regularly measured, at least in Europe and North America. These are duration of sunshine (related to radiation amounts), mean air temperature, mean air humidity and mean wind speed (which limit the losses of heat and vapour from the surface).

The relative roles of the factors which have been mentioned are illustrated by the global pattern of evaporation (fig. 2.3). Losses decrease sharply in high latitudes where there is little available energy. In middle and lower latitudes there are appreciable differences between land and sea (fig. 2.4B). Rates are naturally high over the oceans in view of the unlimited availability of water, and on a seasonal basis the maximum rates occur in winter over the western Pacific and Atlantic, where cold continental air blows across warm ocean currents. On an annual basis maximum oceanic losses occur about 15°–20°N and 10°–20°S in the belts of the constant trade winds. The highest annual losses, estimated to be about 200 cm (80 in), are in the western Pacific and central Indian Ocean near 15°S (cf. fig. 1.34; 100 kcal cm^{-2} is equivalent to an evaporation of 170 cm of water/cm^2). There is a subsidiary equatorial minimum over the oceans mainly as a result of the lower wind speeds in the doldrum belt and the proximity of the vapour pressure in the air

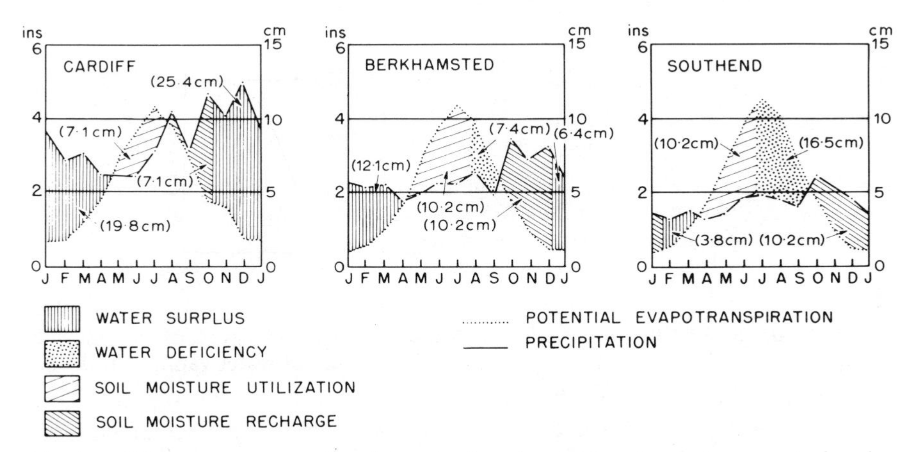

Fig. 2.5. The average annual moisture budget for stations in western, central and eastern Britain determined by Thornthwaite's method (*from Howe 1956*).

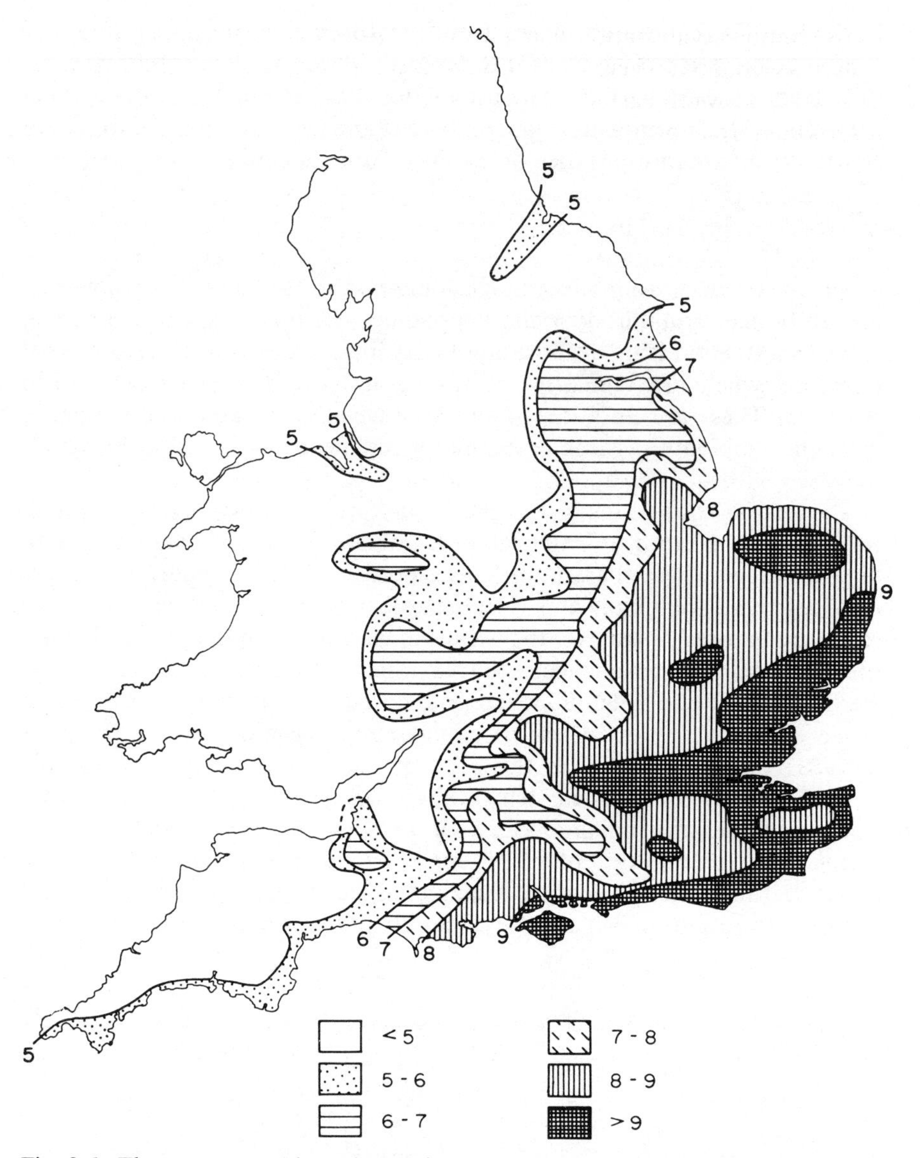

Fig. 2.6. The average number of years in ten when irrigation is theoretically necessary for crops in England and Wales, based on Penman's formula (*from Pearl et al. 1954; Crown Copyright Reserved*).

to its saturation value, but the land maximum occurs more or less at the equator because of the relatively high solar radiation receipts and the large transpiration losses from the luxuriant vegetation of this region. The secondary maximum over land in mid-latitudes is related to the strong prevailing westerly winds. The other parts of fig. 2.4, incorporated here for convenient comparison, are discussed in later sections.

The annual evaporation over Britain, calculated by Penman's formula, ranges from about 38 cm (15 in) in Scotland to about 50 cm (20 in) in parts of south and south-east England. The annual potential evapotranspiration determined by Thornthwaite's method (based on mean temperature) is over 64 cm (25 in) in most of south-eastern England. Since this loss is concentrated in the period May–September there may be seasonal water deficits of 12–15 cm (5–6 in) in these parts of the country (as shown in fig. 2.5 for Southend), necessitating considerable use of irrigation water by farmers. Figure 2.6 indicates that in southern and south-eastern England it is necessary to irrigate in about nine years out of ten during the summer six months (April–September), assuming that the crop can extract 6·4 cm ($2\frac{1}{2}$ in) of moisture from the soil.

In the United States, monthly moisture conditions are commonly evaluated on the basis of the Palmer Drought Severity Index (PDSI). This is determined from accumulated weighted differences between actual precipitation and the calculated amount required for evapotranspiration, soil recharge and runoff. Accordingly, it takes account of the persistence effects of drought situations. The PSDI has a range from ≥ 4 extremely wet to ≤ -4, extreme drought.

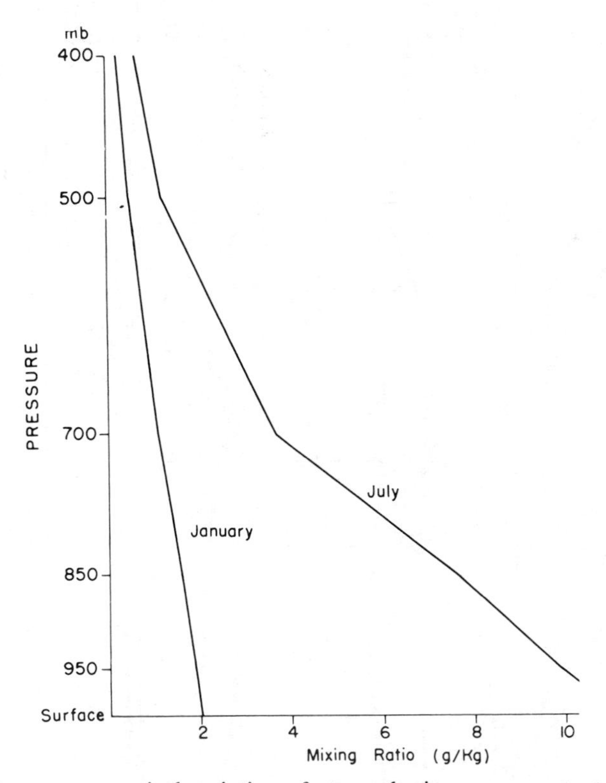

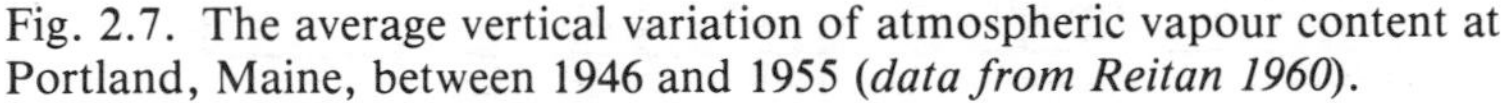
Fig. 2.7. The average vertical variation of atmospheric vapour content at Portland, Maine, between 1946 and 1955 (*data from Reitan 1960*).

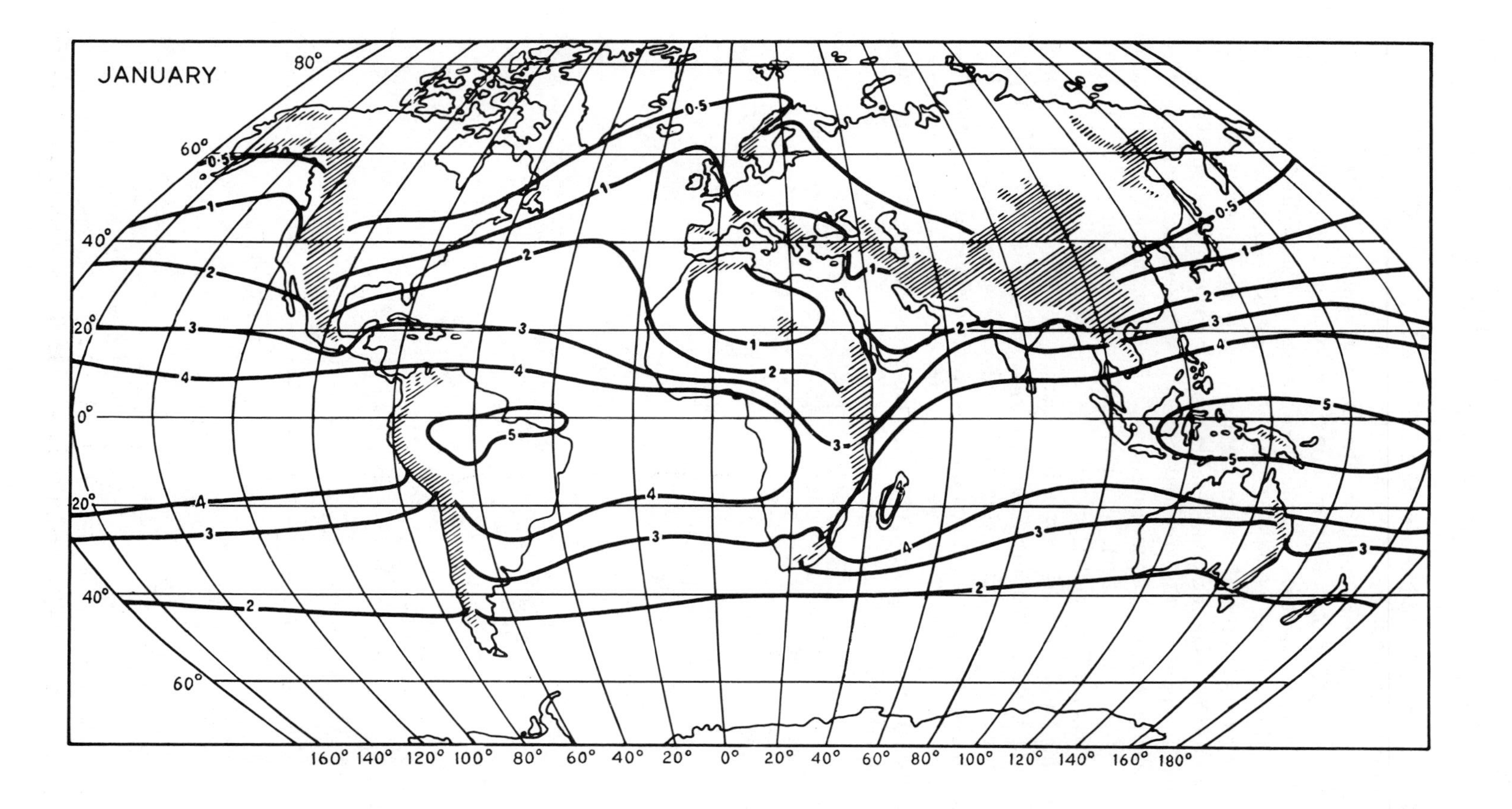
JANUARY
80°
60°
40°
20°
0°
20°
40°
60°
160° 140° 120° 100° 80° 60° 40° 20° 0° 20° 40° 60° 80° 100° 120° 140° 160° 180°

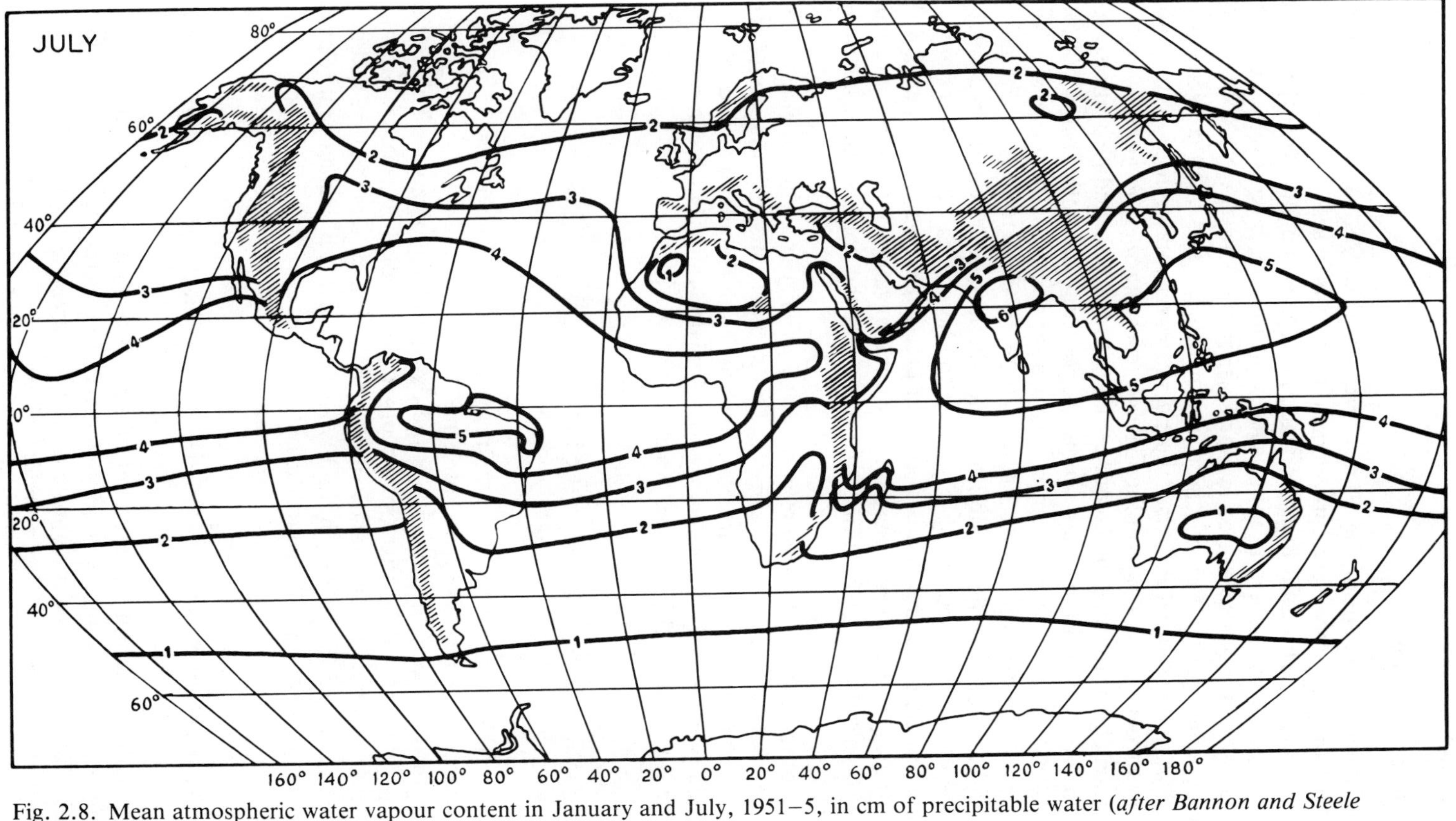

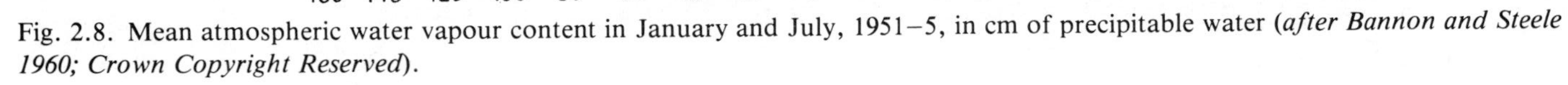
Fig. 2.8. Mean atmospheric water vapour content in January and July, 1951–5, in cm of precipitable water (*after Bannon and Steele 1960; Crown Copyright Reserved*).

B Humidity

1 Moisture content

The moisture content of the atmosphere can be expressed in several ways, apart from the vapour pressure, depending on which aspect the user wishes to emphasize. The total mass of water in a given volume of air, i.e. the density of the water vapour, is one such measure. This is termed the *absolute humidity* (ϱ_w) and is measured in grams per cubic metre (g m^{-3}). Volumetric measurements are not greatly used in meteorology and more convenient is the *mass mixing ratio* (x). This is the mass of water vapour in grams per kilogram of dry air. For most practical purposes the *specific humidity* (q) is identical, being the mass of vapour per kilogram of air including its moisture.

The bulk of the atmosphere's moisture content is contained below 500 mb (5574 m), as fig. 2.7 clearly shows. It is also apparent that the seasonal effect is most marked in the lowest 3000 m (10,000 ft) or so, that is below about 700 mb. The global distribution of atmospheric vapour content in January and July is illustrated in fig. 2.8. Over southern Asia during the summer monsoon an air column holds 5–6 cm of precipitable water, compared with less than 1 cm in tropical desert areas. Minimum values of 0·1–0·2 cm occur over high latitudes and continental interiors of the northern hemisphere in winter.

Another important measure is *relative humidity* (r), which expresses the actual moisture content of a sample of air as a percentage of that contained in the same volume of saturated air at the same temperature. The relative humidity is defined with reference to the mixing ratio, but it can be determined approximately in several ways:

$$r = \frac{x}{x_s} \times 100 \approx \frac{q}{q_s} \times 100 \approx \frac{e}{e_s} \times 100$$

where the subscript s refers to the respective saturation values at the same temperature; e denotes vapour pressure.

A further index of humidity is the dew-point temperature. This is the temperature at which saturation occurs if air is cooled at constant pressure without addition or removal of vapour. When the air temperature and dew point are equal the relative humidity is 100% and it is evident that relative humidity can also be determined from

$$\frac{e_s \text{ at dew point}}{e_s \text{ at air temperature}} \times 100$$

The relative humidity of a parcel of air will obviously change if either its temperature or its mixing ratio is changed. In general the relative humidity varies inversely with temperature during the day, tending to be lower in the early afternoon and higher at night.

Atmospheric moisture can be measured by at least five types of instrument. The most common for routine measurements is the *wet-bulb thermometer* installed in a louvered instrument shelter (Stevenson screen). The bulb of a standard thermometer is wrapped in muslin which is kept moist by a wick from a reservoir of pure water. The evaporative cooling of this wet bulb gives a reading which can be used in conjunction with a simultaneous dry bulb temperature reading to calculate the dew-point temperature. A similar portable device, called an aspirated *psychrometer*, uses a forced flow of air at a fixed rate over the dry and wet bulbs. A sophisticated instrument for determining dew point, based on a different principle, is the *dew-point hygrometer*. This detects when condensation first occurs on a cooled surface. Two other types of instrument are used to determine relative humidity. The *hygrograph* utilizes the expansion/contraction of a bundle of human air, in response to humidity, to record relative humidity continuously by a mechanical coupling to a pen arm marking on a rotating drum. This has an accuracy of about ±5–10%. For upper air measurements, a *lithium chloride* element is used to detect changes in its electrical resistance to vapour pressure differences. Relative humidity changes are accurate to about ±3%.

2 Moisture transport

It is sometimes overlooked that the atmosphere transports moisture horizontally as well as vertically. Figure 2.4c illustrates the quantities which must be transported meridionally in order to maintain the required moisture balance at a given latitude (i.e. precipitation – evaporation = net horizontal transport of moisture into the air column). A prominent feature is the equatorward transport into low latitudes and the poleward transport in middle latitudes. The reader should inspect this diagram again in the light of the discussion of winds belts in ch. 3, E.

At this point it is necessary to stress emphatically the fact that local evaporation is, in general, not the major source of local precipitation. For example, only 6% of the annual precipitation of Arizona and 10% of that over the Mississippi River basin is of local origin, the remainder being transported into these areas (i.e. moisture advection). Even when moisture is available in the atmosphere over a region only a small portion of it is usually precipitated. This depends on the efficiency of the condensation and precipitation mechanisms, both microphysical and large-scale, which we shall now consider.

C Condensation

Condensation, the direct cause of all the various forms of precipitation, occurs under varying conditions which in one way or another are associated with change in one of the linked parameters of air volume, temperature, pressure or humidity. Thus, condensation takes place (i) when the temperature of the air is reduced but its volume remains constant and the air is cooled to dew point; (ii) if the volume of the air is increased without addition of heat; this cooling takes places because adiabatic expansion causes energy to be consumed through work (see ch. 2, D); (iii) when a joint change of temperature and volume reduces the moisture-holding capacity of the air below its existing moisture content; or (iv) by evaporation adding moisture to the air. The key to the understanding of condensation clearly lies in the fine balance that exists between these variables. Whenever the balance between one or more of them is disturbed beyond a certain limit condensation may result.

The most common circumstances favourable to the production of condensation are those producing a drop in air temperature; namely contact cooling, mixing of air masses of different temperatures and dynamic cooling of the atmosphere. Contact cooling is produced, for example, within warm, moist air passing over a cold land surface (pl. 5). On a clear winter's night strong radiation will cool the surface very quickly and this surface cooling will gradually extend to the moist lower air, reducing the temperature to a point where condensation occurs in the form of dew, fog or frost, depending on the amount of moisture involved, the thickness of the cooling air layer and the dew-point value. When the latter is below 0°C it is referred to as the hoar frost-point if the air is saturated with respect to ice.

The mixing of the differing layers within a single air mass or of two differing air masses can also produce condensation. Figure 2.9 indicates how the horizontal mixing of two air masses (A and B), of given temperature and moisture characteristics, may produce an air mass (C) which is oversaturated at the intermediate temperature and consequently forms cloud. Vertical mixing of an air layer, which is discussed below (fig. 2.16) can have the same effect. Fog, or low stratus, with drizzle – known as 'crachin' – which is common along the coasts of South China and the Gulf of Tonkin in

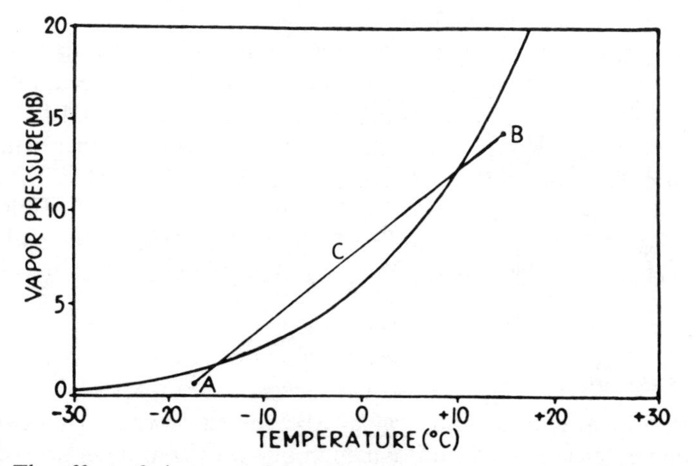

Fig. 2.9. The effect of air-mass mixing (*from Petterssen 1941*). The horizontal mixing of two unsaturated air masses A and B will produce one supersaturated air mass C. The saturation vapour pressure curve is shown (cf. fig. 1.7 which is a semi-logarithmic plot).

February–April, can develop as a result of either air mass mixing or warm advection over a colder surface.

The addition of moisture into the air near the surface by evaporation occurs when cold air moves out over a warm water surface. This can cause steam fog to form, which is common in arctic regions. Attempts at fog dispersal are one area where some progress has been made in local weather modification. Cold fogs can be locally dissipated by the use of dry ice (frozen CO_2) or the release of propane gas through expansion nozzles to produce freezing and the subsequent fallout of ice crystals (cf. p. 89). Warm fogs (i.e. having drops above freezing temperatures) present bigger problems, but attempts at dissipation have shown some limited success in evaporating droplets by artificial heating, the use of large fans to draw down dry air from above, the sweeping out of fog particles by jets of water, and the injection of electrical charges into the fog to produce coagulation.

Undoubtedly the most effective cause of condensation, however, is the dynamic process of adiabatic cooling. This is considered in some detail in the next section.

D Adiabatic temperature changes

The displacement of an air parcel to an environment of lower pressure (without heat exchange with surrounding air) causes an increase in its volume and a consequent lowering of its temperature. A volume increase involves work and the consumption of energy, thus reducing the heat available per unit volume and hence the temperature. Such a temperature change, involving no subtraction or addition of heat, is termed *adiabatic*. Vertical displacements of air are obviously a major cause of adiabatic temperature changes.

Near the earth's surface most processes of change are non-adiabatic (sometimes termed *diabatic*) because of the tendency of air to mix and modify its characteristics by lateral movement, turbulence and related physical processes. When a parcel of air moves vertically the changes that take place often follow an adiabatic pattern because air is fundamentally a poor thermal conductor, and the air parcel as a whole tends to retain its own thermal identity which distinguishes it from the surrounding air masses. In some circumstances, on the other hand, mixing of air with its surroundings must be taken into account.

Let us consider the changes that occur when an air parcel rises and a decrease of pressure is accompanied by volume increase and temperature decrease (see ch. 1, B). The rate at which temperature decreases in a rising, expanding air parcel is called the *adiabatic lapse rate*. If the upward movement of air does not produce condensation then the energy expended by expansion will cause the temperature of the mass to fall at what is called the *dry adiabatic lapse rate* or DALR (9·8°C/km or 5·4°F/1000 ft). However,

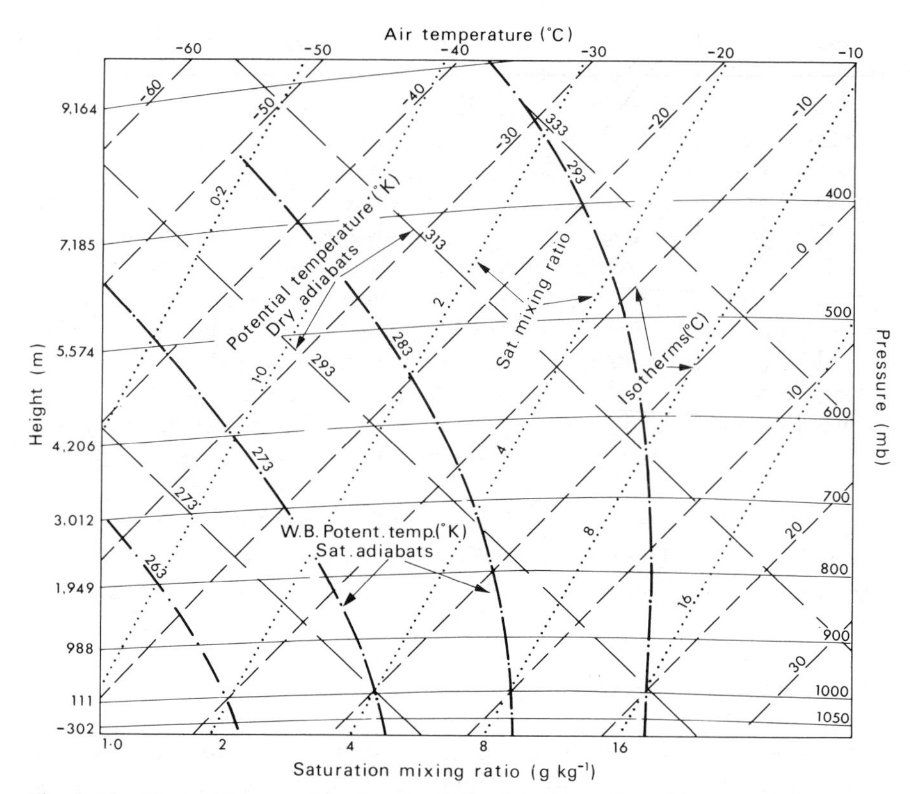

Fig. 2.10. The tephigram, allowing the following properties of the atmosphere to be displayed: temperature, pressure, potential temperature, wet-bulb potential temperature and saturation (humidity) mixing ratio.

prolonged reduction of the temperature invariably produces condensation, and when this happens latent heat is liberated, counteracting the dry adiabatic temperature decrease to a certain extent. It is therefore a distinguishing feature of rising and saturated (or precipitating) air that it cools at a slower rate (i.e. the *saturated adiabatic lapse rate* or SALR) than air which is unsaturated. Another difference between the dry and saturated adiabatic rates is that whereas the former remains constant the latter varies with temperature. This is because air masses at higher temperatures are able to hold more moisture and on condensation therefore to release a greater quantity of latent heat. For high temperatures the saturated adiabatic lapse rate may be as low as 4°C/km (or 2·2°F/1000 ft), but this rate increases with decreasing temperatures, approaching 9°C/km (5°F/1000 ft) at –40°C (–40°F).

In all, three different lapse rates can be differentiated, two dynamic and one static. There is the environmental (or static) rate, which is the actual temperature decrease with height on any occasion, such as an observer ascending with a balloon would record. This is not an adiabatic rate therefore and may assume any form depending on local air temperature con-

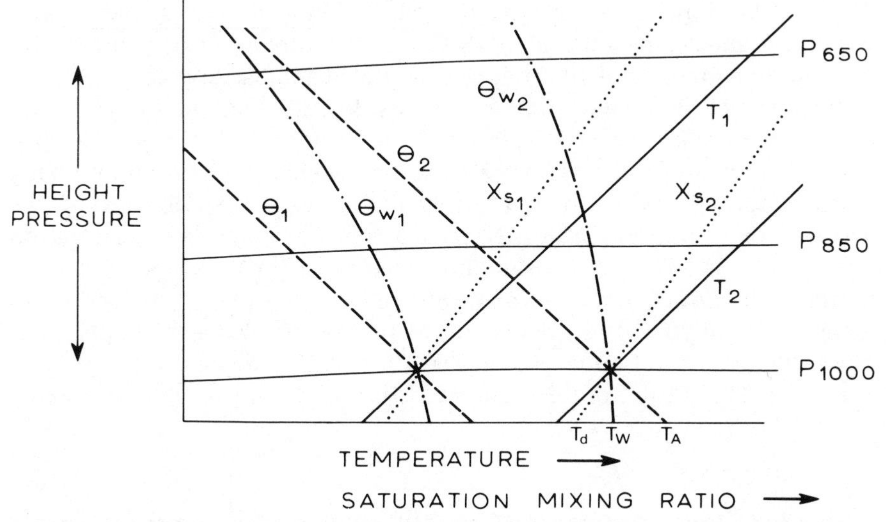

Fig. 2.11. Graph showing the relationships between temperature (T), potential temperature θ, wet-bulb potential temperature (θ_w) and saturation mixing ratio (X_s). T_d = dew point, T_w = wet-bulb temperature and T_A = air temperature.

ditions. There are also the dynamic adiabatic dry and saturated lapse rates (or cooling rates) which apply to rising parcels of air moving through their environment. Close to the surface the vertical temperature gradient sometimes greatly exceeds the dry adiabatic lapse rate, that is, it is superadiabatic. This is particularly common in arid areas in summer (see table 1.4). Over most ordinary dry surfaces the lapse rate approaches the dry adiabatic value at an elevation of 100 m or so.

The changing properties of moving air parcels can be conveniently expressed by plotting them as *path curves* on suitably constructed graphs. One such adiabatic diagram in common use is the tephigram (fig. 2.10). This displays five sets of lines representing properties of the atmosphere:

1 Isotherms – i.e. lines of constant temperature (parallel lines from bottom left to top right).
2 Dry adiabats (parallel lines from bottom right to top left).
3 Isobars – i.e. lines of constant pressure (slightly curved nearly horizontal lines).
4 Saturated adiabats (curved lines sloping up from right to left).
5 Saturation mixing ratio lines (those at a slight angle to the isotherms).

The dry adiabats are also lines of constant potential temperature, θ (or isentropes). Potential temperature is the temperature of an air parcel brought dry adiabatically to a pressure of 1000 mb. Mathematically,

$$\theta = T\left(\frac{1000}{p}\right)^{0.286} \quad \text{where } \theta \text{ and } T \text{ are in degK; } p = \text{pressure (mb).}$$

Figure 2.11 shows schematically the relationship between T and θ; also between T and θ_w, the wet-bulb potential temperature (where the air parcel is brought to a pressure of 1000 mb by a saturated adiabatic process).

Figure 2.11 illustrates that on an aerological diagram (such as the tephigram or adiabatic chart) a line along a dry adiabat (θ) through the dry-bulb temperature of the surface air (T_A), an isopleth of saturation mixing ratio (x_s) through the dew point (T_d) and a saturated adiabat (θ_w) through the wet-bulb temperature (T_w) all intersect at a point corresponding to saturation for the air mass. This relationship, known as Normand's theorem, is used to estimate the *lifting condensation level* (see figs. 2.13 and 2.14). For an air temperature of 20°C and a dew point of 10°C at 1000 mb surface pressure, the lifting condensation level is at 860 mb with a temperature of 8°C (see fig. 2.10). The altitude of this 'characteristic point' can be estimated roughly by

$$h\,(\mathrm{m}) = 120\,(T - T_d)$$

where T = air temperature and T_d = dew-point temperature at the surface in °C).

The lifting condensation level (LCL) formulation does not take account of vertical mixing. A modified calculation defines a *convective condensation*

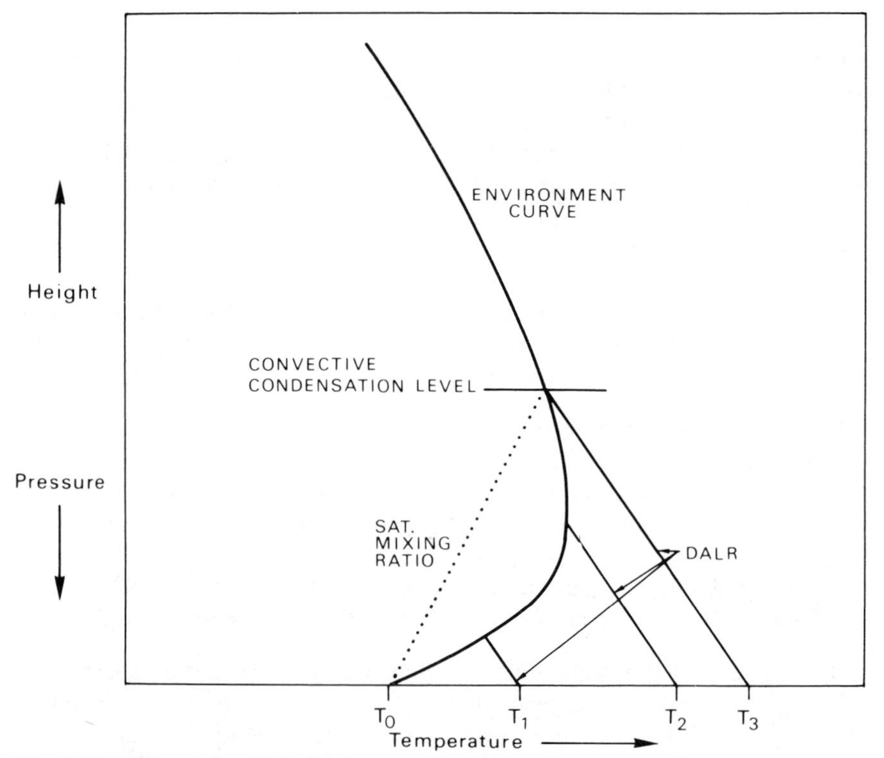

Fig. 2.12. Schematic adiabatic chart used to determine the convective condensation level (see text). T_0 represents the early-morning temperature; T_1, T_2 and T_3 illustrate daytime heating of the surface air.

level (CCL). In the near-ground layer, surface heating may establish a super-adiabatic lapse rate, but convection modifies this to the DALR profile. Day-time heating steadily raises the surface air temperature from T_0 to T_1, T_2 and T_3 (fig. 2.12). Convection also equalizes the humidity mixing ratio, assumed equal to the value for the initial temperature. The CCL is located at the inter-section of the environment temperature curve with a saturation mixing ratio line corresponding to the average mixing ratio in the surface layer (1000–1500 m). Expressed in another way, the surface air temperature is the minimum which will cloud to form as a result of free convection. Because the air near the surface is often well mixed, the CCL and LCL, in practice, are commonly nearly identical.

Experimentation with a tephigram shows that both the convective and lifting condensation levels rise as the surface temperature increases with little change of dew point. This is commonly observed in the early afternoon when the base of cumulus clouds tends to be at higher levels.

Potential temperature provides an important yardstick for air mass characteristics, since if the air is only affected by dry adiabatic processes the potential temperature remains constant. This helps to identify different air masses and indicates when latent heat has been released through saturation of the air mass or when non-adiabatic temperature changes have occurred.

E Air stability and instability

The important characteristic of stable air is that if it is forced up or down it has a tendency to return to its former position once the motivating force ceases. Figure 2.13 shows the reason for this, in that the environmental temperature curve (*a*) lies to the right of any path curve representing the lapse rate of an unsaturated air parcel cooling dry adiabatically when forced to rise. At any level the rising parcel is cooler and more dense than its surround-ings and therefore tends to revert to its former level. Similarly, if the air is forced downwards it will gain in temperature at the dry adiabatic rate, always be warmer and less dense than the surrounding air, and tend to return to its former position (unless prevented from doing so). However, if local surface heating causes the environmental lapse rate near the surface to exceed the dry adiabatic lapse rate (*b*), then the adiabatic cooling of a convective air parcel allows it to remain warmer and less dense than the surrounding air so that it continues to rise through buoyancy. Similarly, if an air parcel is im-pelled downwards under these same conditions from a higher level it will become colder than its surroundings and there will be no check on its down-ward progress until it reaches the surface. The characteristic of unstable air is a tendency to continue moving away from its original level when set in motion.

A further possibility is illustrated in fig. 2.14. The air is stable in the lower layers, but if the air is forced to rise, for example by passage over a mountain range, or by local surface heating, until the path curve crosses the environment curve (the level of free convection) then the air, being warmer than its surroundings, is free to rise. This is termed *conditional instability*, as

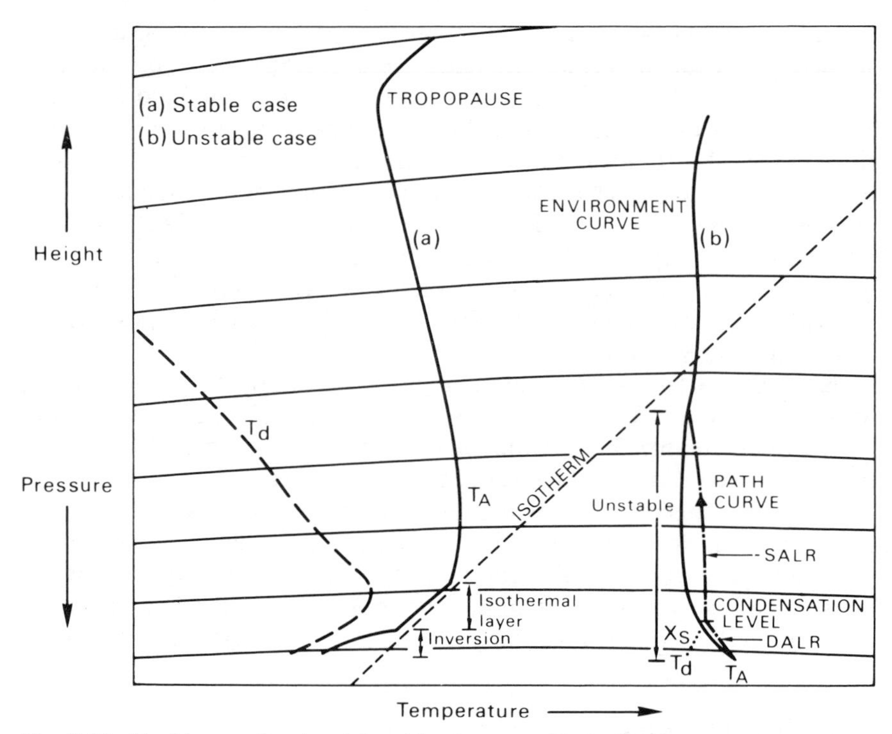

Fig. 2.13. Tephigram showing (a) stable air case – T_A is the air temperature and T_d the dewpoint; and (b) unstable air case. The lifting condensation level is shown, together with the path curve (arrowed) of a rising air parcel. X_s is the saturation humidity mixing ratio line through the dewpoint temperature (see text).

the development of instability is dependent on the air mass becoming saturated. Since the environmental lapse rate is frequently between the dry and saturated adiabatic rates the state of conditional instability is a common one.

The environment curve (*b*) in fig. 2.13 intersects the path curve at a higher level. Above the level of this intersection the atmosphere is stable, but the buoyant energy gained by the rising parcel enables it to move some distance into this region. The theoretical upper limit of cloud development can be estimated from the tephigram by determining an area (B) above the intersection of the environment and path curves equal to that between the two curves from the level of free convection to the intersection (A in fig. 2.14); the tephigram is so constructed that equal areas represent equal energy.

These examples assume that a small air parcel is being displaced without any compensating air motion or mixing of the parcel with its surroundings. These assumptions are rather unrealistic since dilution of an ascending air parcel by mixing of the surrounding air with it through *entrainment* will reduce its buoyant energy. However, the method is generally satisfactory for routine forecasting, especially perhaps since the assumptions approximate conditions in the updraught of cumulonimbus clouds.

A further consideration is that a deep layer of air may be displaced by vertical motion over an extensive topographic barrier. Figure 2.15 shows a

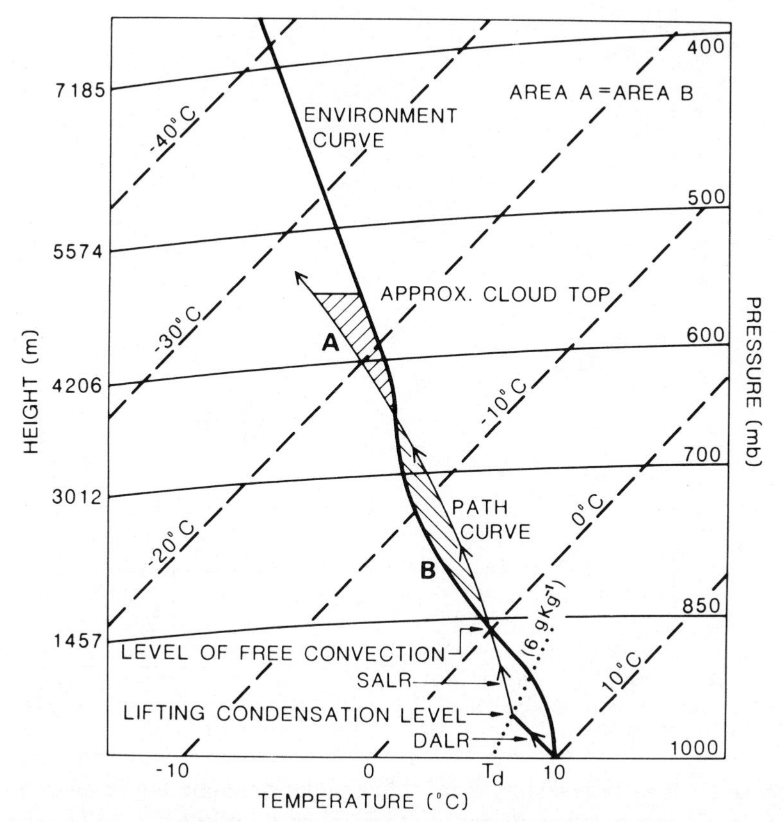

Fig. 2.14. Schematic tephigram illustrating the conditions associated with the conditional instability of an air mass that is forced to rise. The saturation mixing ratio is a broken line and the lifting condensation level (cloud base) is below the level of free convection.

case where the air in the upper levels is less moist than that at lower levels. If the whole layer is forced bodily upwards the drier air at *B* follows the dry adiabatic rate, and so, for a while, will the air about *A*, but there eventually comes a time when the condensation level is reached, after which the lower layers of the rising air mass cool at the saturated adiabatic rate. This has the final effect of increasing the actual lapse rate of the total thickness of the raised layer, and, if this new rate is greater than that of the saturated adiabatic, the air layer becomes unstable, and may overturn. This is termed *convective (or potential) instability.*

Vertical mixing of air was referred to earlier as a possible cause of condensation. This is best illustrated by use of a tephigram. Figure 2.16 shows an initial distribution of temperature and dew point. Vertical mixing has the effect of averaging these conditions through the layer affected. Thus, the *mixing condensation level* is determined from the intersection of the average values of saturation humidity mixing ratio and potential temperature. The areas above and below the points where these average value lines cross the initial environment curves are equal.

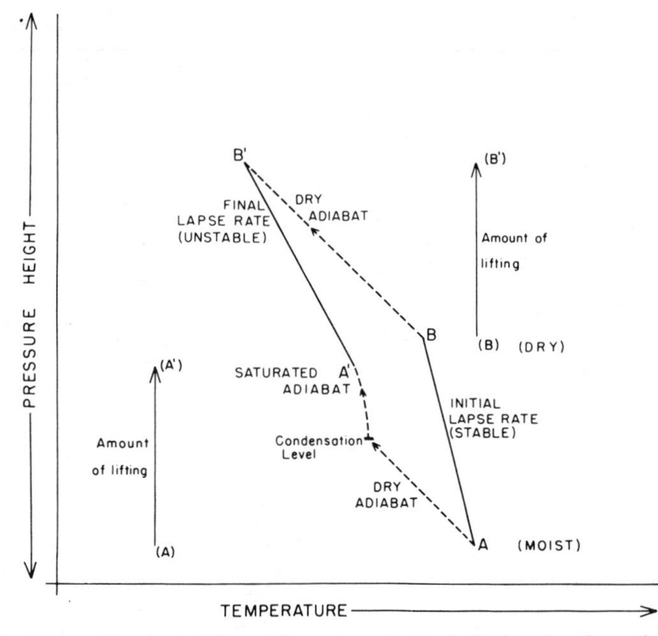

Fig. 2.15. Convective instability. AB represents the initial state of an air column; moist at A, dry at B. After uplift of the whole air column the temperature gradient A′ B′ exceeds the saturated adiabatic lapse rate so that the air column is unstable.

Subsidence usually results from either radiational cooling or an excess of horizontal convergence of air in the upper troposphere. Subsiding air generally moves with a vertical velocity of only 1–10 cm s^{-1} (about 100–1000 ft hr^{-1}), although convectional conditions provide an exception (see ch. 2, H). Subsidence can produce substantial changes in the atmosphere, and, for instance, if an air mass subsides about 300 m (1000 ft) all average-size cloud droplets will usually be evaporated.

F Cloud formation

The formation of clouds is dependent on atmospheric instability and vertical motion but it is also controlled by microscale processes. These are now discussed before we examine larger scale aspects of cloud development and type.

1 Condensation nuclei

It is very important to note that condensation occurs with the utmost difficulty in *clean* air; moisture must generally find a suitable surface upon which it can condense. If pure air is reduced in temperature below its dew point it becomes *supersaturated* (i.e. relative humidity exceeding 100%). To maintain a pure water drop of radius 10^{-7} cm (0·001 μm) requires a relative humidity of 320%, and for one of 10^{-5} cm (0·1 μm) radius 101%.

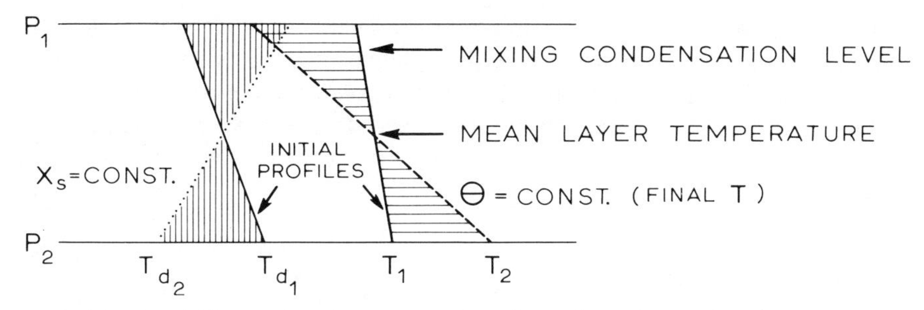

Fig. 2.16. Graph illustrating the effects of vertical mixing in an air mass.

Usually condensation occurs on a foreign surface which can be a land or plant surface, as is the case for dew or frost, while in the free air condensation begins around so-called *hygroscopic nuclei*. These are microscopic particles – aerosols – the surfaces of which (like the weather enthusiast's seaweed!) have the property of *wettability*. Aerosols include dust, smoke, salts and chemical compounds. Sea salts, which are particularly hygroscopic, enter the atmosphere principally by the bursting of air bubbles in foam and are a major component of the aerosol load. Other large contributions are from fine soil particles and various natural, industrial and domestic combustion products raised by the wind. A further source is the conversion of atmospheric trace gas to particles through photochemical reactions, particularly over urban areas. Nuclei range in size from those with a radius of 0·001 μm, which are ineffective because of the high supersaturations required for their activation, to *giants* of over 10 μm which do not remain airborne for very long. On average, oceanic air contains 1 million condensation nuclei per litre (i.e. thousand cm^3) and land air holds some 5 or 6 million.

Hygroscopic aerosols are soluble. This is very important since the saturation vapour pressure is less over a solution droplet (for example, sodium chloride or sulphuric acid) than over a pure water drop of the same size and temperature (fig. 2.17). Indeed, condensation begins on hygroscopic particles before the air is saturated; in the case of sodium chloride nuclei at 78% relative humidity. Figure 2.17 illustrates droplet radii for three sets of solution droplets of sodium chloride (a common sea salt) in relation to their equilibrium relative humidity, called Kohler curves. Droplets in an environment where values are below/above the appropriate line will evaporate/grow. Each curve has a maximum beyond which the droplet can grow in air with less supersaturation.

Once they are initially formed, the process of growth of water droplets is far from simple and much remains to be explained. In the early stages the solution effect is predominant and small drops grow more quickly than large ones, but, as the size of a droplet increases its growth rate by condensation decreases, as shown in fig. 2.18. The radial growth rate obviously slows down as the drop size increases because there is an increasingly greater surface area to add to with every increment of radius. However, the condensation rate is limited by the speed with which the released latent heat can be lost from the drop by conduction to the air and this heat reduces the vapour gradient. Moreover competition between droplets for the available moisture increasingly tends to reduce the degree of supersaturation.

Supersaturation in clouds very rarely exceeds 1% and, because the saturation vapour pressure is greater over a curved droplet surface than over a

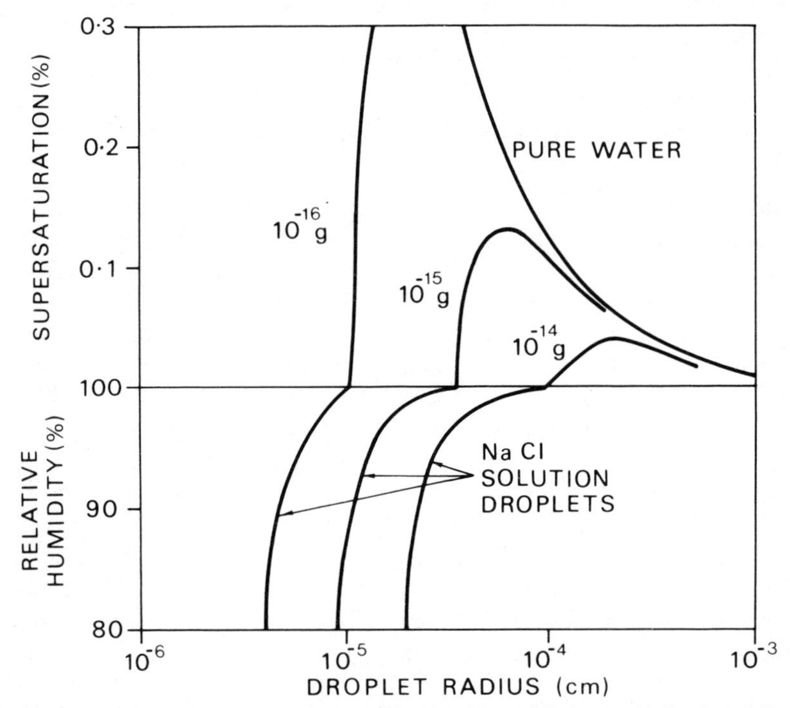

Fig. 2.17. Kohler curves showing the variation of equilibrium relative humidity or supersaturation (%) with droplet radius for pure water and NaCl solution droplets. The numbers show the mass of sodium chloride (a similar family of curves is obtained for sulphate solutions). The pure water droplet line illustrates the curvature effect (*after Mason 1975*).

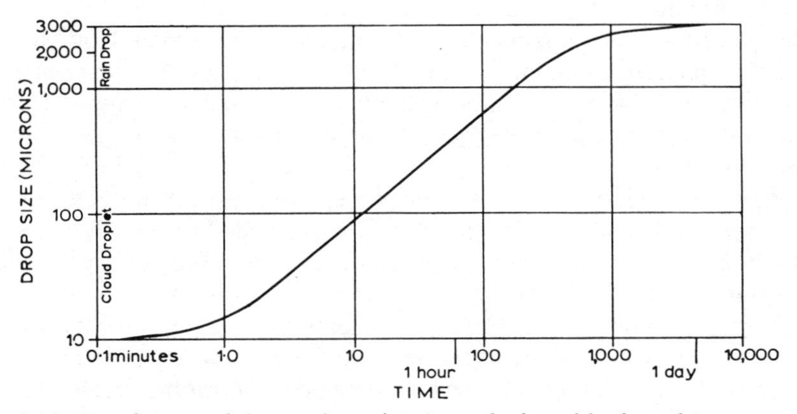

Fig. 2.18. Droplet growth by condensation (note the logarithmic scale).

plane water surface, very small droplets ($<0 \cdot 1\,\mu m$ radius) are readily evaporated (fig. 2.17). In the early stages the nucleus size is important; for supersaturation of 0·05% a droplet of 1 μm radius with a salt nucleus of mass 10^{-13} g reaches 10 μm in 30 min, whereas one with a salt nucleus of 10^{-14} g would take 45 min. Later, when the dissolved salt has ceased to have significant effect, the radial growth rate becomes slow as a result of decreasing supersaturation.

Figure 2.18 illustrates not only the slow growth of droplets but also the immense size difference between cloud droplets (<1 to 50 μm radius) and raindrops (exceeding 1 mm diameter). These facts strongly suggest that the gradual process of condensation is inadequate to explain the rates of formation of raindrops which are often observed. For example, in most clouds precipitation develops within an hour. It must be remembered too that falling raindrops undergo evaporation in the unsaturated air below the cloud base. A droplet of 0·1 mm radius evaporates after falling only 150 m at a temperature of 5°C and 90% relative humidity, but a drop of 1 mm radius would fall 42 km before evaporating. It seems likely then that cloud droplets are not necessarily the immediate source of raindrops. This point is taken up again in section G.

2 Cloud types

The great variety of cloud forms necessitates a classification for purposes of weather reporting. The internationally adopted system is based upon (*a*) the general shape, structure and vertical extent of the clouds, and (*b*) their altitude.

These primary characteristics are used to define the ten basic groups (or genera) as shown in fig. 2.19. High cirriform cloud is composed of ice crystals giving a generally fibrous appearance. Stratiform clouds are layer-shaped, while cumuliform ones have a heaped appearance and usually show progressive vertical development. Other prefixes are *alto-* for middle level

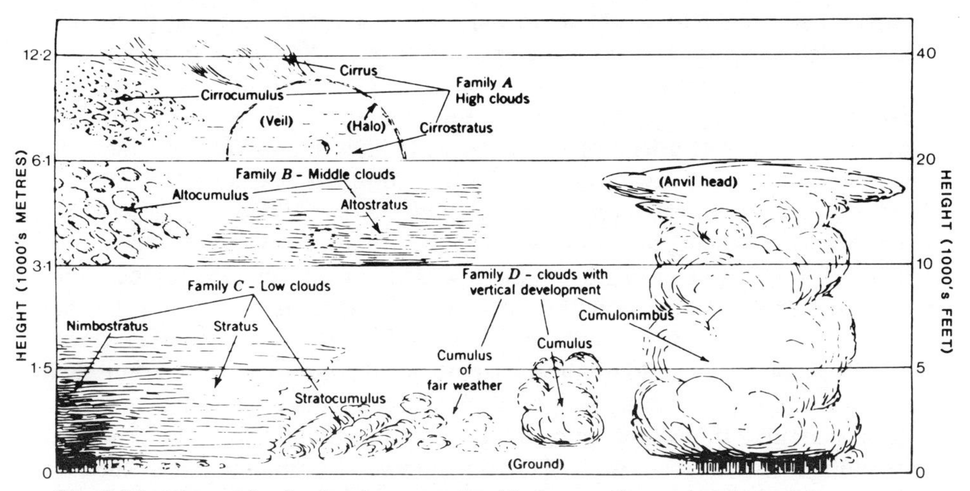

Fig. 2.19. The ten basic cloud groups classified according to height and form (*from Strahler 1965*).

(medium) clouds and *nimbo-* for low cloud of considerable thickness, which appear dark grey and from which continuous rain is falling.

The height of the cloud base may show a considerable range for any of these types and varies with latitude. The approximate limits in thousands of metres for different latitudes are:

	Tropics	*Middle latitudes*	*High latitudes*
High cloud	Above 6	Above 5	Above 3
Medium cloud	2–7·5	2–7	2–4
Low cloud	Below 2	Below 2	Below 2

Following taxonomic practice, the classification subdivides the major groups into species and varieties with Latin names according to details of their appearance. The World Meteorological Organization has produced an *International Cloud Atlas* illustrating all of these types.

Other possible classifications of clouds take into account their mode of origin. For instance, a broad genetic grouping can be made according to the mechanism of vertical motion which produces condensation. Four categories are:

(*a*) gradual uplift of air over a wide area in association with a low-pressure system;
(*b*) thermal convection (on the local cumulus-scale);
(*c*) uplift by mechanical turbulence (*forced convection*);
(*d*) ascent over an orographic barrier.

Group (*a*) includes a wide range of cloud types and will be discussed more fully in ch. 4, D.2. In connection with thermal convection, which forms cumuliform clouds, group (*b*), it is worth noting that upward convection currents (thermals) form plumes of warm air which, as they rise, expand and are carried downwind. Towers in cumulus and other clouds are caused not by thermals of surface origin, but by ones set up *within* the cloud as a result of the release of latent heat by condensation. Thermals gradually lose their impetus as mixing of cooler, drier air from the surroundings dilutes the more buoyant warm air. Cumulus towers also tend to evaporate as updraughts diminish, leaving a shallow oval-shaped 'shelf' cloud (*stratocumulus cumulogenitus*) which may amalgamate with others to produce a high overcast. Group (*c*) includes fog, stratus or stratocumulus and is important whenever air near the surface is cooled to dew point by conduction or night-time radiation and the air is stirred by irregularities of the ground. The final group (*d*) could include stratiform or cumulus clouds produced by forced uplift of air over mountains (see pls. 6 and 7). Hill fog is simply stratiform cloud enveloping high ground. A special and important category is the wave (lenticular) cloud which develops when air flows over hills setting up a wave motion in the air current downwind of the ridge (see ch. 3, C.2). Clouds form in the crest of these waves if the air reaches its condensation level (see pls. 8 and 9).

A great deal of information on cloud amounts, especially in remote areas, and on cloud patterns in relation to weather systems is now being provided by operational weather satellites. These supply direct readout imagery and data that cannot be obtained by ground observations. Special classifications of cloud elements and patterns have been devised in order to analyse satellite imagery. The most common patterns seen on satellite photographs are cellular, or honeycomb-like, with a typical diameter of 30 km. These develop from the movement of cold air over a warmer sea surface. An open cellular pattern, where cumulus clouds are along the cell sides, forms where there is a large air–sea temperature difference, whereas closed polygonal cells occur if this difference is small. In both cases there is subsidence above the cloud layer. Open (closed) cells are more common over warm (cool) ocean currents to the east (west) of the continents. The honeycomb pattern has been attributed to mesoscale convective mixing, but the cells have a width–depth ratio of about 30:1, whereas laboratory thermal convection cells have a corresponding ratio of only 3:1. Thus the true explanation may be more complicated. Occasionally, subsidence over the tropical oceans leads to a modification of the cellular pattern referred to as 'actiniform' (or radiating), illustrated in pl. 10. Another common pattern over oceans and uniform terrain is provided by linear cumulus cloud 'streets'. Helical motion in these two-dimensional cloud cells develops with surface heating, particularly when outbreaks of polar air move over warm seas (see pls. 11 and 28) and there is a capping inversion. Cloud patterns related to cyclonic systems are discussed in ch. 4, D.

G Formation of precipitation

The puzzle of raindrop formation has already been briefly mentioned. The simple growth of cloud droplets is apparently an inadequate mechanism by itself, but if this is the case then more complex processes have to be envisaged.

Numerous early theories of raindrop growth have met with objections. For example, it was proposed that differently charged droplets could coalesce by electrical attraction, but it appears that distances between drops are too great and the difference between the electrical charges too small for this to happen. It was also suggested that large drops might grow at the expense of small ones, but observations show that the distribution of droplet size in a cloud tends to maintain a regular pattern, with the average radius between about 10 to 15 μm and a few larger than 40 μm. Another proposal was based on the variation of saturation vapour pressure with temperature such that if atmospheric turbulence brought warm and cold cloud droplets into close conjunction the supersaturation of the air with reference to the cold drop surfaces and the undersaturation with reference to the warm drop surfaces would cause the warm drops to evaporate and the cold ones to develop at their expense. However, except perhaps in some tropical clouds, the temperature of cloud droplets is too low for this differential mechanism to operate. Figure 2.9 shows that below about −10°C the slope angle of the saturation vapour pressure curve is low. Another theory was that raindrops grow around exceptionally large condensation nuclei (such as have been observed in some tropical storms). Large nuclei, it is known, do experience a more rapid rate of initial condensation, but after this stage they are subject to the same limiting rates of growth that apply to all other atmospheric water drops.

The two main groups of current theories which attempt to explain the rapid growth of raindrops involve the growth of ice crystals at the expense of water drops, and the coalescence of small water droplets by the sweeping action of falling drops.

1 Bergeron–Findeisen theory

This theory forms an important part of the presently accepted mechanism of raindrop growth, and is based on the fact that the relative humidity of air is greater with respect to an ice surface than with respect to a water surface. As the air temperature falls below 0°C the atmospheric vapour pressure decreases more rapidly over an ice surface than over water (see fig. 1.7). This results in the saturation vapour pressure over water becoming greater than that over ice, especially between temperatures of −5°C and −25°C where the difference exceeds 0·2 mb. If ice crystals and supercooled water droplets exist together in a cloud the latter tend to evaporate and direct deposition takes place from the vapour on to the ice crystals (this is often described by meteorologists as *sublimation*, which properly refers to direct evaporation from ice).

Just as the presence of condensation nuclei is necessary for the formation of water droplets, so *freezing nuclei* are necessary before ice particles can form – usually at very low temperatures (about −15° to −25°C). Small water droplets can, in fact, be super-cooled in pure air to −40°C before spontaneous freezing occurs, but ice crystals generally predominate in clouds where temperatures are below about −22°C. Freezing nuclei are far less numerous than condensation nuclei; for example, there may be as few as 10 per litre at −30°C and probably rarely more than 1000. However, some become active at higher temperatures. Kaolinite, a common clay mineral, becomes initially active at −9°C and on subsequent occasions at −4°C. The origin of freezing nuclei has been the subject of much debate but it is generally considered that very fine soil particles are a major source. Another possibility is that meteoric dust provides the nuclei, although there seems to be no firm evidence of a relationship between meteorite showers and rainfall. Volcanic dust ejected into the upper stratosphere and troposphere during eruptions might be an additional terrestrial source. Recent work shows that common biogenic aerosols emitted by decaying plant litter, in the form of complex chemical compounds, serve as freezing nuclei. In the presence of certain associated bacteria, ice nucleation can take place at only −2° to −5°C.

Once minute ice crystals have formed they grow readily by deposition from vapour, with different hexagonal forms of crystal developing at different temperature ranges. The number of ice crystals also tends to increase progressively because small splinters become detached during growth by air currents and act as fresh nuclei. The freezing of super-cooled water drops may also produce ice splinters (see ch. 2, H). Ice crystals readily aggregate upon collision, due to their frequently branched (dendritic) shape, and tens of crystals may form a single snowflake. Temperatures between about 0°C and −5°C are particularly favourable to aggregation, because fine films of water on the crystals' surfaces freeze when two crystals touch, binding them together. When the fall-speed of the growing ice mass exceeds the existing velocities of the air upcurrents the snowflake falls, melting into a raindrop if it passes through a sufficient depth of air warmer than 0°C.

This theory seems to fit most of the observed facts, yet it is not completely satisfactory. Cumulus clouds over tropical oceans can give rain when they are only some 2000 m (6500 ft) deep and the cloud-top temperature is 5°C or more. Even in middle latitudes in summer precipitation may fall from cumuli which have no subfreezing layer (*warm clouds*). A suggested mechanism in such cases is that of 'droplet coalescence', which is discussed below.

Practical rainmaking has been based on the Bergeron theory with some success, which at least supports its principal points. The basis of such experiments is the freezing nucleus. Super-cooled (water) clouds between −5°C and −15°C are *seeded* with especially effective nuclei, such as silver iodide or 'dry ice' (CO_2) from aircraft or ground-based generators in the case of silver iodide, promoting the growth of ice crystals and encouraging precipitation. The seeding of some cumulus clouds at these temperatures probably produces a mean increase of precipitation of 10–15% from clouds which are already precipitating or are 'about to precipitate'. Increases of up to 10% have resulted from seeding winter orographic storms. However, the seeding of depressions has produced no apparent precipitation increases, and it appears likely that clouds with an abundance of natural ice crystals or with above-freezing temperatures throughout are not susceptible to *rainmaking*. Premature release of precipitation may destroy the updraughts and cause dissipation of the cloud, however. This explains why some seeding experiments have actually *decreased* the rainfall! In other instances cloud growth and precipitation have been achieved, with some individually spectacular results having been obtained by such methods in Australia and the United States. Programmes aimed at increasing winter snowfall on the western slopes of the Sierra Nevada and Rocky Mountains by seeding cyclonic storms have been initiated, but it is still too early to regard such rain- (or snow-) making as a routine operation. Its success depends in any case on the presence of suitable super-cooled clouds. When several cloud layers are present in the atmosphere, natural seeding may be important. For example, if ice crystals fall from high-level cirrostratus or altostratus (a 'releaser' cloud) into nimbostratus (a 'spender' cloud) composed of super-cooled water droplets, the latter can grow rapidly by the Bergeron process and such situations may lead to extensive and prolonged precipitation. This is a frequent occurrence in cyclonic systems in winter and is important in orographic precipitation (see ch. 2, I).

Of all current attempts by man to control meteorological events none are more important than those relating to hurricanes. There is some indication that the seeding of the rising air in the cumulus eyewall may widen the ring of condensation and updraught, decrease the angular momentum of the storm, and thus the maximum speed of the winds. Such attempts are still in their infancy.

2 Collision theories

Alternative raindrop theories use collision, coalescence and 'sweeping' as the drop growth generator. It was originally thought that atmospheric turbulence by making cloud particles collide would cause a significant proportion to coalesce. However, it was found that particles might just as easily break up if subjected to collisions and it was also observed that there is often no precipitation from highly turbulent clouds. Langmuir offered a variation of this simple collision theory by pointing out that falling drops have terminal velocities (typically 1–10 cm s^{-1}) directly related to their diameters such that the larger drops might overtake and absorb small droplets and that the latter might also be swept into the wake of the former and be absorbed by them. Figure 2.20 gives experimental results of the rate of growth of water drops by coalescence and also of ice particles by vapour deposition from an initial radius of 20 μm. Although coalescence is initially rather slow, the drop can

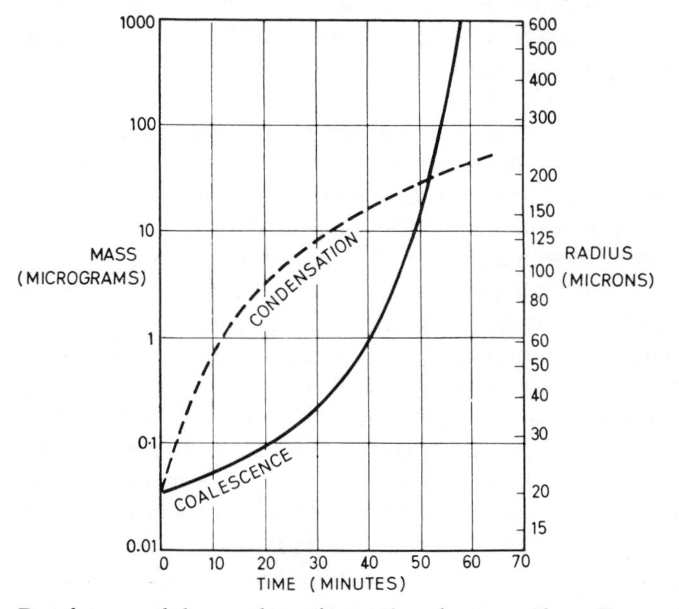

Fig. 2.20. Droplet growth by condensation and coalescence (*from East and Marshall 1954*).

reach 200 μm radius in 50 minutes; moreover, the growth rate is rapid for droplets with radii greater than 40 μm. Calculations show that drops must exceed 19 μm in radius before they can coalesce with other droplets, smaller droplets being swept aside without colliding. The initial presence of a few very large cloud droplets calls for the availability of giant nuclei (e.g. salt particles) if the cloud top does not reach above the freezing level. Observations show that maritime clouds do have relatively few large condensation nuclei (10–50 μm radius) and a high liquid water content, whereas continental air tends to contain many small nuclei (~ 1 μm) and less liquid water. Hence, rapid onset of showers is feasible by the coalescence mechanism in maritime clouds. Alternatively, if a few ice crystals are present at higher levels in the cloud (or if seeding occurs with ice crystals coming from higher cloud layers) they may eventually fall through the cloud as drops and the coalescence mechanism comes into action. Turbulence, especially in cumuliform clouds, serves to encourage collisions in the early stages and cloud electrification also increases the efficiency of coalescence. Thus, the coalescence process allows a more rapid growth than simple condensation can provide and is, in fact, common in clouds in tropical maritime air masses, even in temperate latitudes.

3 Other types of precipitation

Rain has been discussed at length because it is the most common form of precipitation. Snow occurs when the freezing level is so near the surface that aggregations of ice crystals do not have time to melt before reaching the ground. Generally this means that the freezing level must be below 300 m (1000 ft). Mixed snow and rain ('sleet' in British usage) is especially likely when the air temperature at the surface is about 1·5°C (34° to 35°F). Snow rarely occurs with an air temperature exceeding 4°C (39°F).

Soft hail pellets (roughly spherical, opaque grains of ice with much enclosed air) occur when the Bergeron process operates in a cloud with a small liquid water content and ice particles grow mainly by deposition of water vapour. Limited accretion of small super-cooled droplets forms an aggregate of soft, opaque ice particles 1 mm or so in radius. Showers of such pellets are quite common in winter and spring from cumulonimbus clouds.

Ice pellets may develop if the soft hail falls through a region of large liquid water content above the freezing level. Accretion forms a casing of clear ice around the pellet. Alternatively, an ice pellet consisting entirely of transparent ice may result from the freezing of a raindrop or the re-freezing of a melted snowflake.

True hailstones are roughly concentric accretions of clear and opaque ice. The embryo is a raindrop carried aloft in an updraught and frozen. Successive accretions of opaque ice (rime) occur due to impact of super-cooled droplets which freeze instantaneously, whereas the clear ice (glaze) represents a wet surface layer, developed as a result of very rapid collection of super-cooled drops in parts of the cloud with large liquid water content, which has subsequently frozen. A major difficulty in early theories was the necessity to postulate violently fluctuating upcurrents to give the observed banded hailstone structure, but a new thunderstorm model successfully accounts for this by demonstrating that the growing hailstones are re-cycled by the moving storm (see ch. 4, H). On occasions hailstones may reach giant size, weighing up to 0·76 kg each (recorded in September 1970 at Coffeyville, Kansas). In view of their rapid fall speeds, hailstones may fall considerable distances with little melting.

H Thunderstorms

In temperate latitudes probably the most spectacular example of moisture changes and associated energy releases in the atmosphere is the thunderstorm. Unusually great upward and downward movements of air are both the principal ingredients and motivating machinery of such storms. They occur: (*a*) as rising cells of excessively heated moist air; (*b*) in association with the triggering-off of conditional instability by uplift over mountains; or (*c*) along a *squall line* in association with an air-mass discontinuity (see p. 203).

The life cycle of a local storm lasts only 1–2 hours, and begins when a parcel of air is either warmer than the air surrounding it or is actively undercut by colder encroaching air. In both instances the air begins to rise and the embryo thunder cell forms as an unstable updraught of warm air (fig. 2.21). As condensation begins to form cloud droplets, latent heat is released and the initial upward impetus of the air parcel is augmented by an expansion and a decrease in density until the whole mass becomes completely out of thermal equilibrium with the surrounding air. At this stage updraughts may increase from 3 to 5 ms^{-1} at the cloud base to 8–10 ms^{-1} some 2–3 km higher and can exceed 30 ms^{-1}. The constant release of latent heat continuously injects fresh supplies of energy which accelerate the updraught. The rise of the air mass will continue as long as its temperature remains greater (or, in other words, its density less) than that of the surrounding air. Cumulonimbus clouds tend to form where the air is already moist, as a result of previous penetrating towers from a cluster of clouds, and there is persistent ascent.

Raindrops begin to develop rapidly when the ice stage (or freezing stage) is reached by the vertical buildup of the cell, allowing the Bergeron process to operate. They do not immediately fall to the ground because the updraughts are able to support them. The minimum cumulus depth for showers over ocean areas seems to be between 1 and 2 km, but 4–5 km is more typical inland. The corresponding minimum time intervals needed for showers to fall from growing cumulus are about 15 minutes over ocean areas and $\geqslant$30 minutes inland. Falls of hail require the special cloud processes, described in the last section, involving phases of 'dry' (rime accretion) and 'wet' growth on hail pellets. The mature stage of a storm (fig. 2.21B) is usually associated with precipitation downpours and lightning (pl. 13). The precipitation causes frictional downdraughts of cold air. As these gather momentum cold

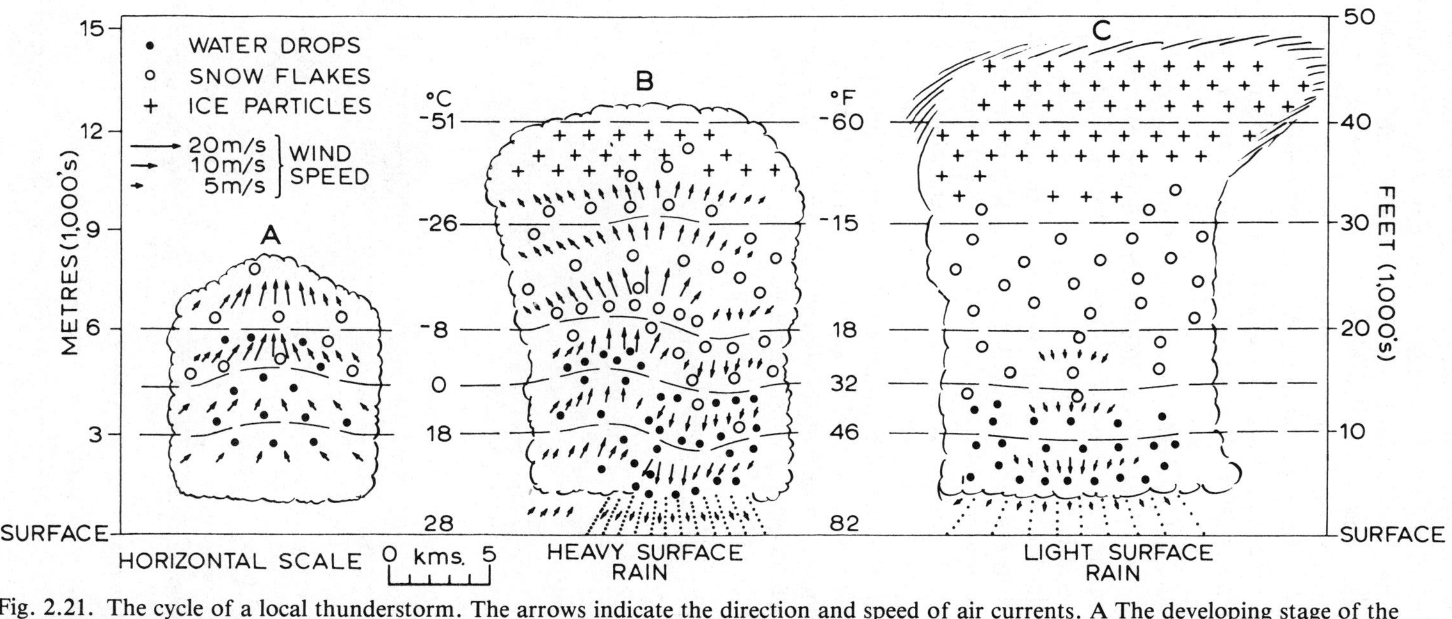

Fig. 2.21. The cycle of a local thunderstorm. The arrows indicate the direction and speed of air currents. A The developing stage of the initial updraught. B The mature stage with updraughts and downdraughts. C The dissipating stage dominated by cool downdraughts. (*After Byers and Braham; adapted from Petterssen 1958.*)

air may eventually spread out below the thunder cell in a wedge. Gradually, as the moisture of the cell is expended, the supply of released latent heat energy diminishes, the downdraughts progressively gain in power over the warm updraughts, and the cell dissipates.

To simplify the explanation, a thunderstorm with only one cell was illustrated. Usually storms are far more complex in structure and consist of several cells arranged in clusters of 2–8 km across, 100 km or so in length and extending up to 10 km or more (see pl. 14). Such systems are known as squall lines (ch. 4, H).

Two general hypotheses have been developed to account for thunderstorm electrification. One involves the induction mechanism, the other non-inductive charge transfer. As an example of the first category, since the ionosphere is positively charged (owing to the action of cosmic and solar ultraviolet radiation in ionization) and the earth's surface is negatively charged during fine weather, cloud droplets can acquire an induced positive charge on their lower side and negative charge on their upper side. Non-inductive charge transfer requires contact between cloud or precipitation particles.

The typically observed distribution of charges in a thundercloud is shown in fig. 2.22. The separation of electrical charges of opposite sign may involve several mechanisms: raindrop breakup (large droplets retaining positive charge, the surface spray carrying negative ions); or the selective capture of

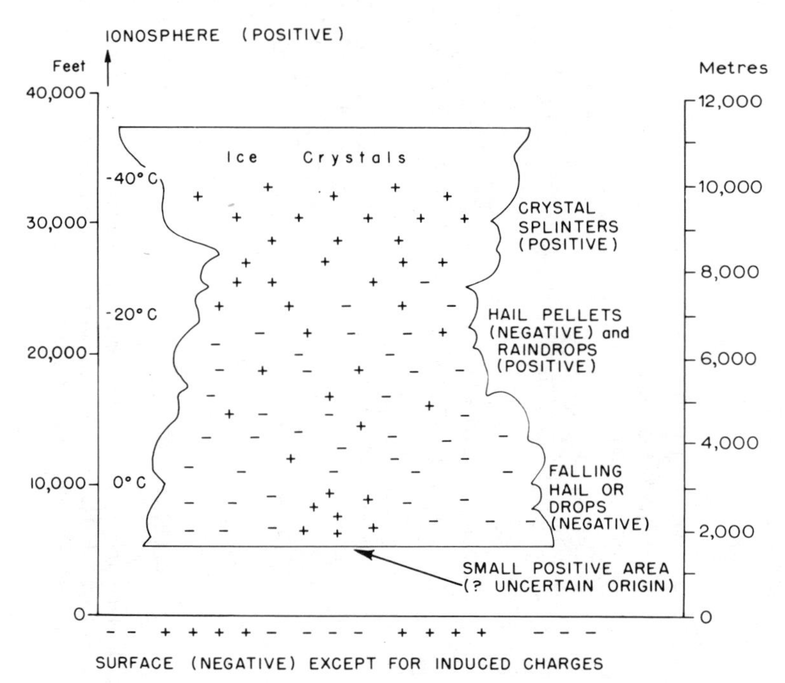

Fig. 2.22. The distribution of electrostatic charges in a thunder-cloud (*after Mason 1962*). See text.

negative atmospheric ions by falling cloud particles are possible factors, but do not appear to create sufficiently large charges. A third mechanism is the splintering of ice crystals during the freezing of cloud droplets. This operates as follows. A super-cooled droplet freezes inwards from its surface and this leads to a negatively-charged warmer core (OH^- ions) and a positively-charged colder surface due to the migration of H^+ ions outwards down the temperature gradient. When this soft hailstone ruptures during freezing small ice splinters carrying a positive charge are ejected by the ice shell and preferentially lifted to the top of the convection cell in updraughts. This theory can explain the charge distribution in fig. 2.22, which shows that the upper part of the cloud (above about the –20°C isotherm) is positively charged. Equally, the negatively charged hail pellets fall towards the cloud base. However, this ice-splintering mechanism appears only to work for a narrow range of temperature conditions and the charge transfer is small.

The processes discussed so far probably play a contributing role in cloud electrification, but according to J. Latham the major factor is non-inductive charge transfer. This involves collisions between splintered ice crystals and warmer pellets of soft hail. Previous accretion of super-cooled droplets on the hail pellets produces an irregular surface which is warmed as the droplets release latent heat on freezing. The impacts of ice crystals on this irregular surface generate negative charge whereas the crystals acquire positive charge. Negative charge is usually concentrated between about –10°C and –20°C in a thundercloud where ice crystal concentrations are large, due to splintering at about the –5°C level and then ascent of the crystals in up-currents. Radar studies show that lightning is associated with both ice particles in clouds and with rising air currents causing upward motion of small hail. The origin of small positive areas near the cloud base (fig. 2.22) is still under discussion. They could arise through the action of convective up-draughts carrying positive charge. It is likely that the very varied electrical properties of thunderclouds (from cloud to cloud and within individual clouds as they develop) cannot be explained by any single theory of charge generation.

Lightning commonly begins more or less simultaneously with precipitation downpours. It may occur between the lower part of the cloud and the ground (which locally has an induced positive charge). The first (leader) stage of the flash bringing down negative charge from the cloud is met near the ground by a return stroke which rapidly takes positive charge upward along the already-formed channel of ionized air. Just as the leader is neutralized by the return stroke, so the latter is neutralized in turn within the clouds. Subsequent leaders and return strokes drain higher regions of the cloud until its supply of negative charge is temporarily exhausted. The total flash typically lasts only about 0·2 seconds. Other more frequent flashes occur within a cloud or between clouds. The extreme heating and explosive expansion of air immediately round the path of the lightning sets up intense sound waves causing thunder to be heard. The sound travels at about $300\,m\,s^{-1}$.

Lightning is only one aspect of the atmospheric electricity cycle. During fine weather the earth's surface is negatively charged, the ionosphere positively charged. The potential gradient of this vertical electrical field in fine weather is about $100\ \mathrm{V\,m^{-1}}$, whereas beneath a thundercloud it may exceed $1000\ \mathrm{V\,m^{-1}}$. The 'breakdown potential' for lightning to occur in dry air is $3 \times 10^6\ \mathrm{V\,m^{-1}}$, but this is ten times the largest observed fields in thunderclouds. Hence, the necessity for localized cloud droplet/ice crystal charging processes, as already described, to initiate flash leaders.

Atmospheric ions conduct electrically from the ionosphere down to the earth, and hence a return supply must be forthcoming to maintain the observed electrical field. One source is the slow *point discharge*, from objects such as buildings and trees, of ions carrying positive charge induced by the negative thundercloud base. Similar upward currents occur above thunderstorm clouds. The other source (estimated to be smaller in its effect over the earth as a whole) is the instantaneous upward transfer of positive charge by lightning strokes, leaving the earth negatively charged. The joint operation of these supply currents, in approximately 1800 thunderstorms over the globe at any instant, is thought to be sufficient to balance the air–earth leakage, and this number seems to agree reasonably well with observations. Lightning is a significant environmental hazard. In the United States alone there are nearly 100 deaths per year, on average, as a result of lightning accidents.

The Russians have claimed some success in dissipating damaging hailstorms by the use of radar-directed artillery shells and rockets to inject silver iodide into high-liquid-water-content portions of clouds which freezes the available super-cooled water, so preventing it from accreting as shells on growing ice crystals. Attempts to 'drain off' lightning charges by seeding clouds with silver iodide or with millions of metallic needles have produced even less certain results.

I Precipitation characteristics and types

Strictly, *precipitation* refers to all liquid and frozen forms of water (p. 91) – rain, snow, hail, dew, hoar-frost, fog-drip and rime (ice accretion on objects through the freezing on impact of super-cooled fog droplets) – but, in general, only rain and snow make significant contributions to precipitation totals. In many parts of the world the term rainfall can be used interchangeably with precipitation. The data in the following section refer to rainfall, since snowfall is less easily measured with the same degree of accuracy.

It must be emphasized that precipitation records are only *estimates*. Factors of site location, gauge height, large- and small-scale turbulence in the air flow, splash-in and evaporation all introduce errors in the catch. Falling snow is particularly subject to wind effects which can result in under-representation of the true amount by 50% or more.

Precipitation data are also limited by the density of gauge networks. On a national basis, per $10{,}000\ \mathrm{km^2}$ area, this ranges from 245 gauges in Britain, to 10 in the United States and only 3 in Canada and Asia. In mountain areas the coverage is usually much worse.

1 Precipitation characteristics

There are many measures by which the various attributes of precipitation can be described. Long-term measures such as mean annual precipitation, annual variability and year-to-year trends have been of great interest to the geographer, and these statistical measures are treated in the concluding chapter (see ch. 8, A). However, particularly in terms of hydrological considerations, the characteristics and relationships of individual rainstorms are being studied increasingly, and it is possible here to point to some of their commonly observed features. Weather observations usually indicate the amount, duration and frequency of precipitation and these enable other derived characteristics to be determined. Three of these are discussed below.

a Rainfall intensity The intensity (= amount/duration) of rainfall during an individual storm, or a still shorter period, is of vital interest to hydrologists and water engineers concerned with flood forecasting and prevention, as well as to conservationists dealing with soil erosion. Chart records of the rate of rainfall (*hyetograms*) are necessary to assess intensity, which varies markedly with the time interval selected. Average intensities for short periods (thunderstorm-type downpours) are much greater than those for longer time intervals as fig. 2.23 illustrates for Washington DC. In the case of extreme rates at different points over the earth (fig. 2.24) the record intensity over 10 minutes is approximately three times that for 100 minutes, and the latter exceeds by as much again the record intensity over 1000 minutes (i.e. 16½ hours). High-intensity rain is associated with increased drop size rather than an increased number of drops. For example, with precipitation intensities of 0·1, 1·3 and 10·2 cm/hr (i.e. 0·05, 0·5 and

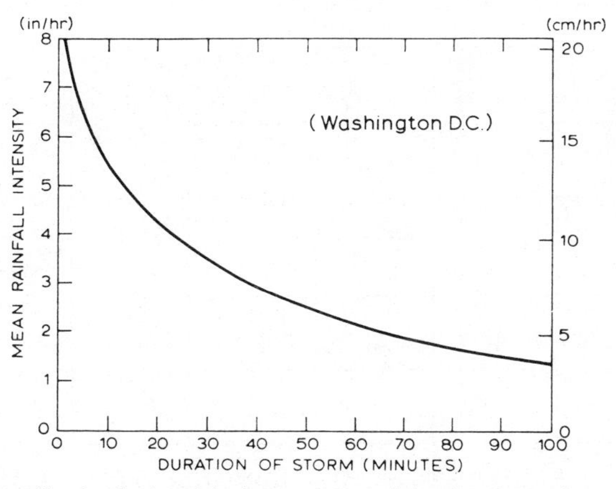

Fig. 2.23. Generalized relationship between precipitation intensity and duration for Washington DC (*after Yarnell 1935*).

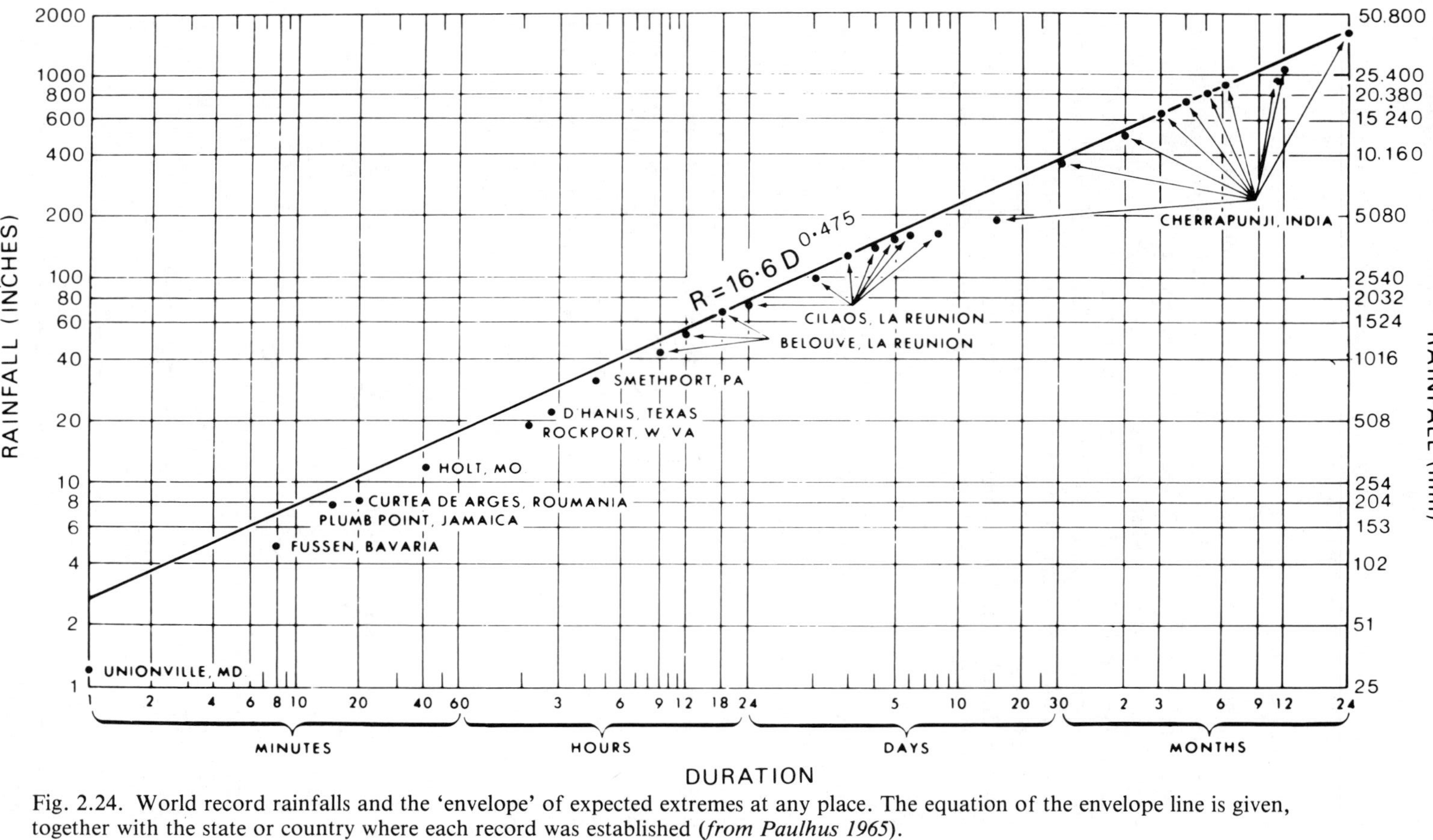

Fig. 2.24. World record rainfalls and the 'envelope' of expected extremes at any place. The equation of the envelope line is given, together with the state or country where each record was established (*from Paulhus 1965*).

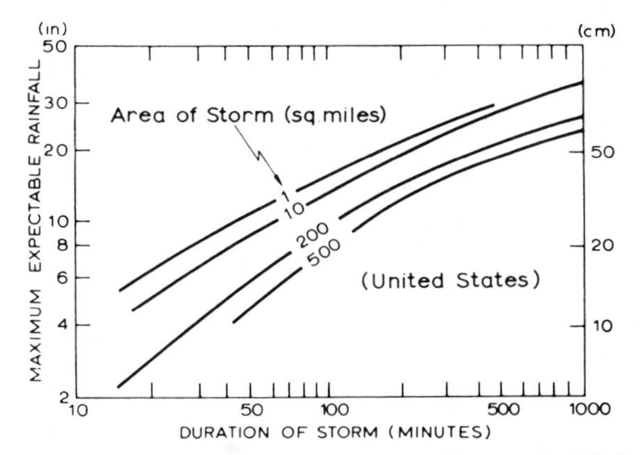

Fig. 2.25. Enveloping depth/duration curves for maximum rainfall for areas of under 500 square miles in the United States (*after Berry, Bollay and Beers 1945*).

4·0 in/hr), the most frequent raindrop diameters are 0·1, 0·2 and 0·3 cm, respectively. The occurrence of daily amounts exceeding 1·3 cm (0·5 in) is considered to be important for gully erosion in North America. Such falls account for 90% of the annual rainfall on the Gulf Coast compared with only 20% in the Great Basin.

b Areal extent of a rainstorm The rainfall totals received in a given time interval vary according to the size of the area which is considered, showing a relationship analogous to that of rainfall duration and intensity. The maximum 24-hour rainfalls over areas of different extent in the United States (up to 1960) were as follows:

Sq. miles	*cm*	*in*
10 (25·9 km^2)	98·3	38·7
10^2	89·4	35·2
10^3	76·7	30·2
10^4	30·7	12·1
10^5	10·9	4·3

(*After Gilman 1964*)

Figure 2.25, based on data of this type, illustrates the maximum rainfall to be expected for a given storm area and given duration in the United States.

c Frequency of rainstorms Another useful item is the average time-period within which a rainfall of specified amount or intensity can be expected to occur once. This is known as the *recurrence interval* or *return period*. Figure 2.26 gives this type of information on rainfall amount and duration for six contrasting stations. From this, for example, it would appear that on average each 20 years a 24-hour rainfall of at least 95 mm (3·75 in) is likely to occur at Cleveland and 216 mm (8·5 in) at Lagos. However, this *average* return period does not mean that such falls necessarily occur in the twentieth

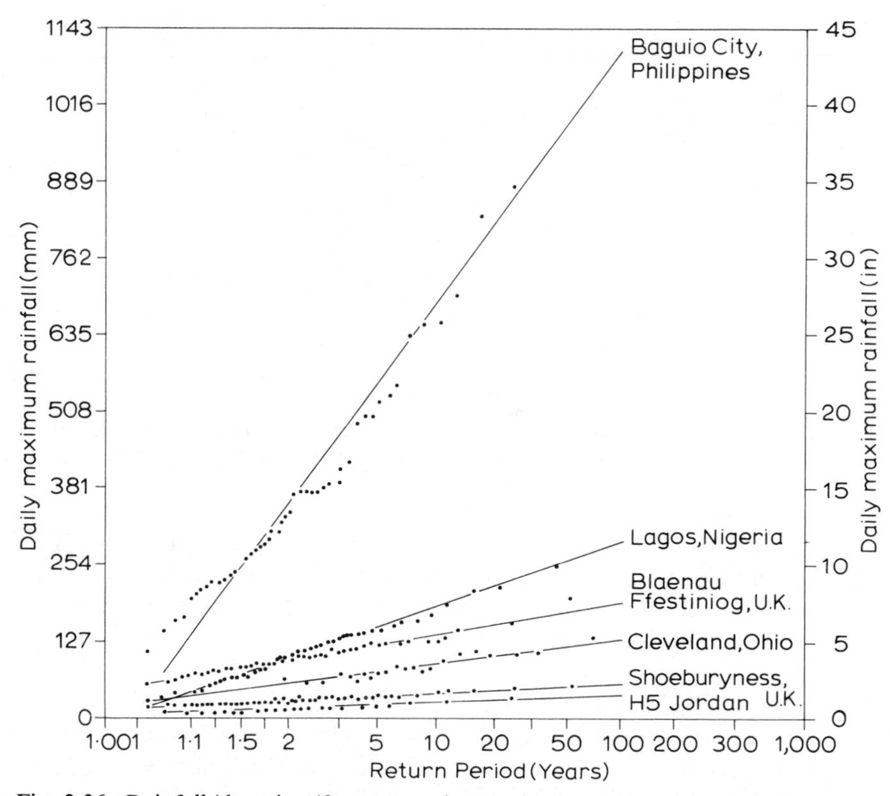

Fig. 2.26. Rainfall/duration/frequency plots for daily maximum rainfalls in respect of a range of stations from the Jordan desert to an elevation of 1482 m in the monsoonal Philippines (*after Rodda 1970; Linsley and Franzini 1964; and Ayoade 1976*).

year of a selected period. Indeed, they might occur in the first! These estimates require long periods of observational data, but the approximately linear relationships shown by such graphs are of great practical significance for the design of flood-control systems.

Numerous case-studies of rainstorm events have been carried out in different climatic areas. Two examples are shown for south-west England and China in fig. 2.27A and B. The former was a 24-hour storm with an estimated 150–200 year return period, the latter was a 100-year storm and the figure shows the 1-hour rainfall. Despite less orographic assistance and the shorter recurrence interval, the Hong Kong storm produced three times the maximum hourly intensity and ten times the core area of the English storm.

2 *Precipitation types*

The above material can now be related to that of the previous sections in a discussion of precipitation types. A convenient starting point is the usual

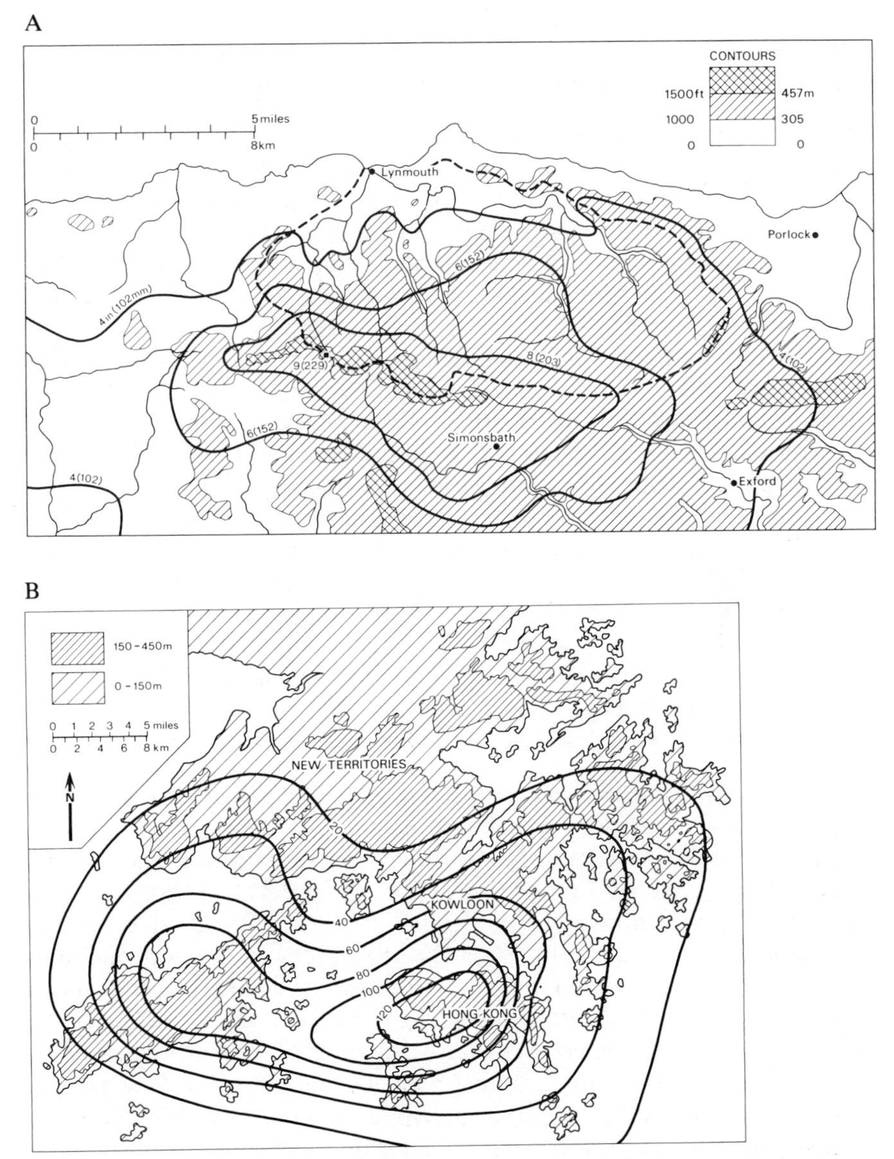

Fig. 2.27. Distribution of rainfall in mm (in parentheses in A, where inches are also given). A Over Exmoor, England, during a 24-hour period on 15 August 1952. 75% of this rainfall occurred during a 7-hour period and 18% in one hour (*from Dobbie and Wolf 1953*). B Over Hong Kong during 0630 to 0730 hours on 12 June 1966 (*from So 1971*).

division into three main types – convective, cyclonic and orographic precipitation – according to the primary mode of uplift of the air. Essential to this analysis is some knowledge of storm systems. These are treated in later chapters and the newcomer to the subject may prefer to read the following in conjunction with them.

a 'Convective type' precipitation This is associated with towering cumulus (cumulus congestus) and cumulonimbus clouds. Three subcategories can be distinguished according to their degree of spatial organization.

(i) Scattered convective cells develop through strong heating of the land surface in summer, especially when low upper tropospheric temperatures facilitate the release of conditional or convective instability (see ch. 2, E). Precipitation, often including hail, is of the thunderstorm type, although thunder and lightning do not necessarily occur. Small areas (20 to 50 km^2) are affected by the individual heavy downpours, which generally last for about $\frac{1}{2}$ to 1 hour.

(ii) Showers of rain, snow or soft hail pellets may form in cold, moist unstable air passing over a warmer surface. Convective cells moving with the wind can produce a streaky distribution of precipitation parallel to the wind direction (pl. 15), although over a period of several days the variable paths and intensities of the showers tend to obscure this pattern.

Two locations in which these cells may occur are parallel to a surface cold front in the warm sector of a depression (sometimes as a squall line) or parallel to and ahead of the warm front (see ch. 4, D). Hence the precipitation is widespread, though of brief duration at any locality.

(iii) In tropical cyclones cumulonimbus cells become organized about the centre in spiralling bands (see ch. 6, C.2). Particularly in the decaying stages of such cyclones, typically over land, the rainfall can be very heavy and prolonged, affecting areas of thousands of square kilometres.

b 'Cyclonic type' precipitation Precipitation characteristics vary according to the type of low-pressure system and its stage of development, but the essential mechanism is ascent of air through horizontal convergence of airstreams in an area of low pressure (see ch. 3, B.1). In extra-tropical depressions this is reinforced by uplift of warm, less-dense air along an air-mass boundary (see ch. 4, D.2). Such depressions give moderate and generally continuous precipitation over very extensive areas as they move usually eastward in the westerly wind belts between about 40° and 65° latitude. The precipitation belt in the forward sector of the storm can affect a locality in its path for 6 to 12 hours, whereas the belt in the rear gives a shorter period of thunderstorm-type precipitation. These sectors are, therefore, sometimes distinguished in precipitation classifications, and a more detailed breakdown is illustrated in table 5.3. Polar lows (see ch. 4, G.3) combine the effects of airstream convergence and convective activity of category *a(ii)*, above, whereas troughs in the equatorial low-pressure area give convective precipitation as a result of airstream convergence in the tropical easterlies (see ch. 6, C.1).

c Orographic precipitation Orographic precipitation is commonly regarded as a distinct type, but this requires careful qualification. Mountains are not especially efficient in causing moisture to be removed from airstreams crossing them, yet because precipitation falls repeatedly in more or less the same locations the cumulative totals are large. Orography, dependent on the alignment and size of the barrier, may (*i*) trigger conditional or convective instability by giving an initial upward motion or by differential heating of the mountain slopes, (*ii*) increase cyclonic precipitation by retarding the rate of movement of the depression system, (*iii*) cause convergence and uplift through the funnelling effects of valleys on airstreams. In mid-latitude areas where precipitation is predominantly of cyclonic origin, orographic effects tend to increase both frequency and intensity of winter precipitation, whereas during summer and in continental climates with a higher condensation level the main effect of relief is the occasional triggering of intense thunderstorm-type precipitation. The orographic influence occurs only in the proximity of high ground in the case of a stable atmosphere. Recent radar studies show that the main effect in this case is one of redistribution, whereas in the case of an unstable atmosphere precipitation appears to be increased, or at least redistributed on a larger scale, since the orographic

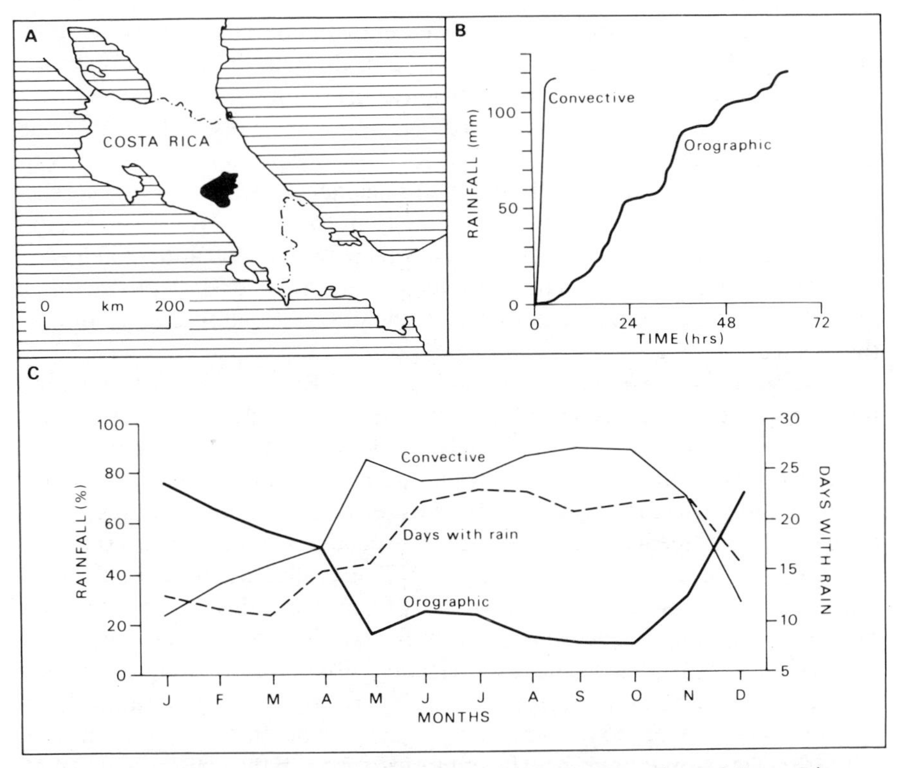

Fig. 2.28. Orographic and convective rainfall in the Cachi region of Costa Rica for the period 1977–80. A The Cachi region, elevation 500–3000 m. B Typical accumulated rainfall distributions for individual convective (duration 1–6 hours, high intensity) and orographic (1–5 days, lower intensity except during convective bursts) rainstorms. C Monthly rainfall divided into percentages of convective and orographic, plus days with rain, for Cachi (1018 m) (*from Chacon and Fernandez 1985*). (*By permission of the Royal Meteorological Society*).

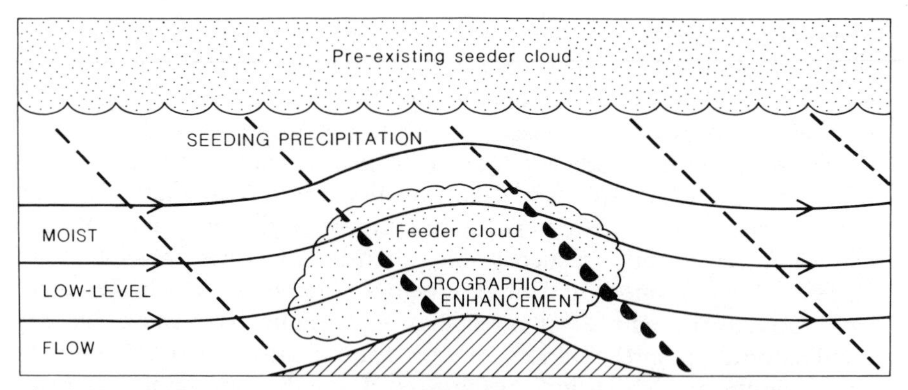

Fig. 2.29. Schematic diagram of T. Bergeron's 'seeder-feeder' cloud model of orographic precipitation over hills (*after Browning and Hill 1981*). (*By permission of the Royal Meteorological Society*).

effects may extend well downwind due to the activation of the meso-scale rain bands (see fig. 4.11).

In tropical highland areas there is a clearer distinction between orographic and convective contributions to total rainfall than in the mid-latitude cyclonic belt, although elements of tropical orographic rainfall are of convective origin due to the orographic release of potential instability. Fig. 2.28 shows that in the mountains of Costa Rica the temporal character of convective and orographic rainfalls and their seasonal occurrences are quite distinguishable. Convective rain occurs mainly in the May–November period when 60% of the rain falls in the afternoons between 1200 and 1800 hours; orographic rain predominates between December and April with a secondary maximum in June and July coinciding with an intensification of the Trades.

Even low hills may have an orographic effect. Research in Sweden shows that low wooded hills rising only 30–50 m above the surrounding lowlands increase precipitation amounts locally by 50–80% during cyclonic spells. Until Doppler radar studies of the motion of falling raindrops became feasible, the processes responsible for such effects were unknown. A principal cause is the 'seeder–feeder' ('releaser–spender') cloud mechanism, proposed by T. Bergeron (described above). This is illustrated in fig. 2.29. In moist stable airflows, shallow-cap clouds form over hilltops. Precipitation falling from an upper layer of altostratus (the seeder cloud) grows rapidly by the washout of droplets in the lower (feeder) cloud. The seeding cloud may release ice crystals, that subsequently melt, or droplets. Precipitation from the upper cloud layer alone would not give significant amounts at the ground as the droplets would have insufficient time to grow in the airflow which may traverse the hills in only 15–30 minutes. Most of the precipitation intensification takes place in the lowest kilometre of the atmosphere in moist, fast airflows.

Two special cases of orographic effects may be mentioned. One is the general influence of surface friction which by turbulent uplift may assist the formation of stratus or stratocumulus layers when other conditions are suitable (see ch. 2, C.2). Only light precipitation (drizzle, light rain or snow grains) is to be expected under these circumstances. The other case arises through frictional slowing down of an airstream moving inland over the coast. A particular instance of the convergence and uplift which this may initiate has been reported by Bergeron. During a 24-hour period in October 1945, a west-south-westerly airstream over Holland produced a belt of precipitation (3 cm or more) – a result of frictional convergence and uplift – in crossing the narrow zone of coastal sand dunes only a few metres high. Over the remainder of that virtually flat country a series of lee waves developed in the tropospheric airflow downwind from the coast (see fig. 3.11, for example), and these gave a series of transverse (north–south) bands of precipitation up to 2 cm in amount. On the following day the surface flow had changed little, but a temperature decrease from –20°C to –28°C at the 500-mb level altered the vertical stability so that the lee waves broke down and the precipitation distribution showed convective streaks, up to 4 cm per day, parallel to the surface wind direction.

3 Regional variations in the altitudinal maximum of precipitation

The increase of precipitation with height on mountain slopes is a world-wide characteristic, although actual profiles of precipitation differ regionally and seasonally. In middle latitudes an increase may be observed up to at least 3000–4000 m, as is the case in the Colorado Rocky Mountains and the Alps. In Western Britain with mountains of about 1000 m, the maximum falls are recorded to leeward of the summits. This probably reflects the general tendency of air to go on rising for a while after it has crossed the crestline and the time-lag involved in the precipitation process after condensation. Over narrow uplands the horizontal distance may allow insufficient time for maximum cloud build-up and the occurrence of precipitation. However, a further factor may be the effect of eddies, set up in the airflow by the mountains, on the catch of rain gauges. Studies in Bavaria at the Hohenpeissenberg observatory show that standard rain gauges may overestimate amounts by about 10% on the lee slopes and underestimate them by 14% on windward slopes.

In the tropics and subtropics the maximum precipitation occurs below the higher mountain summits, from which level it decreases upwards towards the crests. Observations are generally sparse in the tropics, but numerous records from Java show that the average elevation of greatest precipitation is approximately 1200 m (4000 ft). Above about 2000 m the decrease in amounts becomes quite marked. Similar features are reported from Hawaii and, at a rather higher elevation, on mountains in East Africa (see ch. 6, E.3). Figure 2.30A shows that, despite the wide range of records for individual stations, this effect is clearly apparent along the Pacific flank of the Guatemalan Highlands. Further north along the coast, the occurrence of a

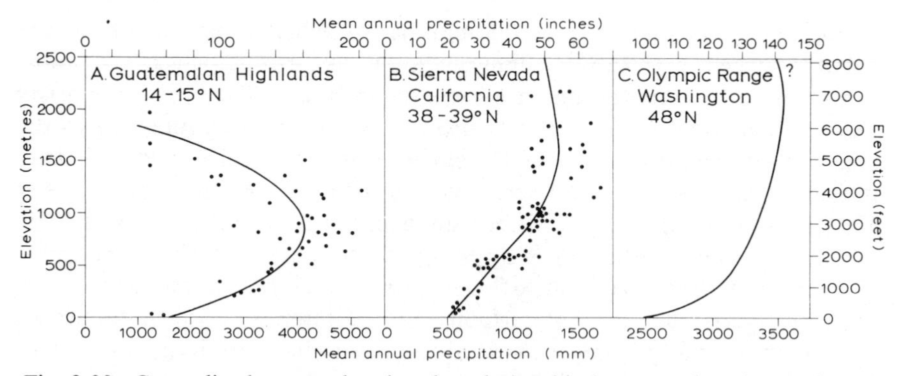

Fig. 2.30. Generalized curves showing the relationship between elevation and mean annual precipitation for west-facing mountain slopes in Central and North America. The dots give some indication of the wide scatter of individual precipitation readings (*adapted from Hastenrath 1967 and Armstrong and Stidd 1967*).

precipitation maximum below the mountain crest is observed in the Sierra Nevadas, despite some complication introduced by the shielding effect of the Coast Ranges (fig. 2.30B), but in the Olympic Mountains of Washington precipitation increases right up to the summits (fig. 2.30C). As has been previously mentioned, precipitation catches on mountain crests may underestimate the actual precipitation due to the effect of eddies, and this is particularly true where much of the precipitation falls in the form of snow which is very susceptible to blowing by the wind.

One explanation of this orographic difference between tropical and temperate rainfall is based on the concentration of moisture in a fairly shallow layer of air near the surface in the tropics (see ch. 6, A). Much of the orographic precipitation seems to derive from warm clouds (particularly cumulus congestus), composed of water droplets, which commonly have an upper limit at about 3000 m. It is probable that the height of the maximum precipitation zone is close to the mean cloud base since the maximum size and number of falling drops will occur at that level. Thus, stations located above the level of mean cloud base will receive only a proportion of the orographic increment. In temperate latitudes much of the precipitation, especially in winter, falls from stratiform cloud which commonly extends through a considerable depth of the troposphere. In this case there tends to be a smaller fraction of the total cloud depth below the station level. These differences according to cloud type and depth are apparent even on a day-to-day basis in middle latitudes, as has been shown by detailed studies in the Bavarian Alps. Seasonal variations in the altitude of the mean condensation level and zone of maximum precipitation are similarly observed. In the mountains of central Asia (the Pamirs and Tien Shan), for instance, the maximum is reported to occur at about 1500 m (5000 ft) in winter and at 3000 m (9900 ft) or more in summer. A further difference between orographic effects on precipitation in the tropics and the mid-latitudes relates to

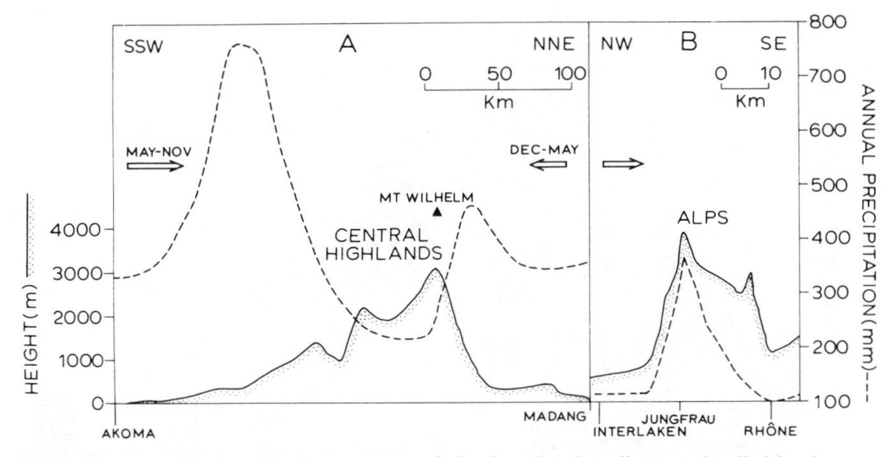

Fig. 2.31. The relationship between precipitation (broken line) and relief in the tropics and the mid-latitudes: A The highly saturated airmasses over the Central Highlands of Papua New Guinea give seasonal maximum precipitations on the windward slopes of the mountains with changes in the monsoonal circulation (*after Barry 1981*); B Across the Jungfrau Massif in the Swiss Alps the precipitation is much less than in A and is closely correlated with the topography on the windward side of the mountains (*after Maurer and Lütschg; from Barry 1981*). The arrows show the prevailing airflow directions.

the great instability of many tropical air masses. Where high mountains obstruct the flow of very moist maritime tropical air masses, the upwind turbulence may be sufficient to trigger off sufficient precipitation to give a rainfall maximum on the windward side of the range. By contrast, in more stable mid-latitude airflows the rainfall maximum is more closely related to the topography (fig. 2.31).

4 The world pattern of precipitation

A glance at the maps of precipitation amount for December–February and June–August (fig. 2.32) indicates that the distributions are considerably more complex than those, for example, of mean temperature (see fig. 1.17B). Comparison of fig. 2.32 with the meridional profile or average precipitation for each latitude (see fig. 2.4A) brings out the marked longitudinal variations that are superimposed on the zonal pattern. The latter has three main features: an equatorial maximum which, like the thermal equator, is slightly displaced into the northern hemisphere; very low totals in high latitudes; and secondary minima in subtropical latitudes. Figure 2.32 demonstrates why the subtropics do not appear as particularly dry on the meridional profile in spite of the known aridity of the subtropical high pressure areas (see ch. 3, D.1 and 5, D.2). In these latitudes the eastern sides of the continents receive considerable rainfall in summer.

In view of the complex controls involved, no brief explanation of these precipitation distributions can be very satisfactory. Various aspects of

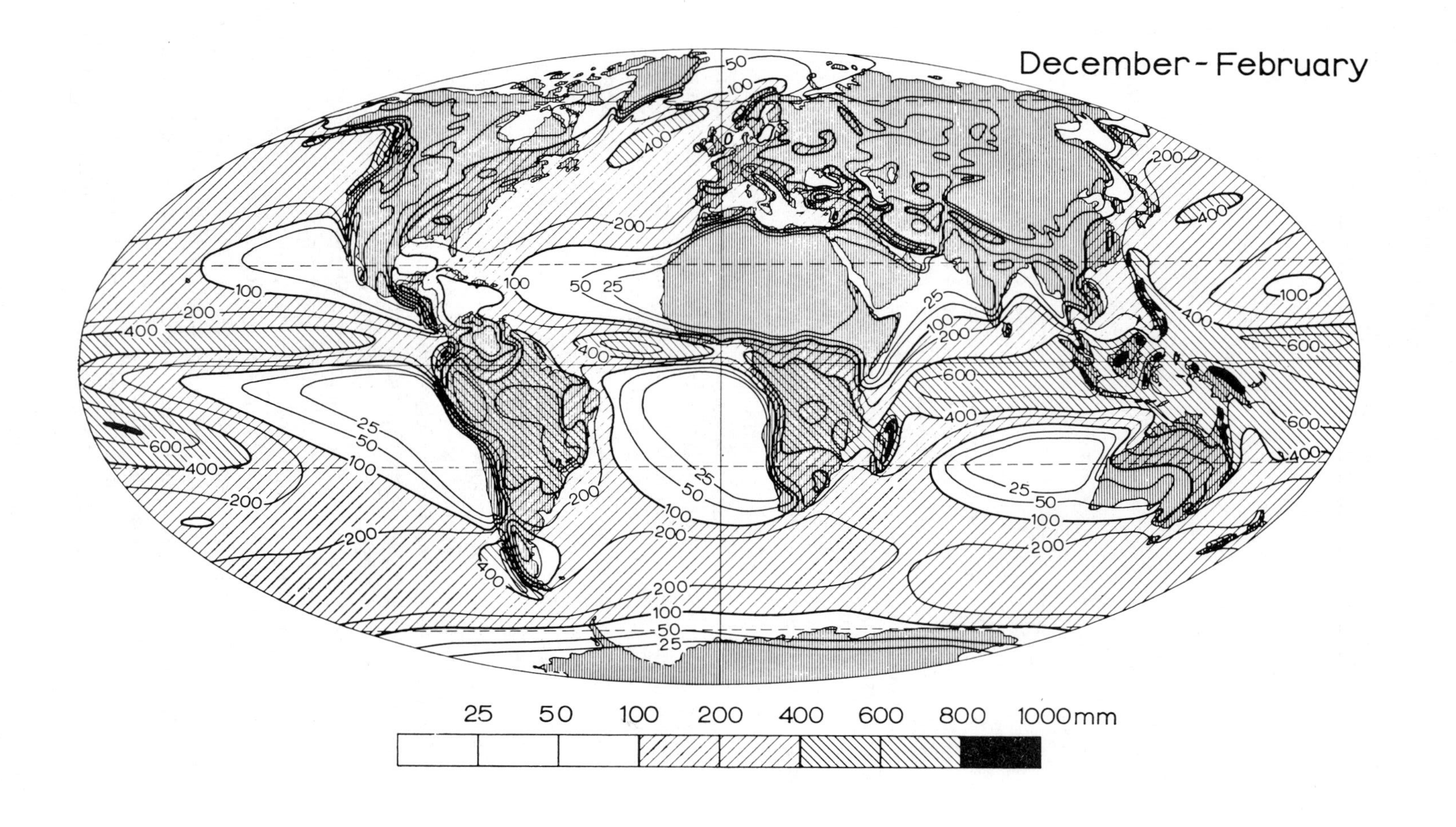
December - February
25
50
100
200
400
600
800
1000mm

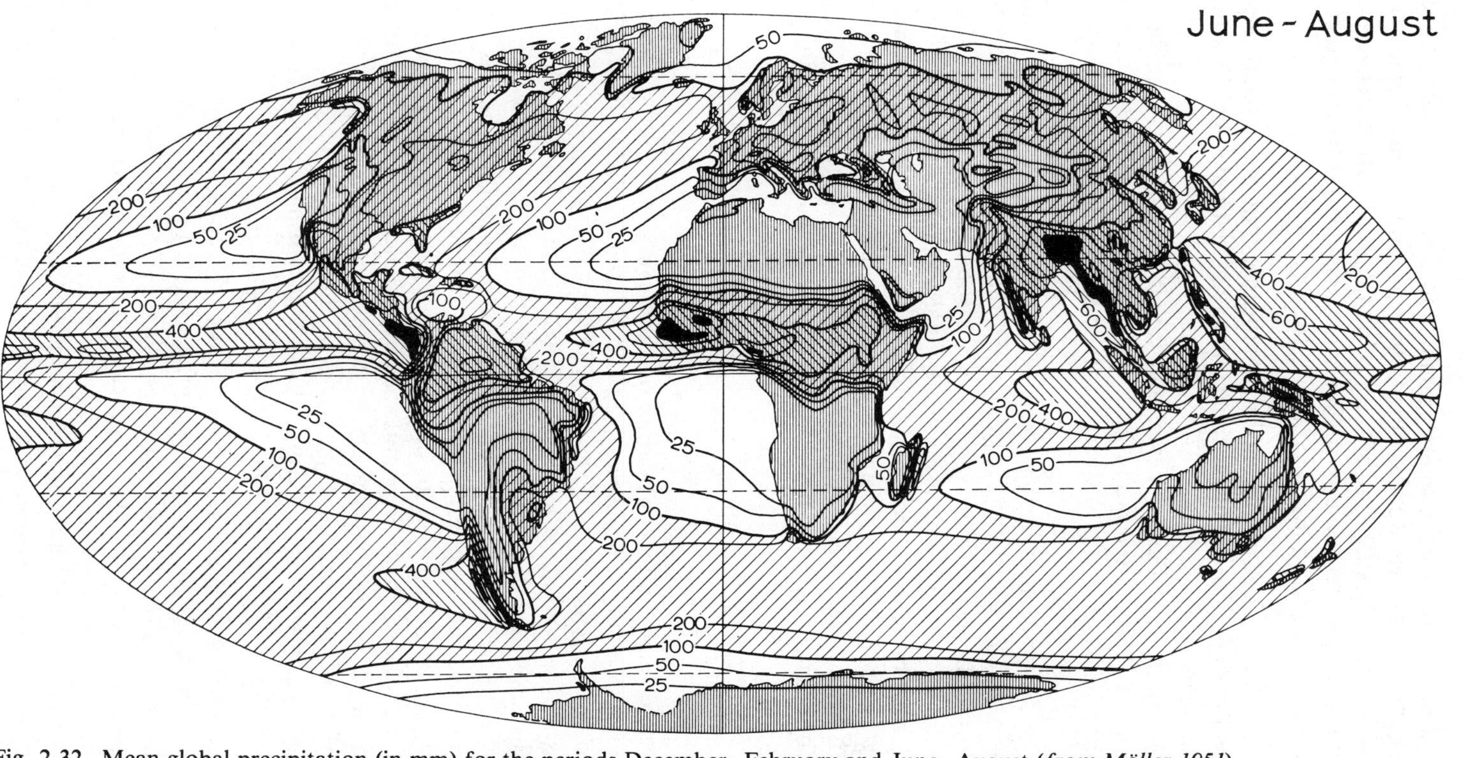

Fig. 2.32. Mean global precipitation (in mm) for the periods December–February and June–August (*from Möller 1951*).

selected precipitation regimes are examined in chs. 5 and 6, after consideration of the fundamental ideas about atmospheric motion, air masses and frontal zones. A classification of wind belts and precipitation characteristics is outlined in App. 1.C. It must suffice at this stage simply to note the factors which have to be taken into account in studying fig. 2.32.

(i) The limit imposed on the maximum moisture content of the atmosphere by air temperature. This is important in high latitudes and in winter in continental interiors.
(ii) The major latitudinal zones of moisture influx due to atmospheric advection. This in itself is a reflection of the global wind systems and their disturbances (i.e. the converging trade wind systems and the cyclonic westerlies, in particular).
(iii) The distribution of the land masses. It is noteworthy that the southern hemisphere lacks the vast, arid, mid-latitude continental interiors of the northern. The oceanic expanses of the southern hemisphere allow the mid-latitude storms to increase the zonal precipitation average for 45°S by about one third compared with that of the northern hemisphere for 50°N (see fig. 2.4A). Another major non-zonal feature is the occurrence of the monsoon regimes, especially in Asia.
(iv) The orientation of mountain ranges with respect to the prevailing winds.

5 Drought

The term drought implies an absence of significant precipitation for a period which is long enough to cause moisture deficits in the soil due to evapotranspiration and decreases in stream flow, thereby disrupting the normal biological and human activities. Thus a drought condition may obtain after only three weeks without rain in parts of Britain, whereas some areas of the tropics regularly experience many successive dry months. There is then no precise, universally applicable definition of drought. At least 150 different definitions can be found in the literature developed by specialists in meteorology, agriculture, hydrology and socio-economic studies having differing perspectives! All regions suffer the temporary but irregularly-recurring condition of drought, but particularly those with marginal climates alternately influenced by differing climatic mechanisms. Drought conditions are especially associated with:

(i) Increases in area and persistence of the subtropical high-pressure cells. Drought in southern Israel has been shown to be significantly related to this mechanism. The major droughts in the African Sahel since the 1970s have also been attributed to an eastward and southward expansion of the Azores anticyclone.
(ii) Changes in the summer monsoonal circulation, causing a postponement or failure of incursions of maritime tropical air, such as may occur in Nigeria or the Punjab of India.

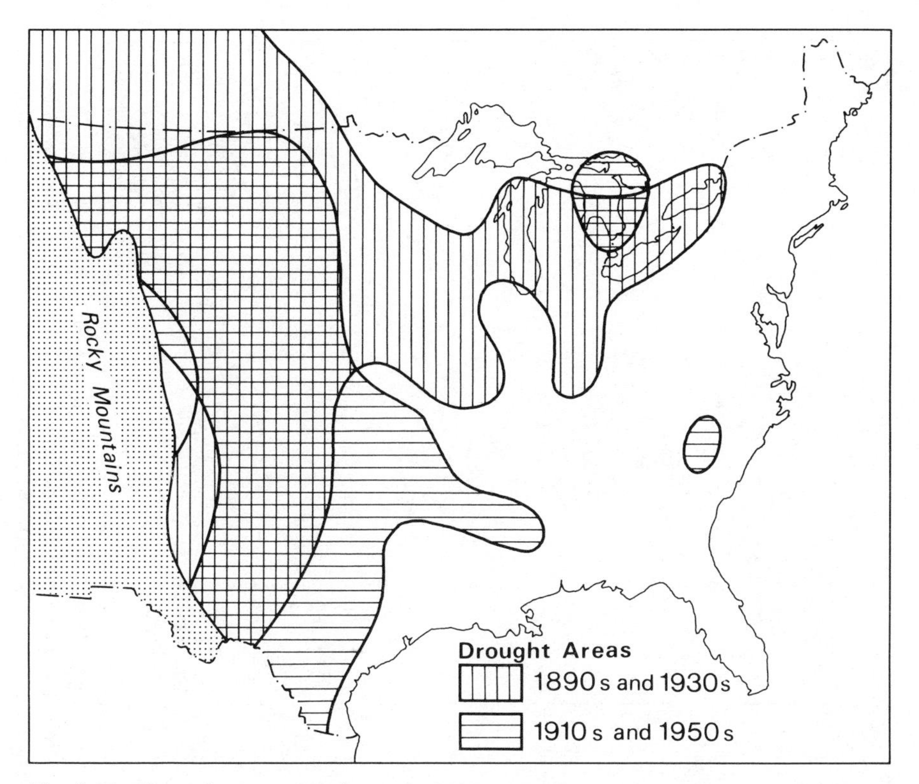

Fig. 2.33. Drought areas for the central USA based on areas receiving less than 80% of the normal July–August precipitation (*after Borchert 1971*).

(iii) Lower ocean surface temperatures produced by changes in currents or increased upwelling of cold waters. Rainfall in California and Chile may be affected by such mechanisms (see p. 161), and adequate rainfall in the drought-prone region of north-east Brazil appears to be strongly dependent on high sea-surface temperatures in the 0°–15°S belt of the South Atlantic.

(iv) Displacement of mid-latitude storm tracks associated with either an expansion of the circumpolar westerlies into lower latitudes, or with the development of persistent blocking patterns of circulation in middle latitudes (see fig. 3.35). It has been suggested that droughts on the Great Plains east of the Rockies in the 1890s and 1930s were due to such changes in the general circulation. However, the droughts of the 1910s and 1950s in this area were caused by persistent high pressure in the south-east and the northward displacement of storm tracks (see fig. 2.33).

Clearly, most severe and prolonged droughts involve combinations of several mechanisms. The prolonged drought in the Sahel – a 3000 × 700 km zone stretching along the southern edge of the Sahara from Mauretania to Chad – which began in 1969 and has continued with interruptions up to the

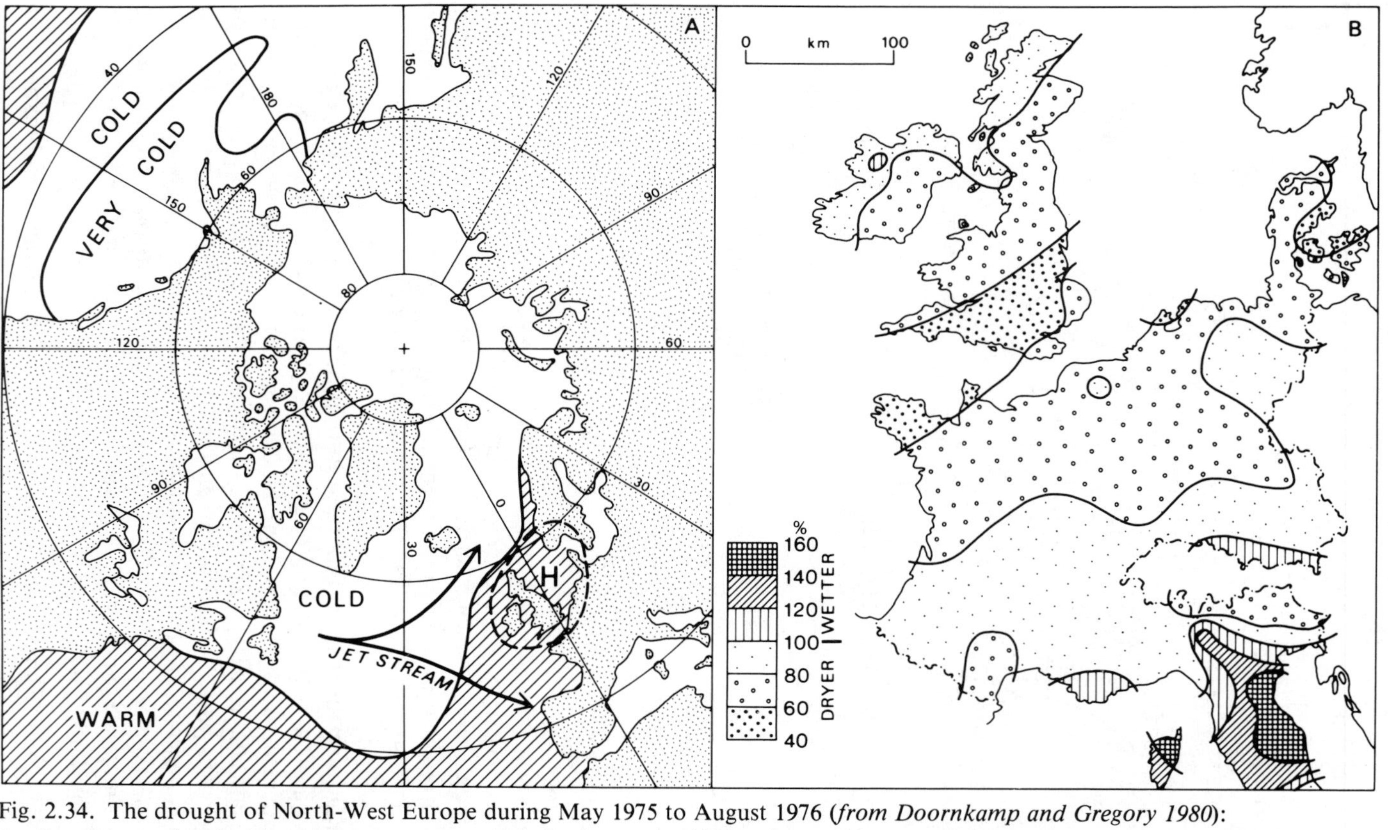

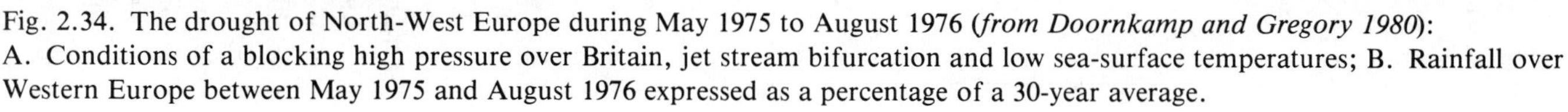

Fig. 2.34. The drought of North-West Europe during May 1975 to August 1976 (*from Doornkamp and Gregory 1980*):
A. Conditions of a blocking high pressure over Britain, jet stream bifurcation and low sea-surface temperatures; B. Rainfall over Western Europe between May 1975 and August 1976 expressed as a percentage of a 30-year average.

present (fig. 8.6), has been attributed to several factors. These include an expansion of the circumpolar westerly vortex, shifting of the subtropical high-pressure belt towards the equator, lower sea-surface temperatures in the eastern North Atlantic, and 'desertification' due to overgrazing. The removal of vegetation, increasing the surface albedo, is thought to result in a reduction of precipitation. There is no evidence that the subtropical high pressure was further south, but dry easterly airflow was stronger across Africa during the drought years.

From May 1975 to August 1976 parts of north-west Europe from Sweden to western France experienced severe drought conditions. Southern England received less than 50% of average rainfall, the most severe and prolonged drought since records began in 1727 (fig. 2.34). The immediate causes of this regime were the establishment of a persistent blocking ridge of high pressure over the area, displacing depression tracks 5°–10° latitude northward over the eastern North Atlantic. Farther afield, the circulation over the North Pacific had changed earlier, with the development of a stronger high-pressure cell and stronger upper-level westerlies, perhaps associated with a cooler-than-average sea surface. The westerlies were displaced northward over both the Atlantic and the Pacific. Over Europe, the dry conditions at the surface increased the stability of the atmosphere, further lessening the possibility of precipitation.

Summary

Atmospheric humidity can be described by the absolute mass of moisture in unit mass (or volume) of air, as a proportion of the saturation value, or in terms of the water vapour pressure. When cooled at constant pressure, air becomes saturated at the dew-point temperature.

The components of the surface moisture budget are total precipitation (including condensation on the surface), evaporation, storage change of water in the soil or in a snow cover, and runoff (on the surface or in the ground). Evaporation rate is determined by the available energy, the surface–air difference in vapour pressure, and the wind speed, assuming moisture supply is unlimited. If moisture supply is limited, the rate is affected by soil water tension, and plant factors. Evapotranspiration is best determined with a lysimeter. Otherwise, it may be calculated by formulae based on the energy budget, or on the aerodynamic profile method using the measured gradients of wind speed, temperature and moisture content near the ground.

Condensation in the atmosphere may occur by continued evaporation into the air; by mixing of air of different temperature and vapour pressure, such that the saturation point is reached; or by adiabatic cooling of the air through lifting until the condensation level is reached.

continued

Air may be lifted through instability due to surface heating or mechanical turbulence, ascent of air at a frontal zone, or forced ascent over an orographic barrier. Instability is determined by the actual rate of temperature decrease with height in the atmosphere relative to the appropriate adiabatic rate. The dry adiabatic lapse rate is 9·8°C/km; the saturated adiabatic rate is less than the DALR due to latent heat released by condensation. It is least (around 5°C/km) at high temperatures, but approaches the DALR at subzero temperatures.

Condensation requires the presence of hygroscopic nuclei such as salt particles in the air. Otherwise, supersaturation occurs. Similarly, ice crystals only form naturally in clouds containing freezing nuclei (clay mineral particles). Otherwise, water droplets may super-cool to −39°C. Both super-cooled droplets and ice crystals may be present at cloud temperatures of −10° to −20°C.

Clouds are classified in ten basic types, according to altitude and cloud form. Satellites are providing new information on spatial patterns of cloudiness.

Precipitation drops do not form directly by growth of cloud droplets through condensation. Two processes may be involved – coalescence of falling drops of differing sizes, and the growth of ice crystals by vapour deposition (the Bergeron–Findeisen process). Low-level cloud may be seeded naturally by ice crystals from upper cloud layers, or by introducing artificial freezing nuclei.

The freezing process appears to be a major element of cloud electrification in thunderstorms. Lightning plays a key role in maintaining the electrical field between the surface and the ionosphere.

Rainfall is described statistically by the intensity, areal extent and frequency (or recurrence interval) of rainstorms. Convective and cyclonic types of precipitation are commonly distinguished; orography generally intensifies the precipitation on windward slopes, but there are geographical differences in this altitudinal effect. Droughts may occur in many different climatic regions due to various causal factors.

3

Atmospheric motion

The atmosphere acts rather like a gigantic heat engine in which the constantly maintained difference in temperature existing between the poles and the equator provides the energy supply necessary to drive the planetary atmospheric circulation. The conversion of the heat energy into kinetic energy to produce motion must involve rising and descending air, but vertical movements are generally much less in evidence than horizontal ones, which may cover vast areas and persist for periods of a few days to several months.

Before considering these global aspects, however, it is important to look at the immediate controls on air motion. The downward-acting gravitational field of the earth sets up the observed decrease of pressure away from the earth's surface that is represented in the vertical distribution of atmospheric mass (fig. 1.6). This mutual balance between the force of gravity and the vertical pressure gradient is referred to as *hydrostatic equilibrium* (p. 9). This state of balance, together with the general stability of the atmosphere and its shallow depth, greatly limits vertical air motion. Average horizontal wind speeds are of the order of one hundred times greater than average vertical movements, though individual exceptions occur – particularly in convective storms.

A Laws of horizontal motion

There are four controls on the horizontal movement of air near the earth's surface: pressure-gradient force, Coriolis force, centripetal acceleration and frictional forces. The primary cause of air movement is the development of a horizontal pressure gradient and the fact that such a gradient can persist (rather than being destroyed by air motion towards the low pressure) results from the effect of the earth's rotation in giving rise to the Coriolis force.

1 The pressure-gradient force

The pressure-gradient force has vertical and horizontal components but, as already noted, the vertical component is more or less in balance with the force of gravity. Horizontal differences in pressure can be due to thermal or mechanical causes (often not easily distinguishable), and these differences control the horizontal movement of an air mass. In effect the pressure gradient serves as the motivating force that causes the movement of air away from areas of high pressure and towards areas where it is lower, although other forces prevent air from moving directly across the isobars (lines of equal pressure). The pressure-gradient force per unit mass is expressed mathematically as

$$-\frac{1}{\varrho}\frac{dp}{dn}$$

where ϱ = air density and dp/dn = the horizontal gradient of pressure. Hence the closer the isobar spacing the more intense is the pressure gradient and the greater the wind speed. The pressure-gradient force is also inversely proportional to air density, and this relationship is of particular importance in understanding the behaviour of upper winds.

2 The earth's rotational deflective (Coriolis) force

The Coriolis force arises from the fact that the movement of masses over the earth's surface is usually referred to a moving co-ordinate system (i.e. the latitude and longitude grid which 'rotates' with the earth). The simplest way to begin to visualize the manner in which this deflecting force operates is to picture a rotating disc on which moving objects are deflected. Figure 3.1 shows the effect of such a deflective force operating on a mass moving outward from the centre of a spinning disc. The body follows a straight path in relation to a fixed frame of reference (for instance, a box which contains a spinning disc), but viewed relative to co-ordinates rotating with the disc the body swings to the right of its initial line of motion. This effect is readily

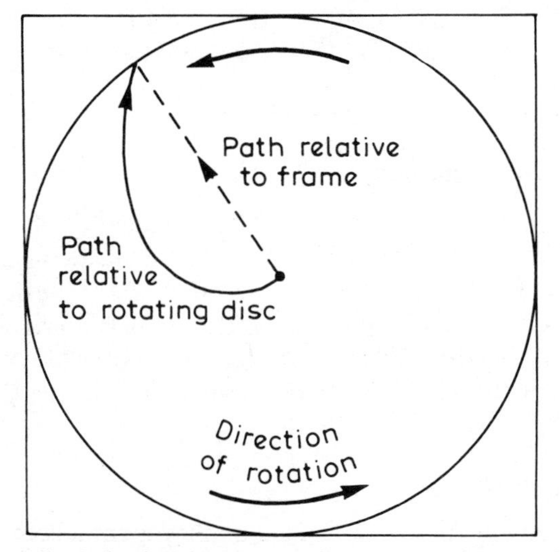

Fig. 3.1. The Coriolis deflecting force operating on an object moving outward from the centre of a rotating turntable.

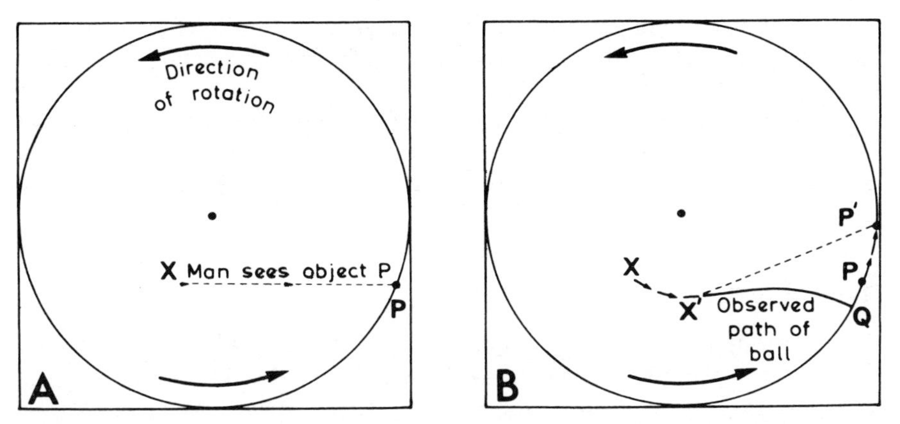

Fig. 3.2. The Coriolis deflecting force on a rotating turntable. A A man at X sees the object P and attempts to throw a ball towards it. Both locations are rotating anticlockwise. B The man's position is now X′ and the object is at P′. To the man, the ball appears to follow a curved path and lands at Q. The man overlooked the fact that P was moving to his left and that the path of the ball would be affected by the initial impetus due to the man's own rotation.

demonstrated if a pencil line is drawn across a white disc on a rotating turntable. Figure 3.2 illustrates a case where the movement is not from the centre of the turntable and the object possesses an initial momentum in relation to its distance from the axis of rotation. In the analogous case of the rotating earth (with rotating reference co-ordinates of latitude and longitude), there is apparent deflection of moving objects to the right of their line of motion in the northern hemisphere and to the left in the southern hemisphere, as viewed by observers on the earth. The deflective force (per unit mass) is expressed by:

$$-2\omega V \sin \phi$$

where ω = the angular velocity of spin (15° hr^{-1} or $2\pi/24$ rad hr^{-1} for the earth = $7 \cdot 29 \times 10^{-5}$ rad s^{-1}); ϕ = the latitude and V = the velocity of the mass. $2\omega \sin \phi$ is referred to as the Coriolis parameter (f).

The magnitude of the deflection is directly proportional to: (a) the horizontal velocity of the air (i.e. air moving at 11 m s^{-1} (25 mph) having half the deflective force operating on it as that moving at 22 m s^{-1} (50 mph); and (*b*) the sine of the latitude ($\sin 0° = 0$; $\sin 90° = 1$). The effect is thus a maximum at the poles (i.e. where the plane of the deflecting force is parallel with the earth's surface) and decreases with the sine of the latitude, becoming zero at the equator (i.e. where there is no component of the deflection in a plane parallel to the surface). Values of f vary with latitude as follows:

Latitude	0°	10°	20°	43°	90°N
$f(10^{-4}\ s^{-1})$	0	0·25	0·50	1·00	1·458

The Coriolis force always acts at right angles to the direction of the air motion to the right in the northern hemisphere (f positive) and to the left in the southern hemisphere (f negative).

The earth's rotation also produces a vertical component of rotation about a horizontal axis. This is a maximum at the equator (zero at the poles), but it is much less important to atmospheric motions due to the existence of hydrostatic equilibrium.

3 *The geostrophic wind*

Observations in the *free atmosphere* (above the level affected by surface friction at about 500 to 1000 m) show that the wind blows more or less at right angles to the pressure gradient (i.e. parallel to the isobars) with, for the northern hemisphere, the high-pressure core on the right and the low-pressure on the left when viewed downwind. This implies that for steady motion the pressure gradient force is exactly balanced by the Coriolis deflection acting in the diametrically opposite direction (fig. 3.3). The wind in this idealized case is called a *geostrophic wind*, the velocity (V_g) of which is given by the following formula:

$$V_g = \frac{1}{2\omega \sin \phi \varrho} \cdot \frac{dp}{dn}$$

where dp/dn = the pressure gradient. The velocity is thus inversely dependent on latitude, such that the same pressure gradient which will be associated with geostrophic wind speeds of 15 m s^{-1} (34 mph) at latitude 43° will produce a velocity of only 10 m s^{-1} (23 mph) at latitude 90°. Except in low latitudes, where the Coriolis deflection approaches zero, the geostrophic wind is a close approximation to the observed air motion in the free atmosphere. Since pressure systems are rarely stationary this fact implies that air motion must continually change towards a new balance. In other words mutual adjustments of the wind and pressure fields are constantly taking place. The common

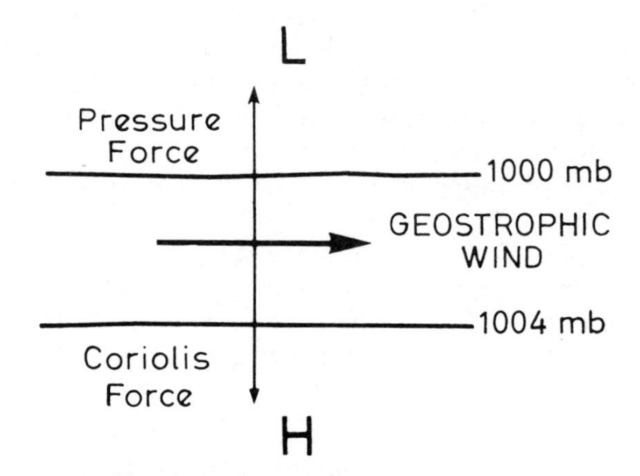

Fig. 3.3. The geostrophic wind case of balanced motion (northern hemisphere).

'cause-and-effect' argument that a pressure gradient is formed and air begins to move towards low pressure before coming into geostrophic balance is an unfortunate oversimplification of reality.

4 The centripetal acceleration

For a body to follow a curved path there must be an inward acceleration towards the centre of rotation. This acceleration (c) is expressed by:

$$c = -\frac{mV^2}{r}$$

where m = the moving mass, V = its velocity and r = the radius of curvature. This factor is sometimes regarded for convenience as a centrifugal force operating radially outward (see note 1). In the case of the earth itself this is valid. The centrifugal effect due to rotation has in fact resulted in a slight bulging of the earth's mass in low latitudes and a flattening near the poles. The small decrease in apparent gravity towards the equator (see note 2) reflects the effect of the centrifugal force working against the gravitational attraction directed towards the earth's centre. It is only necessary therefore to consider the forces involved in the rotation of the air about a local axis of high or low pressure. Here the curved path of the air (parallel to the isobars) is maintained by an inward-acting, or centripetal, acceleration.

Figure 3.4 shows (for the northern hemisphere) that in a low-pressure system balanced flow is maintained in a curved path (referred to as the *gradient wind*) by the Coriolis force being weaker than the pressure force. The difference between the two gives the net centripetal acceleration inward. In the high-pressure case the inward acceleration is provided by the Coriolis force exceeding the pressure force. Since the pressure gradients are assumed to be equal, the different contributions of the Coriolis force in each case imply that the wind speed around the low-pressure must be less than the geostrophic value (*subgeostrophic*) whereas in the high-pressure case it is *supergeostrophic*. In reality this effect is obscured by the fact that the pressure gradient in a high is usually much weaker than in a low. Moreover, the fact that the earth's rotation is cyclonic imposes a limit on the speed of anticyclonic flow. The maximum occurs when the angular velocity is $f/2$

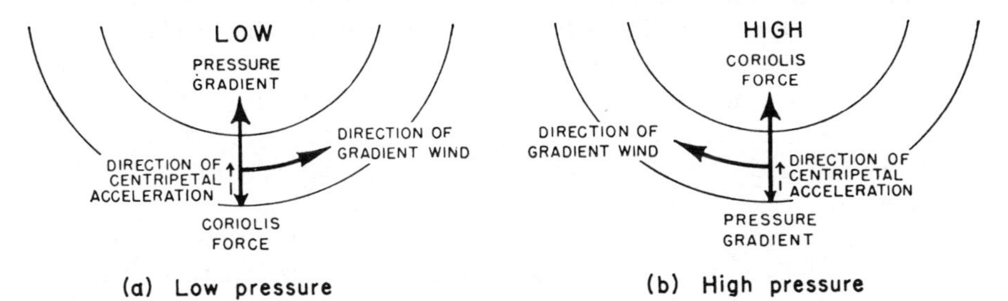

Fig. 3.4. The gradient wind case of balanced motion around a low pressure (a) and a high pressure (b) in the northern hemisphere.

($=\omega \sin \phi$), at which value the absolute rotation of the air (viewed from space) is just cyclonic. Beyond this point anticyclonic flow breaks down ('dynamic instability'). There is no maximum speed in the case of cyclonic rotation.

The magnitude of the centripetal acceleration is generally small, but it becomes important where high-velocity winds are moving in very curved paths (i.e. about an intense low-pressure system). Two cases are of meteorological significance: firstly, in intense cyclones near the equator where the Coriolis force is negligible; and, secondly, in a narrow vortex such as a tornado. Under these conditions, when the large pressure gradient force provides the necessary centripetal acceleration for balanced flow parallel to the isobars, the motion is called *cyclostrophic*.

The above arguments all assume steady conditions of balanced flow. This simplification is useful, but it must be noticed that two factors prevent a continuous state of balance. Latitudinal motion changes the Coriolis parameter and the movement or changing intensity of a pressure system leads to acceleration or deceleration of the air, causing some degree of cross-isobaric flow. Pressure change itself depends on air displacement through the breakdown of the balanced state. If air movement were purely geostrophic there would be no growth or decay of pressure systems. The acceleration of air in moving at upper levels from a region of cyclonic isobaric curvature (subgeostrophic wind) to one of anticyclonic curvature (supergeostrophic wind) causes a fall of pressure at lower levels in the atmosphere through the removal of air aloft. The significance of this fact will be discussed in ch. 4, F. The interaction of horizontal and vertical air motions is outlined in ch. 3, B.2.

5 Frictional forces

The last force which has an important effect on air movement is that due to friction with the earth's surface. If we follow our study of geostrophic wind a little further we find that towards the surface (i.e. below about 500 m for flat terrain) friction begins to decrease the wind velocity below its geostrophic value. This has an effect on the deflective force (which is dependent on velocity) causing it also to decrease. As these two tendencies continue the wind consequently blows more and more obliquely across the isobars in the direction of the pressure gradient. The angle of obliqueness increases with the growing effect of frictional drag due to the earth's surface and it averages about 10°–20° at the surface over the sea and 25°–35° over land. The result is to produce a wind spiral with height (fig. 3.5), analogous to the turning of ocean currents as the effect of wind stress diminishes with increasing depth. Both are referred to as 'Ekman spirals', after Ekman who investigated the variation of ocean currents with depth (see ch. 3, F.3).

In summary, the surface wind (neglecting any curvature effects) represents a balance between the pressure gradient force and friction parallel to the air motion and between the pressure gradient force and the Coriolis force perpendicular to the air motion.

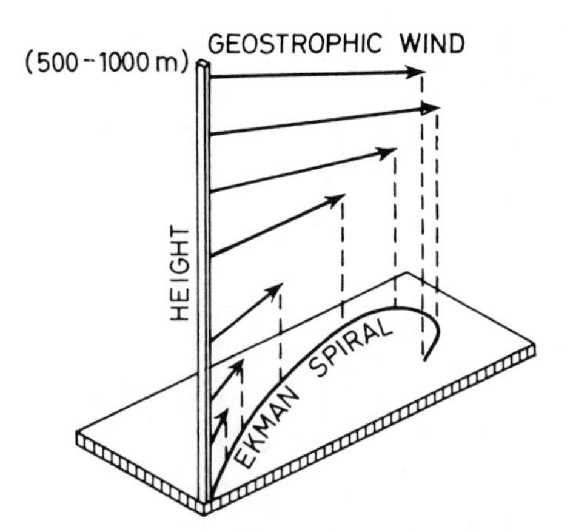

Fig. 3.5. The Ekman spiral of wind with height, in the northern hemisphere. The wind attains the geostrophic velocity at between 500 and 1000 m in the middle and higher latitudes as frictional drag effects become negligible. This is a theoretical profile of wind velocity under conditions of mechanical turbulence.

B Divergence, vertical motion and vorticity

These three terms essentially hold the key to a proper understanding of modern meteorological studies of wind and pressure systems on a synoptic and global scale. Mass uplift or descent of air occurs primarily in response to dynamic factors related to horizontal airflow and is only secondarily affected by air-mass-stability. Hence the significance of these factors for weather processes.

1 Divergence

Different types of horizontal flow are shown in fig. 3.6A. The first panel shows that air may accelerate (decelerate) leading to velocity divergence (convergence). When streamlines (lines of instantaneous air motion) spread out or squeeze together, this is termed diffluence or confluence respectively. If the streamline pattern is strengthened by that of the isotachs (lines of equal wind speed), as shown in the third panels of fig. 3.6A, then there may be mass divergence or convergence at a point (fig. 3.6B). In this case the compressibility of the air causes the density to decrease, or increase, respectively. Usually, however, confluence (diffluence) is associated with an increase (decrease) in air velocity which essentially compensate for each other (the indeterminate cases). Hence, convergence (divergence) may give rise to vertical stretching (shrinking) as illustrated in fig. 3.6C. It is important to note that if all winds were geostrophic, there could be no convergence or divergence and hence no weather!

Other ways in which convergence or divergence can occur are the result of surface friction effects. Onshore winds undergo convergence at low levels

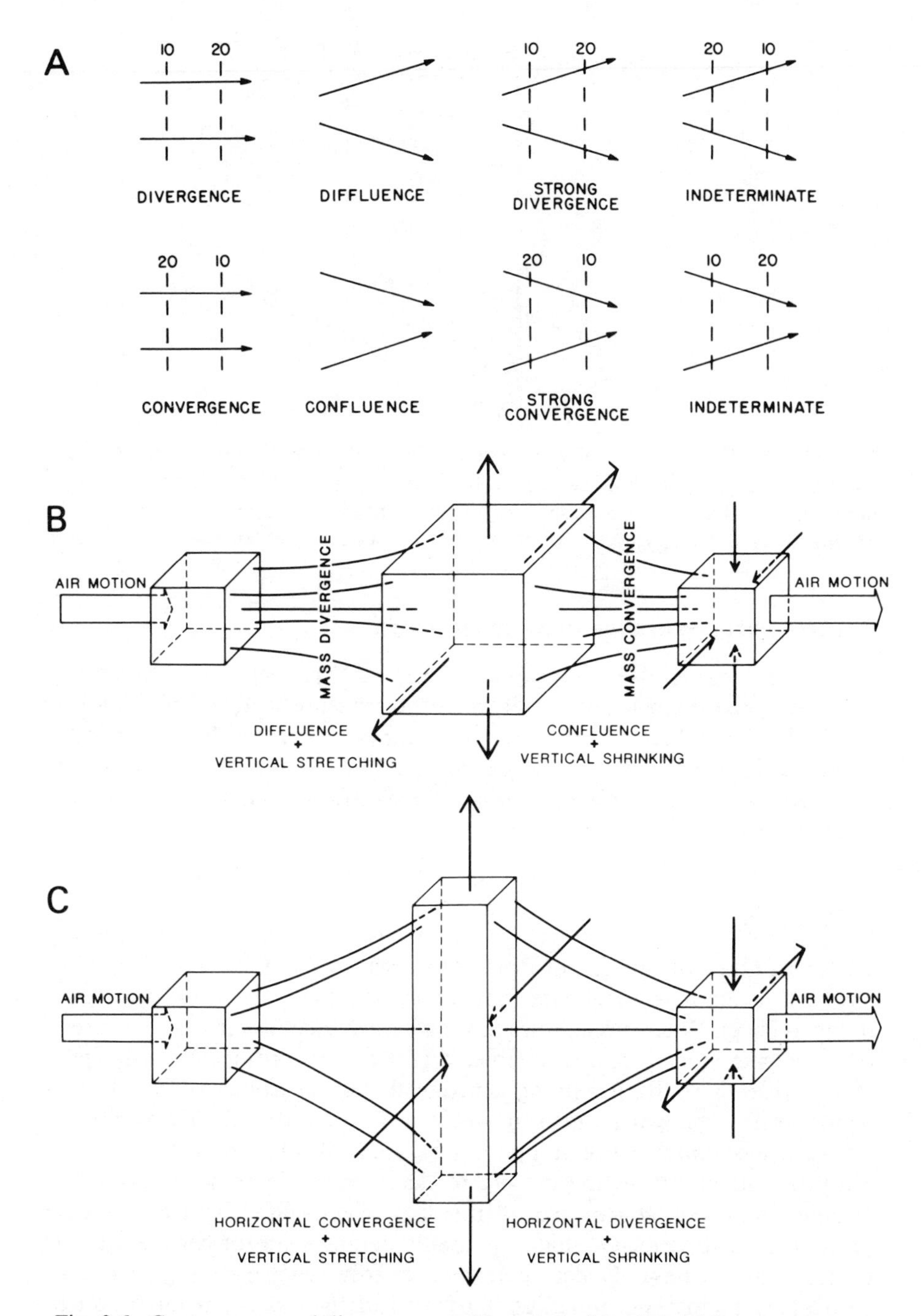

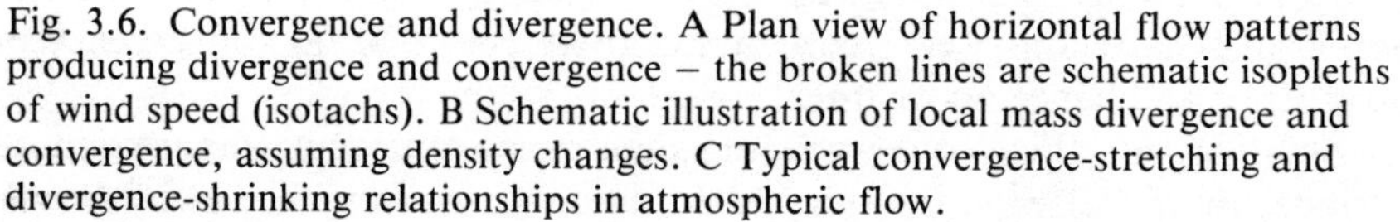

Fig. 3.6. Convergence and divergence. A Plan view of horizontal flow patterns producing divergence and convergence – the broken lines are schematic isopleths of wind speed (isotachs). B Schematic illustration of local mass divergence and convergence, assuming density changes. C Typical convergence-stretching and divergence-shrinking relationships in atmospheric flow.

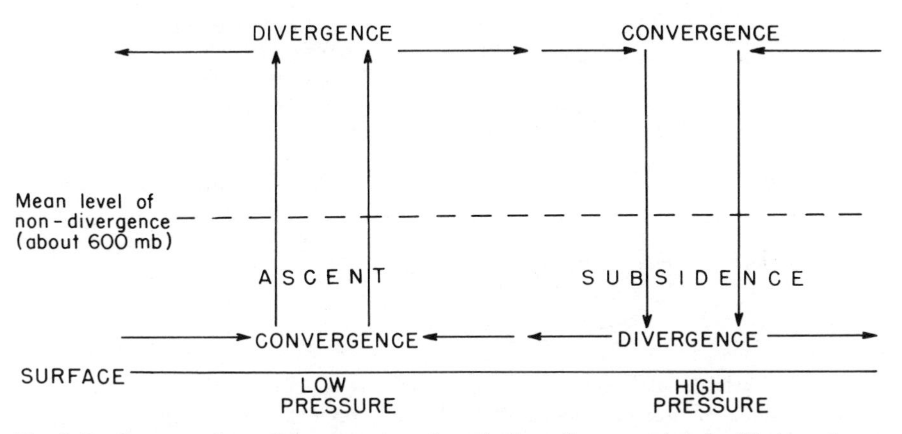

Fig. 3.7. Cross-section of the patterns of vertical motion associated with (mass) divergence and convergence in the troposphere, illustrating mass continuity.

when the air slows down on crossing the coastline owing to the greater friction overland, whereas offshore winds accelerate and become divergent. Frictional differences can also set up coastal convergence (or divergence) if the geostrophic wind is parallel to the coastline with, for the northern hemisphere, land to the right (or left) of the air current, viewed downwind.

2 Vertical motion

Horizontal inflow or outflow near the surface has to be compensated by vertical motion, as illustrated in fig. 3.7, if the low- or high-pressure systems are to persist and there is to be no continuous density increase or decrease. Air rises above a low-pressure cell and subsides over a high-pressure, with compensating divergence and convergence, respectively, in the upper troposphere. In the middle troposphere there must clearly be some level at which horizontal divergence or convergence is effectively zero; the mean 'level of non-divergence' is generally at about 600 mb. Large-scale vertical motion is extremely slow compared with convective and downdraught currents in cumulus, for example. Typical rates in large depressions and anticyclones are of the order of 5–10 cm s^{-1}, whereas updraughts in cumulus may exceed 10 m s^{-1}.

3 Vorticity

Vorticity implies the rotation or angular velocity of minute (imaginary) particles in any fluid system. The air within a depression can be regarded as comprising an infinite number of small air parcels, each rotating cyclonically about an axis vertical to the earth's surface (fig. 3.8). Vorticity has three elements – magnitude (defined as *twice* the angular velocity, ω) (see note 2), direction (the horizontal or vertical axis about which the rotation occurs) and the sense of rotation. Rotation in the same sense as the earth's rotation – cyclonic in the northern hemisphere – is defined as positive. Cyclonic

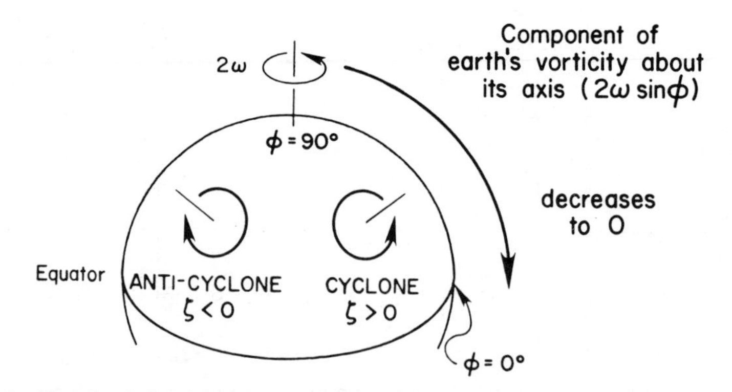

Fig. 3.8. Sketch of the relative vertical vorticity (ζ) about a cyclone and an anticyclone in the northern hemisphere. The component of the earth's vorticity about its axis of rotation (or, the Coriolis parameter, f) is equal to twice the angular velocity (ω) times the sine of the latitude (ϕ). At the pole $f = 2\omega$, diminishing to 0 at the equator. Cyclonic vorticity is in the same sense as the earth's rotation about its own axis, viewed from above, in the northern hemisphere: this cyclonic vorticity is defined as positive ($\zeta>0$).

vorticity may result from cyclonic curvature of the streamlines, from cyclonic shear (stronger winds on the right side of the current, viewed downwind in the northern hemisphere), or a combination of the two (fig. 3.9). Lateral shear (fig. 3.9B) results from changes in isobar spacing. Anticyclonic vorticity occurs with the corresponding anticyclonic situation. The component of vorticity about a vertical axis is referred to as the vertical vorticity. This is generally the most important, but near the ground surface frictional shear causes vorticity about an axis parallel to the surface and normal to the wind direction.

Vorticity is related not only to air motion about a cyclone or anticyclone (*relative vorticity*), but also to the location of that system on the rotating earth. The vertical component of *absolute vorticity* consists of the relative vorticity (ζ) and the latitudinal value of the Coriolis parameter, $f = 2\omega \sin \phi$ (see ch. 3, A.2). At the equator the local vertical is at right angles to the earth's axis so that $f = 0$, but at the north pole cyclonic relative vorticity and the earth's rotation act in the same sense (fig. 3.8).

C Local winds

To the practising meteorologist, local controls over air movement often provide more problems than the effects of the major planetary forces just discussed. Diurnal tendencies are superimposed upon both the large- and small-scale patterns of wind velocity. These are particularly noticeable in the case of local winds and therefore are examined before we consider the major types of local wind regime.

In normal conditions there is a general tendency for wind velocities to be least about dawn, at which time there is little vertical thermal mixing and the lower air does not therefore partake of the velocity of the more freely moving

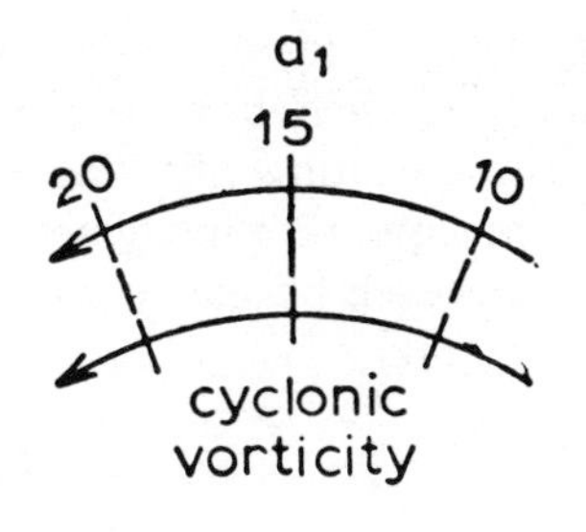

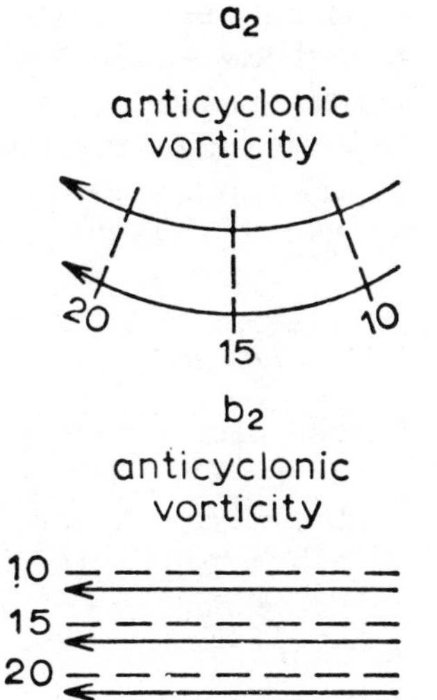

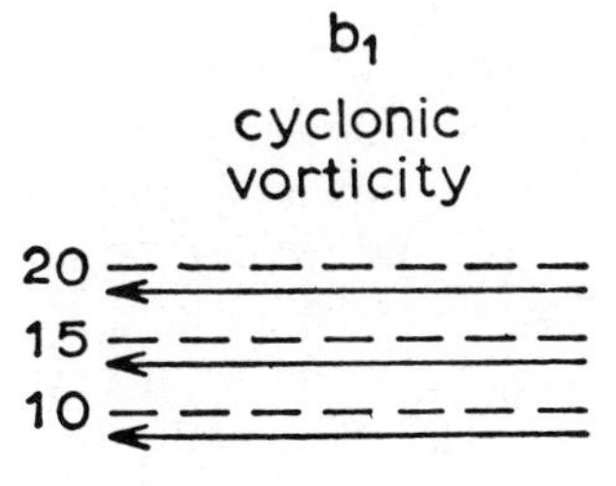

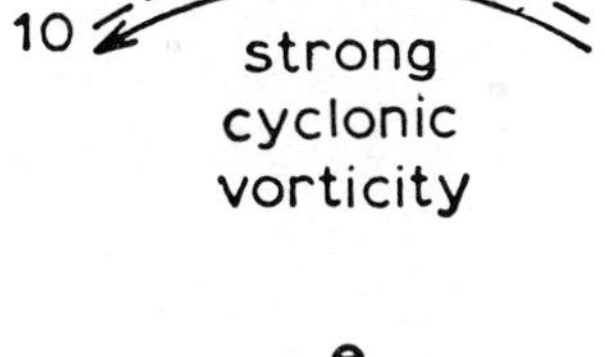

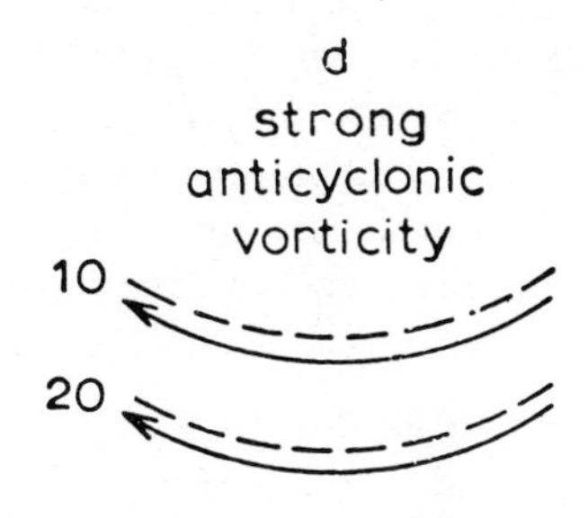

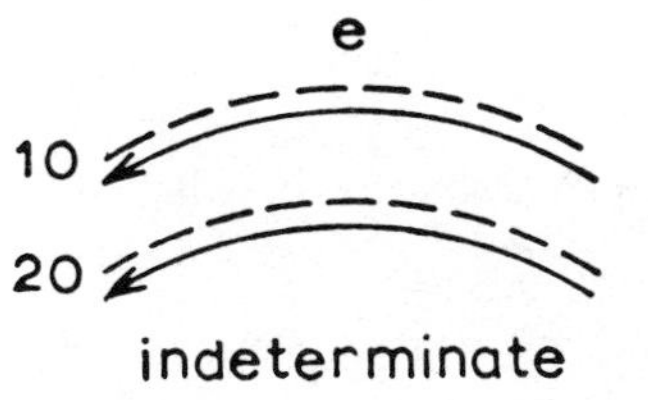

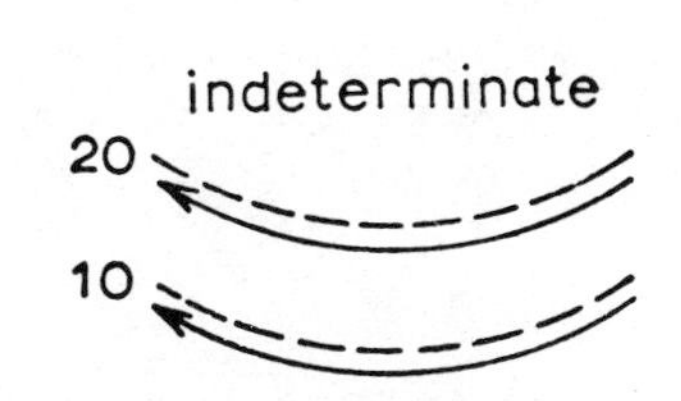

Fig. 3.9. Streamline models illustrating in plan view the flow patterns with cyclonic and anticyclonic vorticity in the northern hemisphere (*after Riehl et al. 1954*). In c and d the effects of curvature (a_1 and a_2) and lateral shear (b_1 and b_2) are additive, whereas in e and f they more or less cancel out. Dashed lines are schematic isopleths of wind speed.

upper air (see ch. 3, D). Conversely, velocities of some local winds are greatest between 1300 and 1400 hours, for this is the time when the air suffers its greatest tendency to move vertically due to terrestrial heating, allowing it, subject to surface frictional effects, to join in the freer upper-air movement. Upper air always moves more freely than air at surface levels because it is not subject to the retarding effects of friction and obstruction.

1 Mountain and valley winds

Terrain irregularities give rise to their own special meteorological conditions. On warm sunny days, the heated air in a valley is laterally constricted, compared with that over an equivalent area of lowland, and so tends to expand vertically. The volume ratio of lowland : valley air is typically about 2 or 3 : 1 and this encourages air to flow from the lowland up the axis of the valley. This valley wind is generally very light and requires a weak regional pressure gradient in order to develop. This flow along the main valley develops more or less simultaneously with *anabatic* (upslope) winds which result from greater heating of the valley sides compared with the valley floor. These slope winds rise above the ridge line and feed an upper return current along the line of the valley to compensate for the valley wind. This feature may be obscured, however, by the regional airflow. Speeds reach a maximum around 1400 hours. At night there is a reverse process as the cold denser air at higher elevations drains into depressions and valleys; this is known as a *katabatic* wind (fig. 3.10).

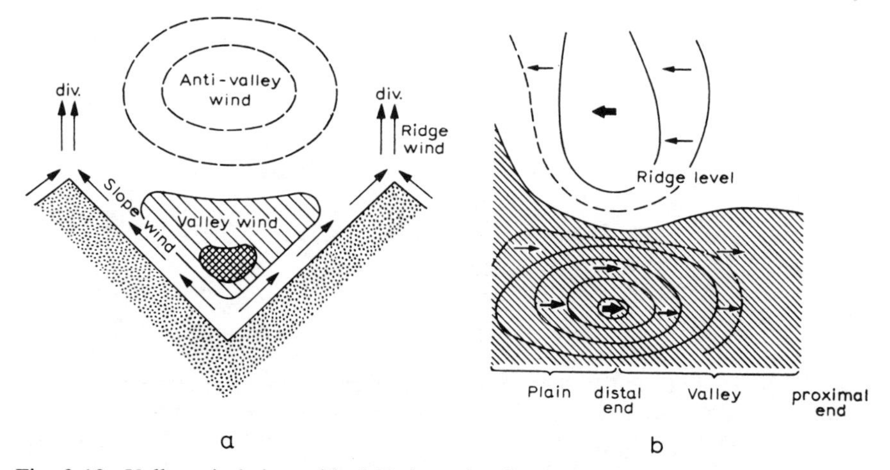

Fig. 3.10. Valley winds in an ideal V-shaped valley. a Section across the valley. The valley wind and anti-valley wind are directed at right angles to the plane of the paper. The arrows show the slope and ridge wind in the plane of the paper, the latter diverging (div.) into the anti-valley wind system. b Section running along the centre of the valley and out on to the adjacent plain, illustrating the valley wind (*below*) and the anti-valley wind (*above*) (*after Buettner and Thyer 1965*).

If the air drains downslope into an open valley, a 'mountain wind' develops more or less simultaneously along the axis of the valley. This flows towards the plain where it replaces warmer, less dense air. The maximum velocity occurs just before sunrise at the time of maximum diurnal cooling. Like the valley wind, the mountain wind is also overlain by an upper return current, in this case up-valley.

Katabatic drainage is usually cited as the cause of frost pockets in hilly and mountainous areas. It is argued that greater radiational cooling on the slopes, especially if they are snow-covered, leads to a gravity flow of cold, dense air into the valley bottoms. Observations in California and elsewhere, however, suggest that the valley air remains colder than the slope air from the onset of nocturnal cooling, so that the air moving downslope slides over the denser air in the valley bottom. Moderate drainage winds will also act to raise the valley temperatures through turbulent mixing. One suggestion is that cold air pockets in valley bottoms and hollows may result from the cessation of turbulent heat transfer to the surface in sheltered locations rather than by cold air drainage. Clearly, the problem requires further study.

2 Winds due to topographic barriers

Mountains ranges have important effects on the airflow across them. The displacement of air upwards over the obstacle may trigger instability if the air is conditionally unstable (see ch. 2, E), whereas stable air returns to its original level in the lee of the barrier and this descent often forms the first of a series of *lee waves* (or *standing waves*) downwind as shown in fig. 3.11. The wave form remains more or less stationary relative to the barrier with the air moving quite rapidly through it. Below the crest of the waves there may be circular air motion in a vertical plane which is termed a *rotor*. The formation of such features is naturally of vital interest to pilots. The development of lee waves is commonly disclosed by the presence of lenticular clouds (pl. 8), and on occasion a rotor causes reversal of the surface wind direction in the lee of high mountains (see pl. 9).

Winds on summits are usually strong, at least in middle and higher latitudes. Average speeds on summits in the Rocky Mountains in winter months are around 12–15 m s^{-1}, for example, and on Mt Washington, New Hampshire, an extreme value of 103 m s^{-1} has been recorded. Peak monthly speeds in excess of 40–50 m s^{-1} are typical in both these areas in winter. The air is constricted and thus accelerated over mountain barriers (the Venturi effect), but friction with the ground also retards the flow, compared with the free air at the same level. Over low hills there may be a considerable speed-up of the wind by comparison with the surrounding lowland.

A wind of local importance in mountain areas is the föhn or chinook. It is a strong, gusty, dry and warm wind which develops on the lee side of a mountain range when stable air is forced to flow over the barrier by the regional pressure gradient. Sometimes, there is a loss of moisture by precipitation on the mountains (fig. 3.12) and the air, having cooled at the

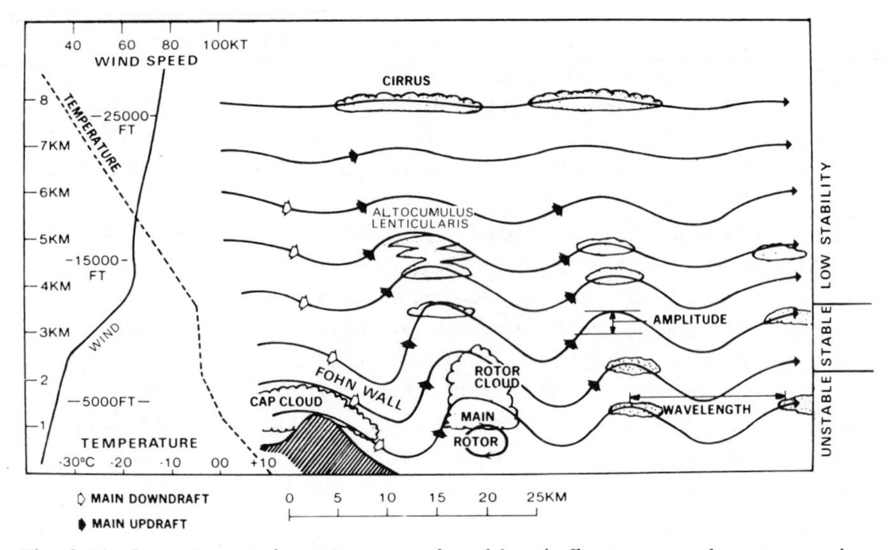

Fig. 3.11. Lee waves and rotors are produced by air flow across a long mountain range. The first wave crest usually forms less than one wavelength downwind of the ridge. There is a strong surface wind down the lee slope. Wave characteristics are determined by the wind speed and temperature relationships, shown schematically on the left of the diagram. The existence of an upper stable layer is particularly important (*after Ernst 1976*).

saturated adiabatic lapse rate above the condensation level, subsequently warms at the greater dry adiabatic lapse rate as it descends on the lee side with a consequent lowering of both the relative and absolute humidity. Other investigations show that in many instances there is no loss of moisture over the mountains. In such cases the föhn effect is the result of the blocking of air to windward of the mountains by a summit-level temperature inversion. This forces air from higher levels to descend and warm adiabatically. Föhn winds are common along the northern flanks of the Alps and the mountains of the Caucasus and central Asia in winter and spring,

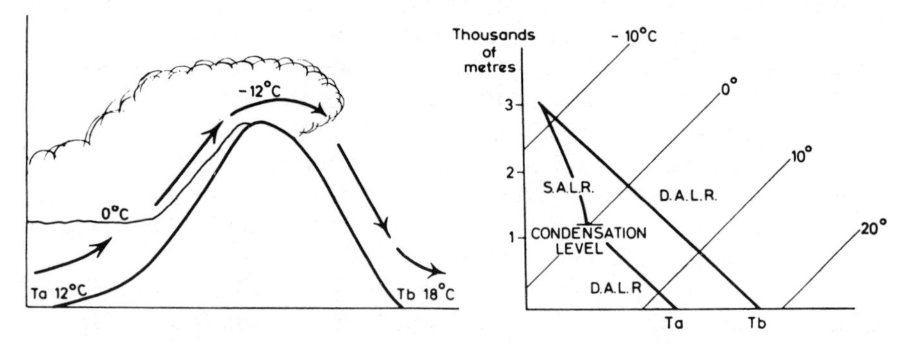

Fig. 3.12. The föhn effect when an air parcel is forced to cross a mountain range. Ta refers to the temperature at the windward foot of the range and Tb to that at the leeward foot.

when the accompanying rapid temperature rise may help to trigger off avalanches on the snow-covered slopes. At Tashkent in central Asia, where the mean winter temperature is about the freezing point, temperatures may rise to more than 21 °C during a föhn. In the same way the chinook is a significant feature at the eastern foot of the New Zealand Alps and of the Rocky Mountains. At Pincher Creek, Alberta, a temperature rise of 21 °C (38 °F) occurred in 4 minutes with the onset of a chinook on 6 January 1966. Less spectacular effects are also noticeable in the lee of the Welsh mountains, the Pennines and the Grampians, where the importance of föhn winds lies mainly in the dispersal of cloud by the subsiding dry air (see also ch. 5, p. 240 and p. 255). This is an important component of so-called 'rain shadow' effects.

In some parts of the world, winds descending on the lee slope of a mountain range are cold. The type example of such 'fall-winds' is the *bora* of the northern Adriatic, although similar winds occur on the northern Black Sea coast, in northern Scandinavia, Novaya Zemlya and in Japan. These winds occur when cold continental air masses are forced across a mountain range by the pressure gradient and, despite adiabatic warming, displace warmer air. They are therefore primarily a winter phenomenon.

On the eastern slope of the Rocky Mountains in Colorado (and probably also in other similar continental locations), winds of either *bora* or chinook type can occur. Locally, at the foot of the mountains, such winds may attain hurricane force with gusts exceeding 45 m s^{-1} (100 mph). A few downslope storms of this type have caused millions of dollars of property damage in Boulder, Colorado and the immediate vicinity. These windstorms develop when a stable layer close to the mountain-crest level prevents airflow to windward from crossing over the mountains. Extreme amplification of a lee wave (fig. 3.11) drags air from above the summit level (about 4000 m) down to the plains (1700 m) in a very short distance, so that the high velocities occur. However, the flow is not simply 'downslope'; the storm may affect the mountain slopes but not the foot of the slope, or vice versa, depending on the location of the lee wavetrough. The high winds are caused by the acceleration of the air towards this pressure minimum.

3 *Land and sea breezes*

Another thermally-induced type of air movement is the land and sea breeze (fig. 3.13). The vertical expansion of the air column which occurs daily during the hours of heating over the more rapidly heated land (see ch. 1, D.5), tilts the isobaric surfaces downwards at the coast, causing onshore winds at the surface and a compensating offshore movement aloft. At night the air over the sea is warmer and the situation is reversed, although this reversal is also the effect of downslope winds blowing off the land. Figure 3.14 shows that these local winds can have a decisive effect on coastal temperature and humidity. The advancing cool sea air may form a distinct line (or *front*, see ch. 4, C) marked by cumulus cloud development, behind which there is a

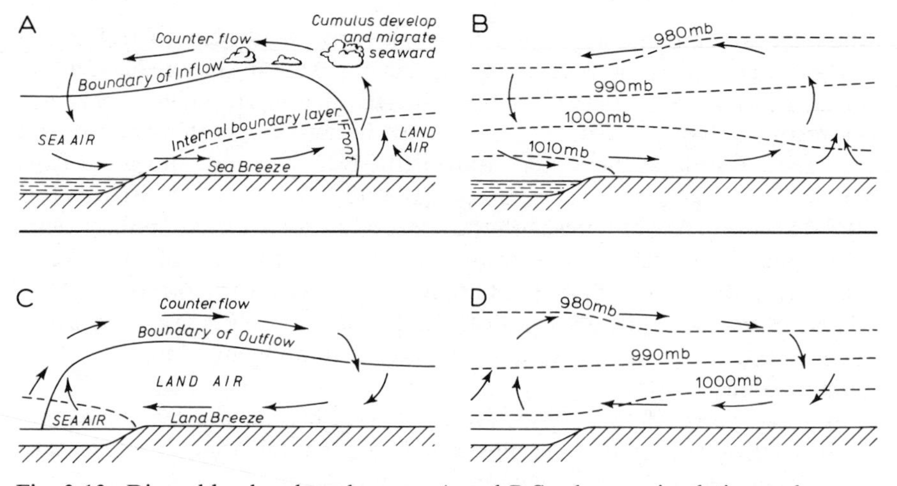

Fig. 3.13. Diurnal land and sea breezes. A and B Sea breeze circulation and pressure distribution in the early afternoon during anticyclonic weather. C and D Land breeze circulation and pressure distribution at night during anticyclonic weather. (A and C *after Oke 1978*).

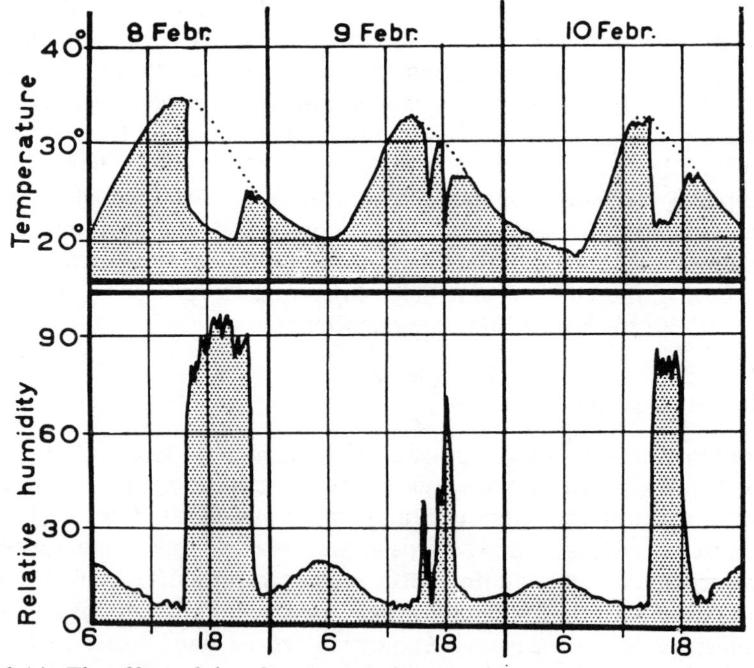

Fig. 3.14. The effect of the afternoon sea breeze on the temperature (°C) and relative humidity (%) at Joal on the Senegal coast, 8–10 February 1893 (*after Angot and De Martonne; from Kuenen 1955*).

distinct wind velocity maximum. This often develops in summer, for example, along the Gulf Coast of Texas. On a smaller scale such features can also be observed in Britain, particularly along the south and east coasts. The sea breeze has a depth of about 1 km (3300 ft), although it thins towards the advancing edge, and may penetrate 50 km (30 miles) or more inland by 2100 hours. Typical wind speeds in such sea breezes are 4–7 m s^{-1} (about 10–15 mph), although these may be greatly increased where a well-marked low-level temperature inversion produces a 'Venturi effect' in constricting and accelerating the flow. The much shallower land breezes are usually only about 2 m s^{-1} (about 5 mph). The counter currents aloft are generally less evident and may be obscured by the regional airflow, but recent work along the Oregon coast has suggested that under certain conditions this upper return flow may be very closely related to the lower sea breeze conditions, even to the extent of mirroring the surges in the latter. It is worth noting that in middle latitudes the Coriolis deflection causes turning of a well-developed onshore sea breeze (clockwise in the northern hemisphere) so that eventually it may blow more or less parallel to the shore. Analogous 'lake breeze' systems develop adjacent to large inland water bodies such as the Great Lakes.

D Variation of pressure and wind velocity with height

Changes of height reveal variations both of pressure and of wind characteristics. Above the level of surface frictional effects (about 500–1000 m) the wind increases in speed and becomes more or less geostrophic. With further height increase the reduction of air density leads to a general increase in wind speed (see ch. 3, A.1). At 45°N a geostrophic wind of 14 m s^{-1} at 3 km is equivalent to one of 10 m s^{-1} at the surface for the same pressure gradient. There is also a seasonal variation in wind speeds aloft, these being much greater during winter months when the meridional temperature gradients are at a maximum. In addition, the persistence of these gradients tends to cause the upper winds to be more constant in direction.

1 The vertical variation of pressure systems

The general relationships between surface and tropospheric pressure conditions are illustrated by the models of fig. 3.15. A low-pressure cell at sea-level with a cold core will intensify with elevation, whereas one with a warm core tends to weaken and may be replaced by high pressure. A warm air column of relatively low density causes the pressure surfaces to bulge upwards and conversely a cold, more dense air column leads to downward contraction of the pressure surfaces. Thus, a surface high-pressure cell with a cold core (a *cold anticyclone*), such as the Siberian winter anticyclone, weakens with increasing elevation and is replaced by low pressure aloft. Cold anticyclones are shallow and rarely extend their influence above about 2500 m (8000 ft). By contrast a surface high with a warm core (a *warm*

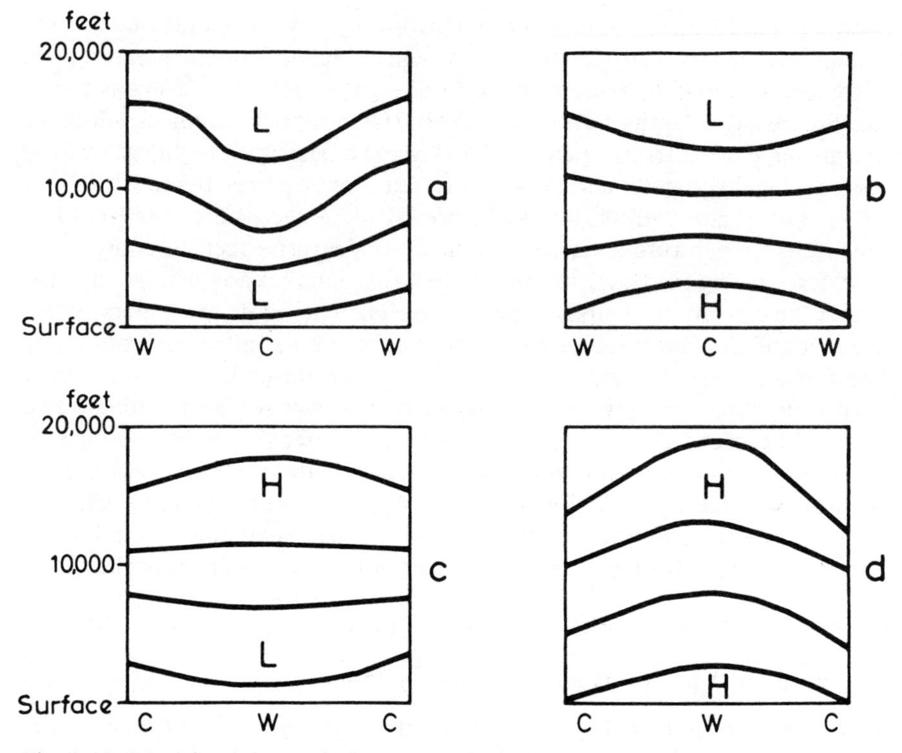

Fig. 3.15. Models of the vertical pressure distribution in cold and warm air columns. a A surface low pressure intensifies aloft in a cold air column. b A surface high pressure weakens aloft and may become a low pressure in a cold air column. c A surface low pressure weakens aloft and may become a high pressure in a warm air column. d A surface high pressure intensifies aloft in a warm air column.

anticyclone) intensifies with height (fig. 3.15D). This is characteristic of the large subtropical cells which maintain their warmth through dynamic subsidence. The warm low (fig. 3.15C) and cold high (fig. 3.15B) are consistent with the vertical motion schemes illustrated in fig. 3.7, whereas the other two types are primarily produced by dynamic processes. The high surface pressure in a warm anticyclone is linked hydrostatically with cold, relatively dense air in the lower stratosphere. Conversely a cold depression (fig. 3.15A) is associated with a warm lower stratosphere.

Mid-latitude low-pressure cells have cold air in the rear and in consequence the axis of low pressure slopes with height towards the colder air to the west. High-pressure cells slope towards the warmest air (fig. 3.16) and in this manner the northern hemisphere subtropical high-pressure cells are displaced 10°–15° south in latitude at the 3000-m level, as well as towards the west (fig. 3.17). Even so, this slope of the high-pressure axes is not constant through time and stations located between the cells may experience widely fluctuating upper winds associated with variations in the inclination of the axes.

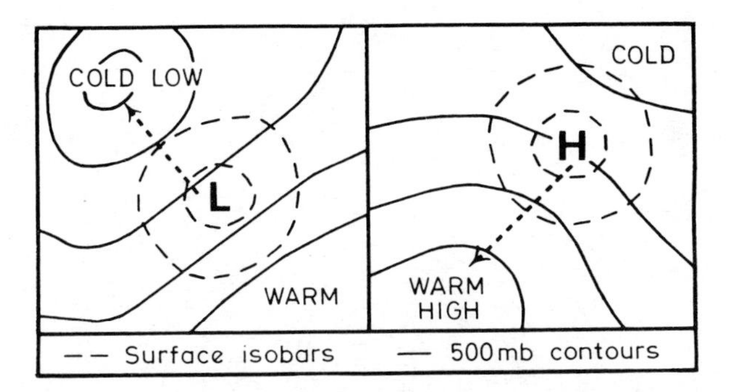

Fig. 3.16. The characteristic slope of the axes of low- and high-pressure cells with height in the northern hemisphere.

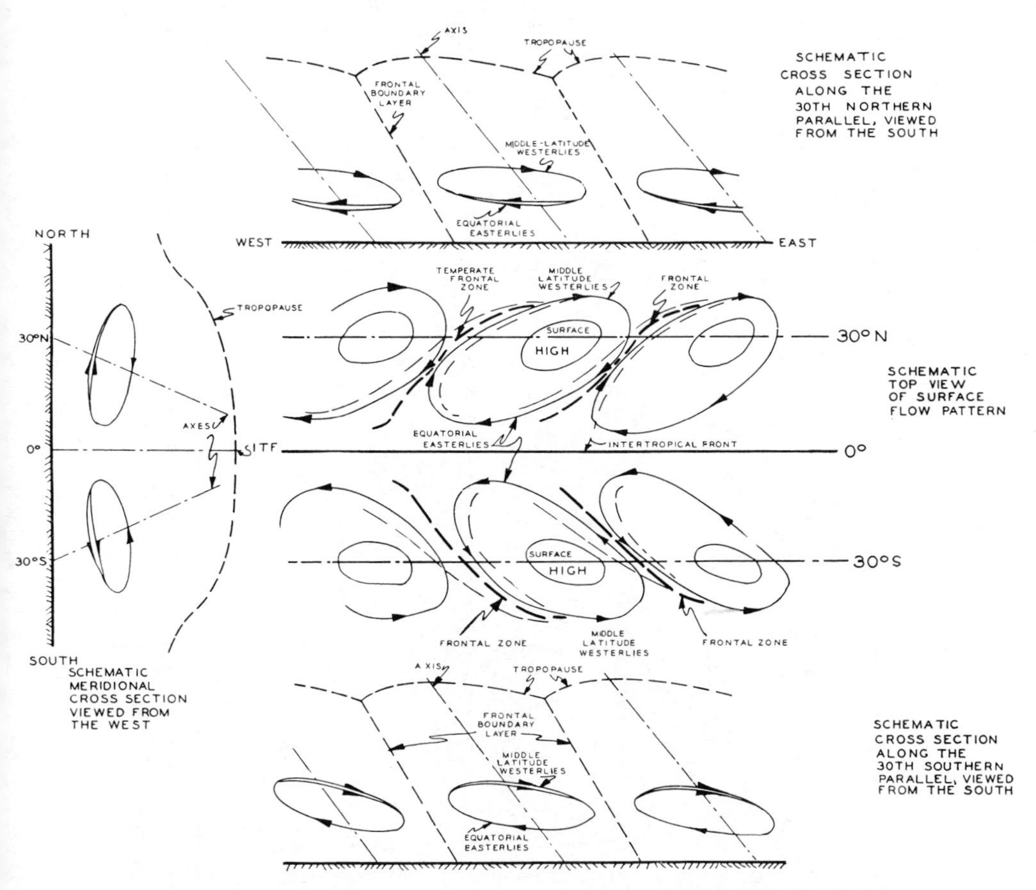

Fig. 3.17. Schematic horizontal and vertical structure of the subtropical high-pressure cells. Note particularly the convergence along the belts between the cells, the slope of the axes with height westward and equatorward, and the inclined spiral of air motion in the middle troposphere – up on the west sides (dynamically unstable air) and down on the east sides (dynamically stable air) (*from Garbell 1947*).

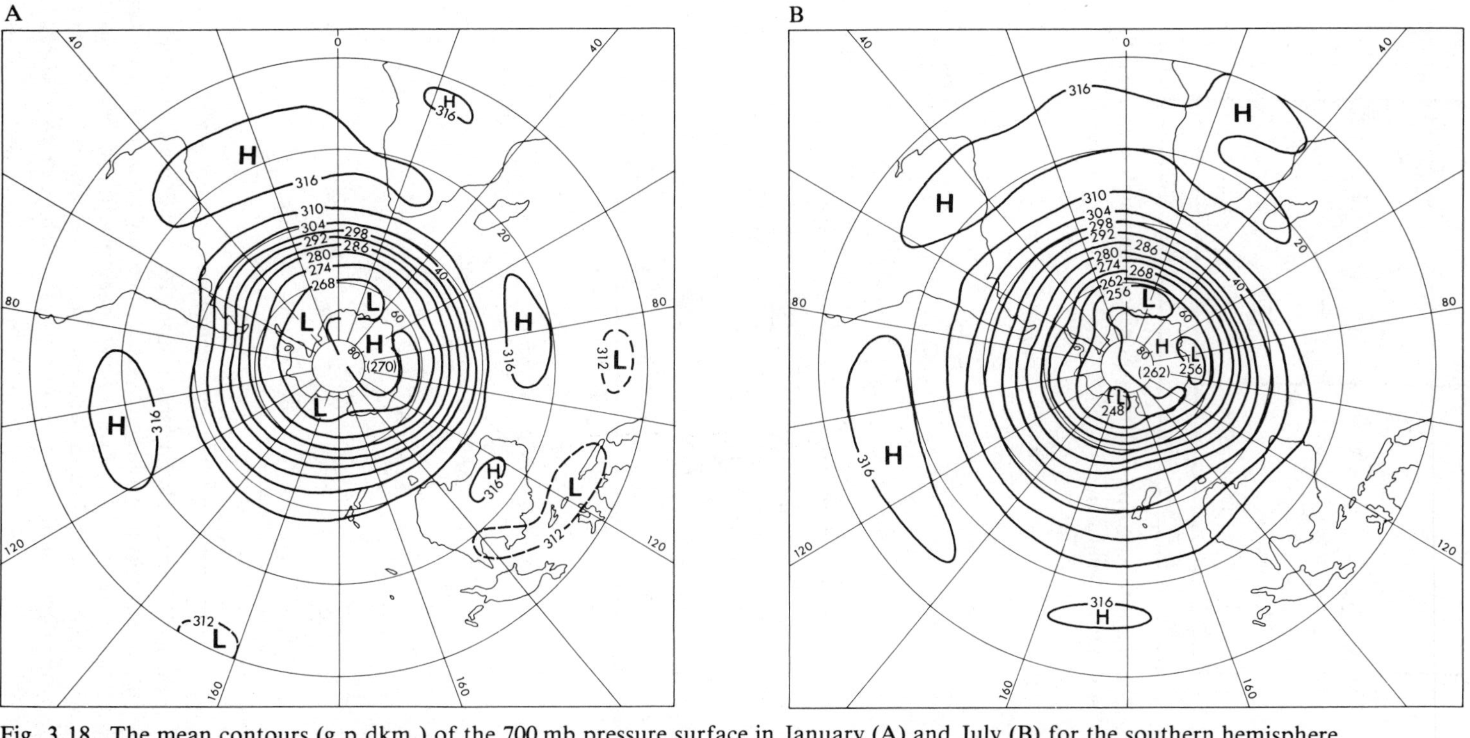

Fig. 3.18. The mean contours (g.p.dkm.) of the 700 mb pressure surface in January (A) and July (B) for the southern hemisphere, 1949–60 (*after Taljaard et al. 1969*).

2 Mean upper-air patterns

It is helpful to begin by considering the patterns of pressure and wind in the middle troposphere. These are less complicated in appearance than surface maps as a result of the diminished effects of the land masses. Rather than using pressure maps at a particular height it is convenient to depict the height of a selected pressure surface; this is termed *contour chart* by analogy with topographic relief maps (see note 4). Figure 3.18 shows that in the middle troposphere of the southern hemisphere there is a vast circumpolar cyclonic vortex poleward of latitude 30°S in summer and winter. The vortex is more or less symmetrical about the pole, although the low centre is towards the Ross Sea sector. Corresponding charts for the northern hemisphere (fig. 3.19) also show an extensive cyclonic vortex, but one which is markedly more asymmetric with a primary centre over the eastern Canadian Arctic and a secondary one over eastern Siberia. The major troughs and ridges form what are referred to as *long waves* (or *Rossby waves*) in the upper flow (see ch. 4, F). The two major troughs at about 70°W and 150°E are thought to be induced by the combined influence on upper-air pressure and winds of large orographic barriers, like the Rockies and the Tibetan Plateau, and heat sources such as warm ocean currents (in winter) or land masses (in summer). It is worth noting that land surfaces occupy over 50% of the northern hemisphere between latitudes 40° and 70°N. The subtropical high-pressure belt has only one clearly distinct cell in January over the eastern Caribbean, whereas in July cells are well developed over the Atlantic and the Pacific. In addition, the July map shows greater prominence of the subtropical high over the Sahara and southern North America. The northern hemisphere shows a marked summer to winter intensification of the mean circulation which is explained below.

In the southern hemisphere, the predominance of ocean surface (which comprises 81% of the hemisphere) considerably reduces the development of long waves in the upper westerlies. Nevertheless, asymmetries are initiated by the effects on the atmosphere of such geographical features as the Andes, the elevated and extensive dome of eastern Antarctica, and ocean currents, particularly the Humboldt and Benguela Currents (see fig. 3.38) and the associated cold coastal upwellings.

3 Upper winds

It is a common observation that clouds at different levels move in different directions. The wind speeds at these levels may also be markedly different, although this is not so evident to the casual observer. The gradient of wind velocity with height is referred to as the (vertical) *wind shear*, and in the free air, above the friction level, the amount of shear depends upon the temperature structure of the air. This important relationship is illustrated in fig. 3.20. The diagram shows hypothetical contours of the 1000 and 500 mb pressure surfaces. The *thickness* of the 1000–500 mb layer is proportional to its mean

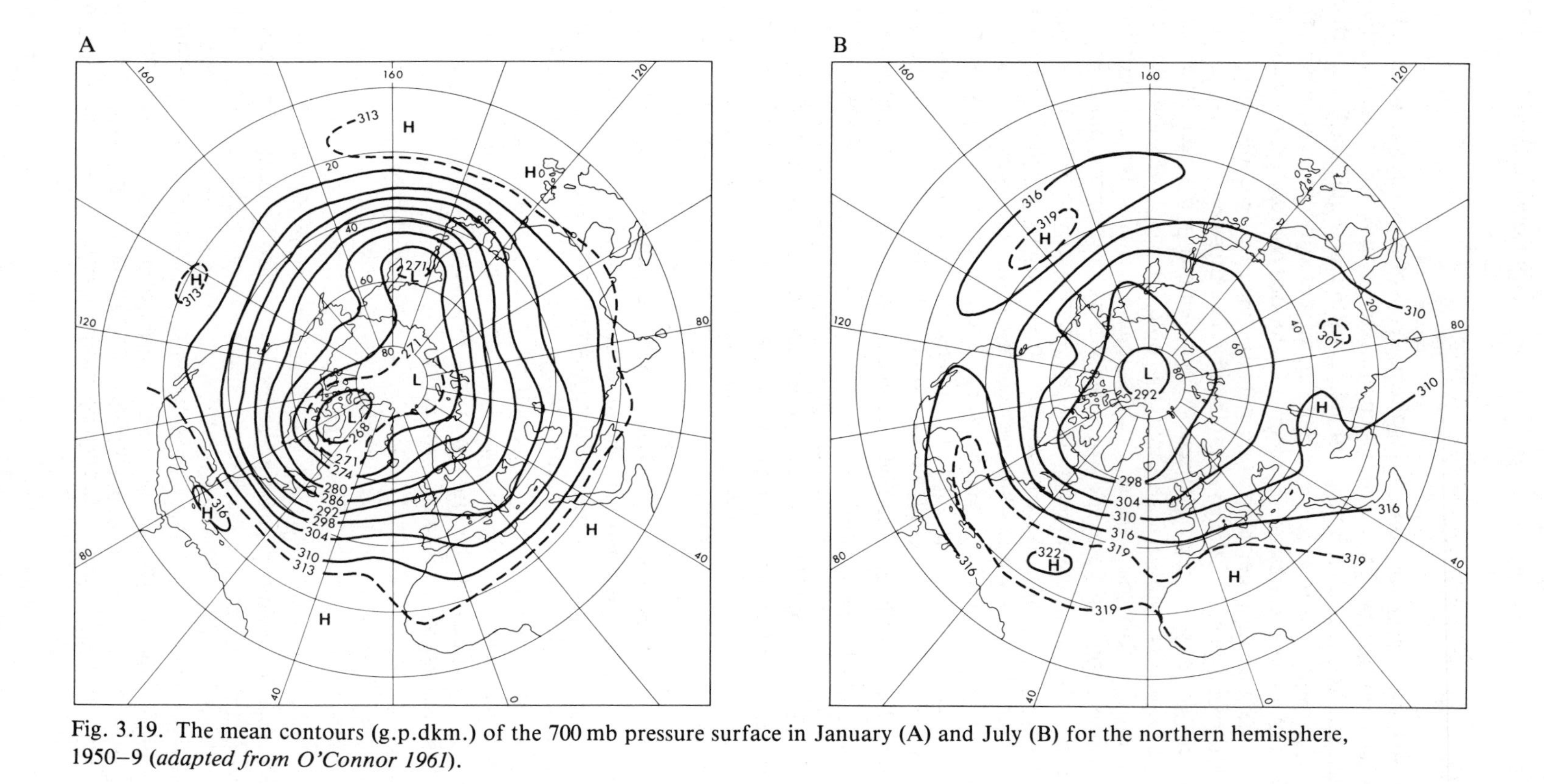

Fig. 3.19. The mean contours (g.p.dkm.) of the 700 mb pressure surface in January (A) and July (B) for the northern hemisphere, 1950–9 (*adapted from O'Connor 1961*).

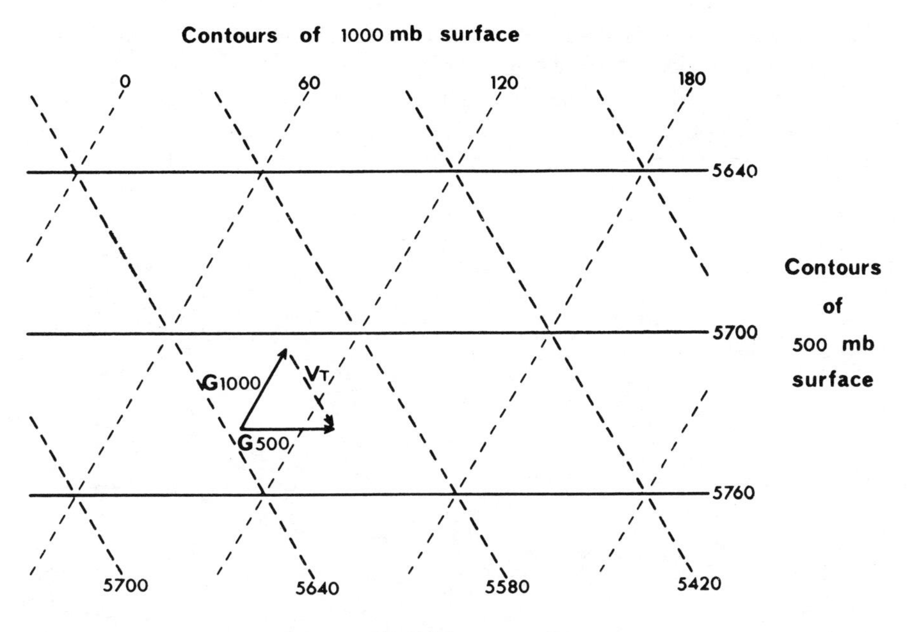

Fig. 3.20. Schematic map of superimposed contours of isobaric height and thickness of the 1000–500 mb layer (in metres). G_{1000} is the geostrophic velocity at 1000 mb, G_{500} that at 500 mb, V_T is the resultant 'thermal wind' blowing parallel to the thickness lines.

temperature – low thickness values correspond to cold air, high thickness values to warm air. This relationship is apparent in the vertical sections of fig. 3.15. The theoretical wind vector (V_T) blowing parallel to the thickness lines, with a velocity proportional to their gradient, is termed the *thermal wind*. The geostrophic wind velocity at 500 mb (G_{500}) is the vector sum of the 1000 mb geostrophic wind (G_{1000}) and the thermal wind (V_T), as shown in fig. 3.20.

Since the thermal wind component blows with cold air (low thickness) to the left in the northern hemisphere, when viewed downwind, it is readily apparent that in the troposphere the poleward decrease of temperature should be associated with a large westerly component in the upper winds. Furthermore, since the meridional temperature gradient is steepest in winter (in the northern hemisphere) the zonal westerlies are most intense at this time.

The total result of the above influences is that in both hemispheres the mean upper geostrophic winds are dominantly westerly between the subtropical high-pressure cells (centred aloft about 15° latitude) and the polar low-pressure centre aloft. Between the subtropical high-pressure cells and the equator they are easterly. This dominant, westerly circulation reaches maximum speeds of 45–67 m s^{-1} (100–150 mph), which even increase to 135 m s^{-1} (300 mph) in winter. These maximum speeds are concentrated in a narrow band often situated at about 30° latitude, between 9000 and 15,000 m, called the *jet stream* (see note 5). Plate 16 shows bands of cirrus cloud which may have been related to jet-stream systems.

This stream, which is essentially a fast-moving mass of laterally concentrated air, is connected with the zone of maximum slope, folding, or fragmentation of the tropopause, which in turn coincides with the latitude of

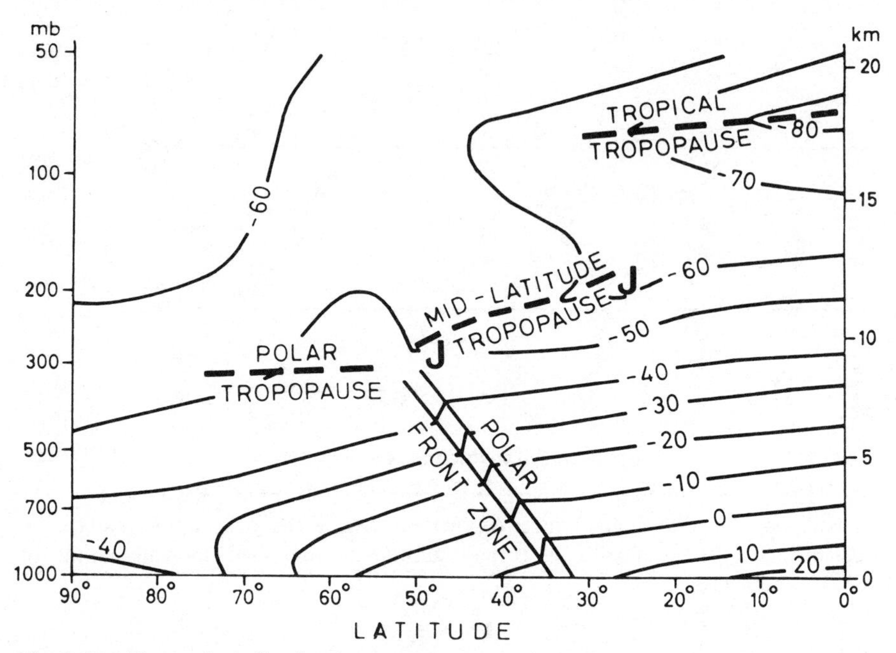

Fig. 3.21. The typical distribution of temperature and the location of the westerly jet streams (J) in the northern hemisphere in winter (*partly after Defant and Taba 1957*).

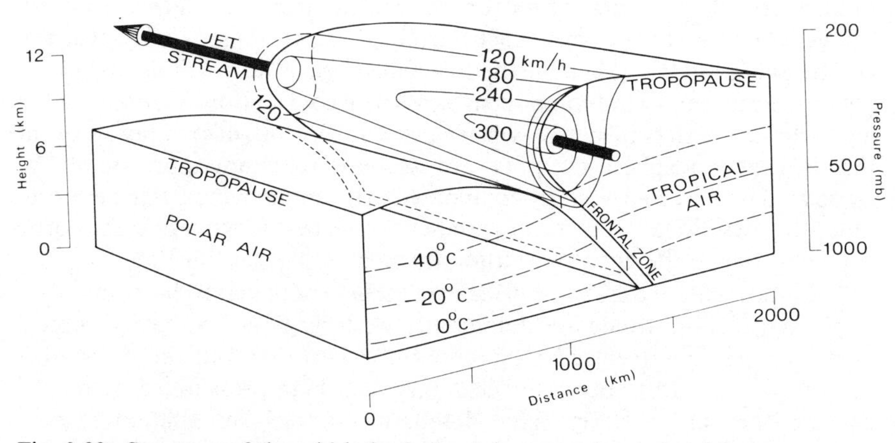

Fig. 3.22. Structure of the mid-latitude frontal zone and associated jet stream showing generalized distribution of temperature, pressure and wind velocity (*after Riley and Spalton 1981*).

maximum poleward temperature gradient and energy transfer. The thermal wind, as described above, is a major component of the jet stream, but the basic reason for the concentration of the meridional temperature gradient in a narrow zone (or zones) is still uncertain. One theory is that the temperature gradient becomes accentuated when the upper wind pattern is confluent (see ch. 3, B.1). Figure 3.21, giving a generalized view of the wind and temperature distribution in the northern hemisphere troposphere in winter, shows that there are two westerly jet streams (see fig. 1.37). The more northerly one, termed the *Polar Front Jet Stream* (cf. ch. 4, E), is associated with the steep temperature gradient where polar and tropical air interact (fig. 3.22), but the *Subtropical Jet Stream* is related to a temperature gradient confined to the upper troposphere. The Polar Front Jet Stream is very irregular in its longitudinal location and is commonly discontinuous, whereas the Subtropical Jet Stream is much more persistent. For these reasons the location of the mean jet stream (fig. 3.23) primarily reflects the position of the Subtropical Jet Stream. The synoptic pattern of jet-stream occurrence may be further complicated in some sectors by the presence of additional frontal zones (see ch. 4, E), each associated with a jet stream. This situation is common in winter over North America. Comparison of figs. 3.19 and 3.23 indicates that the main jet-stream cores are associated with the principal troughs of the Rossby long waves. In summer, an *Easterly Tropical Jet Stream* forms in the upper troposphere over India and Africa due to regional reversal of the S–N temperature gradient (p. 306). The relationships between these upper tropospheric wind systems and surface weather and climate will be considered in chs. 4, 5 and 6.

In the southern hemisphere, the mean jet stream in winter is similar in strength to its northern hemisphere winter counterpart and it weakens less in summer, because the meridional temperature gradient between 30° and 50°S is reinforced by heating over the southern continents. Upper air temperature gradients in the northern hemisphere, in contrast, are much stronger in winter, as noted above.

4 *Surface pressure conditions*

The most permanent features of the mean sea-level pressure maps are the oceanic subtropical high-pressure cells (figs. 3.24 and 3.25). These anticyclones are located at about 30° latitude, suggestively situated below the mean Subtropical Jet Stream. They move a few degrees equatorward in winter and poleward in summer in response to the seasonal expansion and contraction of the two circumpolar vortices. In the northern hemisphere the subtropical ridges of high pressure weaken over the heated continents in summer but are thermally intensified over them in winter. The principal subtropical high-pressure cells are located: (*a*) over the Bermuda–Azores ocean region (aloft the centre of this cell lies over the east Caribbean); (*b*) over the south and south-west United States (the Great Basin or Sonoran cell) – this continental cell is naturally prone to seasonal decline, being

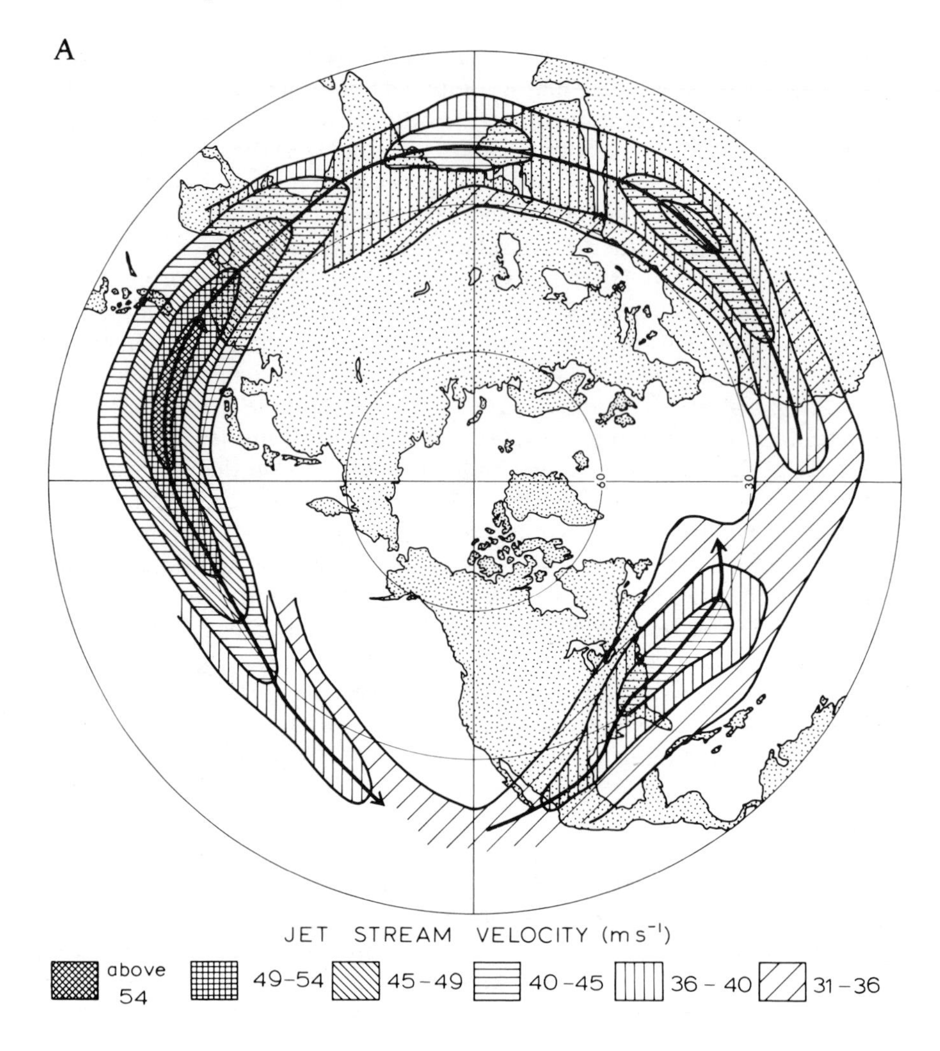

replaced by a thermal surface low in summer; (*c*) over the east and north Pacific – a large and powerful cell (sometimes dividing into two, especially during the summer); and (*d*) over the Sahara – this, like other continental source areas is, seasonally variable both in intensity and extent, being most prominent in winter. In the southern hemisphere the subtropical anticyclones are oceanic, except over southern Australia in winter.

The latitude of the subtropical high-pressure belt depends on the meridional temperature difference between the equator and the pole and on the vertical temperature lapse rate (i.e. vertical stability). The greater the meridional temperature difference the more equatorward is the location of the subtropical high-pressure belt (fig. 3.25).

Equatorward of the subtropical anticyclones there is an equatorial trough of low pressure, associated broadly with the zone of maximum insolation and tending to migrate with it, especially towards the heated continental

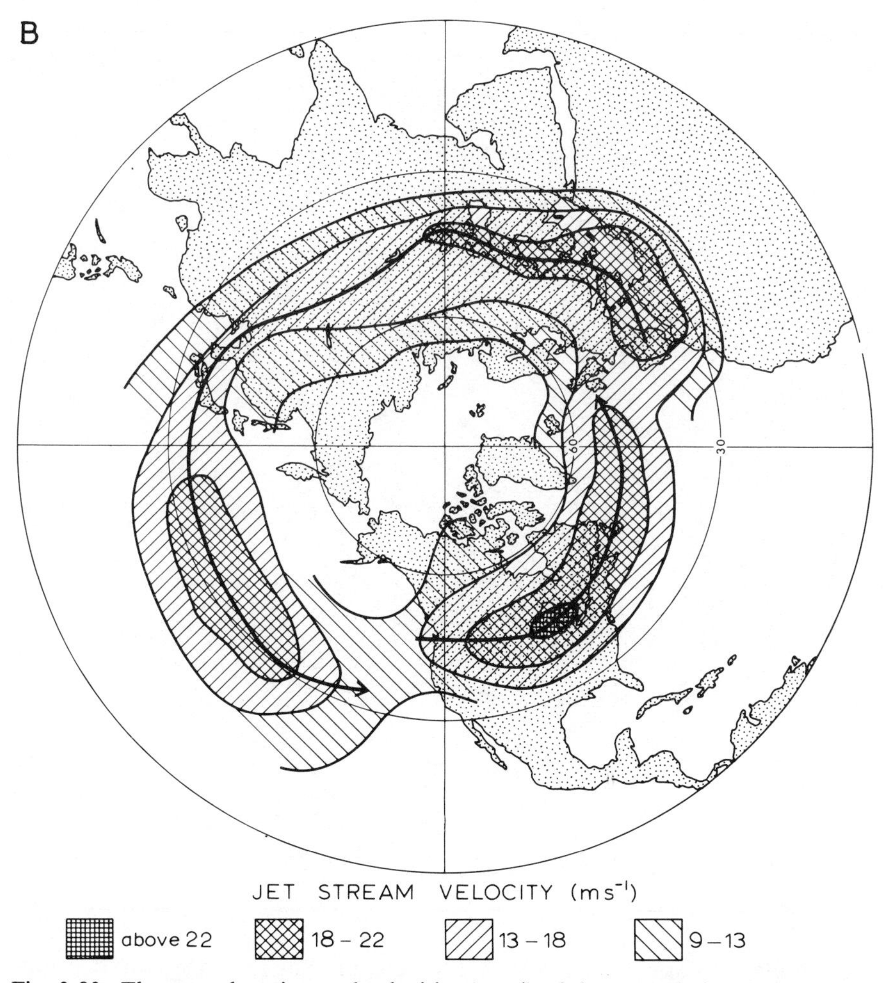

Fig. 3.23. The mean location and velocities (m s^{-1}) of the westerly jet stream in the northern hemisphere in January (A) and July (B) (*after Namias and Clapp; adapted from Petterssen 1958*).

interiors of the summer hemisphere. Poleward of the subtropical anticyclones lies a general zone of subpolar low pressure. In the southern hemisphere this sub-Antarctic Trough is virtually circumpolar (fig. 3.26) whereas in the northern hemisphere the major centres are near Iceland and the Aleutians in winter and primarily over continental areas in summer. It is commonly stated that in high latitudes there is a surface anticyclone due to the cold polar air, but in the Arctic this is true only in spring over the Canadian Arctic Archipelago. In winter the Polar Basin is affected by high and low pressure cells with the major semi-permanent cold air anticyclones over Siberia and, to a lesser extent, north-western Canada. The shallow Siberian high is in part a result of the exclusion of tropical air masses from the interior by the Tibetan massif and the Himalayas. Over Antarctica it is

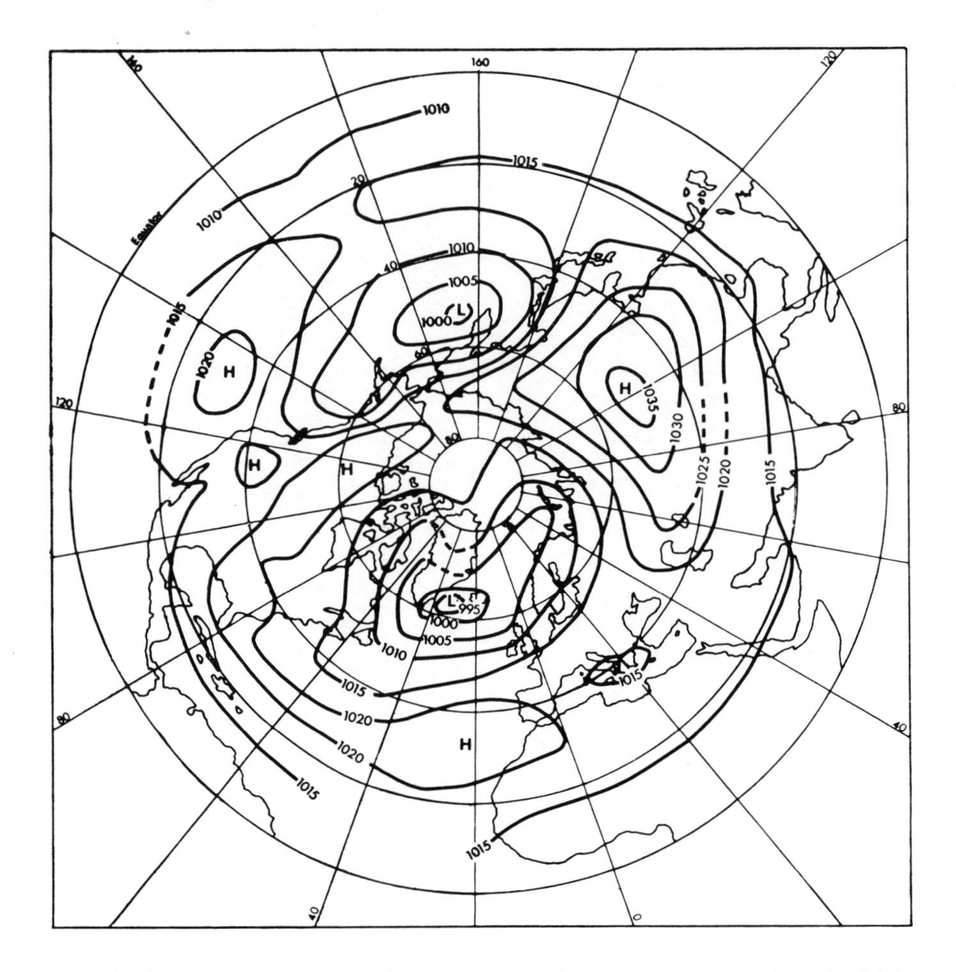

meaningless to speak of sea-level pressure but, on average, there is high pressure over the 3–4-km-high eastern Antarctic plateau.

The main circulation in the southern hemisphere is much more zonal at both 700 mb and sea-level than in the northern hemisphere, due to the limited area and effect of the southern land masses. There is also little difference between summer and winter circulation intensity (figs. 3.18 and 3.19). Indeed, it is important at this point to differentiate between mean pressure patterns and the highs and lows shown on synoptic weather maps. Thus, in the southern hemisphere, the zonality of the mean circulation conceals a high degree of day-to-day variability. The synoptic map is one which shows the principal pressure systems over a very large area at a given time; local wind features, for example, being ignored. The subpolar lows over Iceland and the Aleutians (fig. 3.24) shown on recurrent mean pressure maps represent the passage of deep depressions across these areas downstream of the upper long-wave troughs. The mean high-pressure areas, however, relate to more or less permanent highs. The intermediate areas, such as the zones about 50°–55°N and 40°–60°S, affected by travelling depressions and ridges of high pressure,

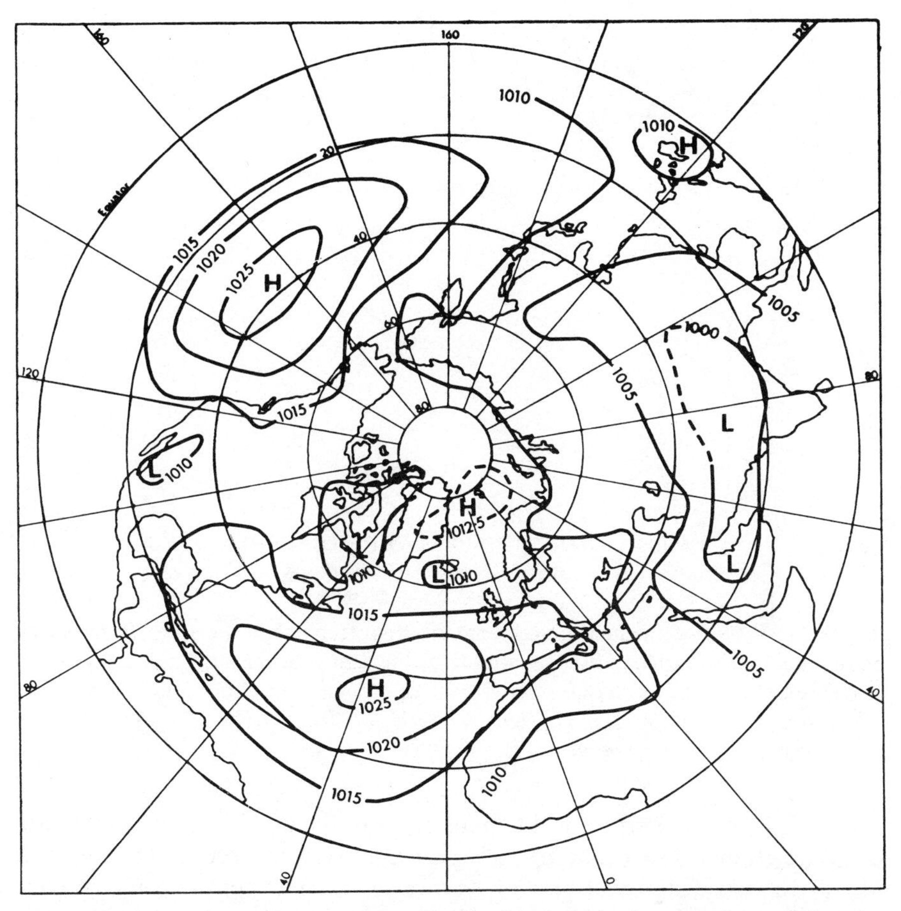

Fig. 3.24 The mean sea-level pressure distribution (mb) in January (opposite) and July (above) for the northern hemisphere, 1950–59 (*after O'Connor 1961*).

appear on the mean maps as being of neither markedly high nor markedly low pressure. The movement of depressions is considered in ch. 4, F.

On comparing the surface and tropospheric pressure distributions for January (figs. 3.19 and 3.24) it will be noticed that only the subtropical high-pressure cells extend to high levels. The reasons for this are evident from fig. 3.15B and D. In summer the equatorial low-pressure belt is also evident aloft over southern Asia. The subtropical cells are still discernible at 300 mb, showing them to be a fundamental feature of the global circulation and not merely a response to surface conditions.

E The global wind belts

One fact that emerges from the preceding discussion is the importance of the subtropical high-pressure cells. Dynamic, rather than immediately thermal, in origin, and situated between 20° and 30° latitude, they seem to provide the

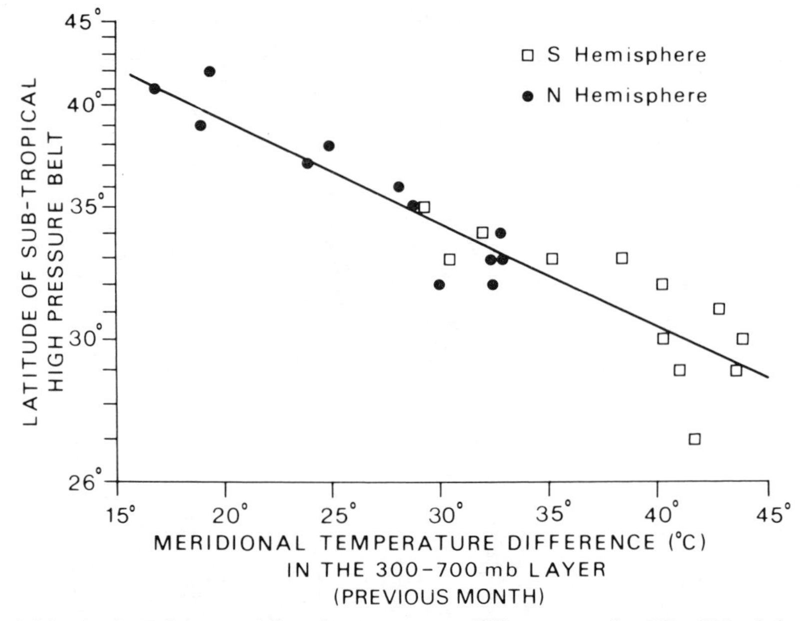

Fig. 3.25. A plot of the meridional temperature difference at the 300–700 mb level in the previous month against the latitude of the centre of the subtropical high-pressure belt, assuming a constant vertical tropospheric lapse rate (*after Flohn, in Flohn and Fantechi 1984*). (*Copyright © 1980/1982 by D. Reidel Publishing Company. Reprinted by permission*).

key to the world's surface wind circulation. In the northern hemisphere the pressure gradients surrounding these cells are strongest between October and April. In terms of actual pressure, however, oceanic cells experience their highest pressure in summer, the belt being counterbalanced at low levels by thermal low-pressure conditions over the continents. Their strength and persistence clearly mark them as the dominating factor which controls the position and activities of both the trades and westerlies.

1 The trade winds

The trades (or tropical easterlies) are important because of the great extent of their activity; they blow over nearly half the globe (fig. 3.27). They originate at low latitudes on the margins of the subtropical high-pressure cells, and their constancy of direction and speed (about 7 m s^{-1}) is remarkable. Trade winds, like the westerlies, are strongest during the winter half-year, which suggests they are both controlled by the same fundamental mechanism.

The two trade-wind systems tend to converge in the *Equatorial Trough* (of low-pressure). Over the oceans, particularly the central Pacific, the convergence of these air streams is often pronounced and in this sector the term *Inter-Tropical Convergence Zone* (ITCZ) is applicable. Generally, however, the convergence is discontinuous in space and time (pl. 12). Equatorward of

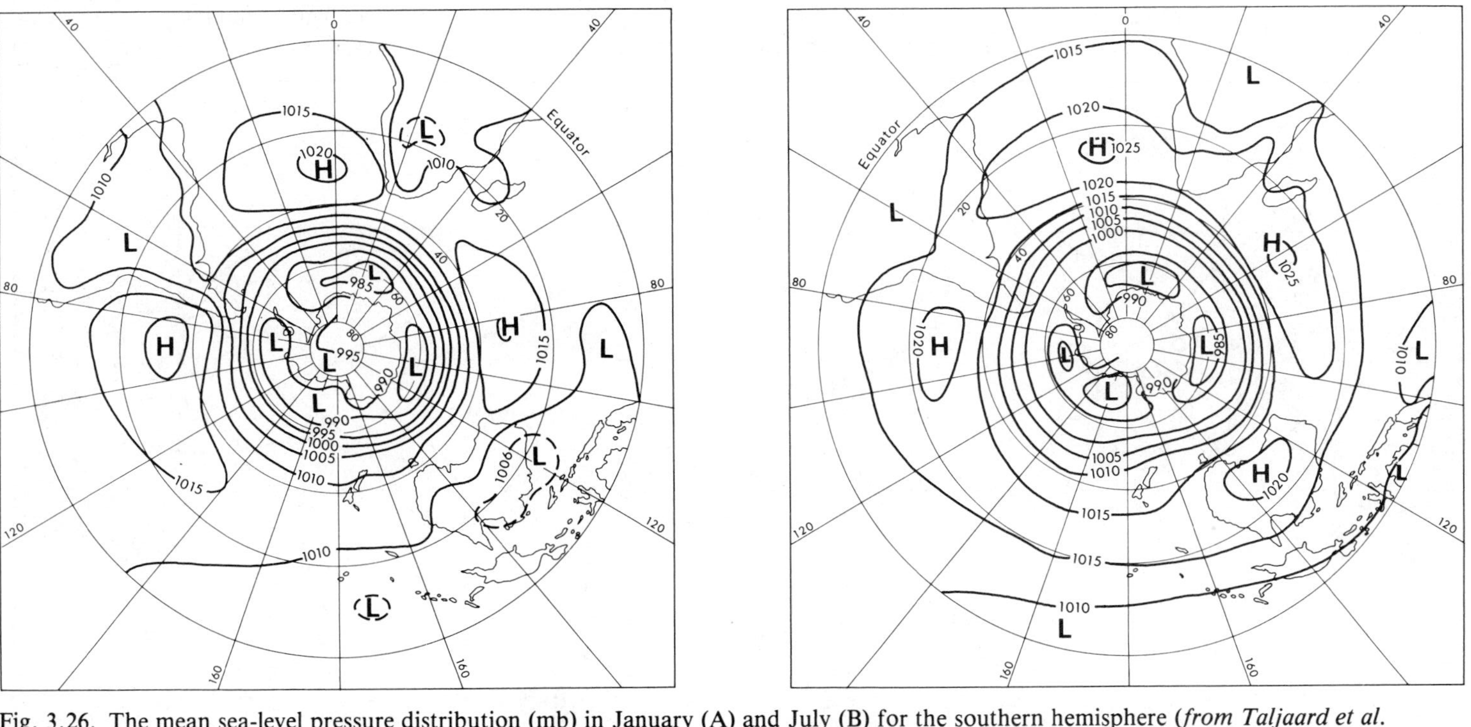

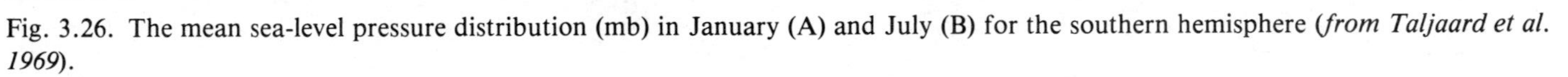
Fig. 3.26. The mean sea-level pressure distribution (mb) in January (A) and July (B) for the southern hemisphere (*from Taljaard et al. 1969*).

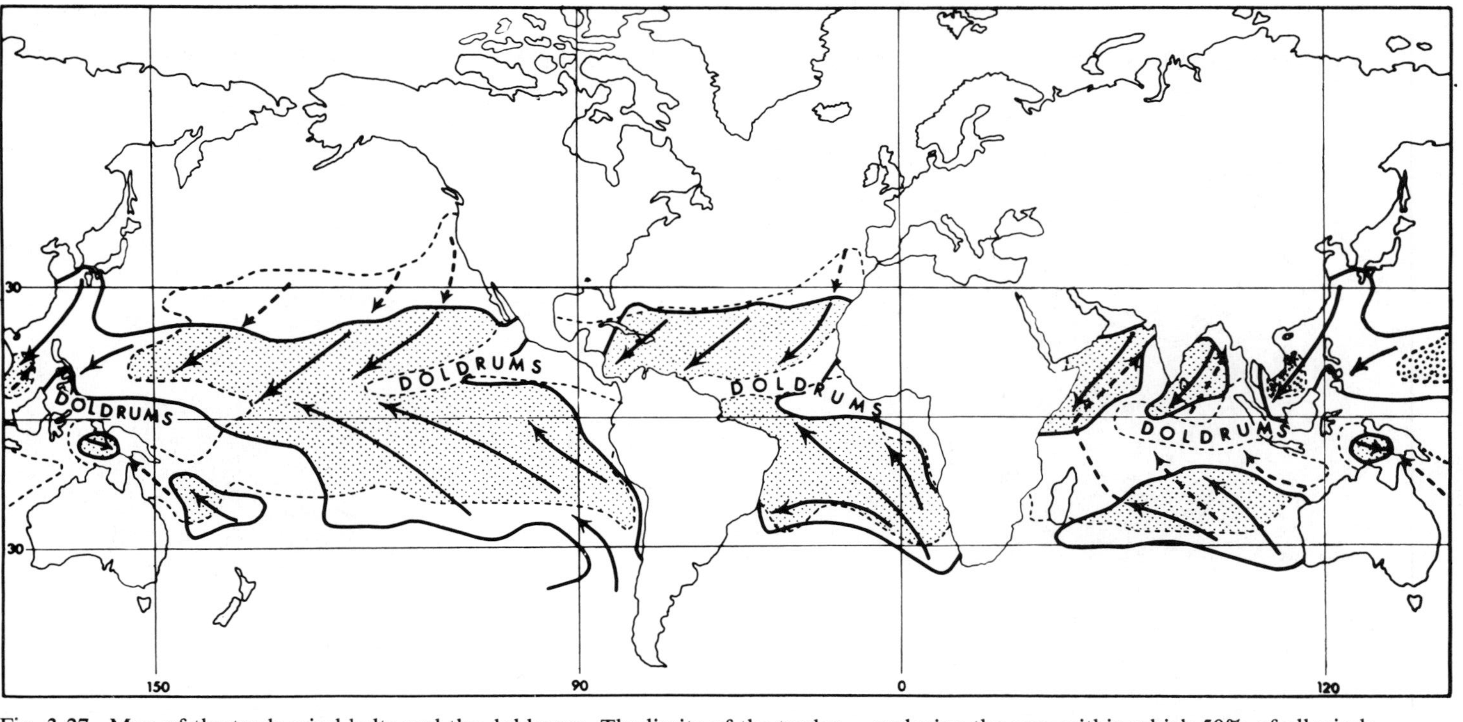

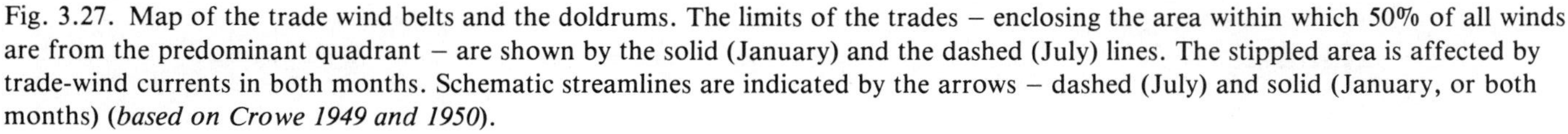

Fig. 3.27. Map of the trade wind belts and the doldrums. The limits of the trades – enclosing the area within which 50% of all winds are from the predominant quadrant – are shown by the solid (January) and the dashed (July) lines. The stippled area is affected by trade-wind currents in both months. Schematic streamlines are indicated by the arrows – dashed (July) and solid (January, or both months) (*based on Crowe 1949 and 1950*).

the main *root zones* of the trades over the eastern Pacific and eastern Atlantic are regions of light, variable winds, known traditionally as the *doldrums* and much feared in past centuries by the crews of sailing ships. Their seasonal extent varies considerably; from July to September they spread westward into the central Pacific while in the Atlantic they extend to the coast of Brazil. A third major doldrum zone is located in the Indian Ocean and western Pacific. In March–April it stretches 16,000 km from east Africa to 180° longitude and it is again very extensive during October–December.

2 The equatorial westerlies

In the summer hemisphere, and over continental areas especially, there is a zone of generally westerly winds intervening between the two trade-wind belts (fig. 3.28). This westerly system is well marked over Africa and southern Asia in the northern hemisphere summer, when thermal heating over the continents assists the northward displacement of the Equatorial Trough (fig. 3.27). Over Africa the westerlies reach to 2–3 km and over the Indian Ocean to 5–6 km. In Asia these winds are known as the 'Summer Monsoon' but this is now recognized to be a complex phenomenon the cause of which is partly global and partly regional in origin (see ch. 6, D). The equatorial westerlies are not simply trades of the opposite hemisphere which recurve (due to the changed direction of the Coriolis deflection) on crossing the equator, since there is *on average* a westerly component in the Indian Ocean at 2°–3°S in June and July and at 2°–3°N in December and January. Over the Pacific and Atlantic Oceans the ITCZ does not shift sufficiently far from the equator to permit the development of this westerly wind belt.

3 The mid-latitude (Ferrel) westerlies

These are the winds of the mid-latitudes emanating from the poleward sides of the subtropical high-pressure cells. They are far more variable than the trades both in direction and intensity, for in these regions the path of air movement is frequently affected by cells of low and high pressure which travel generally eastwards within the basic flow (pl. 1). Also in the northern hemisphere the preponderance of land areas with their irregular relief and changing seasonal pressure patterns tend to obscure the generally westerly airflow. The Scilly Islands, lying in the south-westerlies, record 46% of winds from between south-west and north-west, but fully 29% from the opposite sector between north-east and south-east.

The westerlies of the southern hemisphere are stronger and more constant in direction than those of the northern hemisphere because the broad expanses of ocean rule out the development of stationary pressure systems (fig. 3.29). Kerguelen Island (49°S, 70°E) has an annual frequency of 81% of winds from between south-west and north-west and the comparable figure of 75%

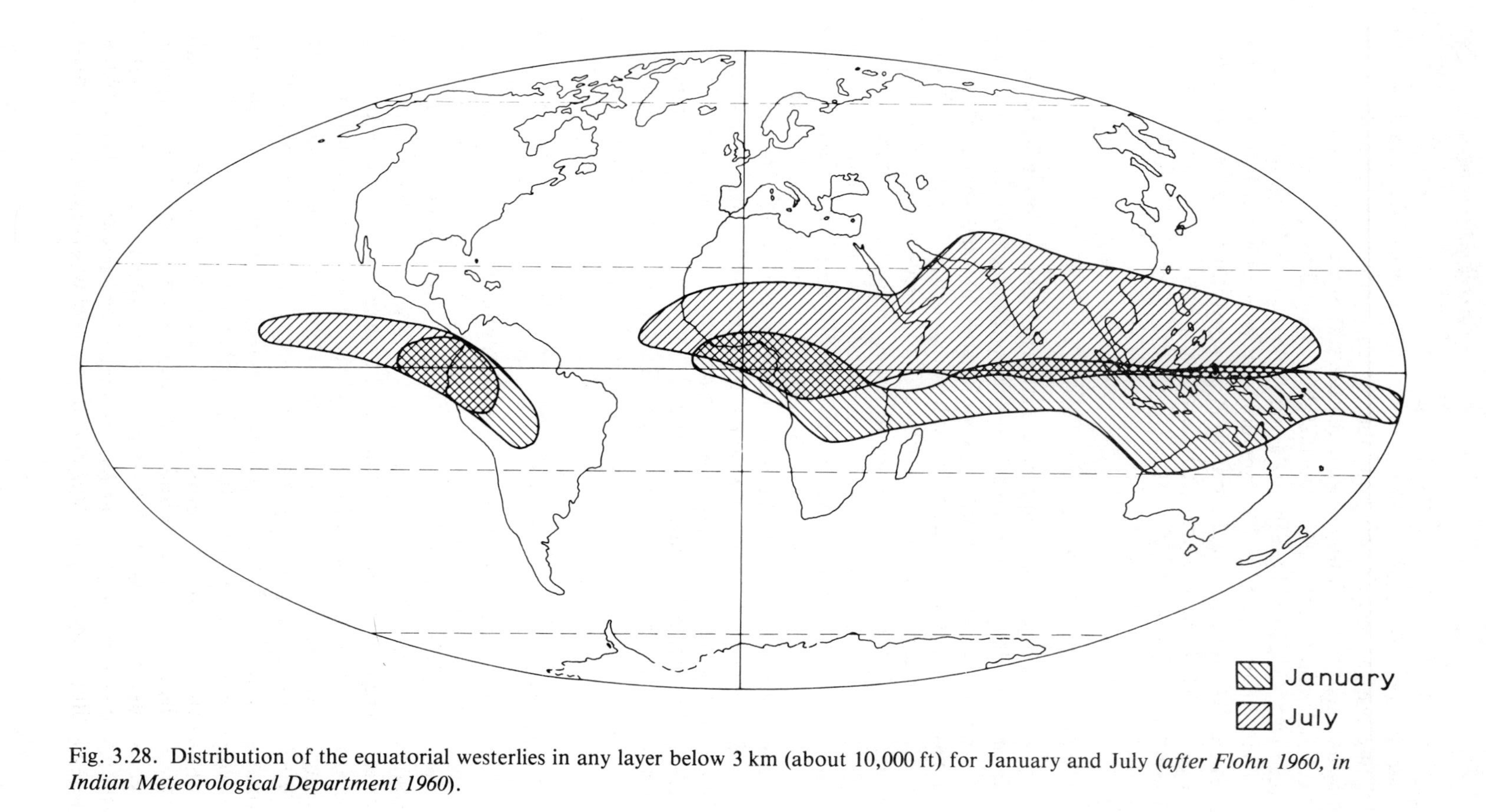

Fig. 3.28. Distribution of the equatorial westerlies in any layer below 3 km (about 10,000 ft) for January and July (*after Flohn 1960, in Indian Meteorological Department 1960*).

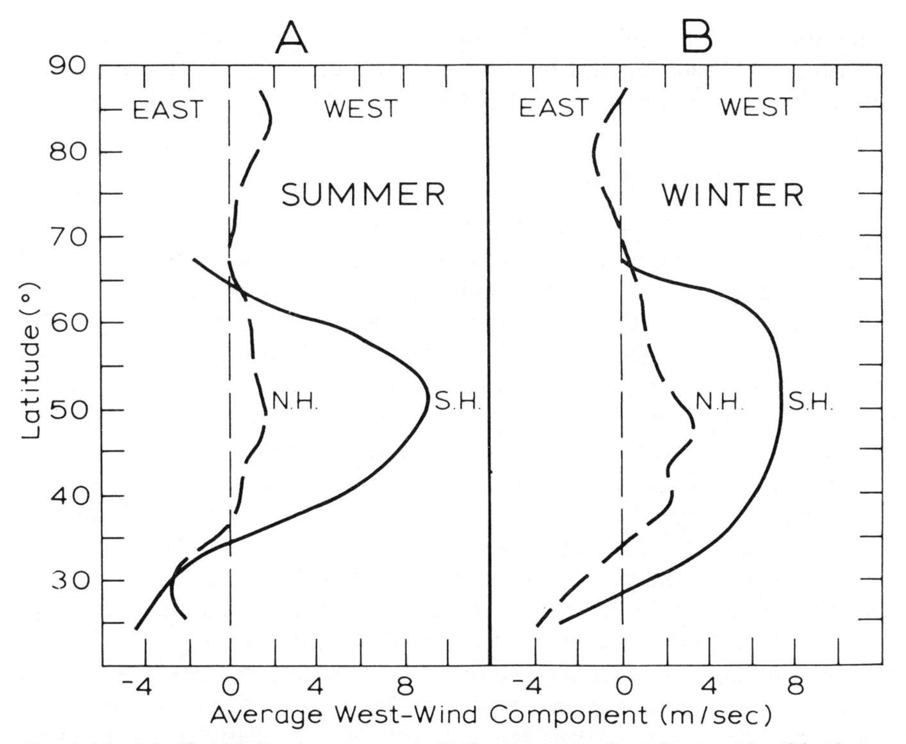

Fig. 3.29. Profiles of the average west-wind component (m s^{-1}) at sea-level in the northern and southern hemispheres during their respective summer (A) and winter (B) seasons (*after Van Loon 1964*).

for Macquarie Island (54°S, 159°E) shows that this predominance is widespread over the southern oceans. However, the apparent zonality of the southern circumpolar vortex (fig. 3.26) conceals considerable synoptic variability of wind velocity.

4 The polar easterlies

This term is applied to winds that are supposed to occur between a polar high pressure and the belt of low pressure of the higher mid-latitudes. The polar high, as has already been pointed out, is by no means a quasi-permanent feature of the arctic circulation. Easterly winds occur mainly on the poleward sides of depressions over the North Atlantic and North Pacific and if average wind directions are calculated for entire latitude belts in high latitudes there is found to be little sign of a coherent system of polar easterlies. The situation in high latitudes of the southern hemisphere is complicated by the presence of Antarctica, but anticyclones appear to be frequent over the high plateau of eastern Antarctica and easterly winds prevail over the Indian Ocean sector of the Antarctic coastline. For example, in 1902–3 the expedition ship *Gauss* at 66°S, 90°E observed winds between north-east and south-east for 70% of the time, and at many coastal stations the constancy of easterlies may be compared with that of the trades. However, westerly components predominate over the sea areas off west Antarctica.

F The general circulation

The observed patterns of wind and pressure prompt consideration of the mechanisms maintaining the *general circulation* of the atmosphere – the large-scale patterns of wind and pressure which persist throughout the year or recur seasonally. Reference has already been made to one of the primary driving forces, the imbalance of radiation between lower and higher latitudes (see ch. 1, G.1), but it is important also to appreciate the significance of energy transfers in the atmosphere. Energy is continually undergoing changes of form as shown schematically in fig. 3.30. Unequal heating of the earth and its atmosphere by solar radiation generates potential energy, some of which is converted into kinetic energy by the rising of warm air and the sinking of cold air. Ultimately, the kinetic energy of atmospheric motion on all scales is dissipated by friction and small scale turbulent eddies (i.e. internal viscosity). In order to maintain the general circulation, the rate of generation of kinetic energy must obviously balance its rate of dissipation. These rates are estimated to be about 2 W m^{-2}, which amounts only to some 1% of the average global solar radiation absorbed at the surface and in the atmosphere. In other words the atmosphere is a highly inefficient heat engine (see ch. 1, G).

A second controlling factor is the angular momentum of the earth and its atmosphere. This is the tendency for the earth's atmosphere to move, with the earth, around the axis of rotation. Angular momentum is proportional to the rate of spin (that is the angular velocity) and the square of the distance of the air parcel from the axis of rotation. With a uniformly rotating earth and atmosphere, the total angular momentum must remain constant (in other words, there is a *conservation of angular momentum*). If, therefore, a large mass of air changes its position on the earth's surface such that its distance from the axis of rotation also changes, then its angular velocity must change

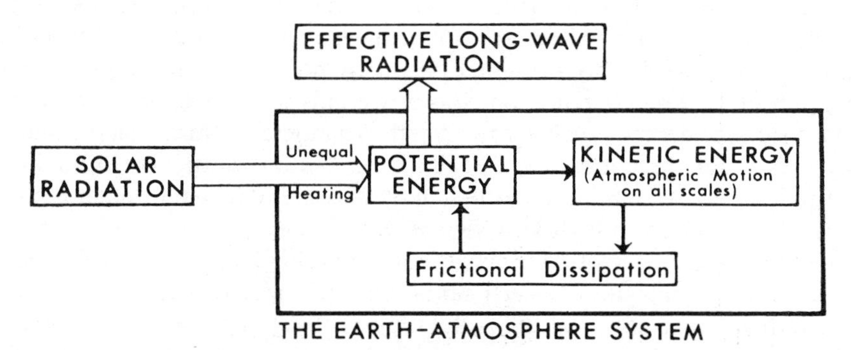

Fig. 3.30. Schematic changes of energy involving the earth-atmosphere system.

in a manner so as to allow the angular momentum to remain constant. Naturally absolute angular momentum is high at the equator (see note 6) and decreases with latitude to become zero at the pole (that is, the axis of rotation), so air moving poleward tends to acquire progressively higher eastward velocities. For example, air travelling from 42° to 46° latitude and conserving its angular momentum, would increase its speed relative to the earth's surface by 29 m s^{-1}. This is the same principle which causes an ice skater to spin more violently when her arms are progressively drawn into the body. In practice this increase of air-mass velocity is countered or masked by the other forces affecting air movement (particularly friction), but there is no doubt that many of the important features of the general atmospheric circulation result from this poleward transfer of angular momentum.

The necessity for a poleward momentum transport is readily appreciated in terms of the maintenance of the mid-latitude westerlies. These winds continually impart westerly (eastward) relative momentum to the earth by friction and it has been estimated that they would cease altogether due to this frictional dissipation of energy in little over a week if their momentum were not continually replenished from elsewhere. In low latitudes the extensive tropical easterlies are gaining westerly relative momentum by friction, as a result of the earth rotating in a direction opposite to their flow (see note 7), and this excess is transferred polewards with the maximum poleward transport occurring, significantly, in the vicinity of the mean subtropical jet stream at about 250 mb at 30°N and 30°S.

1 Circulations in the vertical and horizontal planes

There are two possible ways in which the atmosphere can transport heat and momentum. One is by circulation in the vertical plane as indicated in fig. 3.31 which shows three meridional cells. The low-latitude (or Hadley) cell and its counterpart in the southern hemisphere were considered to be analogous to the convective circulations set-up when a pan of water is heated over a flame and are referred to as *thermally direct* cells. Warm air near the

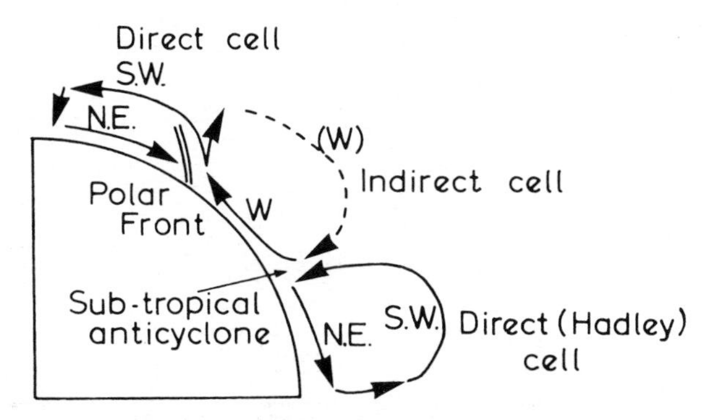

Fig. 3.31. Three-cell model of the northern hemisphere meridional circulation (*after Rossby 1941; from Barry 1967*).

equator was thought to rise and generate a low-level flow towards the equator, the earth's rotation deflecting these currents which thus form the north-east and south-east trades. This explanation was put forward by G. Hadley in 1735, although in 1856 W. Ferrel pointed out that the conservation of angular momentum would be a more effective factor in causing easterlies because the Coriolis force is small in low latitudes. The low-latitude cell, according to the above scheme, would be completed by poleward counter-currents aloft with the air sinking at about 30° latitude as it is cooled by radiation. However, this scheme is not entirely correct since the atmosphere does not have a simple heat source at the equator, the trades are not continuous around the globe (fig. 3.27) and poleward upper flow is restricted mainly to the western ends of the subtropical high-pressure cells aloft (see figs. 3.19 and 3.26).

Figure 3.31 shows another thermally direct cell in high latitudes with cold dense air flowing out from a polar high pressure. The reality of this is doubtful, but it is in any case of limited importance to the general circulation in view of the small mass involved. It is worth noting at this point that a single direct cell in each hemisphere is not possible, because the easterly winds near the surface would slow down the earth's rotation. On average the atmosphere must rotate with the earth, requiring a balance between easterly and westerly winds over the globe.

The mid-latitude cell in fig. 3.31 is thermally indirect and it would need to be driven by the other two. Momentum considerations indicate the necessity for upper easterlies in such a scheme, yet observations with upper-air balloons during the 1930s and 1940s demonstrated the existence of strong westerlies in the upper troposphere (see ch. 3, D.3). Rossby modified the three-cell model to incorporate this fact, proposing that westerly momentum

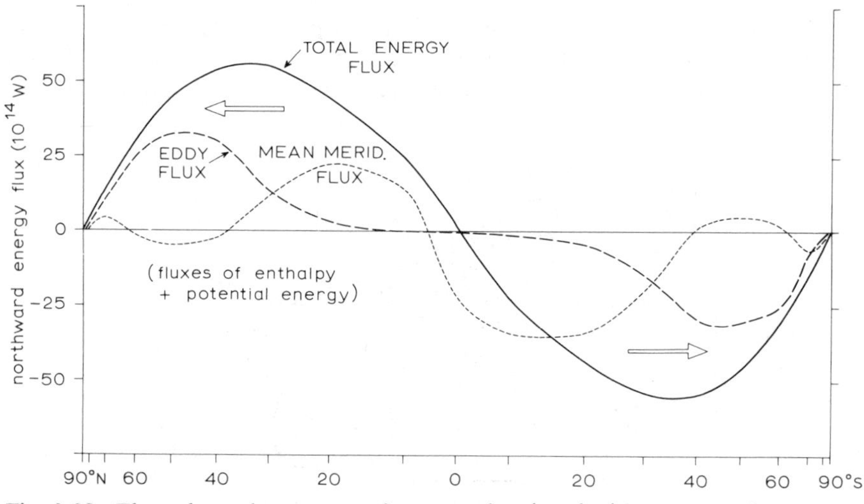

Fig. 3.32. The poleward transport of energy, showing the importance of horizontal eddies.

was transferred to middle latitudes from the upper branches of the cells in high and low latitudes. Such horizontal mixing could, for example, be accomplished by troughs and ridges in the upper flow.

These views underwent radical amendment from about 1948 onwards. The alternative means of transporting heat and momentum – by horizontal circulations – had been suggested in the 1920s by A. Defant and H. Jeffreys but could not be tested until adequate upper-air data became available. Calculations for the northern hemisphere by V. P. Starr and R. M. White at Massachusetts Institute of Technology showed that in middle latitudes horizontal cells transport most of the required heat and momentum polewards. This operates through the mechanism of the quasi-stationary highs and the travelling highs and lows near the surface acting in conjunction with their related wave patterns aloft. The importance of such horizontal eddies for energy transport is shown in fig. 3.32 (see also fig. 1.31B). The modern concept of the general circulation therefore views the energy of the zonal winds as being derived from travelling waves, not from meridional circulations. In lower latitudes, however, this mechanism may be insufficient by itself to account for the total energy transport estimated to be necessary for energy balance. For such reasons the mean Hadley cell still features in current representations of the general circulation, as fig. 3.33 shows, but the low-latitude circulation is recognized as being complex. In particular, the vertical heat transport in the Hadley cell is effected by giant cumulonimbus clouds in disturbance systems associated with the Equatorial Trough (of low pressure), which is located on average at 5°S in January and at 10°N in July (see ch. 6, B). The Hadley cell of the winter hemisphere is by far the most important and it gives rise to low-level transequatorial flow into the summer hemisphere. The traditional model with twin cells, symmetrical about the equator, is found only in spring/autumn. Longitudinally, the Hadley cells are linked with the monsoon regimes of the summer hemisphere. Rising air over southern Asia (and also South America and Indonesia) is associated with east-west (zonal) outflow and these are systems known as 'Walker circulations'. The poleward return transport of the meridional Hadley cells

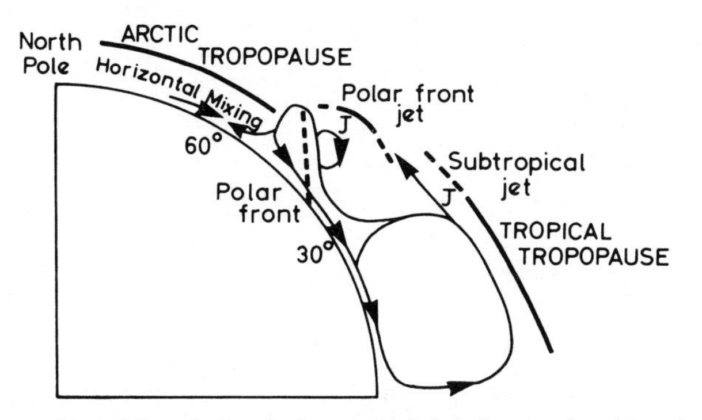

Fig. 3.33. General meridional circulation model for the northern hemisphere in winter (*after Palmén 1951; from Barry 1967*).

takes place in troughs which extend into low latitudes from the mid-latitude westerlies. This tends to occur at the western ends of the upper tropospheric subtropical high pressure cells (see fig. 3.17). Horizontal mixing predominates in middle and high latitudes although it is also thought that there is a weak indirect mid-latitude cell in much reduced form (fig. 3.33). The relationship of the jet streams to regions of steep meridional temperature gradient has already been noted (see fig. 3.21). A complete explanation of the two wind maxima and their role in the general circulation is still lacking, but they undoubtedly form an essential part of the story.

In the light of these theories the origin of the subtropical anticyclones which play such an important role in the world's climates may be re-examined. Their existence has been variously ascribed to the piling-up of poleward-moving air as it is increasingly deflected eastwards through the earth's rotation and the conservation of angular momentum; to the sinking of poleward currents aloft by radiational cooling; to the general necessity for high pressure near 30° latitude separating approximately equal zones of east and west winds; or to combinations of such mechanisms. An adequate theory must account not only for their permanence but also for their cellular nature and the vertical inclination of the axes. The preceding discussion shows that ideas of a simplified Hadley cell and momentum conservation are only partially correct. Moreover, recent studies rather surprisingly show no relationship, on a seasonal basis, between the intensity of the Hadley cell and that of the subtropical highs.

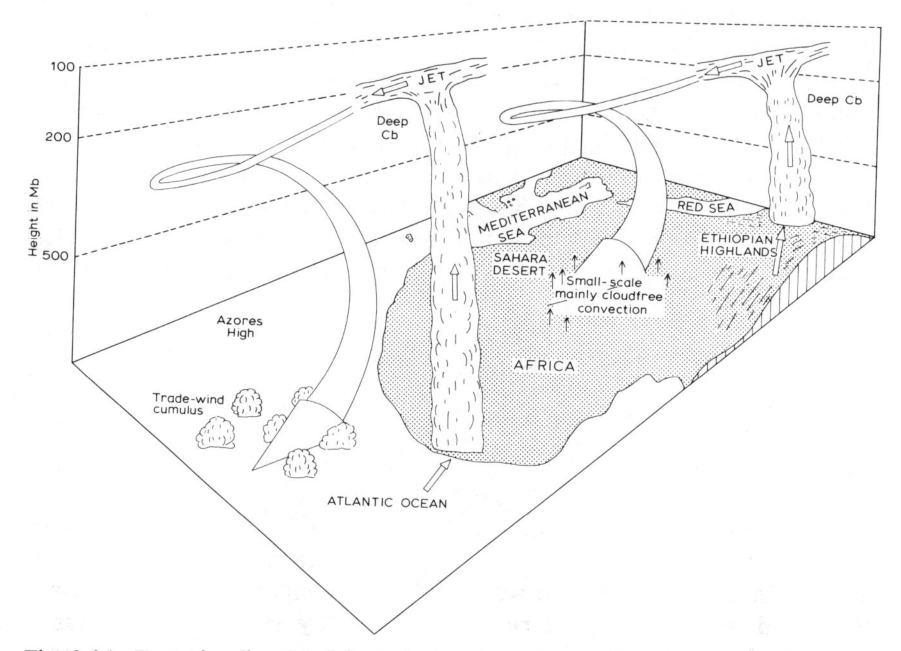

Fig. 3.34. Tentative flow model relating summer convection, the easterly jet stream and high-pressure subsidence over northern Africa and the eastern North Atlantic (*adapted from Walker 1972; Crown Copyright Reserved*).

It is probable that the high-level anticyclonic cells which are evident on *synoptic* charts (these tend to merge on mean maps) are related to anticyclonic eddies on the equatorward side of jet streams. Theoretical and observational studies show that, as a result of the latitudinal variation of the Coriolis parameter, cyclones in the westerlies tend to move poleward and anticyclonic cells equatorward. Hence the subtropical anticyclones are constantly regenerated. There is a statistical relationship between the latitude of the subtropical highs and the mean meridional temperature gradient (see fig. 3.25); a stronger gradient causes an equatorward shift of the high pressure, and vice versa. This shift is evident on a seasonal basis. The cellular pattern at the surface clearly reflects the influence of heat sources. The cells are stationary and elongated north–south over the northern hemisphere oceans in summer when continental heating creates low pressure and also the meridional temperature gradient is weak. In winter, on the other hand, the zonal flow is stronger in response to a greater meridional temperature gradient and continental cooling produces east–west elongation of the cells. Undoubtedly surface and high-level factors reinforce one another in some sectors and tend to cancel out in others. Indeed, it has been suggested that the Azores high-pressure cell, in particular, owes part of its summer intensification and its tendency to extend eastward to air masses which rise locally in areas of high monsoonal rainfall over Africa, enter the tropical easterly jet stream circulation (see ch. 6, p. 314) and then subside over the western Sahara and the eastern North Atlantic (fig. 3.34).

2 *Variations in the circulation of the northern hemisphere*

The pressure and contour patterns during certain periods of the year may be radically different from those indicated by the mean maps (see figs. 3.19, 3.36 and 3.37). These variations, of 3 to 8 weeks' duration, occur irregularly but are rather more noticeable in the winter months when the general circulation is strongest. The nature of the changes is illustrated schematically in fig. 3.35. The zonal westerlies over middle latitudes develop waves and the troughs and ridges become accentuated, ultimately splitting up into a cellular pattern with pronounced meridional flow at certain longitudes. The strength of the westerlies between 35° and 55°N is termed the *zonal index*; strong zonal westerlies are representative of high index and marked cellular patterns occur with low index (pl. 17). A relatively low index may also occur if the westerlies are well south of their usual latitudes and, paradoxically, such expansion of the zonal circulation pattern is associated with strong westerlies in lower latitudes than usual. Figures 3.36 and 3.37 illustrate the mean 700 mb contour patterns and zonal wind speed profiles for two contrasting months. In December 1957 the westerlies were stronger than normal north of 40°N and the troughs and ridges were weakly developed, whereas in February 1958 there was low zonal index and an expanded circumpolar vortex, giving rise to strong low-latitude westerlies. The 700 mb pattern shows very weak subtropical highs, deep meridional troughs and blocking

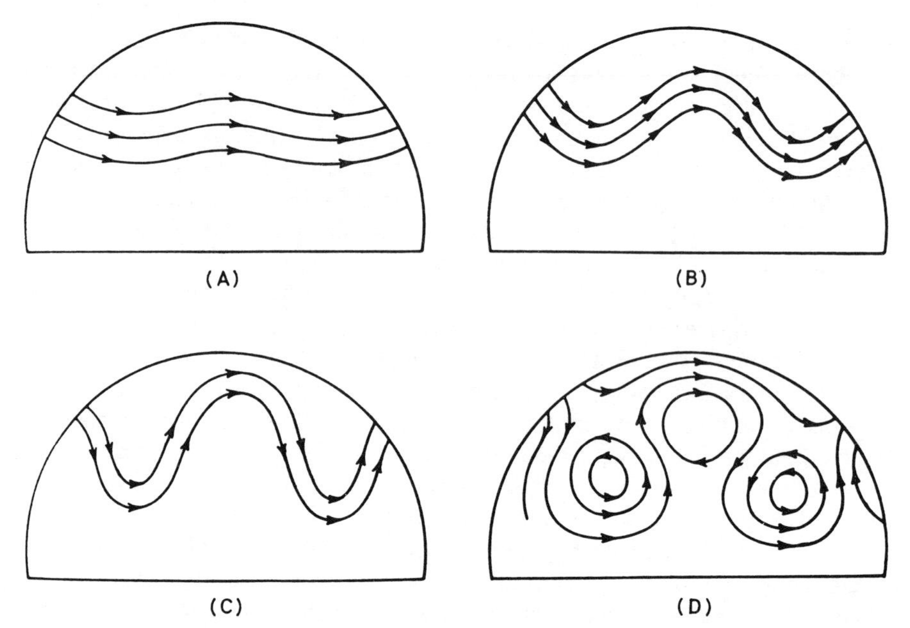

Fig. 3.35. The index cycle. A schematic illustration of the development of cellular patterns in the upper westerlies, usually occupying 3–8 weeks and being especially active in February and March in the northern hemisphere. Statistical studies indicate no regular periodicity in this sequence. A High zonal index. The jet stream and the westerlies lie north of their mean position. The westerlies are strong, pressure systems have a dominantly east–west orientation, and there is little north–south air-mass exchange. B and C The jet expands and increases in velocity, undulating with increasingly larger oscillations. D Low zonal index. Complete break-up and cellular fragmentation of the zonal westerlies. Formation of stationary deep occluding cold depressions in lower mid-latitudes and deep warm blocking anticyclones at higher latitudes. This fragmentation commonly begins in the east and extends westward at a rate of about 60° of longitude per week (*after Namias; from Haltiner and Martin 1957*).

anticyclone off Alaska (see fig. 3.35D). The cause of these variations is still uncertain although it would appear that fast zonal flow is unstable and tends to break down. This tendency is certainly increased in the northern hemisphere by the arrangement of the continents and oceans.

Detailed studies are now beginning to show that the irregular index fluctuations, together with secondary circulation features, such as cells of low and high pressure at the surface or long waves aloft, play a major role in redistributing momentum and energy. Laboratory experiments with rotating 'dishpans' of water to simulate the atmosphere, and computer studies using numerical models of the atmosphere's behaviour, demonstrate that a Hadley circulation cannot provide an adequate mechanism for transporting heat poleward. In consequence, the meridional temperature gradient increases and eventually the flow becomes unstable in the Hadley mode, breaking down into a number of cyclonic and anticyclonic eddies. This phenomenon is referred to as *baroclinic instability*. In energy terms, the potential energy in

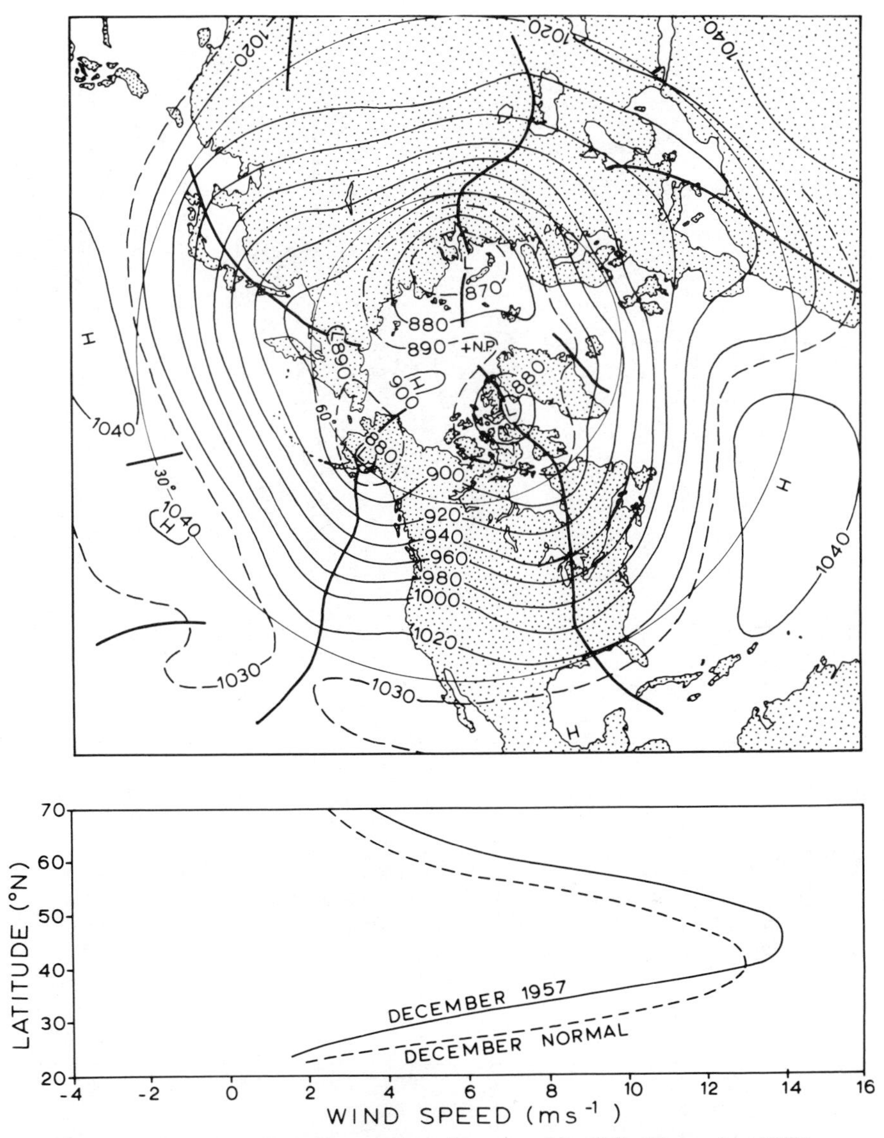

Fig. 3.36. *Above* Mean 700 mb contours (in tens of feet) for December 1957, showing a fast, westerly, small-amplitude flow typical of a high zonal index. *Below* Mean 700 mb zonal wind speed profiles (m s^{-1}) in the Western Hemisphere for December 1957, compared with those of a normal December. The westerly winds were stronger than normal and displaced to the north (*after Monthly Weather Review 85, 1957, 410–11*).

the zonal flow is converted into potential and kinetic energy of the eddies. It is also now known that the kinetic energy of the zonal flow is derived *from* the eddies, the reverse of the classical picture which viewed the disturbances within the global wind belts as superimposed detail. The significance of atmospheric disturbances and the variations of the circulation are becoming

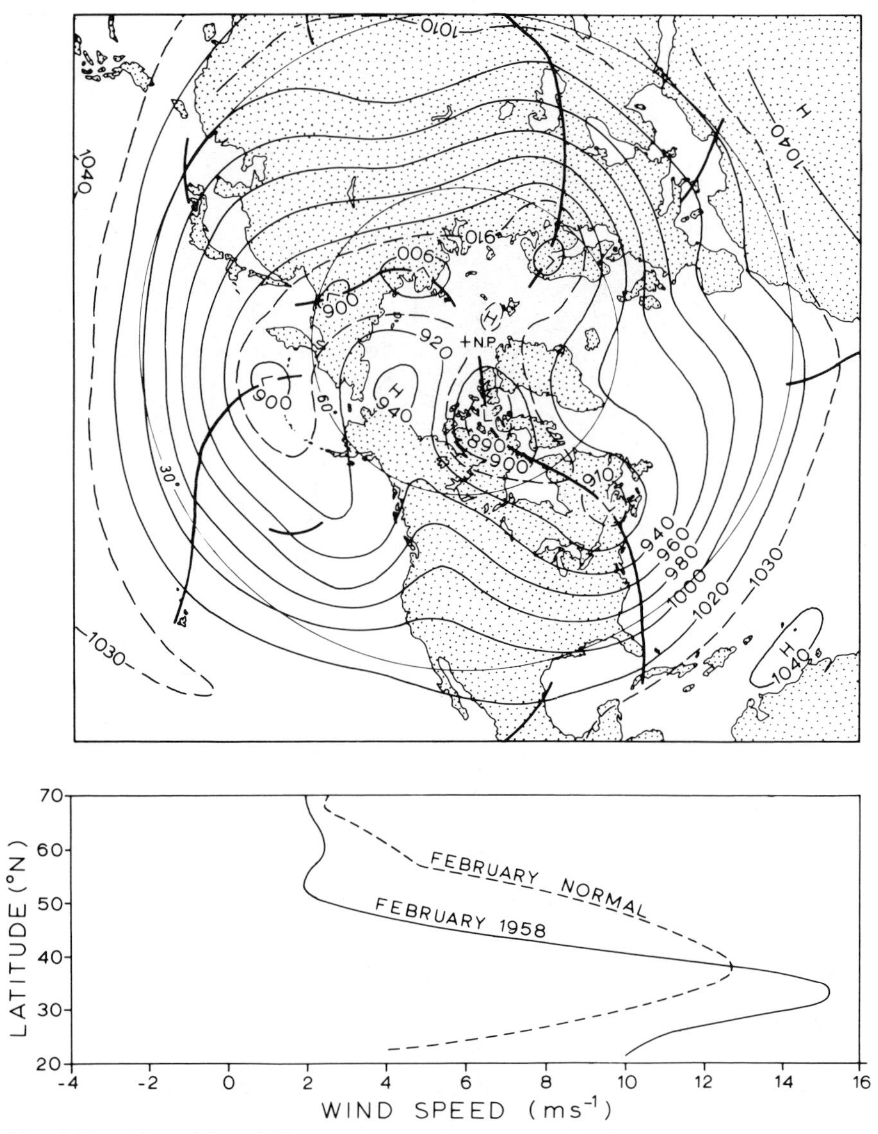

Fig. 3.37. *Above* Mean 700 mb contours (in tens of feet) for February 1958. *Below* Mean 700 mb zonal wind speed profiles (m s^{-1}) in the Western Hemisphere for February 1958, compared with those of a normal February. The westerly winds were stronger than normal at low latitudes, with a peak at about 33°N (*after Monthly Weather Review 86, 1958, 62–3*).

increasingly evident. The mechanisms of the circulation are, however, greatly complicated by numerous interactions and *feedback* processes, one of the most important of which involves the oceanic circulation as outlined below. The significance of interactions between the oceanic and atmospheric heat and moisture budgets has already been discussed in ch. 1, G, and ch. 2, A.

3 The circulation of the ocean surface

The most obvious feature of the surface oceanic circulation is the control exercised over it by the low-level planetary wind circulation, especially by the subtropical oceanic high-pressure circulations and the westerlies. The oceanic circulation even partakes of the seasonal reversals of flow in the monsoonal regions of the northern Indian Ocean, off east Africa and off northern Australia (fig. 3.38). The Ekman effect (see ch. 3, A.5) causes the flow to be increasingly deflected to the right (in the northern hemisphere) and to decrease in velocity as the influence of the wind stress diminishes with depth. However, the rate of change of flow direction with depth increases with latitude, such that near the equator there are no flow reversals at depth which are characteristic of higher latitudes. The depth at which this reversal occurs decreases poleward, but averages about 50 m over large areas of the ocean. In addition, as water moves meridionally the conservation of angular momentum implies changes in relative vorticity (see p. 123 and p. 150), with poleward-moving currents acquiring anticyclonic vorticity and equatorward-moving currents acquiring cyclonic vorticity.

Equatorward of the subtropical high-pressure cells the persistent trade winds generate the broad North and South Equatorial Currents (fig. 3.38). On the western sides of the oceans most of this water swings poleward with the airflow and thereafter increasingly comes under the influence of the Ekman deflection and of the anticyclonic vorticity effect. However, some water tends to pile up near the equator on the western sides of oceans, partly because here the Ekman effect is virtually absent with little poleward deflection and no reverse current at depth. To this is added some of the water which is displaced northward into the equatorial zone by the especially active subtropical high-pressure circulations of the southern hemisphere. This accumulated water flows back eastward down the hydraulic gradient as compensating narrow surface Equatorial Counter Currents, unimpeded by the weak surface winds. As the circulations swing poleward round the western margins of the oceanic subtropical high-pressure cells there is the tendency for water to pile up against the continents giving, for example, an appreciably higher sea-level in the Gulf of Mexico than that along the Atlantic coast of the United States. This accumulated water cannot escape by sinking because of its relatively high temperature and resulting vertical stability, and it consequently continues poleward in the dominant direction of surface airflow. As a result of this movement the current gains anticyclonic vorticity which reinforces the similar tendency imparted by the winds, leading to relatively narrow currents of high velocity (for example, the Kuro Shio, Brazil, Mozambique-Agulhas and, to a less-marked extent, the East Australian Current). In the North Atlantic the configuration of the Caribbean Sea and Gulf of Mexico especially favours this pile-up of water, which is released poleward through the Florida Straits as the particularly narrow and fast Gulf Stream.

These poleward currents are opposed both by their friction with the

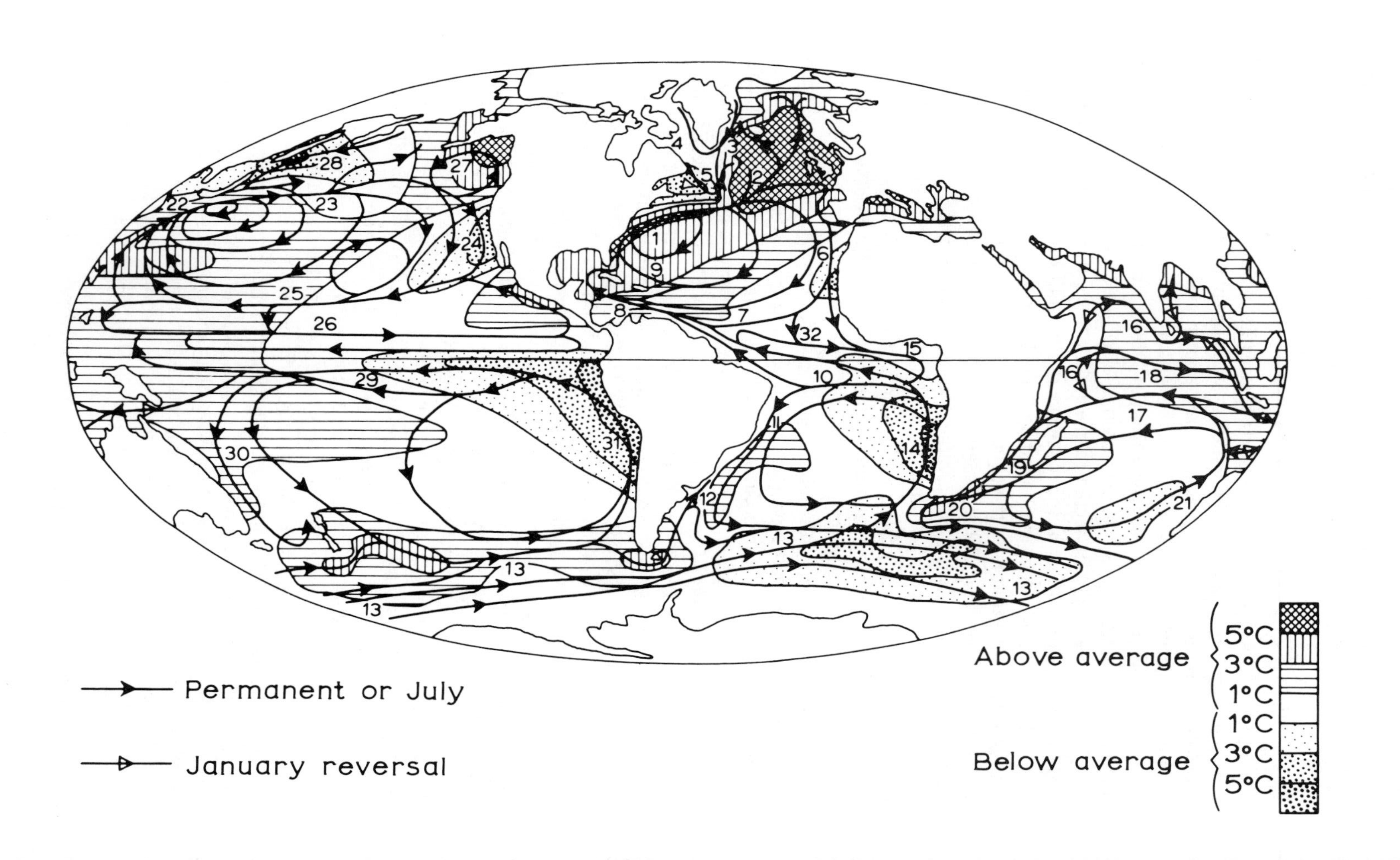
Permanent or July
January reversal
Above average
5°C
3°C
1°C
Below average
1°C
3°C
5°C

nearby continental margins and by energy losses due to turbulent diffusion, such as those accompanying the formation and cutting off of meanders in the Gulf Stream. On the poleward sides of the subtropical high-pressure cells westerly currents dominate, and where they are unimpeded by land masses in the southern hemisphere they form the broad and swift West Wind Drift. In the northern hemisphere a great deal of the eastward-moving current in the Atlantic swings northward, leading to very anomalously high sea temperatures, and is compensated for by a southward flow of cold arctic water at depth. However, more than half of the water mass comprising the North Atlantic Drift, and almost all that of the North Pacific Drift, swings south round the east sides of the subtropical high-pressure cells, forming the Canary and California Currents. Their southern-hemisphere equivalents are the Benguela, Humboldt or Peru, and West Australian Currents. In contrast with the currents on the west sides of the oceans, these currents acquire cyclonic vorticity which is in opposition to the anticyclonic wind tendency, leading to relatively broad flows of low velocity. In addition the deflection due to the Ekman effect causes the surface water to move westward away from the coasts, leading to upwelling of cold water from depths of 100–300 m. Although the band of upwelling may be quite narrow (about 200 km wide for the Benguela Current) the Ekman effect spreads this cold water westward. On the poleward margins of these cold-water coasts the meridional swing of the wind belts imparts a strong seasonality to the upwelling, the California Current up-welling, for example, being particularly well marked during the period March–July.

G Modelling the atmospheric circulation and climate

Better understanding of the complex behavior of the atmosphere and climate processes has been obtained in recent years through numerical modelling

---

Fig. 3.38. The general ocean current circulation of the globe, showing the mean temperature anomalies of surface ocean temperatures.

1 Gulf Stream
2 North Atlantic Drift
3 East Greenland Current
4 West Greenland Current
5 Labrador Current
6 Canary Current
7 North Equatorial Current
8 Caribbean Current
9 Antilles Current
10 South Equatorial Current
11 Brazil Current
12 Falkland Current
13 West Wind Drift
14 Benguela Current
15 Guinea Current
16 South-west and North-east Monsoon Drift
17 South Equatorial Current
18 Equatorial Counter Current
19 Mozambique Current
20 Agulhas Current
21 West Australian Current
22 Kuro Shio Current
23 North Pacific Drift
24 California Current
25 North Equatorial Current
26 Equatorial Counter Current
27 Alaska Current
28 Kamchatka Current
29 South Equatorial Current
30 East Australian Current
31 Peru or Humboldt Current
32 Equatorial Counter Current

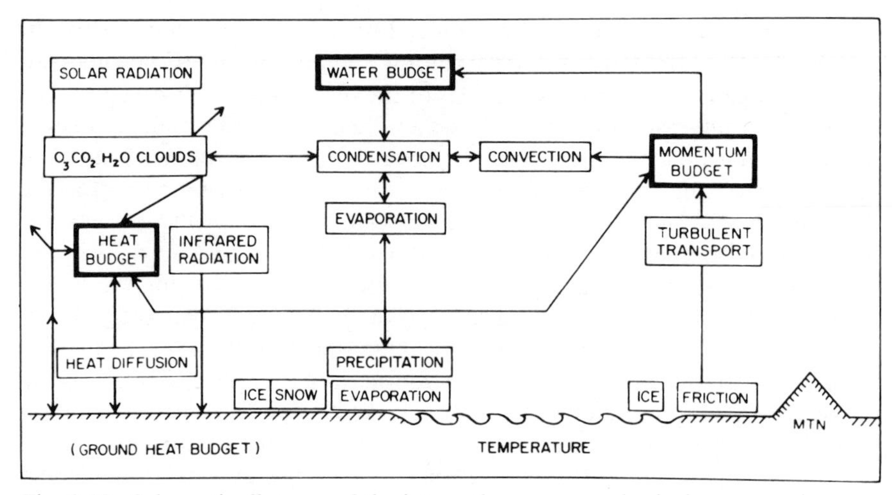

Fig. 3.39. Schematic diagram of the interactions among physical processes in a general circulation model (*from Druyan et al. 1975*).

studies. Here we can only sketch the essential features of this approach. There are three basic categories of model: the general circulation model (GCM), the energy budget model (EBM) and the radiative-convective model (RCM).

In the GCM, all dynamic and thermodynamic processes and the radiative and mass exchanges that have been treated in chs. 1–3 are modelled using five basic sets of equations. These account for the horizontal and vertical components of motion (ch. 3A, B), the conservation of energy (ch. 3, F) and of water substance (ch. 2), the continuity of mass (p. 123) and the equation of state (p. 8). In addition, radiative processes, including a diurnal and seasonal cycle, surface friction and cloud formation are usually represented. These are coupled in the manner shown schematically in fig. 3.39. Beginning with a set of initial atmospheric conditions, the equations are solved repeatedly for time steps of a few minutes at a large number of grid points over the earth and at several levels in the atmosphere. The horizontal grid is usually of the order of 5° latitude and longitude (except near the poles). Another computationally faster approach is to represent the horizontal fields by a series of two-dimensional sine and cosine functions (a spectral model). In the vertical, 2–15 levels may be used. Coastlines and mountains as well as essential elements of the surface vegetation (albedo, roughness) and soil (moisture content) may be incorporated. Sea-ice extent and sea-surface temperatures are usually specified for each month (unless the GCM is coupled to an interactive ocean model where the ocean circulation and heat transports are calculated). Snow cover on the ground is often computed as part of a hydrological cycle.

Model experiments depend upon the nature of the problem to be studied. These include weather forecasting (see ch. 4, I), the sensitivity of the climate to different conditions, short-term effects such as ocean temperature anomalies and volcanic eruptions as well as long-term factors like changes in atmospheric composition or solar radiation. Figure 8.9 illustrates the components of the earth's climate system that are candidates for such climate experiments. The usual procedure is first to develop a model so that it adequately represents present-day conditions. This major task normally involves a group of scientists and computer programmers for several years and necessitates powerful computers. When satisfactory 'control cases' of present conditions are available for a series of model years of data, it is then feasible to examine the effect of changing individual variables. This is important because in the real atmosphere changes will seldom if ever occur in isolation. In this way, the relative importance of different climatic variables can be determined quantitatively.

Because GCMs require massive computer resources, other approaches to modelling climate have developed. A variant of the GCM is the statistical-dynamical model (SDM) in which only zonally averaged features are analysed and north–south energy and momentum exchanges are not treated explicitly, but are represented statistically through parameterization. Simpler still are the EBM and RCM. The EBM assumes a global radiation balance and describes the integrated north–south transports of energy in terms of the poleward temperature gradients; EBMs can be one-dimensional (latitude variations only), two-dimensional (latitude–longitude, with simple land/ocean weightings or simplified geography) and even zero-dimensional (averaged for the globe). They are used particularly in climate change studies. The RCMs represent a single, globally averaged vertical column. The vertical temperature structure is analysed in terms of radiative and convective exchanges. These less-complete models complement the GCMs because, for example, the RCM allows study of complex cloud-radiation interactions and atmospheric composition on lapse rates in the absence of large-scale dynamical processes.

Summary

Air motion is described by its horizontal and vertical components; the latter are much smaller than the horizontal velocities. Horizontal motions compensate for vertical imbalances between gravitational acceleration and the vertical pressure gradient.

Horizontal wind velocity is determined by the horizontal pressure gradient, the earth's rotational effect (Coriolis force), and the curvature of the isobars (centripetal acceleration). All three factors are accounted for in the gradient wind equation, but this can be approximated in large scale flow by the geostrophic wind relationship. Below 1500 m, the wind speed and direction are affected by surface friction.

Air ascends (descends) in association with surface convergence (divergence) of air. Air motion is also subject to relative vertical vorticity as a

continued

result of curvature of the streamlines and/or lateral shear; this together with the earth's rotational effect make up the absolute vertical vorticity.

Local winds occur as a result of diurnally varying thermal differences setting-up local pressure gradients (mountain–valley winds and land–sea breezes) or due to the effect of a topographic barrier on airflow crossing it (examples are the lee-side föhn and bora winds).

The vertical change of pressure with height depends on the temperature structure. High (low) pressure systems intensify with altitude in a warm (cold) air column; thus warm lows and cold highs are shallow features. This 'thickness' relationship is illustrated by the upper level subtropical anticyclones and polar vortex in both hemispheres. The intermediate mid-latitude westerly winds thus have a large 'thermal wind' component. They become concentrated into upper tropospheric jet streams above sharp thermal gradients, such as fronts.

The upper flow displays a large scale long-wave pattern, especially in the northern hemisphere, related to the influence of mountain barriers and land/sea differences. The surface pressure field is dominated by semi-permanent subtropical highs, subpolar lows and, in winter, shallow cold continental highs in Siberia and north-western Canada. The equatorial zone is predominantly low pressure. The associated global wind belts are the easterly trade winds and the mid-latitude westerlies. There are more variable polar easterlies and over land areas in summer a band of equatorial westerlies representing the monsoon systems. This mean zonal (west–east) circulation is intermittently interrupted by 'blocking' highs; an idealized sequence is known as the index cycle.

The atmospheric general circulation which transfers heat and momentum polewards, is predominantly in a vertical meridional plane in low latitudes (the Hadley cell), but in middle and high latitudes it is accomplished by horizontal waves and eddies (cyclones/anticyclones). Substantial energy is also carried poleward by ocean current systems. Various types of numerical models are now used to study the mechanisms of the atmospheric circulation and climate processes.

Notes

1 The centrifugal force is equal in magnitude and opposite in sign to the centripetal acceleration.

2 Apparent gravity, $g = 9{\cdot}78\ \mathrm{m\ s^{-2}}$ at the equator, $9{\cdot}83\ \mathrm{m\ s^{-2}}$ at the poles.

3 The vorticity, or circulation, about a rotating circular fluid disc is given by the product of the rotation on its boundary (ωR) and the circumference ($2\pi R$) where R = radius of the disc. The vorticity is then $2\omega\pi R^2$, or 2ω per unit area.

4 The geostrophic wind concept is equally applicable to contour charts. Heights on these charts are given in geopotential metres (g.p.m.) or dekametres (g.p.dkm.).

5 The World Meteorological Organization recommends an arbitrary lower limit of 30 m s^{-1}.
6 Equatorial speed of rotation is 465 m s^{-1}.
7 Note that, at the equator, an east/west wind of 5 m s^{-1} represents an absolute motion of 460/470 m s^{-1} towards the east.

4

Air masses, fronts and depressions

An air mass may be defined as a large body of air whose physical properties, especially temperature, moisture content and lapse rate, are more or less uniform horizontally for hundreds of kilometres. The theoretical ideal is an atmosphere where surfaces of constant pressure are not intersected by isosteric (constant-density) surfaces, so that in any vertical cross-section, as shown in fig. 4.1, isobars and isotherms are parallel. Such an atmosphere is referred to as *barotropic*.

Three main factors tend to determine the nature and degree of uniformity of air-mass characteristics. They are: (*a*) the nature of the source area (from which the air mass obtains its original qualities): (*b*) the direction of movement and changes that occur in the constitution of an air mass as it moves over long distances; and (*c*) the age of the air mass. The physical properties of all air masses are classified according to the way in which they compare

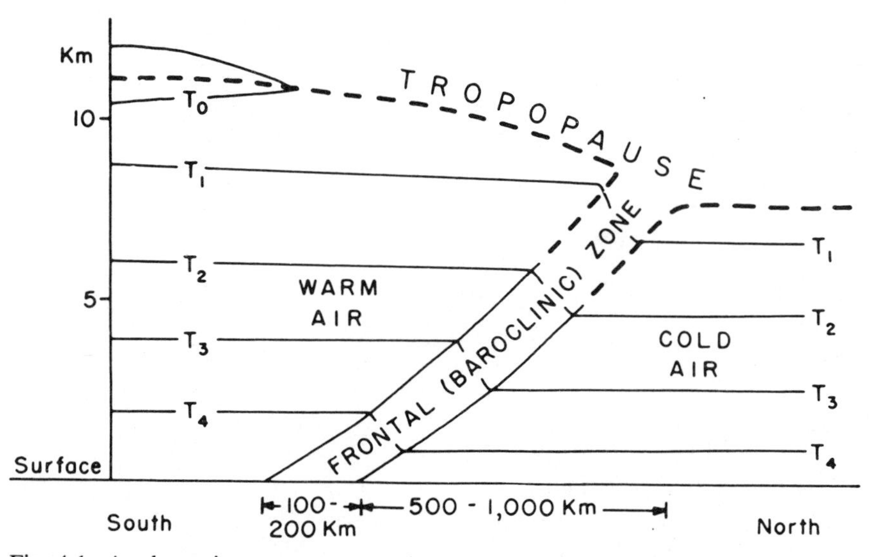

Fig. 4.1. A schematic temperature section for the northern hemisphere showing barotropic air masses and a baroclinic frontal zone (assuming that density decreases with height only).

Table 4.1 Air-mass characteristics in winter

1 *Typical values in North America, 45°–50°N (after Godson 1950)*
2 *Monthly means over the British Isles, using surface data at Kew in place of the 1000 mb values (after Belasco 1952)*
3 *Typical values in the Mediterranean (after 'Weather in the Mediterranean', MO 391, 1962)*
4 *Typical values in Australia, 33°S*
5 *Typical values in Antarctica, 75°S* } *(after Taljaard et al. 1969)*
6 *Typical values in the Southern Oceans, 50°S*

T = air temperature (°C) x = humidity mixing ratio (g/kg)

Air mass	*Level (mb)* 1000	850	700	500
cA (1) T	—	−31	−33	−42
(3) T	1	−8	−21	−36
(3) x	2·4	1·7	0·4	0·2
(5) T	(0·33)††	−28	−30	−42
(5) x	(0·2)††	0·3	0·2	0·1
mA (1) T	—	−10	21	−38
† (2) T	1	−9	−20	−40
(2) x	3·1	1·7	0·7	0·6
(3) T	4	−6	−14	−33
(3) x	4·6	2·2	1·3	0·3
(6) T	0	−10	−20	−35
(6) x	3·0	1·6	0·8	0·2
cP (1) T	—	−18	−20	−33
* (2) T	−2	−12	−22	−41
(2) x	2·6	1·5	0·6	0·1
(3) T	7	−2	−13	−24
(3) x	4·5	2·6	1·3	0·4
mPw (1) T	—	5	−4	−23
** (2) T	8	1	−9	−27
(2) x	5·8	4·0	2·1	0·6
(3) T	12	2	−7	−23
(3) x	7·8	4·0	1·6	0·4
(4) T	10	2	−7	−25
(4) x	5·5	3·4	1·8	0·4
mT (1) T	—	10	0	−17
‡ (2) T	11	6	−2	−17
(2) x	6·8	5·6	3·5	1·2
(3) T	—	10	2	−14
(3) x	—	6·0	2·5	1·0
(4) T	14	6	−2	−18
(4) x	7·8	5·3	2·5	0·9
cT (3) T	—	19	5	−17
(3) x	—	1·8	1·3	0·6
Med (3) T	14	3	−3	−19
(3) x	7·0	3·7	2·5	0·9

Belasco's classification: † P_1, * A_1, ** P_7, ‡ T_1, †† 950 mb level.

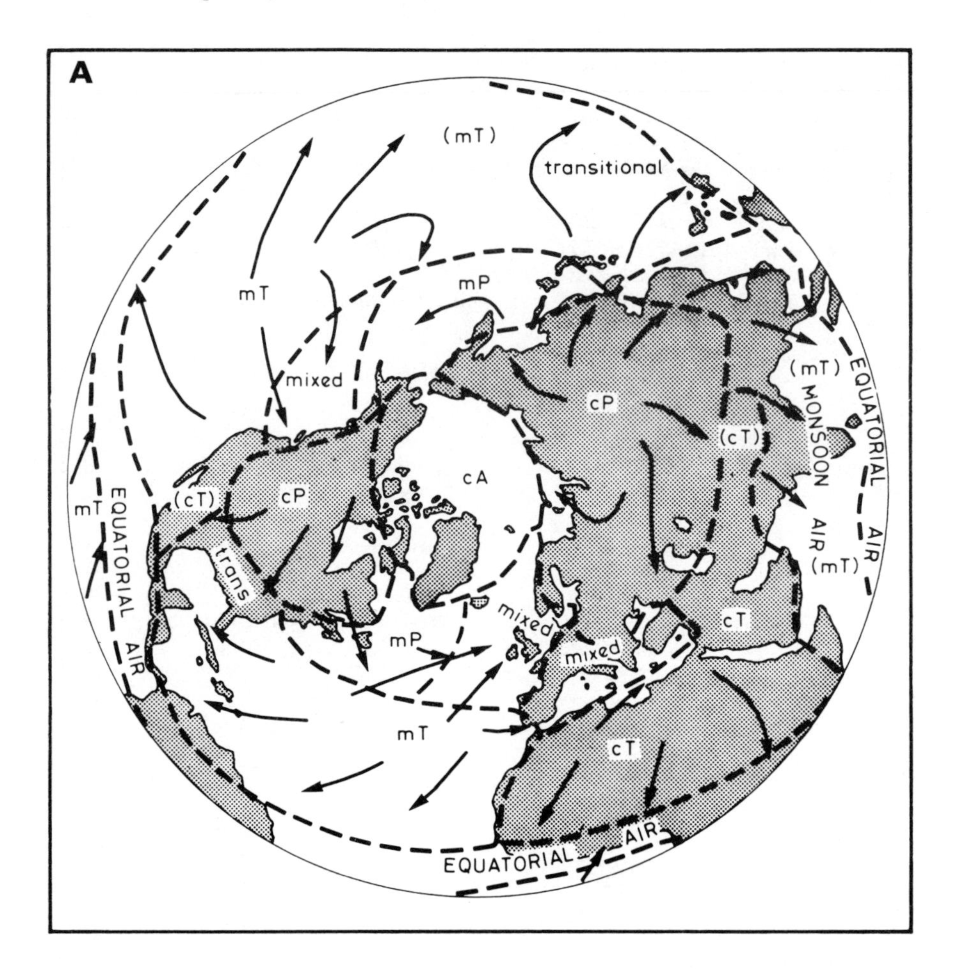

with the corresponding properties of the underlying surface region or with those of adjacent air masses.

Study of the contrasting properties of different air masses leads naturally on to a consideration of air-mass boundaries or *fronts*. Their relationship to low-pressure centres and to the patterns of airflow aloft are also discussed in this chapter, and this is followed by a brief examination of the various approaches adopted in weather forecasting.

A Nature of the source area

We have already observed how most of the physical processes of our atmosphere result from self-regulating attempts to equalize the major differences that arise from inequalities in the world distribution of heat, moisture and pressure. On the world scale the heat and momentum balances refer only to long-term average conditions. However, on a smaller scale, radiation and vertical mixing can produce some measure of equilibrium between the surface conditions and the properties of the overlying air mass if air remains

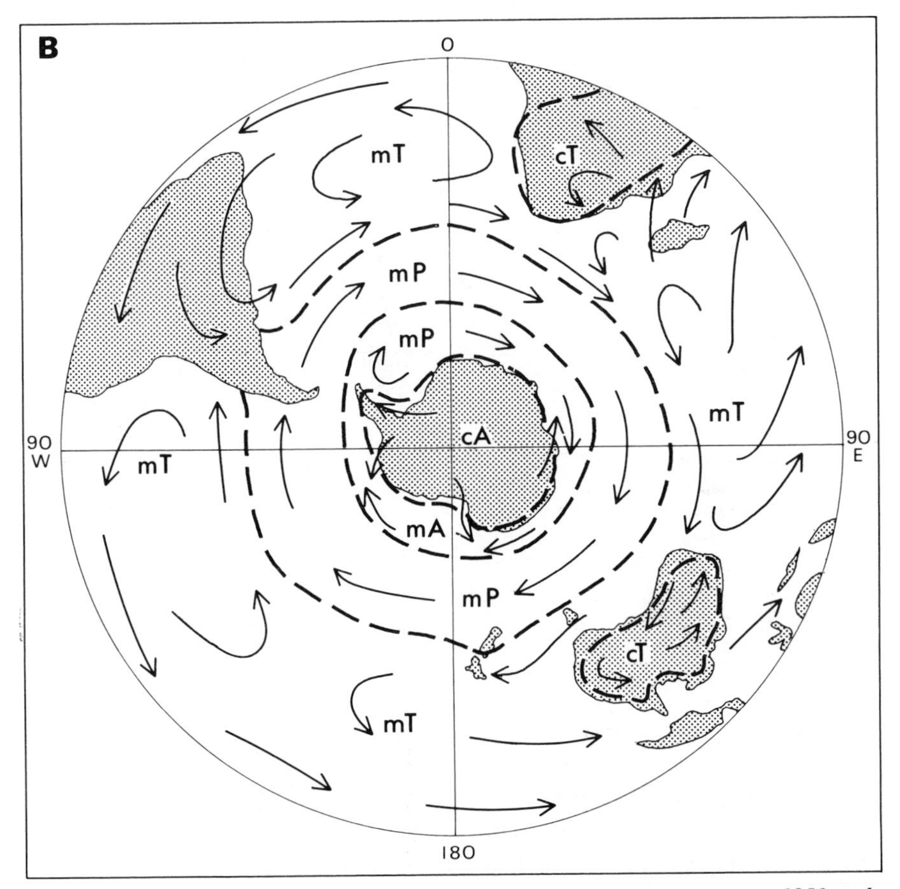

Fig. 4.2. Air masses in winter. A Northern hemisphere (*after Petterssen 1950 and Crowe 1965*). B Southern hemisphere (*after Taljaard et al. 1969*).

over a given geographical region for a period of about 3 to 7 days. Naturally the chief source regions of air masses are areas of extensive, uniform surface type which are normally overlain by quasi-stationary pressure systems. These requirements are fulfilled where there is slow divergent flow from the major thermal and dynamic high-pressure cells, whereas low-pressure regions are zones of convergence into which air masses move (see ch. 4, E).

Air masses are classified on the basis of two primary factors. The first is the temperature, giving arctic, polar and tropical air, and the second is the type of surface in their region of origin, giving maritime and continental categories. The major cold- and warm-air masses will now be discussed.

1 Cold-air masses

The principal sources of cold air in the northern hemisphere are: (*a*) the continental anticyclones of Siberia and northern Canada, which originate

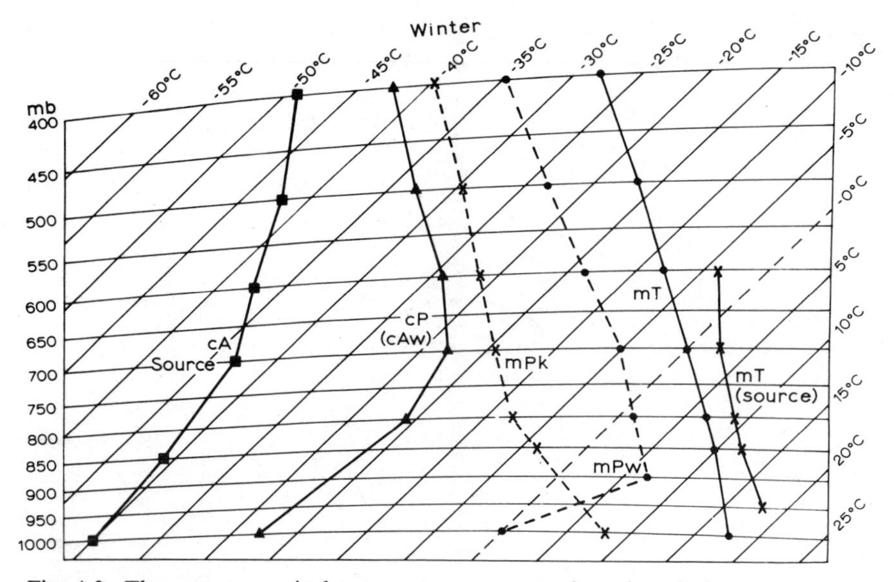

Fig. 4.3. The average vertical temperature structure for selected air masses affecting North America at about 45°–50°N, recorded over their source areas or over North America in winter (*after Godson, Showalter and Willett*).

continental polar (cP) air masses, and (*b*) the Arctic Basin, when it is dominated by high pressure (fig. 4.2). Some classifications designate air of the latter category as continental arctic (cA), but the differences between cP and cA air masses are limited mainly to the middle and upper troposphere, where temperatures are lower in the cA air (table 4.1).

The snow-covered source regions of these two air masses lead to marked cooling of the lower layers (see fig. 4.3) and, since the vapour content of cold air is very limited, the air masses generally have a mixing ratio of only 0·1–0·5 g/kg near the surface. The stability produced by the effect of surface cooling prevents vertical mixing so that further cooling occurs more slowly by radiation losses only. The effect of this radiative cooling and the tendency for air-mass subsidence in high-pressure regions combine to produce a prominent temperature inversion from the surface up to about 850 mb in typical cA and cP air. In view of their extreme dryness these air masses are characterized by small cloud amounts and only occasional light snowfalls. In summer continental heating over northern Canada and Siberia causes the virtual disappearance of their sources of cold air. The Arctic Basin source remains (see fig. 4.4A), but the cold air here is very limited in depth at this time of year. In the southern hemisphere, the Antarctic continent and the ice shelves are a source of cA air in all seasons (figs. 4.2B and 4.4B). There are no sources of cP air, however, due to the dominance of ocean areas in middle latitudes. At all seasons cA or cP air is greatly modified by a passage over the ocean. Secondary types of air mass are produced by such means and these will be considered in ch. 4, B.

Plate 1. Visible image of Africa, Europe and the Atlantic Ocean taken by METEOSAT on 19 August 1978 at 1155 hours GMT. An anticyclone is associated with clear skies over Europe and the Mediterranean, while frontal-wave cyclones are evident in the North Atlantic. Cloud clusters appear along the oceanic Intertropical Convergence Zone and there are extensive monsoon cloud masses over equatorial West Africa. Less-organized cloud cover is present over East Africa. The subtropical anticyclone areas are largely cloud-free but possess trade wind cumulus, particularly in the South-East Trade Wind belt of the South Atlantic. The highly reflective desert surfaces of the Sahara are prominent (*METEOSAT image supplied by the European Space Agency*).

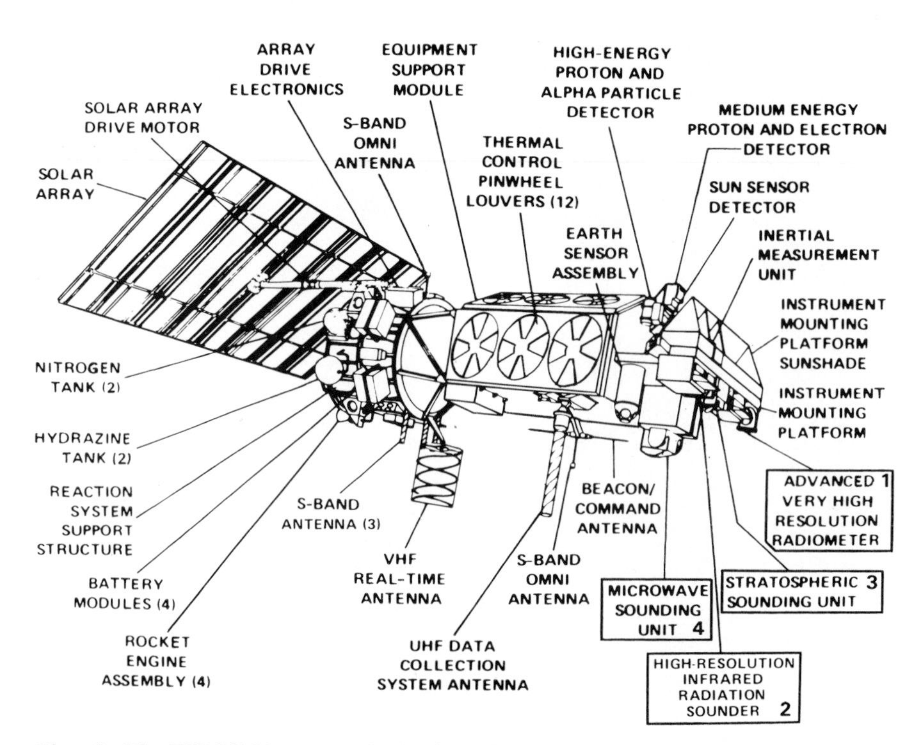

Plate 2. The TIROS-N spacecraft, having a length of 3·71 m and a weight of 1421 kg. The four instruments of particular meteorological importance are shown in the numbered boxes: 1 Visible and infrared detector – discerns clouds, land–sea boundaries, snow and ice extent and temperatures of clouds, earth's surface and sea surface. 2 Infrared detector – permits calculation of temperature profile from the surface to the 10 mb level, as well as the water vapour and ozone contents of the atmosphere. 3 Device for measuring radiation from the top of the atmosphere. 4 Device for measuring microwave radiation from the earth's surface (*NOAA: National Oceanic and Atmospheric Administration*).

Plate 4. (on facing page) Night-time (0900 GMT) infrared photograph showing sea-surface temperatures off the south-east coast of the United States on 15 February 1971 (see fig. 1.24). G Coldest shelf water. H Shelf water of intermediate temperature. I Warm Gulf Stream, clearly showing meanders associated with cold water intrusions (J) (*World Meteorological Organization 1973*).

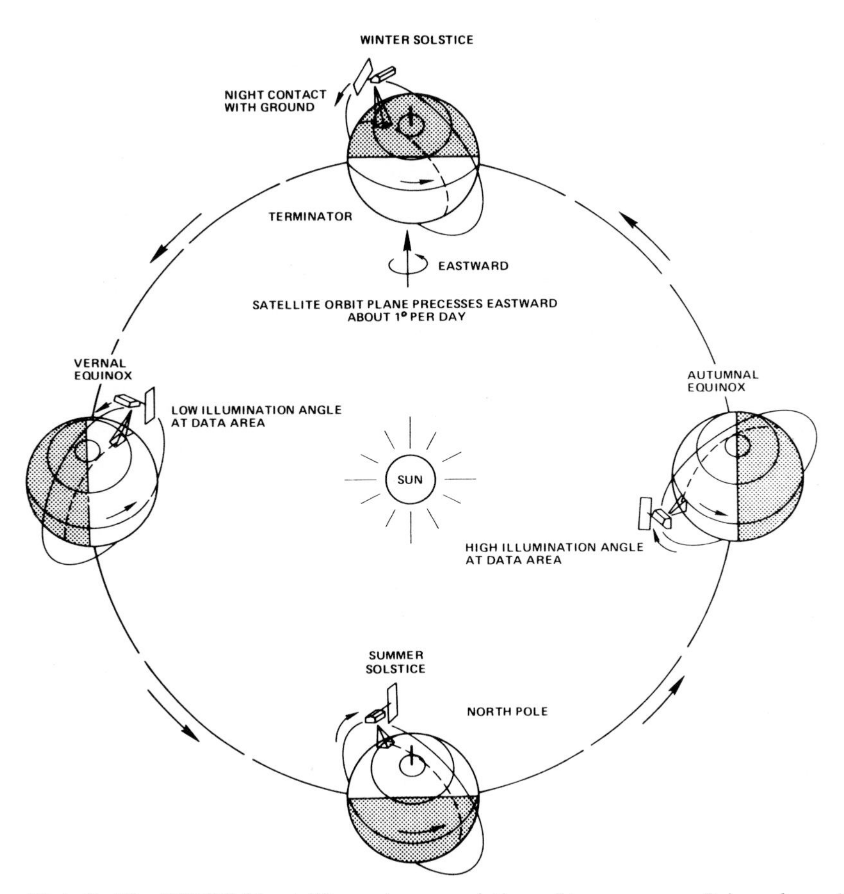

Plate 3. The TIROS-N satellite system consisting of two spacecraft in polar orbit at 833 and 870 km, respectively. The orbital plane of the second satellite lags 90° longitude behind that of the first and the orbital plane of each precesses eastward at about 1° longitude per day. Each satellite transmits data from a circular area of the earth's surface 6200 km in diameter. The satellites make 14·18 and 14·07 orbits of the earth per day, respectively, such that each point on the earth is sensed for 13–14 minutes at a time (*NOAA*).

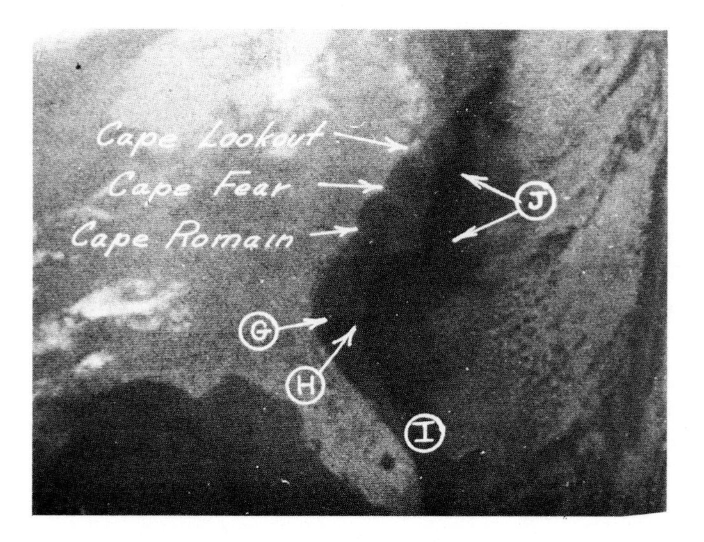

Plate 5. Cold, fog-laden air draining over the southern rim of the Grand Canyon, Arizona, elevation 2075 m (6800 ft) in the early morning (*photograph Ernst Haas; Courtesy Time/Life Publications*).

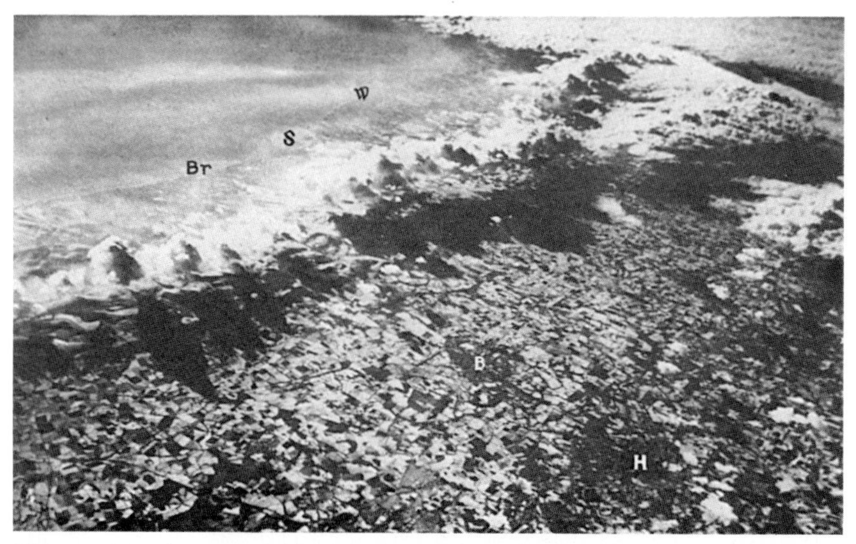

Plate 6. Cumulus orographic clouds developed over the dip-slope of the South Downs in Sussex, England. To the west (*right*), southern Hampshire is covered by stratiform clouds. The English Channel is in the upper left. This infrared photograph was taken from an elevation of about 12,000 m (40,000 ft) (B = Burgess Hill; Br = Brighton; H = Haywards Heath; S = Shoreham; W = Worthing) (*P.A.—Reuter Ltd*).

Plate 7. Layered clouds formed as a moist, stable airflow is forced to rise over a mountain barrier (*from A Field Guide to the Atmosphere by Vincent J. Schaefer and John A. Day. Copyright © 1981 by Vincent J. Schaefer and John A. Day. Reprinted by permission of Houghton Mifflin Co.*)

Plate 8. View north along the eastern front of the Colorado Rockies, showing lee-wave clouds (*NCAR photograph by Robert Bumpas*).

Plate 9. View looking south-south-east from about 9000 m (30,000 ft) along the Owen's Valley, California, showing a roll cloud developing in the lee of the Sierra Nevada mountains. The lee-wave crest is marked by the cloud layer and the vertical turbulence is causing dust to rise high into the air (W = Mount Whitney, 4418 m (14,495 ft); I = Independence) (*photograph by Robert F. Symons; Courtesy R. S. Scorer*).

Plate 11. DMSP visible image of the coastal area off New England at 1433 hours GMT, 17 February 1979. A northerly airflow averaging 10 m s^{-1}, with surface air temperatures of about −15°C, moves offshore where sea-surface temperatures increase to 9°C within 250 km of the coast. Convective cloud streets are visible, also ice in James Bay, *upper left*, and in the Gulf of St Lawrence. (This case is analysed in a study by Atlas *et al.*, *Monthly Weather Review*, vol. 111 (1983), p. 265) (*image courtesy of National Snow and Ice Data Center, University of Colorado, Boulder*).

Plate 10. (on facing page) Radiating or dendritic cellular (actiniform) cloud pattern. These complex convective systems, some 150–250 km (90–150 miles) in diameter, were only discovered as a result of satellite photography. They usually occur in groups over areas of subsidence inversions intensified by cold ocean currents (e.g. in the low latitudes of the eastern Pacific) (*Environmental Science Service Administration*).

Plate 12. Four cumulus towers protruding through a stable altostratus cloud sheet over Montana. Tower tops are displaced by a 70 m s^{-1} (150 mph) jet stream (*from A Field Guide to the Atmosphere by Vincent J. Schaefer and John A. Day. Copyright © 1981 by Vincent J. Schaefer and John A. Day. Reprinted by permission of Houghton Mifflin Co.*)

Plate 13. Isolated desert thunderstorm, 25 km (15 miles) west of Mt Lemmon, near Tucson, Arizona, on 31 July 1973 (*photograph courtesy Dr Brooks Martner, University of Wyoming*).

Plate 14. Thunderstorm approaching Östersund, Sweden, during late afternoon on 23 June 1955. Ahead of the region of intense precipitation there are rings of cloud formed over the squall-front (*copyright F. H. Ludlam; originally published in Weather, vol. XV, no. 2, 1960, p. 63*).

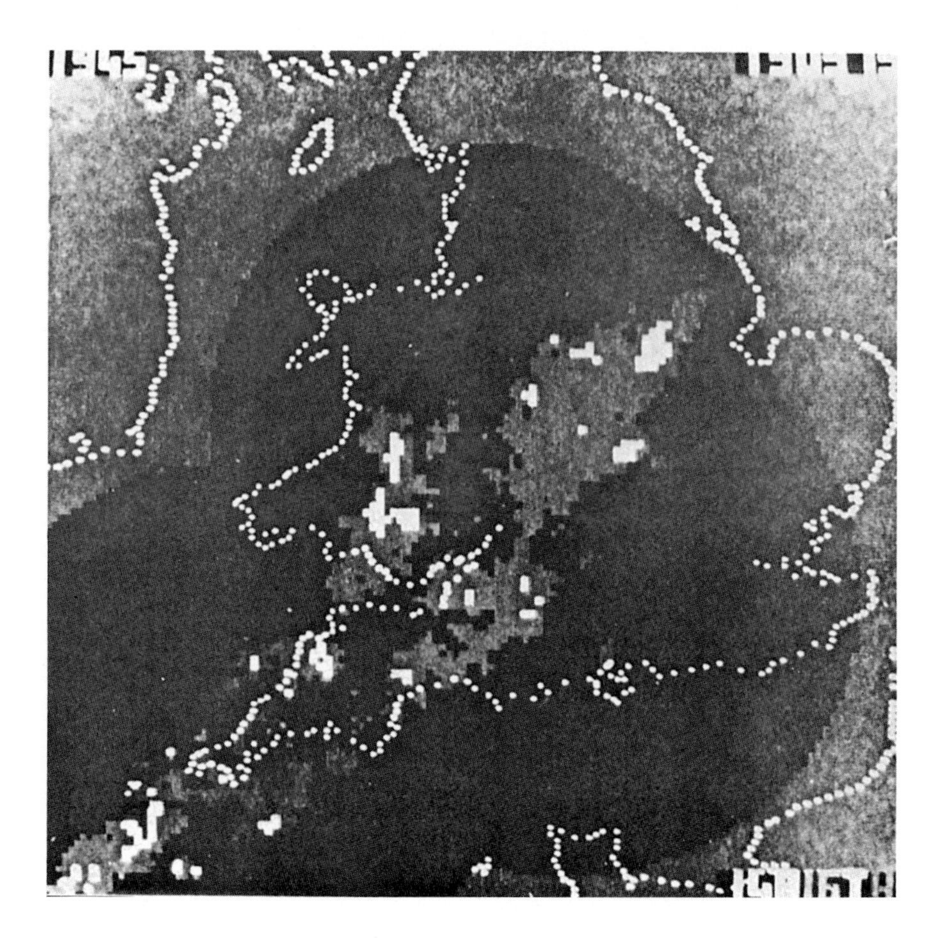

Plate 16. Photograph by an astronaut from Gemini XII manned spacecraft from an elevation of some 180 km (112 miles) looking south-east over Egypt and the Red Sea. The band of cirrus clouds is associated with strong upper winds, possibly concentrated as a jet stream (*NASA photograph*).

Plate 15. (on facing page) Photograph of a composite radar display from three radars (intersecting dark circles) showing the streaky rainfall distribution over southern Britain at 1945 hours on 19 September 1979. The display consists of a 128 × 128 matrix of 5-km squares, with light squares representing heavy rain and grey squares low to moderate rainfall intensity (*from Browning 1980*).

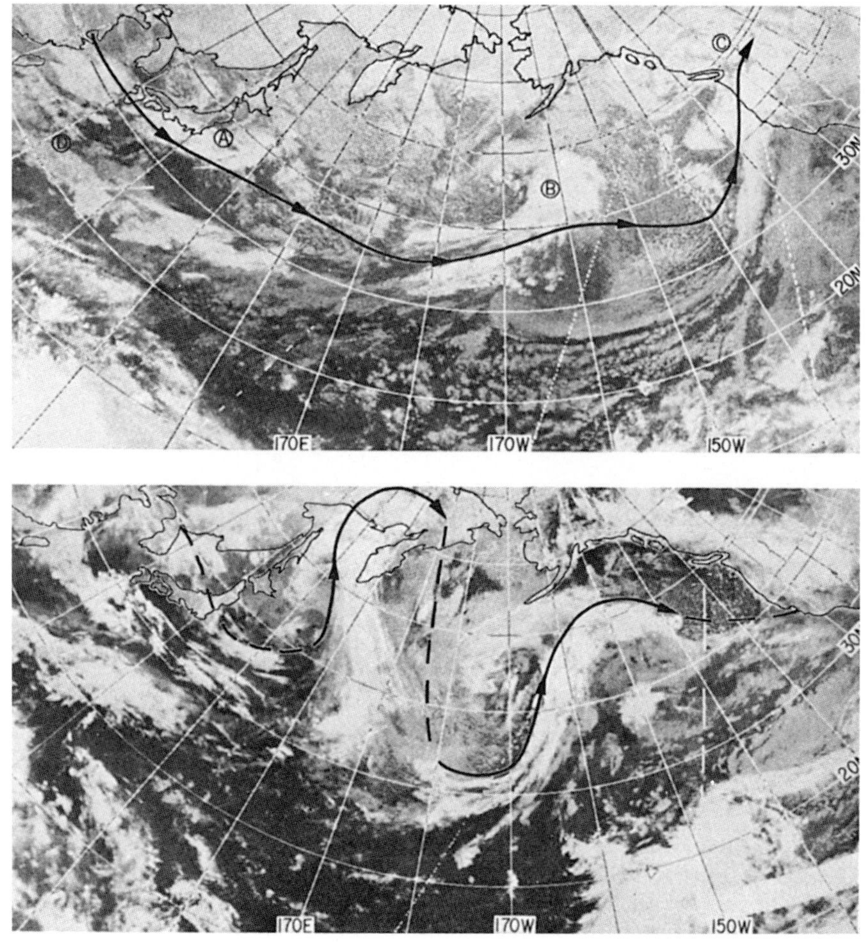

Plate 17. Infrared photographs of the North Pacific, with the 200 mb jet stream inserted. *Above:* general zonal flow associated with a high zonal index, 12 March 1971; three major cloud systems (A, B, C) occur along the belt of zonal flow, and the large east–west belt of cloud (D) to the south of Japan is also characteristic of accentuated zonal flow. *Below:* large amplitude flow regime associated with a lower zonal index, 23 April 1971 (*World Meteorological Organization 1973*).

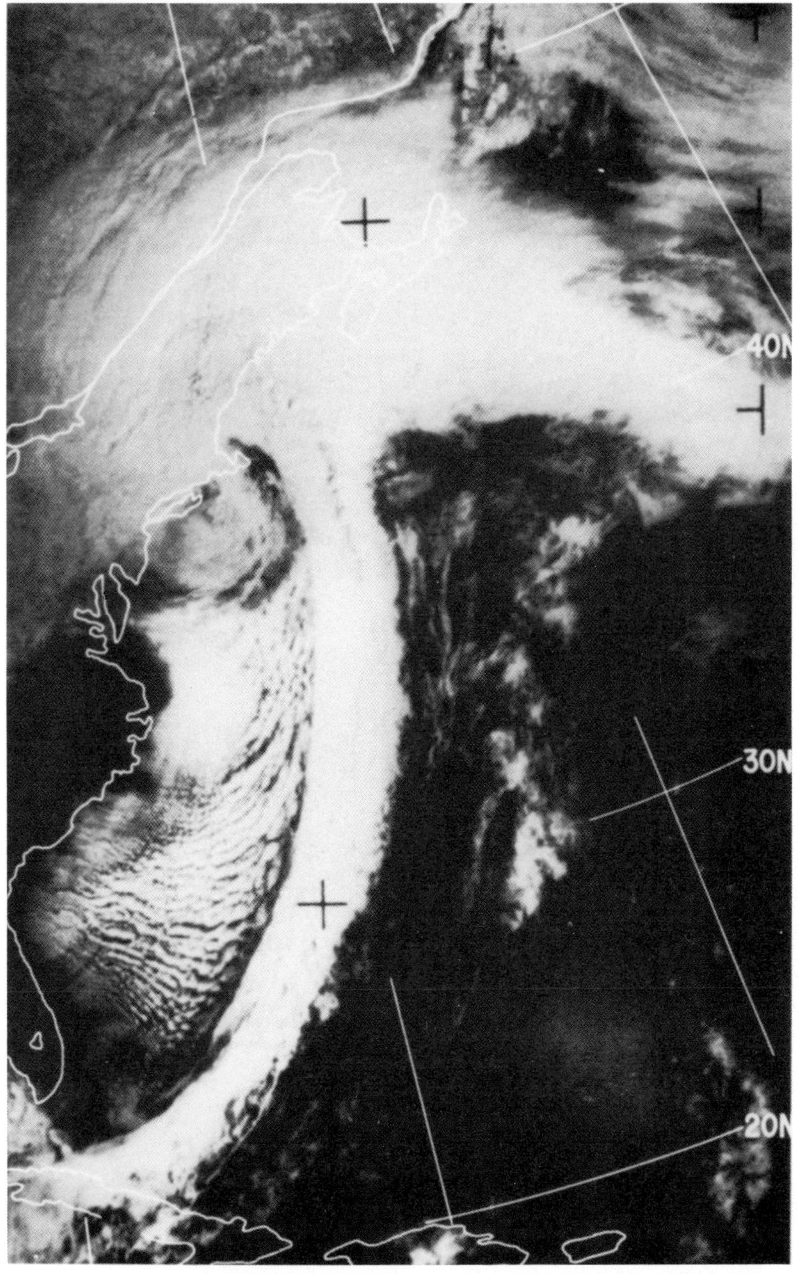

Plate 18. A well-developed, open-stage depression centred over the Maritime Provinces of Canada observed on 4 March 1971 from an ESSA 8 satellite. A broad warm front extends eastwards into the Atlantic and a narrow cold front parallels the east coast of the United States. The cooler north-westerly airflow behind the cold front is generating unstable cloud bands over the warmer ocean (*from A Field Guide to the Atmosphere by Vincent J. Schaefer and John A. Day. Copyright © 1981 by Vincent J. Schaefer and John A. Day. Reprinted by permission of Houghton Mifflin Co.*).

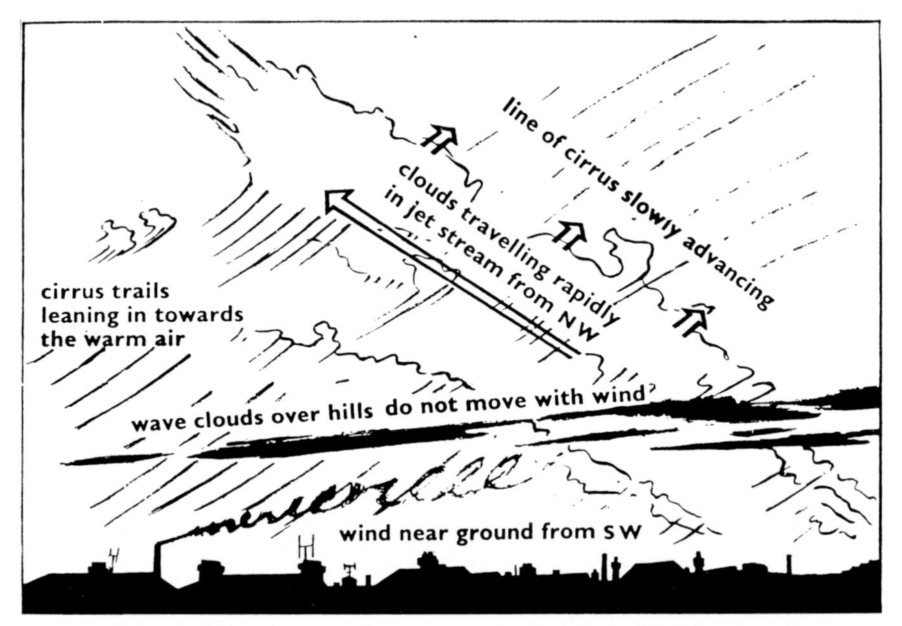

Plate 19. View looking westward towards an approaching warm front, with lines of jet-stream clouds extending from the north-west, from which trails of ice crystals are falling. In the middle levels are dark wave clouds formed in the lee of small hills by the south-westerly airflow, whereas the wind direction at the surface is more southerly – as indicated by the smoke from the chimney (*photograph copyright by F. H. Ludlam; diagram by R. S. Scorer; both published in Weather, vol. 18, no. 8, 1963, pp. 266–7*).

Plate 20. Photograph of a depression to the south-west of the British Isles taken by the ESSA 2 satellite at 1018 hours GMT on 12 November 1966, showing a well-marked spiral cloud structure and open cell cloud areas behind the cold fronts. The depression began as hurricane 'Lois' in the eastern Caribbean and moved north-eastwards until its winds fell below gale force on 10 November. Thereafter it began to deepen again and the pressure at the centre had dropped to 962 mb when this photograph was taken. The frontal zones are well developed, although the 1000–500 mb wind shear was unusually weak for such a well-defined system (*courtesy Weather, vol. XXIV, no 6, 1969, p. 222; Crown Copyright Reserved*).

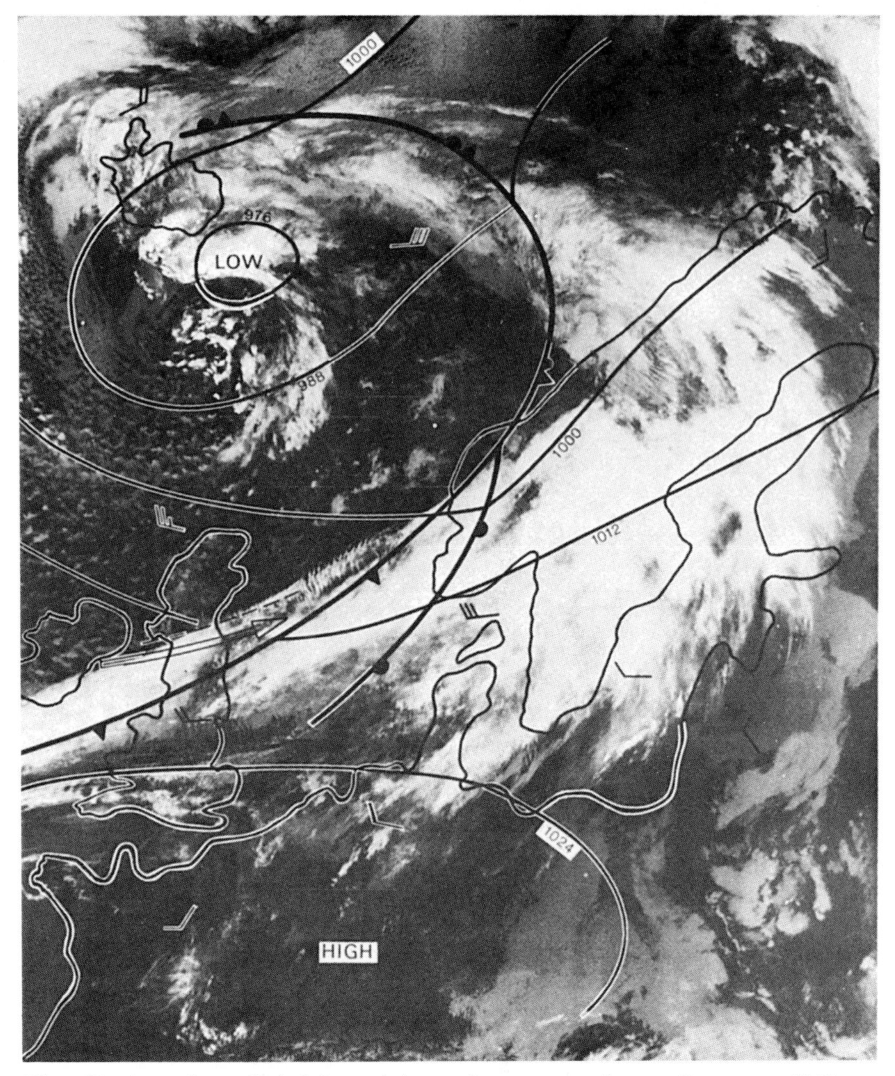

Plate 21. A partly occluded frontal depression over north-west Europe at 1347 hours GMT on 19 October 1979. This TIROS infrared image was processed to show the coldest surfaces (e.g. cloud tops) as white and the warmest (e.g. land surfaces) as black. Surface fronts, isobars and winds are shown, together with the upper tropospheric jet stream (arrowed), lying on the warm side of the cold front aloft (see fig. 3.22) (maximum velocity 65 m s^{-1} at about 250 mb; 11·3 km); also note the break (dashed) between the mid-latitude tropopause (200 mb; 11·8 km) and the polar tropopause (270 mb; 10·2 km). This added synoptic information relates to 1200 hours GMT (*image courtesy of P. E. Baylis, Department of Electrical Engineering and Electronics, University of Dundee; synoptic interpretation courtesy of Dr Ross Reynolds, Department of Meteorology, University of Reading*).

Table 4.2 Air-mass characteristics in summer (*key as for Table 4.1*)

Air mass	*Level (mb)* 1000	850	700	500
cA (5) *T*	(−9)‡	−13	−20	−33
(5) *x*	(1·8)‡	1·1	0·7	0·7
mA (1) *T*	—	−4	−14	−33
(2) *T*	14	2	−7	−25
(2) *x*	6·3	4·3	2·5	0·1
mP (1) *T*	—	11	0	−19
* (2) *T*	16	4	−6	−24
(2) *x*	8·4	3·9	2·2	0·4
(3) *T*	—	18	−2	−19
(3) *x*	—	6·0	2·5	0·8
(4) *T*	17	8	0	−14
(4) *x*	8·0	6·0	3·1	1·0
cP (3) *T*	26	13	4	−14
(3) *x*	16·1	6·7	3·4	0·9
mT (1) *T*	—	18	8	−8
(2) *T*	19	12	4	−11
(2) *x*	10·8	8·1	4·5	2·4
(4) *T*	22	16	5	−11
(4) *x*	13·4	8·0	4·8	1·7
cT (1) *T*	—	22	10	−11
(2) *T*	21	16	6	−11
(2) *x*	12·1	3·9	3·4	1·1
† (3) *T*	—	26	13	−10
(3) *x*	—	4·5	2·5	0·5
(4) *T*	27	20	7	−12
(4) *x*	8·0	4·7	3·6	1·2
Med (3) *T*	29	19	12	−6
(3) *x*	14·1	7·4	3·0	0·9

Belasco's classification: * P_3, † cT originating over Africa, ‡ 950 mb.

2 Warm-air masses

These have their origins in the subtropical high-pressure cells and, during the summer season, in the great accumulations of warm surface air which characterize the heart of large land areas.

The tropical (T) sources are either maritime (mT), originating in the oceanic subtropical high-pressure cells, or continental (cT), originating either from the continental parts of these subtropical cells (e.g. as does the North African *Harmattan*) or simply associated with regions of generally

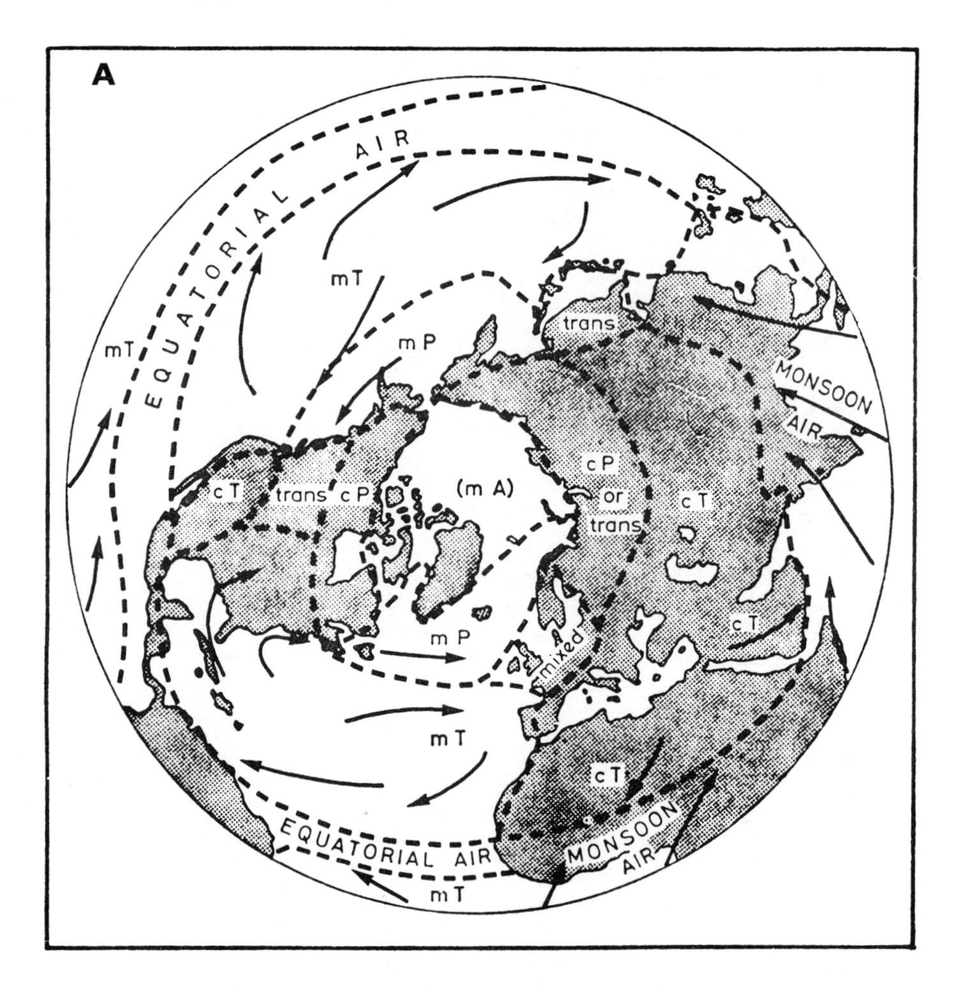

light variable winds, assisted by upper tropospheric subsidence, over the major continents in summer (e.g. Central Asia). In the southern hemisphere, the source area of mT air covers about half of the hemisphere. There is no zone of significant temperature gradient between the equator and the oceanic Subtropical Convergence about 40°S.

The maritime type is characterized by high temperatures (accentuated by the warming action to which the descending air is subjected), high humidity of the lower layers over the oceans, and stable stratification. Since the air is warm and moist near the surface, stratiform cloud commonly develops as the air moves polewards from its source. The continental type in winter is restricted mainly to north Africa (fig. 4.2, table 4.1), where it is a warm, dry and stable air mass. In summer, warming of the lower layers by the heated land generates a steep lapse rate, but despite its instability the low relative and specific humidity prevent the development of cloud and precipitation. In the southern hemisphere, cT air is rather more prevalent in winter over the subtropical continents except South America. In summer, much of southern

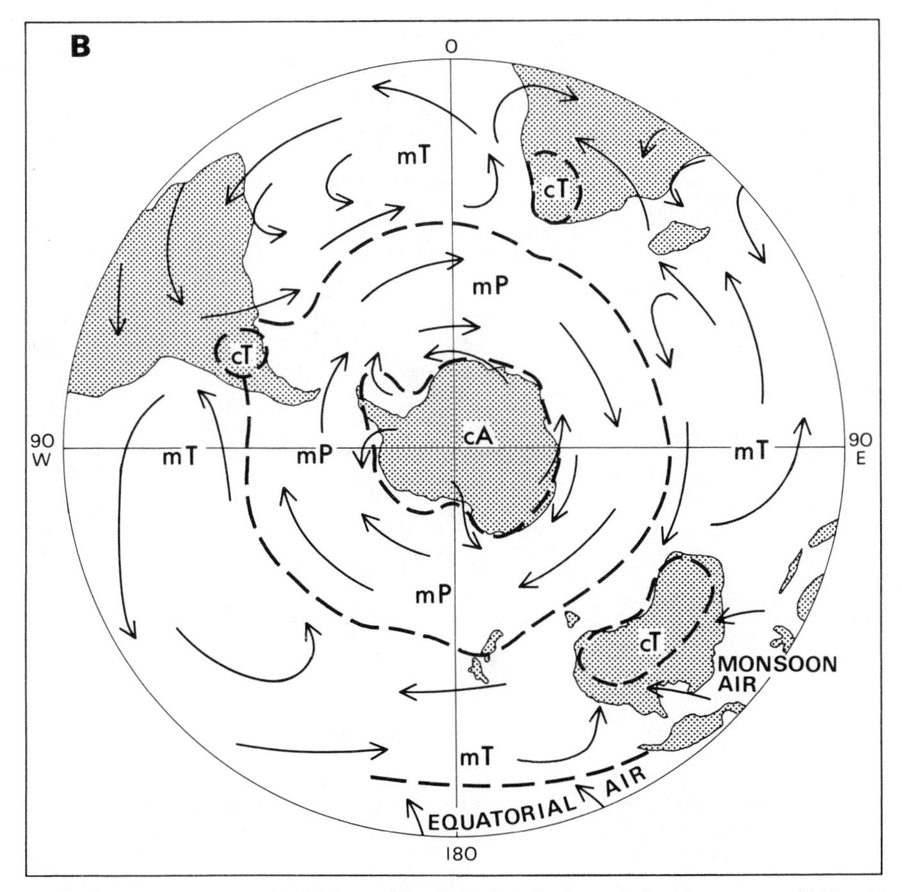

Fig. 4.4. Air masses in summer. A Northern hemisphere (*after Petterssen 1950 and Crowe 1965*). B Southern hemisphere (*after Taljaard et al. 1969*).

Africa and northern Australia is affected by mT air, while there is a small source of cT air over Argentina (fig. 4.4B).

The characteristics of the primary air masses are illustrated in figs. 4.3 and 4.5 and tables 4.1 and 4.2. In some cases their properties have been considerably affected by movement away from the source region, and this question is discussed below.

A different view of source regions can be obtained from analysis of airstreams. Streamlines of the mean resultant winds (see note 1) in individual months can be used to analyse areas of divergence representing air-mass source regions, downstream airflow and the confluence zones between different airstreams. Figure 4.6A shows the sources in the northern hemisphere and their annual duration. Four sources are dominant: the subtropical North Pacific and North Atlantic anticyclones, and their southern hemisphere counterparts. For the entire year air from these sources covers at least 25% of the northern hemisphere; for 6 months of each year they affect almost three-fifths of the hemisphere. In the southern hemisphere, in contrast, the

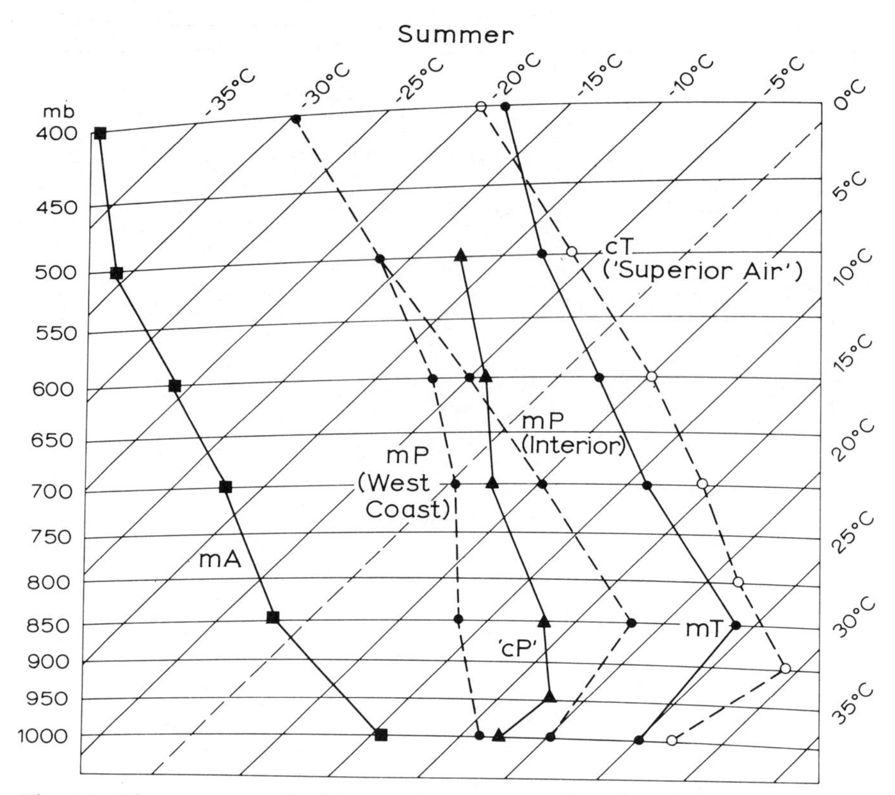

Fig. 4.5. The average vertical temperature structure for selected air masses affecting North America in summer (*after Godson, Showalter and Willett*).

airstream climatology is much simpler as a result of the dominance of ocean surfaces (fig. 4.6B). The source areas are associated with the three oceans, and especially their subtropical anticyclones. Antarctica is the major continental source, with another mainly in winter over Australia.

B Air-mass modification

As air masses move away from their source region they are affected by different heat and moisture exchanges with the ground surface and by dynamic processes in the atmosphere. Thus an initially barotropic air mass is gradually changed into a moderately *baroclinic* airstream in which isosteric and isobaric surfaces intersect one another. The presence of horizontal temperature gradients means that air cannot travel as a solid block maintaining an unchanging internal structure. The trajectory (i.e. actual path) followed by an air parcel in the middle or upper troposphere will normally be quite different from that of a parcel nearer the surface, due to the increase of westerly wind velocity with height in the troposphere. The actual structure of an airstream at a given instant is determined to a large extent by the past history of air-mass modification processes. In spite of these qualifications the air-mass concept nevertheless remains of considerable practical value.

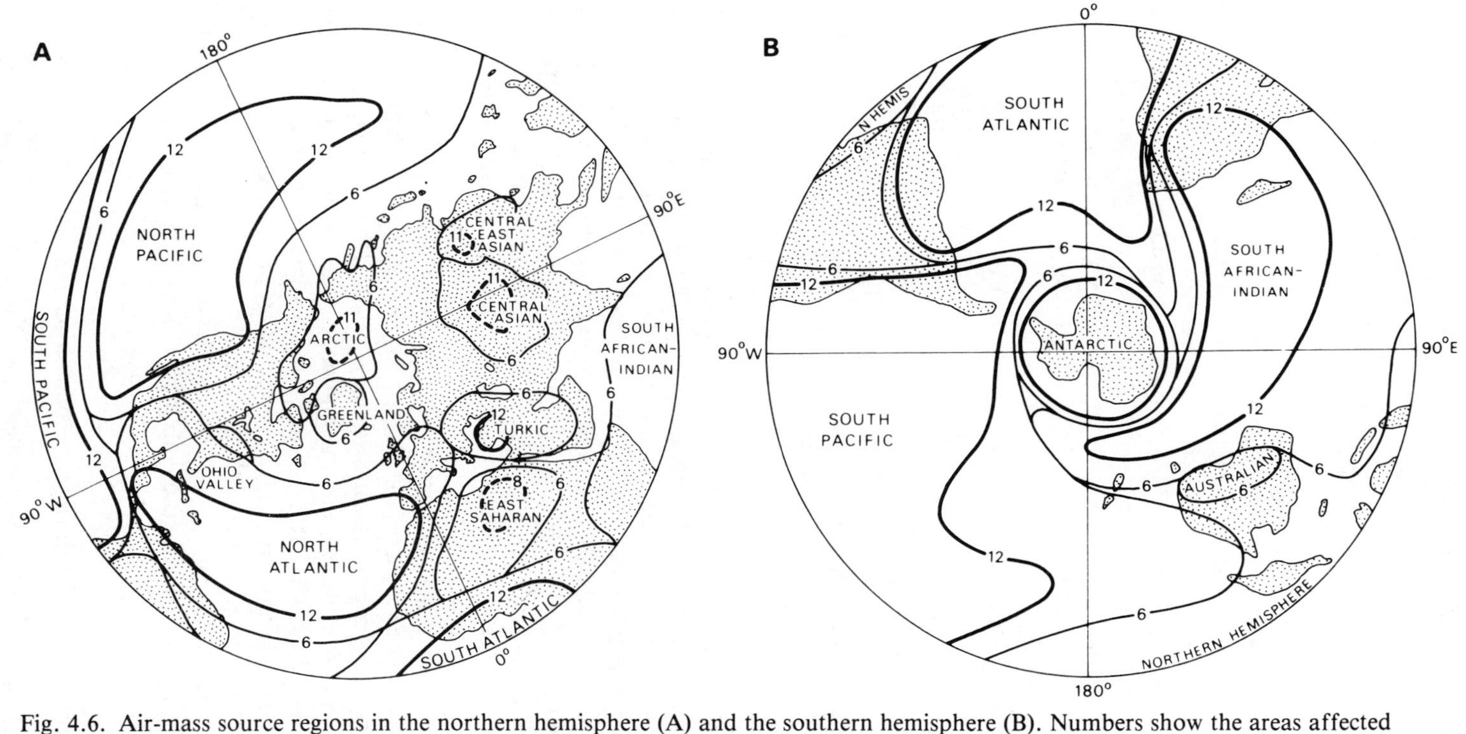

Fig. 4.6. Air-mass source regions in the northern hemisphere (A) and the southern hemisphere (B). Numbers show the areas affected by each air mass in months per year (*after Wendland and Bryson 1981, and Wendland and McDonald 1986*).

1 Mechanisms of modification

The mechanisms by which air masses are modified are, for convenience, treated separately, although this rigid distinction is often not justified in practice.

a Thermodynamic changes An air mass may be heated from below either by passing from a cold to a warm surface or by solar heating of the ground over which the air is located. Similarly, but in reverse, it can be cooled from below. Heating from below acts to increase air-mass instability so that the effect may be spread rapidly through a considerable thickness of air, whereas surface cooling produces a temperature inversion which greatly limits the vertical extent of the cooling. For this reason cooling mainly occurs through radiative heat loss by the air, a process which takes place only very gradually.

Changes can also occur through increased evaporation, the moisture being supplied either from the underlying surface or by precipitation from an overlying air-mass layer. In reverse, the abstraction of moisture by condensation or precipitation can also cause changes. A parallel, and most important, change is the respective addition or loss of latent heat accompanying this condensation or evaporation. The distribution of latent and sensible heat transfers to the atmosphere are illustrated in figs. 1.34 and 1.35, although it must be noticed that these refer to annual values.

b Dynamic changes Dynamic (or mechanical) changes are, superficially at any rate, different from thermodynamic changes because they involve mixing or pressure changes associated with the actual movement of the air mass. The distribution of the physical properties of air masses has been shown to be considerably modified, for example, by a prolonged period of turbulent mixing (see fig. 2.16). This process is particularly important at low levels where surface friction intensifies the natural turbulence of airflow, providing a ready mechanism for the upward transfer of the effects of thermodynamic processes.

The radiative and advective exchanges discussed previously are non-adiabatic, but the ascent or descent of air causes adiabatic changes of temperature. Large-scale lifting may result from forced ascent by a mountain barrier or from airstream convergence. Conversely, sinking may occur when high-level convergence sets up subsidence or when stable air, having been forced up over high ground by the pressure gradient, descends in its lee. Dynamic processes in the middle and upper troposphere are in fact a major cause of air-mass modification. The decrease in stability aloft, as air moves away from the areas of subsidence, is a common example of this type of mechanism.

2 The results of modification: secondary air masses

The study of the ways in which air masses change in character tells us a great deal about our weather, for many common meteorological phenomena are the product of such modification.

a Cold air Continental polar air frequently streams out from Canada over the western Atlantic in winter, where it undergoes rapid transformation. Heating over the Gulf Stream Drift rapidly makes the lower layers unstable and evaporation into the air leads to sharp increases of moisture content (see fig. 1.34) and cloud formation (pl. 11). The turbulence associated with the convective instability is marked by gusty conditions. By the time the air has reached the central Atlantic it has become a cool, moist, maritime polar (mP) air mass. Analogous processes occur with outflow from Asia over the north Pacific (see fig. 4.2). Over middle latitudes of the southern hemisphere the circumpolar ocean gives rise to a continuous zone of mP air which, in summer, extends to the margin of Antarctica. In this season, however, there is a considerable gradient of ocean temperatures associated with the Antarctic Convergence that makes the zone far from uniform in its physical properties. The weather in cP airstreams is typically that of bright periods and squally showers, with a variable cloud cover of cumulus and cumulonimbus. As the air moves eastward towards Europe the cooler sea surface may produce a neutral or even stable stratification near the surface, especially in summer, but subsequent heating over land will again regenerate unstable conditions. Similar conditions, but with lower temperatures (table 4.1), result if cA air crosses sea areas in high latitudes producing maritime arctic (mA) air.

When cP air moves southward over land in winter, in central North America for example, it acquires higher temperatures and a greater tendency towards instability, but there is little gain in moisture content. The cloud type is scattered shallow cumulus which only rarely gives showers even in the afternoon when convectional instability is at a maximum. Exceptions occur in early winter around the eastern and southern shores of Hudson Bay and the Great Lakes. Until these water bodies freeze over, cold airstreams which cross them are rapidly warmed and supplied with moisture leading to locally heavy snowfalls.

Over Eurasia and North America cP air may move southwards and later recurve northwards. Some schemes of air-mass classification cater for such possibilities by specifying whether the air is colder (k), or warmer (w), than the surface over which it is passing. For example, cPk refers to cold, dry continental polar air which is moving over a warmer surface and is thereby likely to become unstable. Likewise, mPw indicates moist, maritime polar air which is being progressively cooled near the surface and, hence, becoming more stable.

In general, a 'k' air mass has gusty, turbulent winds which make for good visibility as smoke and haze are dispersed. The instability leads to cumuliform clouds. A 'w' air mass typically has stable or inversion conditions with stratiform cloud. Limited vertical mixing allows the concentration of smoke, haze and fog at low levels. Clearly, these and similar symbols provide a convenient shorthand description of the important parameters which characterize different air masses.

Many parts of the globe must be regarded as transitional regions where the surface and air circulation produce air masses with intermediate characteristics. Northern Asia and northern Canada fall into this category in summer. In a general sense the air has affinities with continental polar air masses but these land areas, particularly the Canadian shield, have extensive bog and water surfaces so that the air is moist and cloud amounts are quite high. In a similar manner, melt-water pools and leads in the arctic pack-ice make it more appropriate to regard the area as a source of maritime arctic (mA) air in summer (fig. 4.4A). This designation is also applied to air over the antarctic pack-ice in winter that is much less cold in its lower levels than the air over the continent itself.

b Warm air The modification of warm air masses is usually a gradual process. Air moving poleward over progressively cooler surface becomes increasingly stable in the lower layers. In the case of mT air with high moisture content, surface cooling may produce advection fog and this is particularly common, for example, in the south-western approaches to the English Channel during spring and early summer when the sea is still cool. Similar development of advection fog in mT air occurs along the South China coast in February–April, and also off Newfoundland and over the coast of northern California in spring and summer. If the wind velocity is sufficient for vertical mixing low stratus cloud forms in the place of fog, and drizzle may result. In addition, forced ascent of the air by high ground, or by overriding of an adjacent air mass, can produce heavy rainfall.

The cT air originating in those parts of the subtropical anticyclones situated over the arid subtropics in summer is extremely hot and dry (table 4.2). It is typically unstable at low levels and dust-storms may occur, but the dryness and the subsidence of the upper air limit cloud development. In the case of north Africa this cT air may move out over the Mediterranean rapidly acquiring moisture, with the consequent release of potential instability triggering off showers and thunderstorm activity.

The air masses in low latitudes present considerable problems of interpretation. The temperature contrasts found in middle and high latitudes are virtually absent and what differences do exist are due principally to moisture content and, more particularly, to the presence or absence of subsidence. *Equatorial air* is usually cooler than that subsiding in the subtropical anticyclones, for example. *Tropical air* masses can only be differentiated meaningfully in terms of moisture content and the effects of subsidence on the lapse rate. On the equatorial sides of the subtropical anticyclones in summer the air is moving westward from areas with cool sea surfaces (e.g. off north-west Africa and California) towards those of higher sea-surface temperatures. Moreover the south-western parts of the high-pressure cells are affected only by weak subsidence due to the vertical structure of the cells (see fig. 3.17). As a result of these circumstances the mT air moving westwards around the equatorward sides of the subtropical highs becomes much less stable than that on the north-eastern margin of the cells. Eventually such air forms the very warm, moist, unstable 'equatorial air' of the Inter-Tropical Convergence Zone (see figs. 4.2 and 4.4). *Monsoon air* is indicated separately in these figures, although there is no basic difference between it and mT air. The special difficulties of treating tropical climatology in terms of air masses are discussed in ch. 6.

3 The age of the air mass

Eventually the mixing and modification necessarily accompanying the movement of an air mass away from its source will cause the rate of energy exchange with the surroundings to diminish, and the various weather phenomena associated with these changes will dissipate. This process will be associated with the loss of its original identity, until finally its features merge with those of surrounding airstreams and the air may again become subject to the influence of a new source region.

North-west Europe is shown as an area of 'mixed' air masses in figs. 4.2 and 4.4. This is intended to refer to the variety of sources and directions from which air may invade the region, since weather processes associated with air-mass modification and with the frontal zones separating air masses are very much in evidence. The same is also true of the Mediterranean in winter, although the area does impart its own particular characteristics to polar and other air masses which stagnate over it. Such air is termed *Mediterranean*; typical temperature and humidity values are listed in tables 4.1 and 4.2. In winter it is convectively unstable (see fig. 2.15) as a result of the moisture picked up over the Mediterranean Sea.

The length of time during which an air mass retains its original characteristics depends very much on the extent of the source area and the type of pressure pattern affecting the area. In general the lower air is changed much more rapidly than that at higher levels, although dynamic modifications aloft, which are sometimes overlooked by climatologists, are no less significant in terms of weather processes. Modern air-mass concepts must, therefore, be flexible from the point of view of both synoptic and climatological studies.

C Frontogenesis

The first real advance in our detailed understanding of mid-latitude weather variations was made with the discovery that many of the day-to-day changes are associated with the formation and movement of boundaries, or *fronts*, between different air masses. Observations of the temperature, wind directions, humidity and other physical phenomena during unsettled periods showed that discontinuities often persist between impinging air masses of differing characteristics. The term 'front', for these surfaces of air-mass conflict, was a logical one to be proposed during the First World War by a group of meteorologists (including V. and J. Bjerknes, H. Solberg and T. Bergeron) working in Norway, and their ideas are still an integral part of most weather analysis and forecasting particularly in middle and high latitudes.

1 Frontal waves

It was observed that the typical geometry of the air-mass interface, or front, resembles a wave form (see fig. 4.7). Similar wave patterns are, in fact, found to occur on the interfaces between many different media, for example, waves on the sea surface, ripples on beach sand, aeolian sand dunes, etc. Unlike these wave forms, however, the frontal waves in the

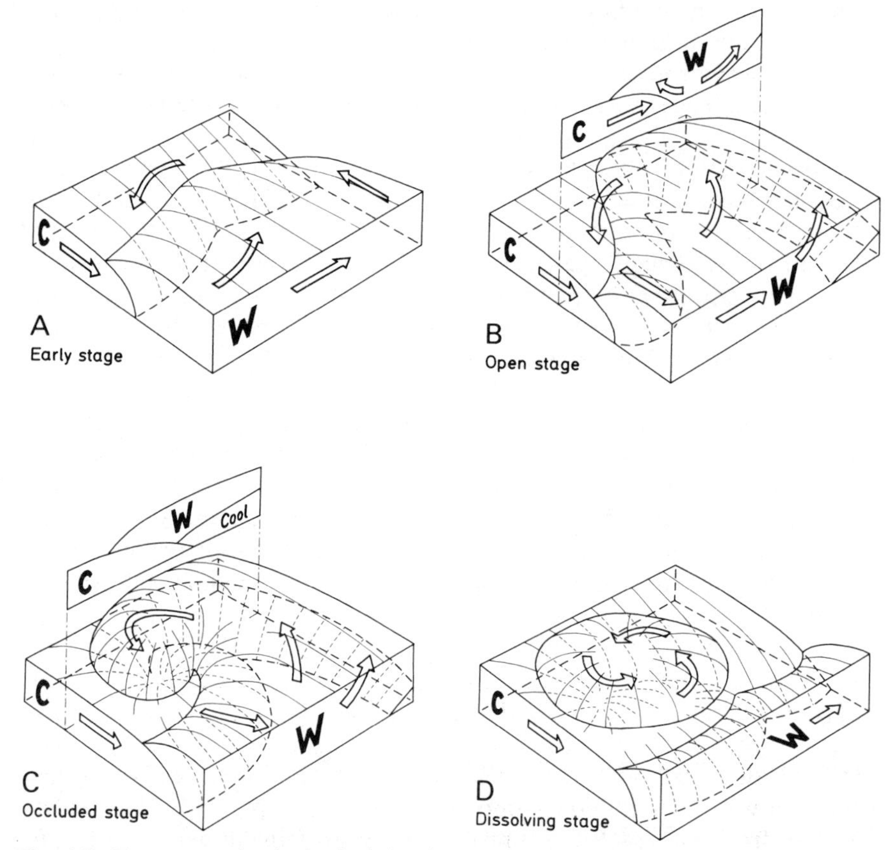

Fig. 4.7. Four stages in the typical development of a mid-latitude depression (*mostly after Strahler 1951, modified after Beckinsale.*) Satellite views of the cloud systems corresponding to these stages are shown in fig. 4.8. C = cold air; W = warm air.

atmosphere are commonly unstable; that is, they suddenly originate, increase in size, and then gradually dissipate. Numerical model calculations show that in middle latitudes waves in a baroclinic atmosphere are unstable if their wavelength exceeds a few thousand kilometres. Frontal wave cyclones are typically 1500–3000 km in wavelength. The initially attractive analogy between atmospheric wave systems and waves formed on the interface of other media is, therefore, an insufficient basis on which to develop explanations of frontal waves. In particular, the circulation of the upper troposphere plays a key role in providing appropriate conditions for their development and growth, as will be shown below.

2 The frontal wave depression

A depression (also termed a low or cyclone) (see note 2) is an area of relatively low pressure, with a more or less circular isobaric pattern. It covers an area 1500–3000 km in diameter and usually has a life-span of 4–7 days. Systems with these characteristics, which are prominent on daily weather maps, are referred to as *synoptic scale* features. The depression, in mid-latitudes at least, is usually associated with a convergence of contrasting air masses. The interface between these air masses develops into a wave form with its apex located at the centre of the low-pressure area. The wave encloses a mass of warm air between modified cold air in front and fresh cold air in the rear. The formation of the wave also creates a distinction between the two sections of the original air-mass discontinuity for, although each section still marks the boundary between cold and warm air, the weather characteristics found in the neighbourhood of each section are very different. The two sections of the frontal surface are distinguished by the names *warm front* for the leading edge of the wave and *cold front* for that of the cold air to the rear (fig. 4.7).

The boundary between two adjacent air masses is marked by a strongly baroclinic zone of large temperature gradient 100–200 km wide (see ch. 4, B and fig. 4.1). Sharp discontinuities of temperature, moisture and wind properties at fronts, especially the warm front, are rather uncommon. Such discontinuities are usually the result of a pronounced surge of fresh, cold air in the rear sector of a depression, but in the middle and upper troposphere they are often caused by subsidence and may not coincide with the location of the baroclinic zone.

On satellite imagery, active cold fronts in a strong baroclinic zone commonly show marked spiral cloud bands formed as a result of the thermal advection (see fig. 4.8B, C). Warm fronts, however, are typically covered by a cirrus shield. As fig. 3.21 shows, an upper tropospheric jet stream is closely associated with the baroclinic zone, blowing roughly parallel to the line of the upper front (see pl. 19). This relationship is examined further in ch. 4, F.

Air behind the cold front, away from the low centre, commonly has an anticyclonic trajectory and hence moves at a greater than geostrophic speed (see ch. 3, A.4) impelling the cold front to acquire a supergeostrophic speed

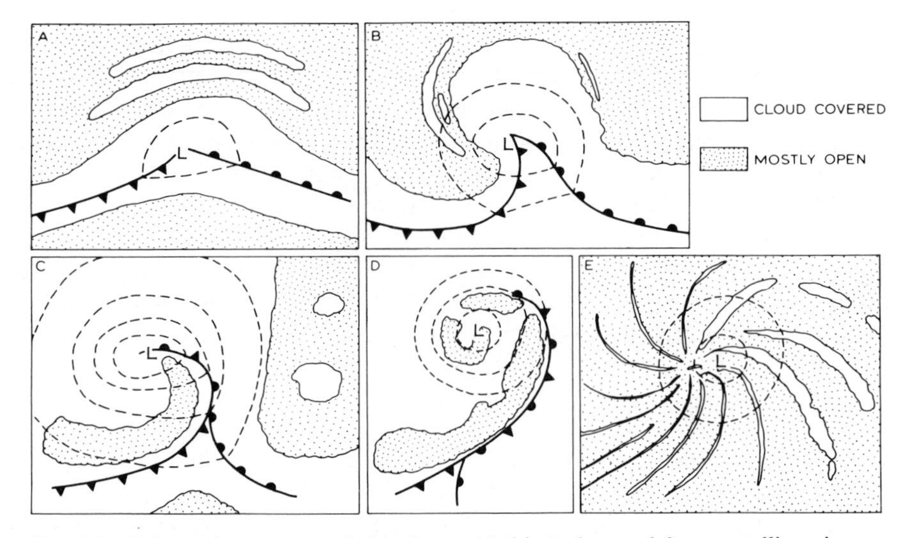

Fig. 4.8. Schematic patterns of cloud cover (white) observed from satellites, in relation to surface fronts and generalized isobars (*after Boucher and Newcomb 1962*). A, B, C and D correspond to the four stages in fig. 4.7.

also. The wedge of warm air is pinched out at the surface and lifted bodily off the ground. This stage of *occlusion* eliminates the wave form at the surface (fig. 4.7). The occlusion gradually works outward from the centre of the depression along the warm front. Sometimes, the cold air wedge advances so rapidly that, in the friction layer close to the surface, cold air overruns the warm air and generates a *squall line* (see below, ch. 4, H, p. 203).

The depression usually achieves its maximum intensity 12–24 hours after the beginning of occlusion.

By no means all frontal lows follow the idealized life cycle discussed above (cf. the caption for pl. 20). It is generally characteristic of oceanic cyclogenesis, but over North America many lows forming east of the Rocky Mountains in the lee pressure trough develop occluded fronts almost immediately. In winter months, the absence of moisture sources in this region greatly reduces the intensity of frontogenesis until the system moves eastward and draws in warm moist air from the south.

D Frontal characteristics

The activity of a front in terms of weather depends upon the vertical motion in the air masses. If the air in the warm sector is rising relative to the frontal zone the fronts are usually very active and are termed *ana-fronts*, whereas sinking of the warm air relative to the cold air masses gives rise to less intense *kata-fronts* (see fig. 4.9).

1 The warm front

The warm front represents the leading edge of the warm sector in the wave. The frontal zone here has a very gentle slope, of the order $\frac{1}{2}°–1°$, so that the

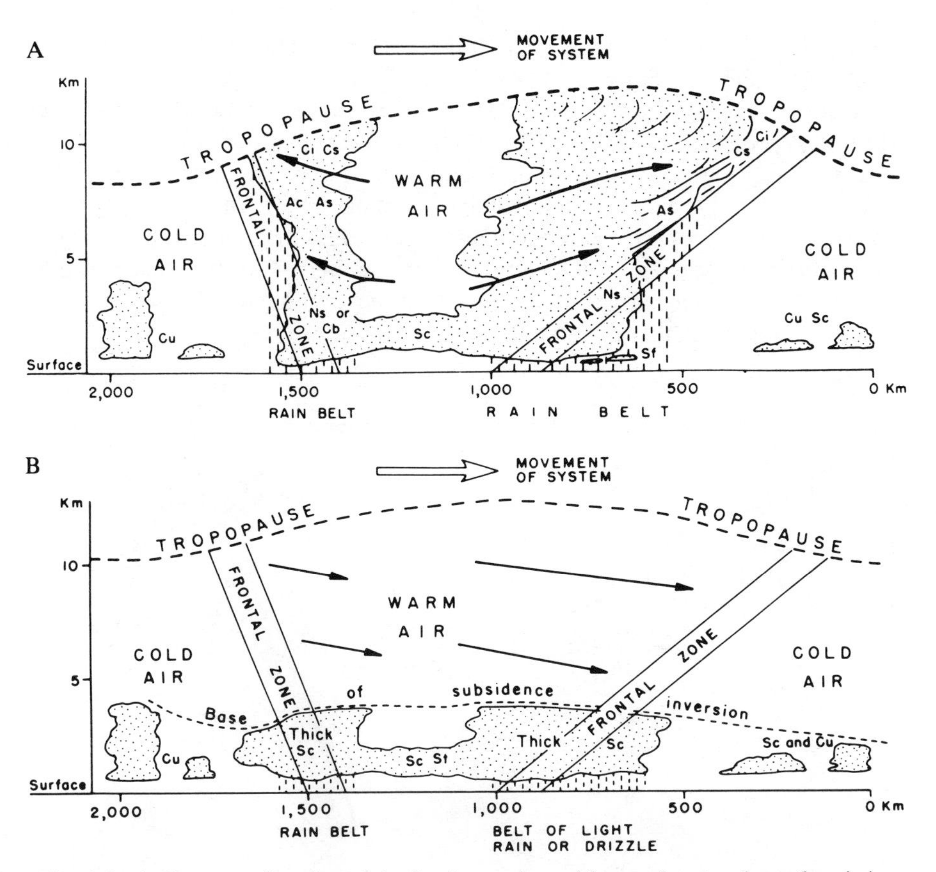

Fig. 4.9. A Cross-sectional model of a depression with ana-fronts where the air is rising relative to each frontal surface. Note that an ana-warm front may occur with a kata-cold front and vice versa. B Model of a depression with kata-fronts where the air is sinking relative to each frontal surface (*after Pedgley 1962; Crown Copyright Reserved*).

cloud systems associated with the upper portion of the front herald its approach some 12 hours or more before the arrival of the surface front (pl. 19). The ana-warm front, with rising warm air, has multi-layered cloud which steadily thickens and lowers towards the surface position of the front. The first clouds are thin, wispy cirrus, followed by sheets of cirrus and cirrostratus, and altostratus (see fig. 4.9A). The sun is obscured as the altostratus layer thickens and drizzle or rain begins to fall. The cloud often extends through most of the troposphere and with continuous precipitation occurring is generally designated as nimbostratus. Patches of stratus may also form in the cold air as rain falling through this air undergoes evaporation and quickly saturates it.

The descending warm air of the kata-warm front greatly restricts the development of medium- and high-level clouds. The frontal cloud is mainly stratocumulus, with a limited depth as a result of the subsidence inversions in both air masses (see fig. 4.9B). Precipitation is usually light rain or drizzle formed by coalescence since the freezing level tends to be above the inversion level, particularly in summer.

At the passage of the warm front the wind veers, the temperature rises and the fall of pressure is checked. The rain becomes intermittent or ceases in the warm air and the thin stratocumulus cloud sheet may break up.

Forecasting the extent of rain belts associated with the warm front is complicated by the fact that most fronts are not ana- or kata-fronts throughout their length or even at all levels in the troposphere. For this reason, radar is being used increasingly to determine by direct means the precise extent of rain belts and to detect differences in rainfall intensity.

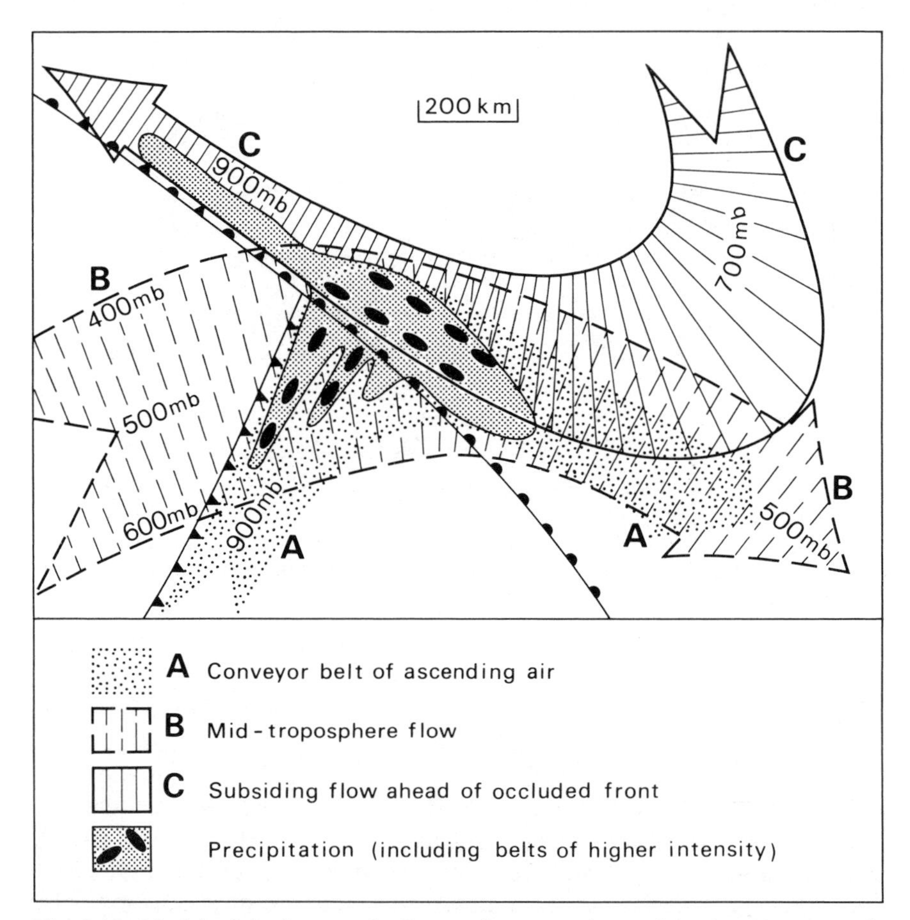

Fig. 4.10. Model of the large-scale flow and meso-scale precipitation structure of a partially-occluded depression typical of those affecting the British Isles. It shows the 'conveyor belt' (A) rising from 900 mb ahead of the cold front over the warm front. This is overlaid by a mid-tropospheric flow (B) of potentially colder air from behind the cold front. Most of the precipitation occurs in the well-defined region shown, within which it exhibits a cellular and banded structure (*after Harrold 1973*).

Such studies have shown that most of the production and distribution of precipitation is controlled by a broad airflow a few hundred kilometres across and several kilometres deep, which flows parallel to and ahead of the surface cold front (fig. 4.10).

Just ahead of the cold front the flow occurs as a low-level jet with winds up to 25–30 m s^{-1} at about 1 km above the surface. The air, which is warm and moist, rises over the warm front and turns south-eastward ahead of it as it merges with the mid-tropospheric flow (B in fig. 4.10). This flow has been termed a 'conveyor belt' (for large-scale heat and momentum transfer in mid-latitudes). Broad-scale convective (potential) instability is generated by the over-running of this low-level flow by potentially colder, drier air in the middle troposphere. Instability is released mainly in small-scale convection

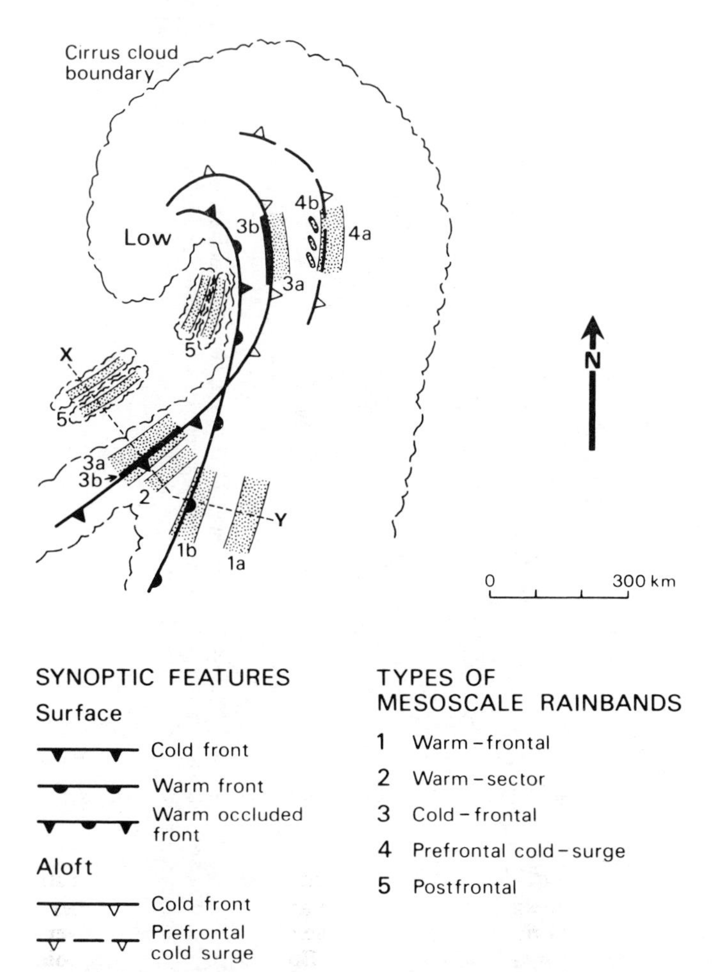

Fig. 4.11. Fronts and associated rain bands typical of a mature depression. The broken line X–Y shows the location of the cross-section given in fig. 4.12 (*after Hobbs; from Houze and Hobbs 1982*).

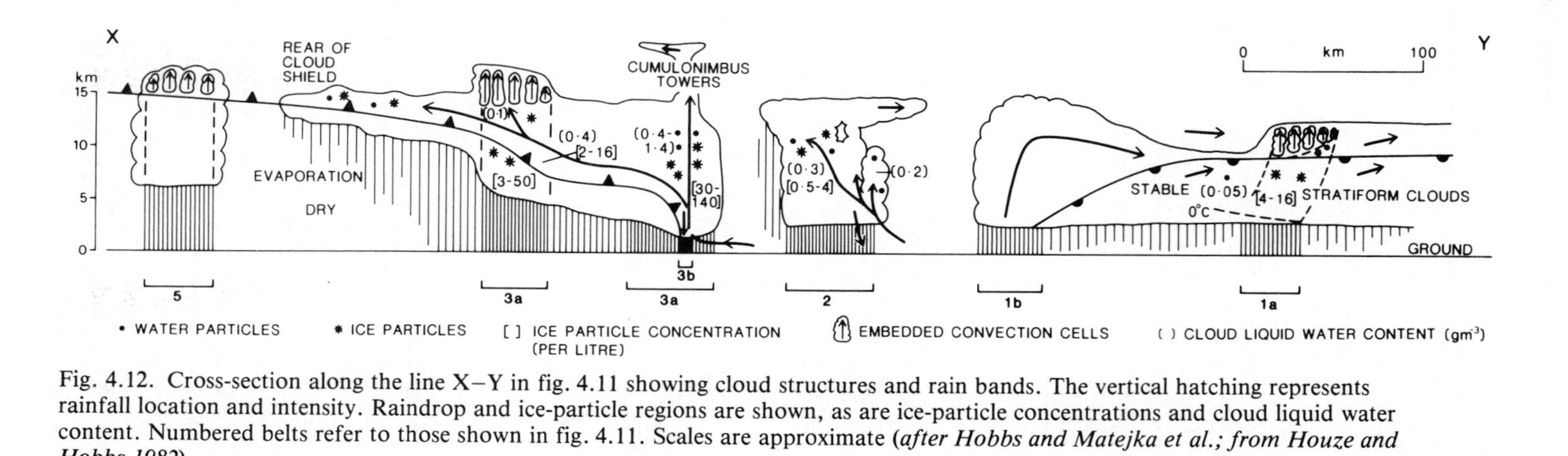

Fig. 4.12. Cross-section along the line X–Y in fig. 4.11 showing cloud structures and rain bands. The vertical hatching represents rainfall location and intensity. Raindrop and ice-particle regions are shown, as are ice-particle concentrations and cloud liquid water content. Numbered belts refer to those shown in fig. 4.11. Scales are approximate (*after Hobbs and Matejka et al.; from Houze and Hobbs 1982*).

cells that are organized into clusters, known as meso-scale precipitation areas (MPAs). These MPAs are further arranged in bands, 50–100 km wide (fig. 4.11). Ahead of the warm front, the bands are broadly parallel to the airflow in the rising section of the conveyor belt, whereas in the warm sector they parallel the cold front and the low-level jet. In some cases, cells and clusters are further arranged in bands within the warm sector and ahead of the warm front (figs. 4.11 and 4.12). Precipitation from warm front rain-bands often involves 'seeding' by ice particles falling from the upper cloud layers. It has been estimated that 20–35% of the precipitation originates in the 'seeder' zone and the remainder in the lower clouds (see also fig. 2.29). Some of the cells and clusters are undoubtedly set up through orographic effects and these influences may extend well down-wind when the atmosphere is unstable.

2 *The cold front*

The weather conditions observed at cold fronts are equally variable, depending upon the stability of the warm sector air and the vertical motion relative to the frontal zone. The classical cold-front model is of the ana-type, and the cloud is usually cumulonimbus. Over the British Isles air in the warm sector is rarely unstable, so that nimbostratus occurs more frequently at the cold front (fig. 4.9A). With the kata-cold front the cloud is generally strato-cumulus (fig. 4.9B) and precipitation is light. With ana-cold fronts there are usually brief, heavy downpours sometimes accompanied by thunder. The steep slope of the cold front, roughly 2°, means that the bad weather is of shorter duration than at the warm front. With the passage of the cold front the wind veers sharply, pressure begins to rise and temperature falls. The sky may clear very abruptly, even before the passage of the surface cold front in some cases, although with kata-cold fronts the changes are altogether more gradual.

3 *The occlusion*

Occlusions are classified as either *cold* or *warm*, the difference depending on the relative states of the cold-air masses lying in front and to the rear of the warm sector (fig. 4.13). If the air is colder than the air following it then the occlusion is warm, but if the reverse is so (which is more likely over the British Isles) it is termed a cold occlusion. The air in advance of the depression is most likely to be coldest when depressions occlude over Europe in winter and very cold cP air is affecting the continent.

The line of the warm air wedge aloft is associated with a zone of layered cloud (similar to that found with a warm front) and often of precipitation. Hence its position is indicated separately on some weather maps and it is referred to by Canadian meteorologists as a *trowal* (trough of warm air aloft). The passage of an occluded front and trowal brings a change back to polar air-mass weather.

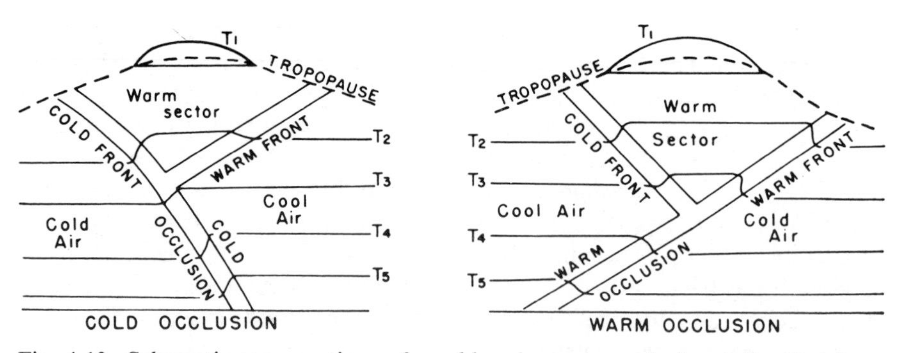

Fig. 4.13. Schematic cross-sections of a cold and a warm occlusion (*after Pedgley 1962; Crown Copyright Reserved*).

A different process occurs when there is interaction between a polar trough and the main polar front, giving rise to an instant occlusion. A warm conveyer belt on the polar front ascends as an upper tropospheric jet forming a stratiform cloud band (fig. 4.14), while a low-level polar trough conveyor belt at right angles to it produces a convective cloud band and precipitation area poleward of the main polar front (pl. 23) on the leading edge of the cold pool.

The occurrence of *frontolysis* (frontal decay) is not necessarily linked with occlusion, although it represents the final phase of a front's existence. Decay occurs when differences no longer exist between adjacent air masses. This may arise in four ways: through their mutual stagnation over a similar surface, as a result of both air masses moving on parallel tracks at the same speed, as a result of their movement in succession along the same track at the same speed, or by the system incorporating into itself air of the same temperature.

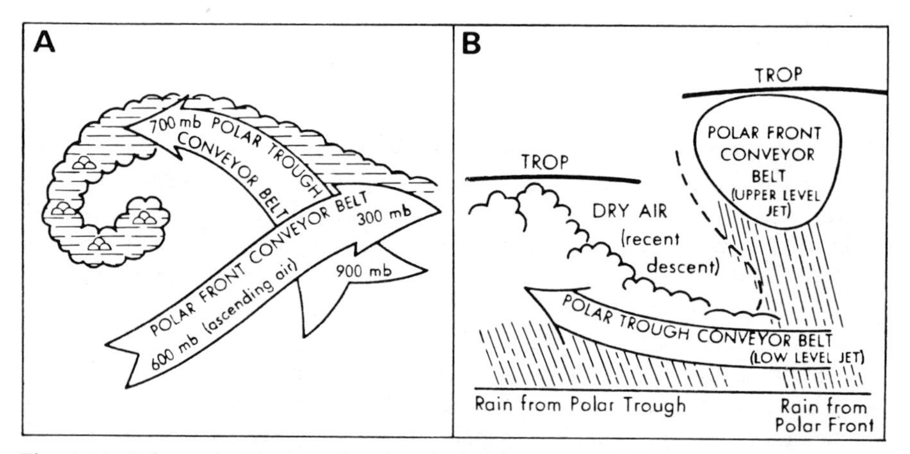

Fig. 4.14. Schematic diagram showing the interaction between a polar trough and the polar front: (A) plan view and (B) vertical section along the polar trough axis (*from Browning 1985; Reproduced by permission of the Controller of Her Majesty's Stationery Office*).

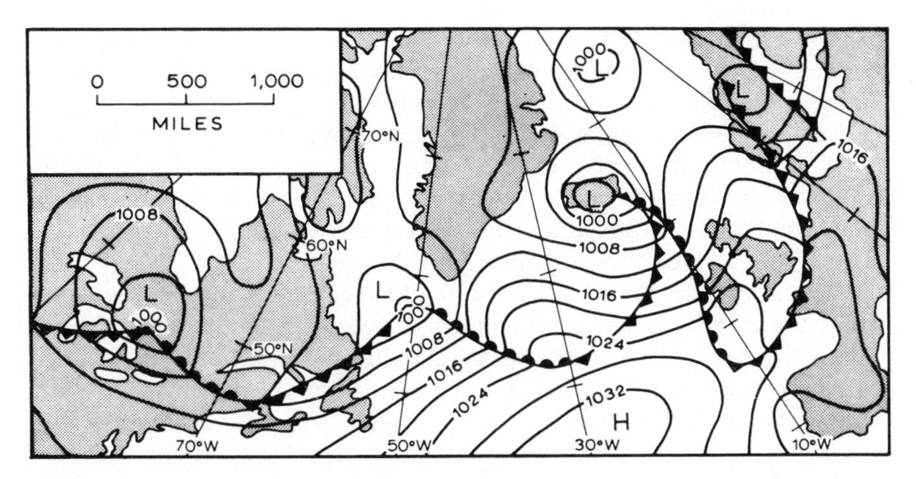

Fig. 4.15. A depression family in the North Atlantic, 22 June 1954 (*after Taylor and Yates 1967; Crown Copyright Reserved*).

4 Frontal wave families

Observation has shown that frontal waves, or depressions, do not generally occur as separate units but in *families* of three or four (fig. 4.15, pl. 32) with the depressions which succeed the original one forming as *secondaries* along the trailing edge of an extended cold front. Each new member follows a course which is south of its progenitor as the polar air pushes farther south to the rear of each depression in the series. Eventually the front trails far to the south and the cold polar air forms an extensive meridional wedge of high pressure terminating the sequence.

Another pattern of development may take place on the warm front, particularly at the point of occlusion, as a separate wave forms running ahead of the parent depression. This type of secondary is more likely with very cold (cA, mA or cP) air ahead of the warm front, and its formation is encouraged when the eastward movement of the occlusion is barred by mountains. This situation often occurs when a primary depression is situated in the Davis Strait and a break-away wave forms south of Cape Farewell (the southern tip of Greenland), moving away eastwards. Analogous developments take place in the Skagerrak–Kattegat area when the occlusion is held up by the Scandinavian mountains.

E Zones of wave development and frontogenesis

Fronts and associated depressions do not form everywhere and their development is restricted to well defined areas. Frontal formation in the temperate latitudes has been studied intensively for some years and knowledge of the general weather conditions to be expected is reasonably accurate. Not nearly so much is known about the nature of tropical fronts but the conditions of their formation and development are unlike those normally associated with higher-latitude fronts. Increasing world air travel and the need of accurate forecasts for tropical routes is fast filling this gap. At the moment it seems that Arctic and Polar Fronts are caused primarily by gross differences in air-mass characteristics, whereas tropical discontinuities within and between somewhat similar air masses are produced mainly by the nature of the large-scale air motion and especially by confluence within an airstream or between two air currents of different humidity.

The major zones of frontal wave development are naturally those areas which are most frequently baroclinic as a result of airstream confluence. This is the case, for instance, off eastern Asia and eastern North America, especially in winter when there is a sharp temperature gradient between the snow-covered land and warm offshore currents. These zones are referred to respectively as the Pacific Polar and Atlantic Polar Fronts (fig. 4.16). Their position is quite variable, but they show a general equatorward shift in winter, when the Atlantic Frontal Zone may extend into the Gulf of Mexico. In this area there is convergence of air masses of different stability between adjacent subtropical high-pressure cells (this frontal zone is sometimes misleadingly termed 'temperate'). Depressions developing here commonly move north-eastwards, sometimes following or amalgamating with others of the northern part of the Polar Front proper or of the Canadian Arctic Front. Frontal frequency remains high across the North Atlantic, but it decreases eastward in the North Pacific, perhaps owing to a less pronounced gradient of sea-surface temperature. Frontal activity is most common in the central North Pacific when the subtropical high is split into two cells with converging air currents between them.

Another section of the Polar Front, often referred to as the *Mediterranean Front*, is located over the Mediterranean–Caspian Sea areas in winter. At intervals, fresh Atlantic mP air, or cool cP air from south-east Europe, converges with warmer air masses often of North African origin, over the Mediterranean Basin and initiates frontogenesis. In summer the area lies under the influence of the Azores subtropical high-pressure cell and the frontal zone is absent.

The summer locations of the Polar Front over the western Atlantic and Pacific are some 10° further north than in winter (fig. 4.16) although the frontal zone is rather weak at this time of year. There is now a frontal zone over Eurasia and a corresponding one over middle North America. These reflect the general meridional temperature gradient and probably also the large-scale influence of orography on the general circulation (see ch. 4, F).

In the southern hemisphere the Polar Front is on average about 45°S in January (summer) with branches spiralling poleward towards it from about 32°S off eastern South America and from 30°S, 150°W in the South Pacific (fig. 4.17). In July (winter) there are two Polar Frontal Zones spiralling towards Antarctica from about 20°S; one starts over South America and the other at 170°W. They terminate some 4°–5° latitude further poleward than in summer. It is noteworthy that the southern hemisphere has more cyclonic

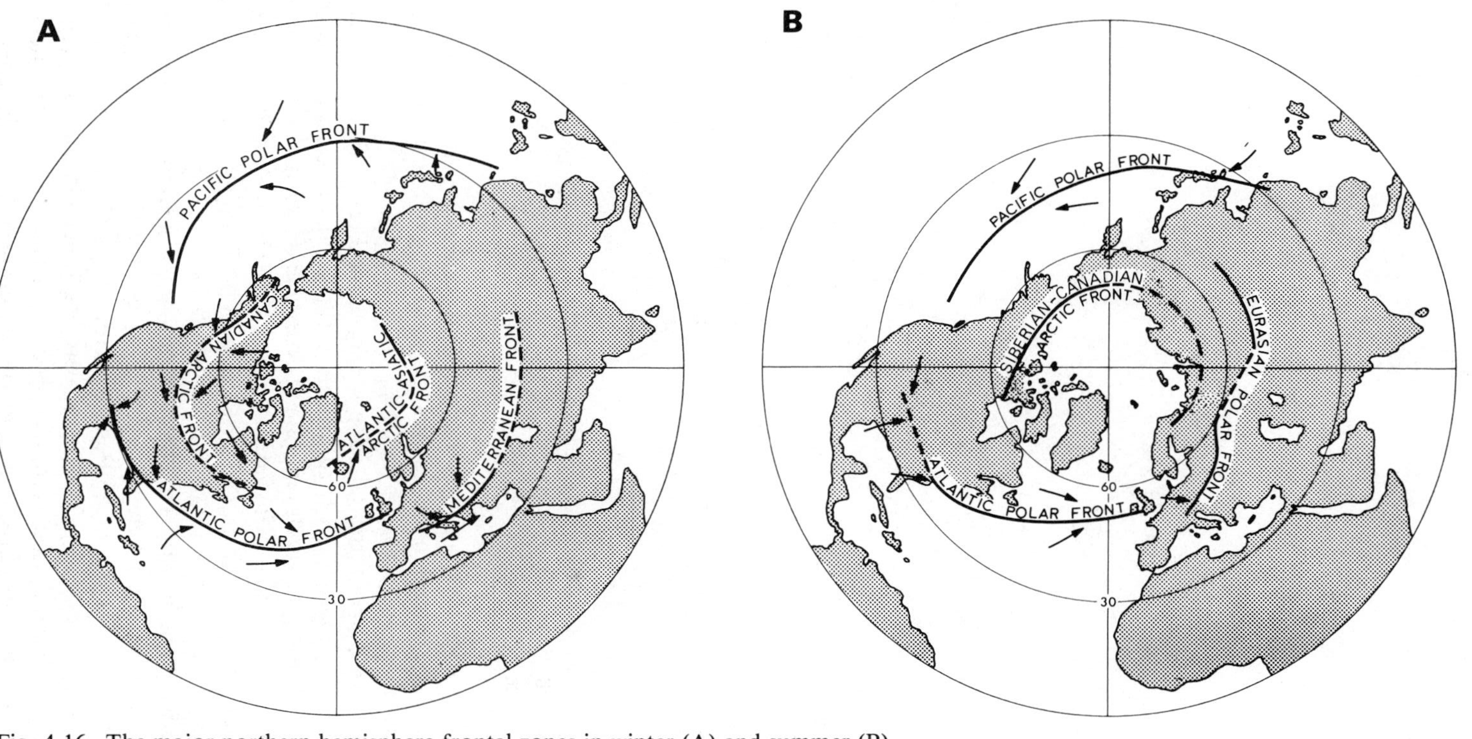

Fig. 4.16. The major northern hemisphere frontal zones in winter (A) and summer (B).

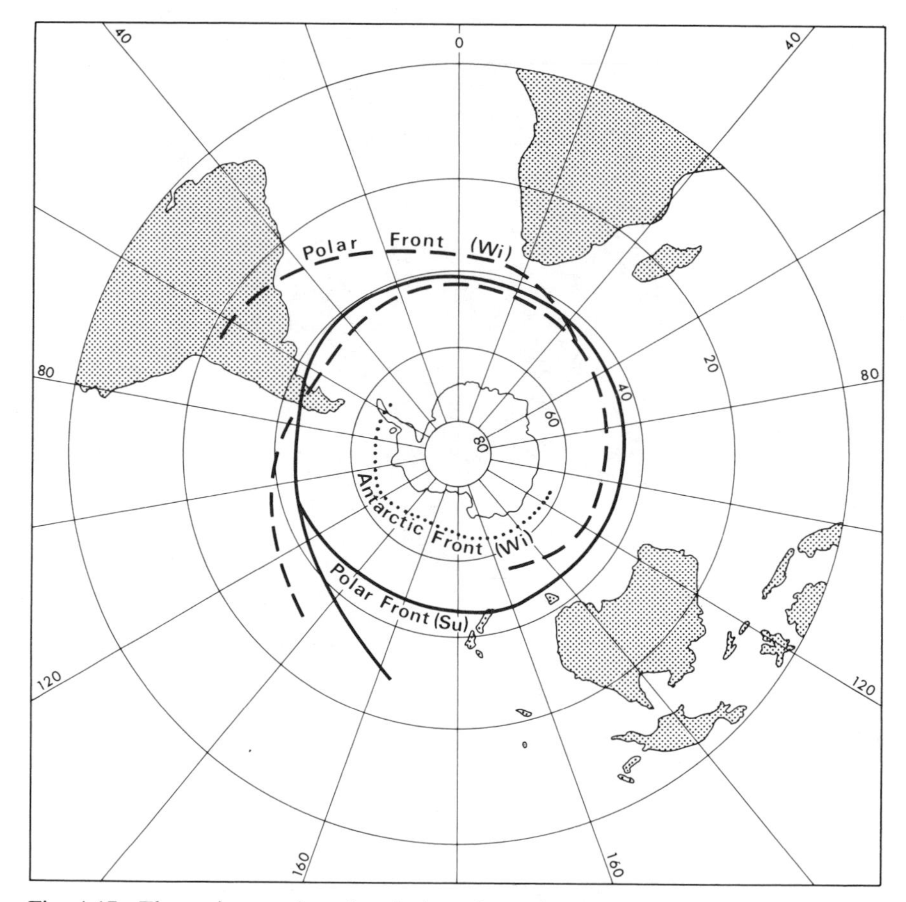

Fig. 4.17. The major southern hemisphere frontal zones in winter (Wi) and summer (Su).

activity in summer than does the northern hemisphere in its summer. This appears to be related to the seasonal strengthening of the meridional temperature gradient noted earlier (p. 139).

The second major frontal zone is the Arctic Front, associated with the snow and ice margins of high latitudes (fig. 4.16). In summer this zone is developed along the tundra margin in Siberia and North America. In winter over North America it is formed between cA (or cP) air and Pacific maritime air modified by crossing the Coast Ranges and Rockies. There is also a less pronounced arctic frontal zone in the North Atlantic–Norwegian Sea area, extending along the Siberian coast. A similar weak frontal zone is found in winter in the southern hemisphere. It is located at 65°–70°S near the edge of the antarctic pack-ice in the Pacific sector (fig. 4.17), although rather few cyclones form there. Zones of airstream confluence in the southern hemisphere (cf. figs. 4.2 and 4.4) are less numerous and more persistent, particularly in coastal regions, than in the northern hemisphere.

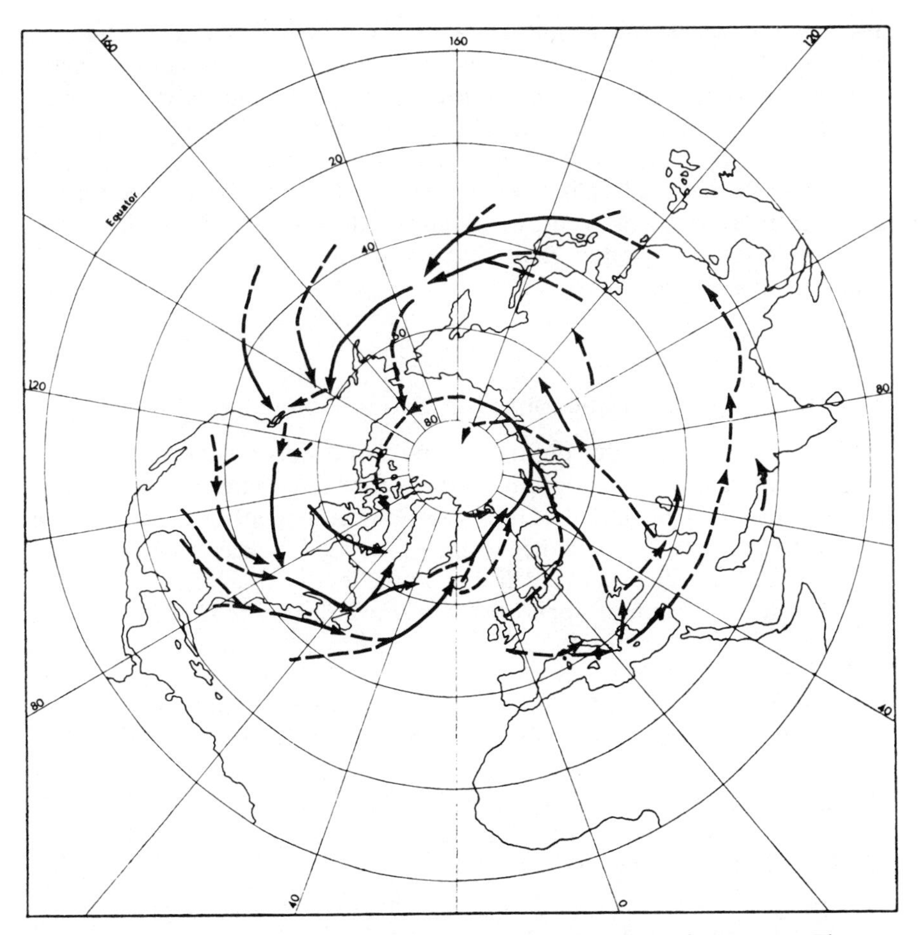

Fig. 4.18. The principal northern hemisphere depression tracks in January. The full lines show major tracks, the dashed lines secondary ones which are less frequent and less well defined. The frequency of lows is a local maximum where arrow-heads end. An area of frequent cyclogenesis is indicated where a secondary track changes to a primary one or where two secondary tracks merge to form a primary (*after Klein 1957*).

The principal tracks of depressions in the northern hemisphere in January are shown in fig. 4.18. The major ones reflect the primary frontal zones already discussed. In summer the Mediterranean route is absent and lows move across Siberia, but the other tracks are similar although generally more zonal at this season and located in higher latitudes (around 60°N).

Between the two hemispherical belts of subtropical high pressure there is a further major world convergence zone, the Inter-Tropical Convergence Zone (or ITCZ). Formerly this was referred to as the Inter-Tropical Front (ITF), but air-mass contrasts only occur in limited sectors. This zone moves seasonally north and south away from the equator, as the subtropical high-pressure cell activity alternates in opposite hemispheres. The contrast between the converging air masses obviously increases with the distance of the ITCZ from the equator, and the degree of difference in their characteristics is naturally associated with considerable variation in activity along the convergence zone. Activity is most intense in June–July over southern Asia and west Africa, when the contrast between the maritime and continental air masses which are involved is at a maximum. In these sectors the zone merits the term Inter-Tropical Front, although this does not imply that it behaves like a mid-latitude frontal zone. The nature of the ITCZ and its role in tropical weather are discussed in ch. 6.

F Surface/upper-air relationships and the formation of depressions

It has already been pointed out that a wave depression is associated with air-mass convergence, yet the barometric pressure at the centre of the low may decrease by 10–20 mb in 12–24 hours as the system intensifies. The explanation of this apparent discrepancy is that upper air divergence removes rising air more quickly than convergence at lower levels replaces it. The superimposition of a region of upper divergence over a frontal zone is the prime motivating force of *cyclogenesis* (i.e. depression formation).

The long (or *Rossby*) waves in the middle and upper troposphere, which were mentioned in ch. 3, D.2, are particularly important in this respect, and it is worth considering first the reason why the hemispheric westerlies show this large-scale wave motion. The key to this problem lies in the rotation of the earth and the latitudinal variation of the Coriolis parameter (ch. 3, A.2). It can be shown that for large-scale motion the absolute vorticity about a vertical axis $(f + \zeta)$ tends to be conserved, i.e.

$$\frac{d(f + \zeta)}{dt} = 0$$

The symbol d/dt denotes a rate of change following the motion (a total differential). Consequently, if air moves poleward so that f increases, the cyclonic vorticity tends to decrease. The curvature thus becomes anticyclonic and the current returns towards lower latitudes. If the air moves equatorward of its original latitude f tends to decrease (fig. 4.19), requiring ζ to increase, and the resulting cyclonic curvature again deflects the current polewards. In this manner large-scale flow tends to oscillate in a wave pattern.

Rossby related the motion of these waves to their wavelength (L) and the speed of the zonal current (u). The speed of the wave (or phase speed, c) is:

$$c = u - \beta \left(\frac{L}{2\pi}\right)^2$$

where $\beta = \partial f/\partial y$, i.e. the variation of the Coriolis parameter with latitude (a local, partial differential). For stationary waves, where $c = 0$, $L = 2\pi\sqrt{u/\beta}$. At 45° latitude this stationary wavelength is 3120 km for a zonal velocity of 4 m s^{-1}, increasing to 5400 km at 12 m s^{-1}. The wavelengths, at 60° latitude

where $\beta = 2f/2y$, i.e. the variation of the Coriolis parameter with latitude (a

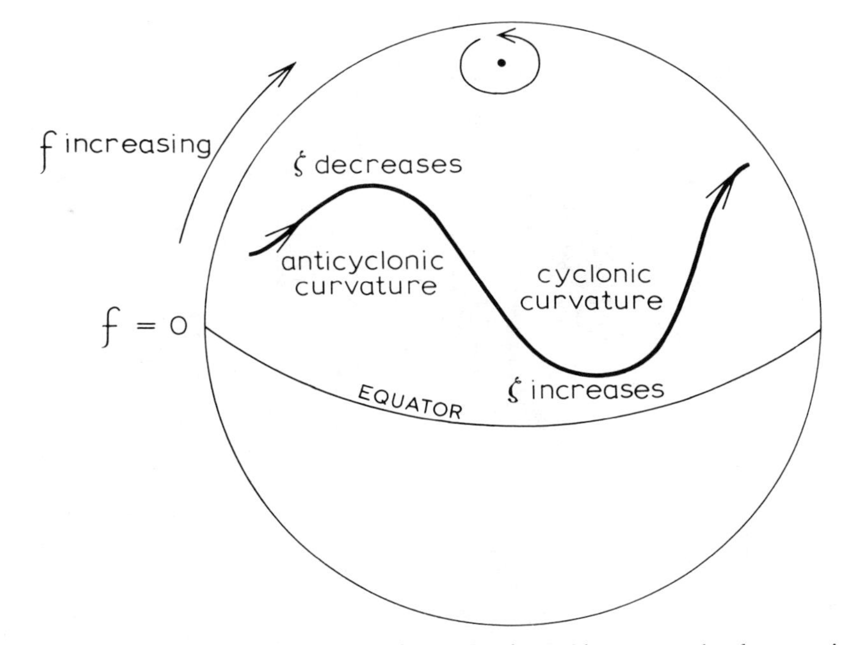

Fig. 4.19. A schematic illustration of the mechanism of long-wave development in the tropospheric westerlies.

for zonal currents of 4 and 12 m s^{-1} are, respectively, 3170 and 6430 km. Long waves tend to remain stationary, or even to move westward against the current, so that $c \leqslant 0$. Shorter waves travel eastward with a speed close to of the zonal current and tend to be steered by the quasi-stationary long waves.

The latitudinal circumference limits the circumpolar westerly flow to between three and six major Rossby waves, and these affect the formation and movement of surface depressions. It has been pointed out that the main stationary waves tend to be located about 70°W and 150°E in response to the influence on the atmospheric circulation of orographic barriers, such as the Rocky Mountains and the Tibetan plateau, and of heat sources. On the eastern limb of troughs in the upper westerlies of the northern hemisphere the flow is normally divergent, since the gradient wind is subgeostrophic in the trough but supergeostrophic in the ridge (see ch. 3, A.4). Thus the sector ahead of an upper trough is a very favourable location for a surface depression to form or deepen, and it will be noted that the mean upper troughs are significantly positioned just west of the Atlantic and Pacific Polar Front Zones in winter.

With these ideas in mind we may now consider further the three-dimensional nature of depression development and the important links existing between upper and lower tropospheric flow. The basic theory relates to the vorticity equation which states that, for frictionless horizontal motion, the rate of change of the vertical component of absolute vorticity

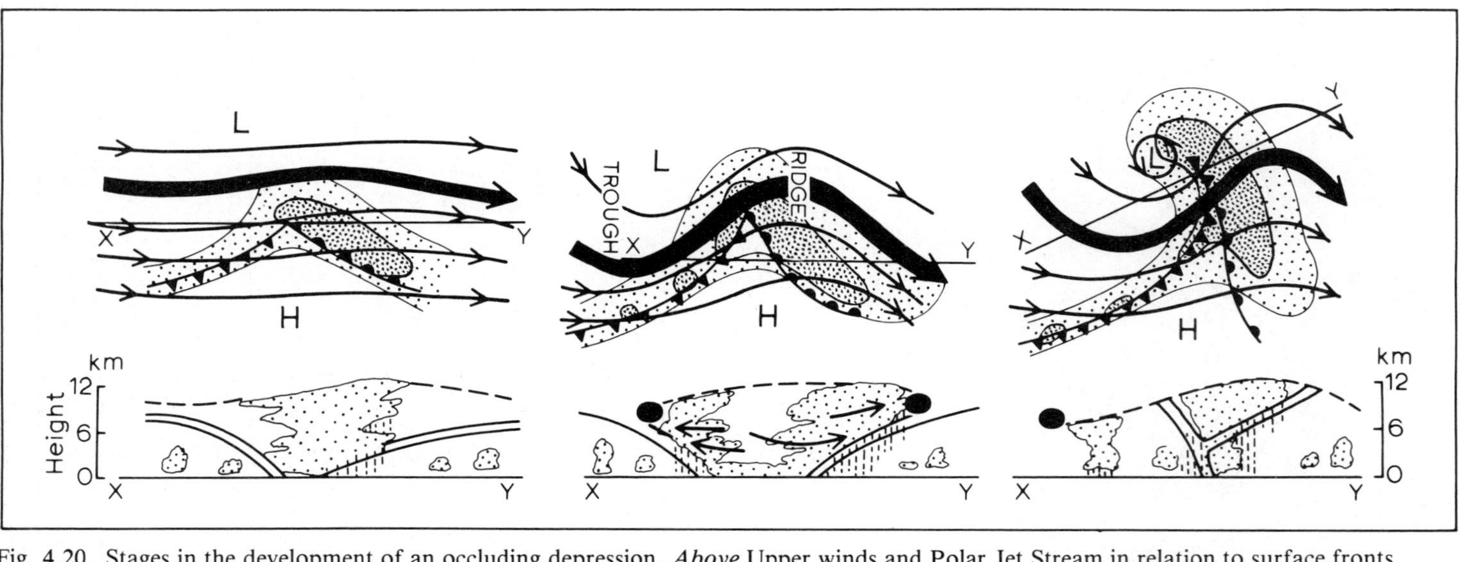

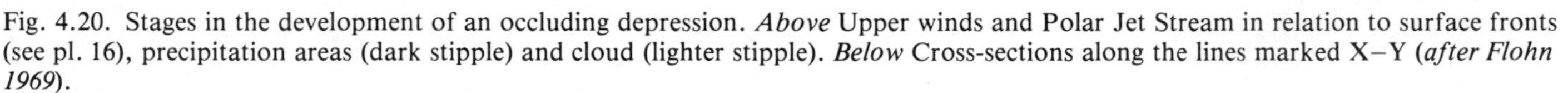
Fig. 4.20. Stages in the development of an occluding depression. *Above* Upper winds and Polar Jet Stream in relation to surface fronts (see pl. 16), precipitation areas (dark stipple) and cloud (lighter stipple). *Below* Cross-sections along the lines marked X–Y (*after Flohn 1969*).

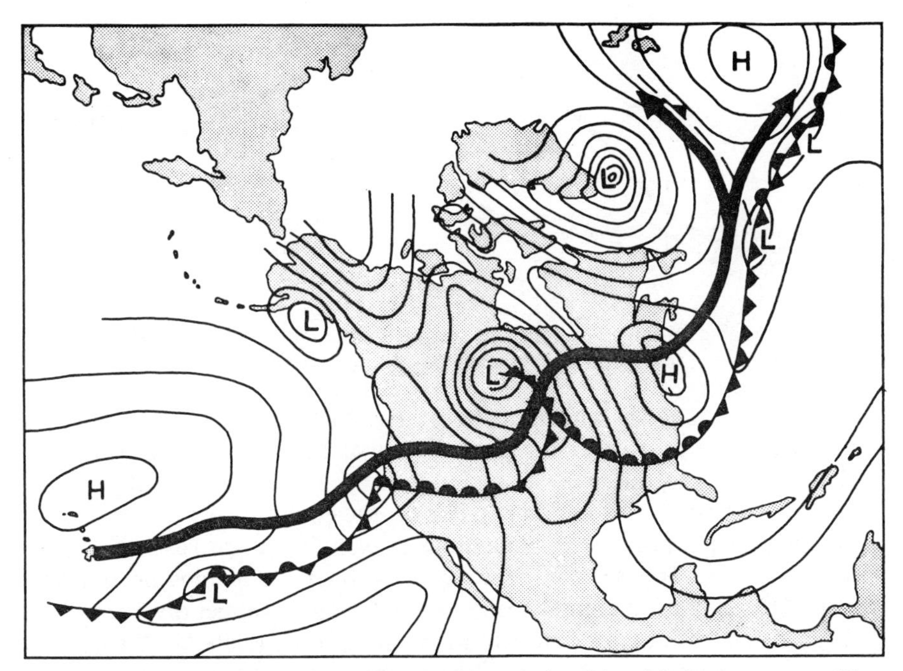

Fig. 4.21. A typical depression family and its relationship with the jet stream. The thin lines are sea-level isobars (*after Vederman 1954*).

(dQ/dt or $d(f+\zeta)/dt$) is proportional to air-mass convergence ($-D$, i.e. negative divergence):

$$\frac{dQ}{dt} = -DQ \quad \text{or} \quad D = -\frac{1}{Q}\frac{dQ}{dt}$$

The relationship implies that a converging (diverging) air column has increasing (decreasing) absolute vorticity. The conservation of vorticity equation, which we have already discussed, is in fact a special case of this relationship.

In the sector ahead of an upper trough the decreasing cyclonic vorticity causes divergence (i.e. D positive), since the change in ζ outweighs that in f, thereby favouring surface convergence and low-level cyclonic vorticity. Once the surface cyclonic circulation has become established vorticity production is increased through the effects of thermal advection. Poleward transport of warm air in the warm sector and the eastward advance of the cold upper trough act to sharpen the baroclinic zone, strengthening the upper jet stream through the thermal wind mechanism (see p. 137). The vertical relationship between jet stream and front has already been shown (fig. 3.21); a model depression sequence is demonstrated in plan view in fig. 4.20. The actual relationship may depart from this idealized case, although the jet is commonly located in the cold air as illustrated by the synoptic example of fig. 4.21. Velocity maxima (core zones) occur along the

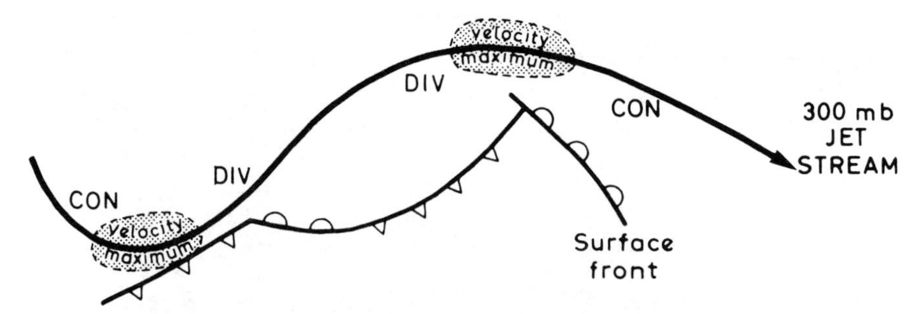

Fig. 4.22. Model of the jet stream and surface fronts, showing zones of upper tropospheric divergence and convergence and the jet-stream cores.

jet stream, as shown schematically in fig. 4.22, and the distribution of vertical motion upstream and downstream of these cores is known to be quite different. In the area of the jet entrance (i.e. upstream of the core) divergence causes lower-level air to rise on the equatorward (i.e. right) side of the jet, whereas in the exit zone (downstream of the core) ascent is on the poleward side. The second depression in fig. 4.22 is moving eastward towards the area of maximum cyclogenetic tendency (upper divergence).

Figure 4.23 shows how precipitation is often more related to the position of the jet stream than to that of surface fronts; maximum precipitation areas are in the right entrance sector of the jet core. This vertical motion pattern is also of basic importance in the initial deepening stage of the depression. If the upper-air pattern is unfavourable (e.g. beneath left entrance and right exit zones, where there is convergence) the depression will fill. Note that to the rear of the second depression in fig. 4.22 upper convergence will encourage a polar outbreak through subsidence.

The development of a depression can also be considered in terms of energy transfers. A cyclone requires the conversion of potential into kinetic energy

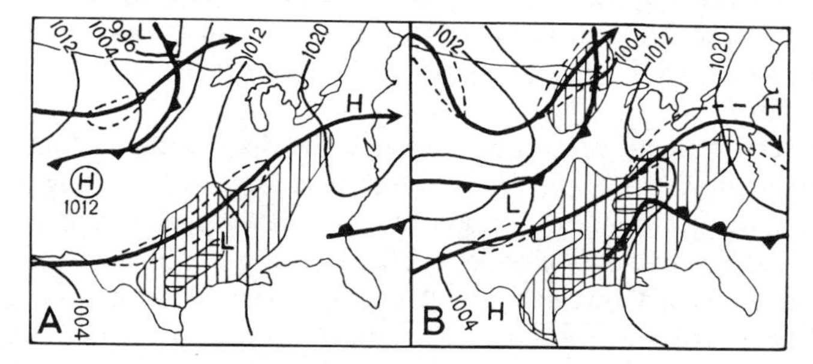

Fig. 4.23. The relations between surface fronts and isobars, surface precipitation ($\lesssim$ 25 mm vertical hatching; > 25 mm cross-hatching), and jet streams (wind speeds in excess of about 100 mph (45 m s^{-1}) occur within the dashed lines) over the United States on (A) 20 September 1958 and (B) 21 September 1958. This illustrates how the surface precipitation area is related more to the position of the jets than to that of the surface fronts. The air over the south-central United States was close to saturation, whereas that associated with the northern jet and the maritime front was much less moist (*after Richter and Dahl 1958*).

and this achieved by the upward (and poleward) motion of *warm* air. The rising warm air is driven by the vertical wind shear and by the superimposition of upper tropospheric divergence over a baroclinic zone. Intensification of this zone further strengthens the upper winds. The upper divergence allows surface convergence and pressure fall to occur simultaneously. Modern theory relegates the fronts to a quite subordinate role. They develop within depressions as narrow zones of intensified ascent, probably through the effects of cloud formation.

The movement of depressions is determined essentially by the upper westerlies and, as a rule of thumb, a depression centre travels at about 70% of the surface geostrophic wind speed in the warm sector. Records for the United States indicate that the average speed of depressions is 32 km h^{-1} (20 mph) in summer and 48 km h^{-1} (30 mph) in winter. The higher speed in winter reflects the stronger westerly flow in response to a greater meridional temperature gradient. Shallow depressions are mainly steered by the direction of the thermal wind in the warm sector and hence their path closely follows that of the upper jet stream (see ch. 3, D.3). Deep depressions may greatly distort the thermal pattern, however, as a result of northward transport of warm air and the southward transport of cold air. In such cases the depression usually becomes slow-moving. The movement of a depression may be additionally guided by energy sources such as a warm sea surface, which generates cyclonic vorticity, or by mountain barriers. The depression may cross obstacles, such as the Rocky Mountains or the Greenland Ice Sheet, as an upper low or trough and subsequently redevelop, aided by the lee-effects of the barrier or by fresh injections or contrasting air masses.

The location and intensity of storm tracks appears to be critically influenced by ocean surface temperatures. Investigations for the North Pacific and the North Atlantic demonstrate interactions with the atmospheric circulation on a near-hemispheric scale. Figure 4.24B suggests, for example, that an extensive relatively warm surface in the north-central Pacific in the winter of 1971–2 caused a northward displacement of the westerly jet stream together with a compensating southward displacement over the western United States, bringing in cold air there. This pattern contrasts with that observed during the 1960s (fig. 4.24A) when a persistent cold anomaly in the central Pacific, with warmer water to the east, led to frequent storm development in the intervening zone of strong temperature gradient. The associated upper airflow produced a ridge over western North America with warm winters in California and Oregon. Models of the global atmospheric circulation support the view that persistent anomalies of sea-surface temperature exert an important control on local and large-scale weather conditions.

G Non-frontal depressions

Not all depressions originate as frontal waves. Tropical depressions are indeed mainly non-frontal and these are considered in ch. 6. In middle and high latitudes four types which develop in distinctly different situations are

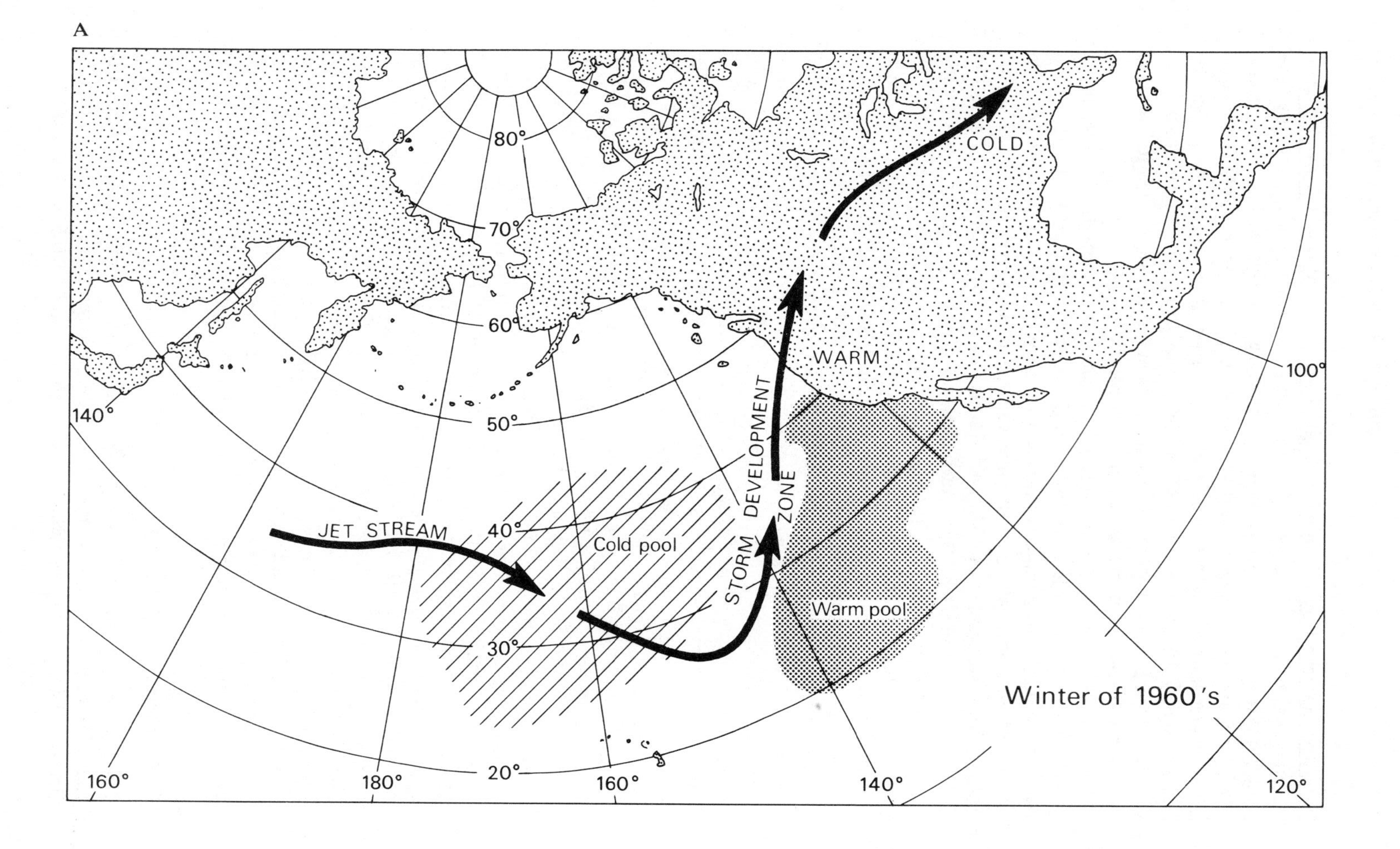
A
COLD
WARM
STORM DEVELOPMENT ZONE
JET STREAM
Cold pool
Warm pool
Winter of 1960's
80°
70°
60°
50°
40°
30°
20°
160°
180°
160°
140°
120°
100°
140°

B

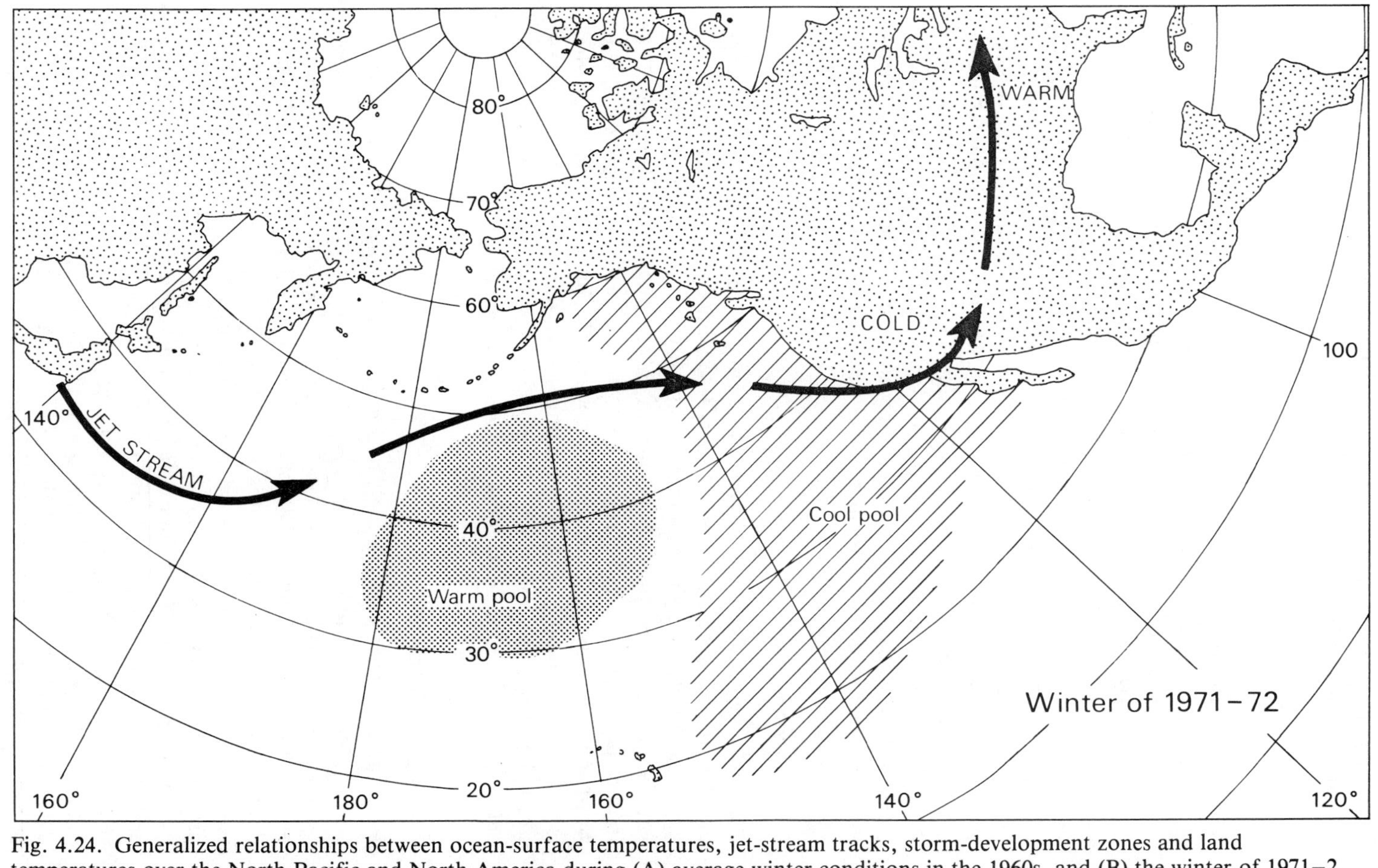

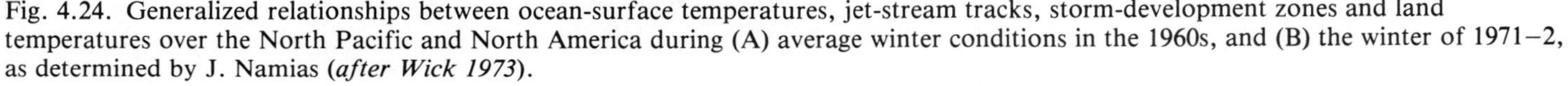
Fig. 4.24. Generalized relationships between ocean-surface temperatures, jet-stream tracks, storm-development zones and land temperatures over the North Pacific and North America during (A) average winter conditions in the 1960s, and (B) the winter of 1971–2, as determined by J. Namias (*after Wick 1973*).

of particular importance and interest: the lee depression, the thermal low, the polar air depression and the cold low.

1 The lee depression

Westerly airflow that is forced over a north–south mountain barrier undergoes vertical contraction over the ridge and expansion on the lee side. This vertical movement creates compensating lateral expansion and contraction, respectively. Hence there is a tendency for divergence and anticyclonic curvature over the crest, convergence and cyclonic curvature in the lee of the barrier. Wave troughs may be set up in this way on the lee side of low hills (see fig. 3.11) as well as major mountain chains. The airflow characteristics and the size of the barrier determine whether or not a closed low-pressure system actually develops. Such depressions, which at least initially tend to remain 'anchored' by the barrier, are frequent in winter to the south of the Alps when the mountains block the low-level flow of north-westerly airstreams. Fronts may occur in these depressions but it is important to recognize that the low does not form as a wave along a frontal zone.

2 The thermal low

These lows occur almost exclusively in summer, resulting from intense daytime heating of continental areas. Figure 3.15C illustrates their vertical structure. The most impressive examples are the summer low-pressure cells over Arabia, the northern part of the Indian sub-continent and Arizona. The Iberian peninsula is another region commonly affected by such lows. The weather accompanying them is usually hot and dry, but if sufficient moisture is present the instability caused by heating may lead to showers and thunderstorms. Thermal lows normally disappear at night when the heat source is cut off, but in fact those of India and Arizona persist.

3 Polar air depressions

Polar air depressions are a loosely defined class of mesocale to subsynoptic scale systems (a few hundred kilometres across) with a lifetime of 1–2 days. On satellite imagery they appear as a cloud spiral with one or several cloud bands, as a comma cloud (pl. 24), or as a swirl in cumulus cloud streets. They develop mainly in winter months when unstable mP or mA air currents stream equatorward along the eastern side of a north–south ridge of high pressure, commonly in the rear of an occluding primary depression. They usually form within a baroclinic zone, e.g. near sea-ice margins where there are strong sea-surface temperature gradients, and their development may be stimulated by an initial upper-level disturbance.

In the northern hemisphere, the comma cloud type (which is mainly a cold core disturbance of the middle troposphere) is more common over the North Pacific, while the spiral-form polar low occurs more often in the Norwegian

Sea. The latter is a low-level warm core disturbance which may have a closed cyclonic circulation up to about 800 mb or may consist simply of one or more troughs embedded in the polar airflow. A key feature is the presence of an ascending, moist, south-westerly flow *relative* to the low centre. This organization accentuates the general instability of the cold airstream to give considerable precipitation often as snow. Heat input to the cold air from the sea continues by night and day so that in exposed coastal districts showers may occur at any time.

4 The cold low

The cold low (or *cold pool*) is usually most evident in the circulation and temperature fields of the middle troposphere. Characteristically it displays symmetrical isotherms about the low centre. Surface charts may show little or no sign of these persistent systems which are frequent over north-eastern North America and north-eastern Siberia. They probably form as the result of strong vertical motion and adiabatic cooling in occluding baroclinic lows along the arctic coastal margins. Such lows are especially important during the arctic winter in that they bring large amounts of medium and high cloud which offset radiational cooling of the surface. Otherwise they usually cause no 'weather' in the Arctic during this season. It is important to emphasize that tropospheric cold lows may be linked with either low- or high-pressure cells at the surface.

In middle latitudes cold lows may also form during periods of low-index circulation pattern (see fig. 3.37) by the cutting-off of polar air from the main body of cold air to the north (these are sometimes referred to as *cut-off lows*). This gives rise to weather of polar air-mass type, although rather weak fronts may also be present. Such lows are commonly slow-moving and give persistent unsettled weather with thunder in summer. Heavy precipitation over Colorado in spring and autumn is often associated with cold lows.

H Meso-scale convective systems

Meso-scale convective systems are intermediate in size and life span between synoptic disturbances and individual cumulonimbus cells (fig. 4.25). They occur in both middle latitudes and the tropics as either nearly circular clusters of convective cells or linear squall lines. The *squall line* consists of a narrow line of thunderstorm cells which may extend for hundreds of kilometres. It is marked by a sharp veer of wind direction and very gusty conditions. The squall line often occurs ahead of a kata-cold front maintained either as a self-propagating disturbance or by thunderstorm downdraughts. It may form a pseudo-cold front between rain-cooled air and a rainless zone within the same air mass. In frontal cyclones, cold air in the rear of the depression may overrun air in the warm sector. The intrusion of this nose of cold air sets up great instability and the subsiding cold wedge tends to act as a scoop forcing up the slower-moving warm air (pl. 14).

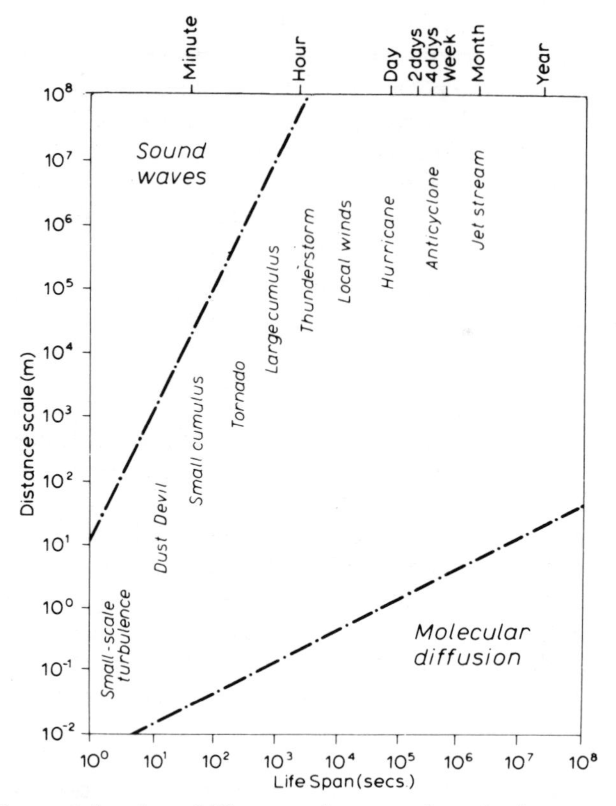

Fig. 4.25. The spatial scale and life span of meso-scale and other meteorological systems (*after Smagorinsky 1974*).

Figure 4.26 shows the movement of clusters of convective cells, each cell about 1 km in diameter, as they crossed southern Britain with a cold front. Each individual cell may be short-lived, but cell clusters may persist for hours, strengthening or weakening due to orographic and other factors.

Figure 4.27 shows that the *relative* motion of the warm air is towards the squall line. Such conditions generate severe frontal thunderstorms like that which struck Wokingham, England, in September 1959. This moved from the south-west at about 20 m s^{-1}, steered by strong south-westerly flow aloft. The cold air subsided from high levels as a violent squall and the updraught ahead of this produced an intense hailstorm. The hailstones grow by accretion in the upper part of the updraught, where speeds in excess of 50 m s^{-1} are not uncommon, are blown ahead of the storm by strong upper winds, and begin to fall. This causes surface melting but the stone is caught up again by the advancing squall line and re-ascends. The melted surface freezes, giving glazed ice as the stone is carried above the freezing level and further growth occurs by the collection of super-cooled droplets (see also ch. 2, pp. 89 and 91).

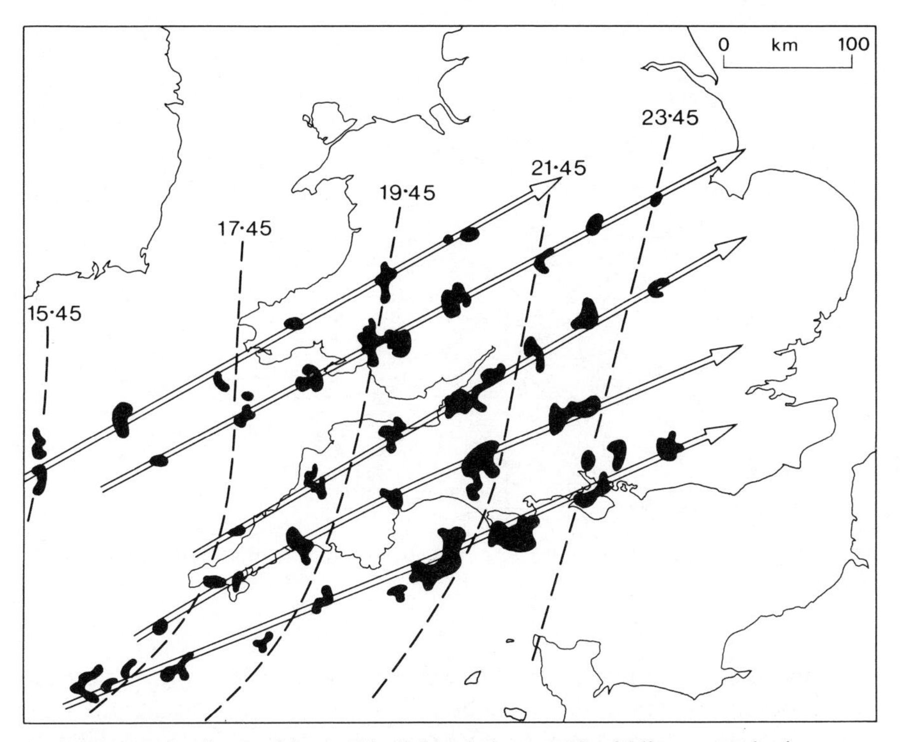

Fig. 4.26. Successive positions of individual clusters of middle-tropospheric convective cells moving across southern Britain at about 50 km h^{-1} with a cold front. Cell location and intensity were determined by radar (see pl. 15) (*after Browning 1980*).

The *meso-scale convective complex* is common over the central United States in spring and summer, where it brings widespread severe weather. It develops from initially isolated cumulonimbus cells. The sequence, illustrated in fig. 4.28, appears to be as follows. As rain falls from the thunderstorm clouds, evaporative cooling of the air beneath the cloud bases sets up cold downdraughts (see fig. 2.21) and when these become sufficiently extensive they create a local high pressure of a few millibars' intensity. The downdraughts trigger the ascent of displaced warm air and a general warming of the middle troposphere results from latent heat release. Inflow develops towards this warm region, above the cold outflow, causing additional convergence of moist unstable air. In at least some cases, this inflow is provided by a low-level jet (see fig. 4.32). As individual cells become organized in a cluster along the leading edge of the surface high, new cells tend to form on the right flank (fig. 4.28) through interaction of cold downdraughts with the surrounding air. Through this process and the decay of older cells on the left flank, the storm system tends to move 10°–20° to the right of the mid-tropospheric wind direction. As the thunderstorm high intensifies a 'wake low', associated with clearing weather, develops to the rear of it. The system is now producing violent winds, intense downpours of

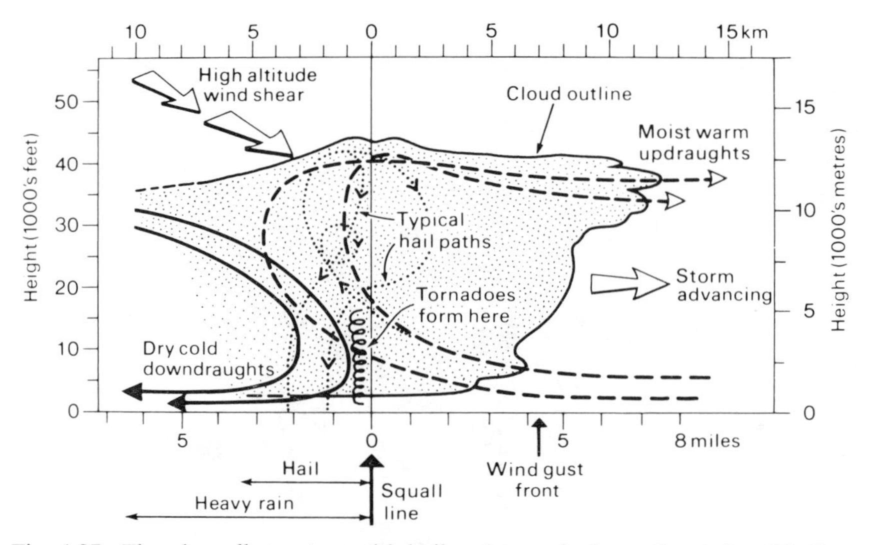

Fig. 4.27. Thunder cell structure with hail and tornado formation (*after Hindley 1977*).

rain and hail accompanied by thunder. During the triggering of new cells tornadoes may form, as discussed below. As the complex reaches maturity, during the evening and night hours over the Great Plains of the United States, the meso-scale circulation is capped by an extensive (> 100,000 km^2) cold upper cloud shield, readily identified on infrared satellite images. Statistics for 43 systems over the Great Plains in 1978 showed that the systems lasted on average 12 hours, with initial meso-scale organization occurring in the early evening (1800–1900 LST) and maximum extent 7 hours later. During their life cycle, systems may travel from the Colorado–Kansas border to the Mississippi River or the Great Lakes, or from the Missouri–Mississippi river valley to the east coast. The complex usually decays when synoptic scale features inhibit its self-propagation. The production of cold air is shut off when new convection ceases, so that the meso-high and -low are steadily weakened and the rainfall becomes light and sporadic, eventually stopping altogether.

On 14 September 1968 a belt of intense thunderstorms moved northwards into south-east England. Strong convergence of air from the south and north-east occurred along this trough in association with jet streams aloft allowing intense surface convection of hot moist air. The intense rainstorms during the period 1000–1700 hours (fig. 4.29) were the first phase of a 2-day period of instability, yielding 200 mm (7·87 in) of rainfall at Tilbury just east of London, which proved to be the outstanding rainfall event of the century in south-east England.

In regions of great potential instability and with strong vertical wind shear supercell thunderstorms may develop (fig. 4.30). These are about the same size as thundercell clusters but are dominated by one giant updraught and

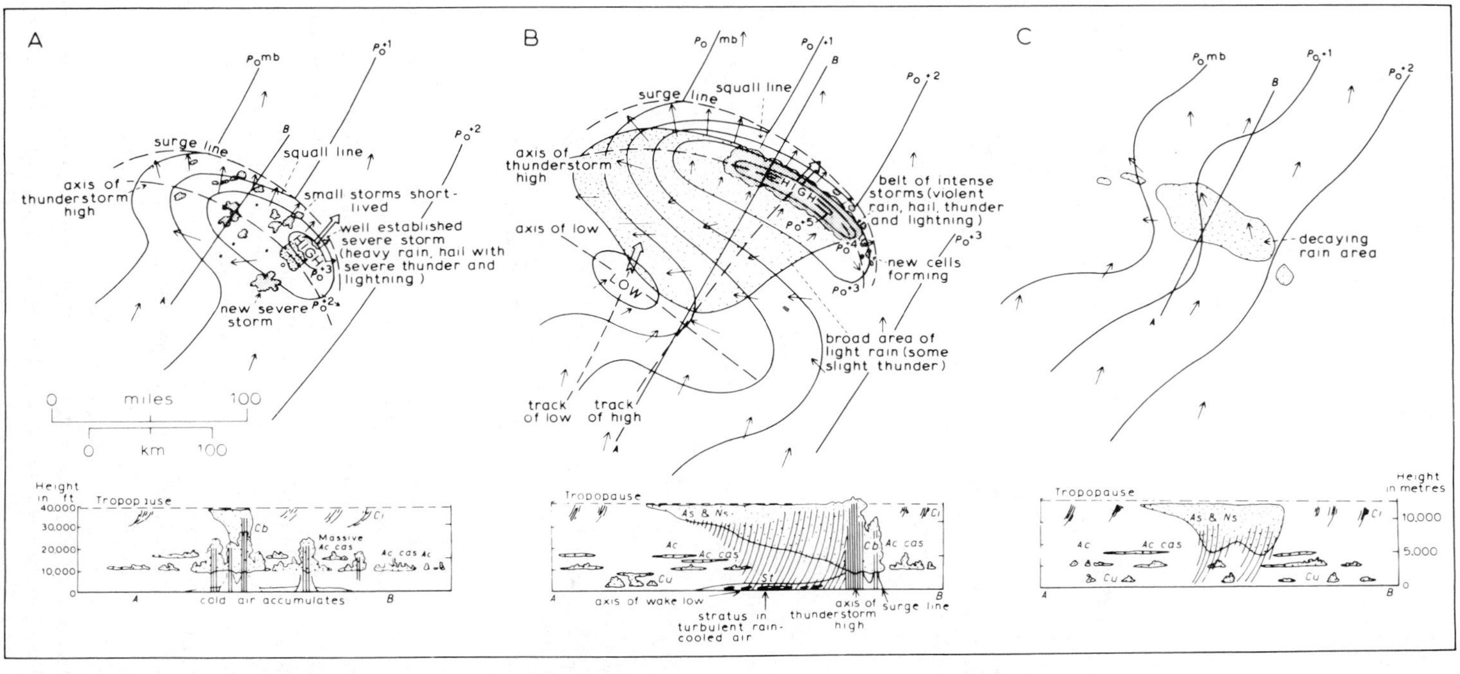

Fig. 4.28. Stages in the development of a meso-scale thunderstorm system: growth (A), maturity (B) and decay (C). The large arrows show the directions of movement of parts of the system, and the small arrows the surface winds. Areas of heavy and light precipitation are delimited (*after Pedgley 1962; Crown Copyright Reserved*).

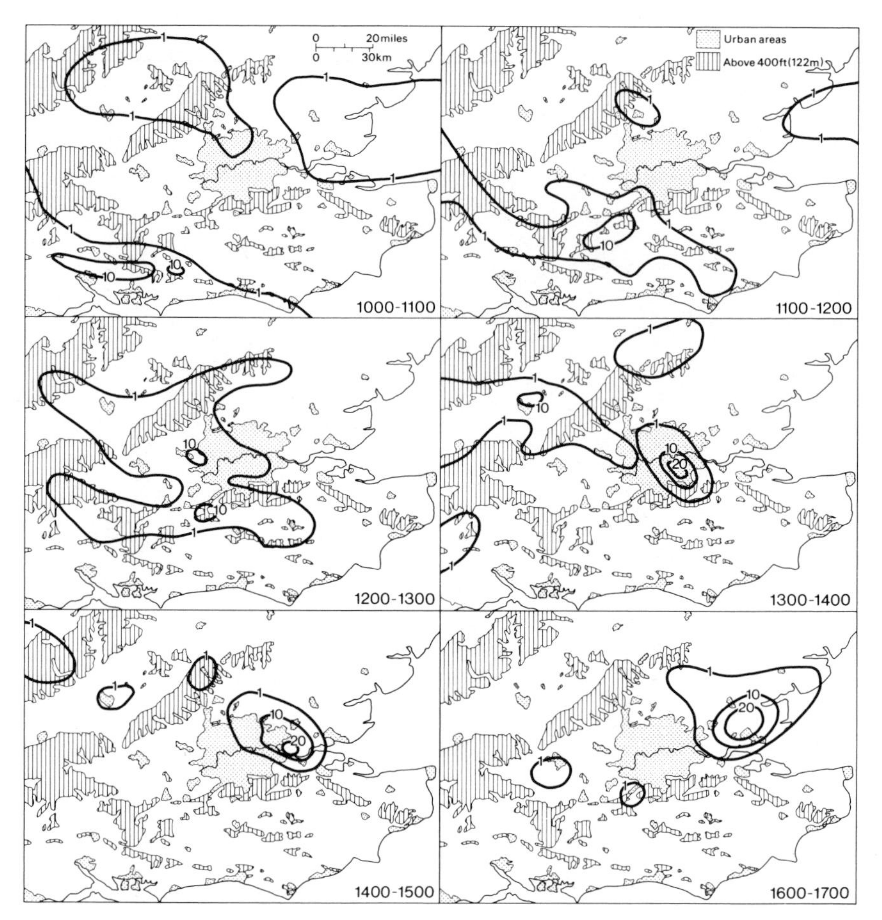

Fig. 4.29. Hourly rainfalls (mm) over south-east England during the periods 1000–1500 hours and 1600–1700 hours on 14 September 1968 (*after Jackson 1977*).

localized strong downdraughts. They are often associated with the production of large hailstones and tornadoes.

Tornadoes which may develop from such squall-line thunderstorms, are common over the Great Plains of the United States, especially in spring and early summer (fig. 4.31) (pl. 25). During this period cold, dry air from the high plateaux may override maritime tropical air (see note 3). Subsidence beneath the upper troposphere westerly jet (fig. 4.32) caps the low-level moist air forming an inversion at about 1500–2000 m. The moist air is extended northward by a low-level southerly jet (cf. p. 185) and through continuing advection the air beneath the inversion becomes progressively more warm and moist. Eventually the general convergence and ascent in the depression trigger the potential instability of the air generating large cumulus clouds which penetrate the inversion. The convective trigger is sometimes provided by the approach of the cold front towards the western edge of the

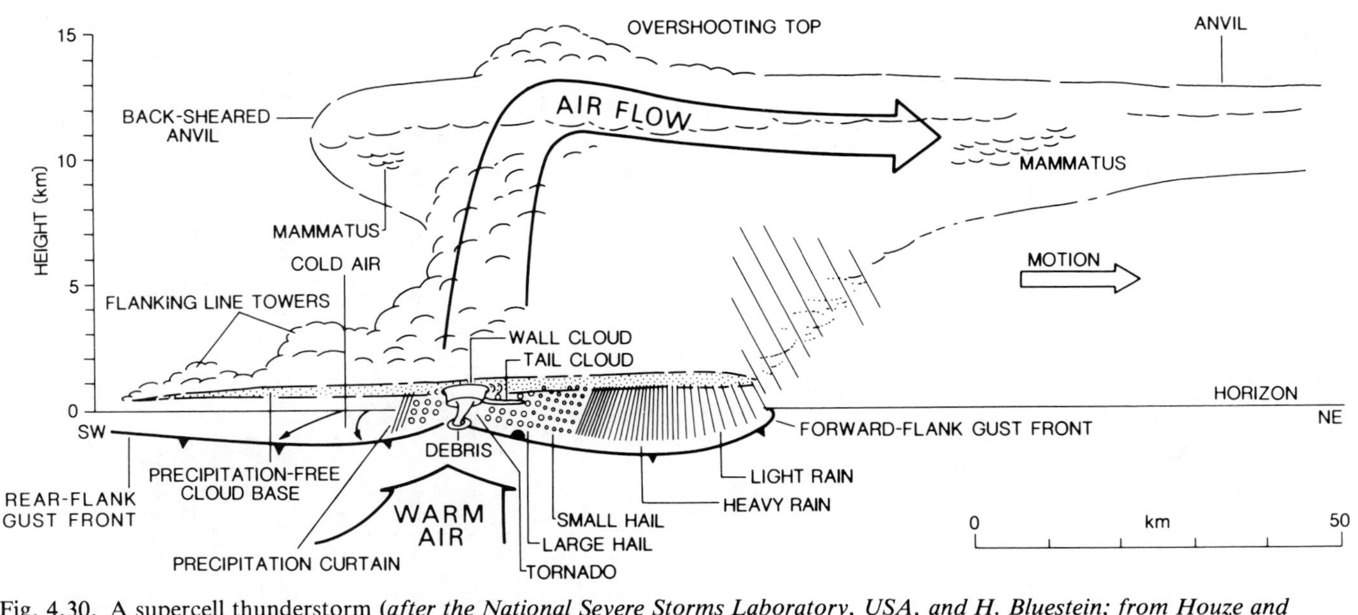

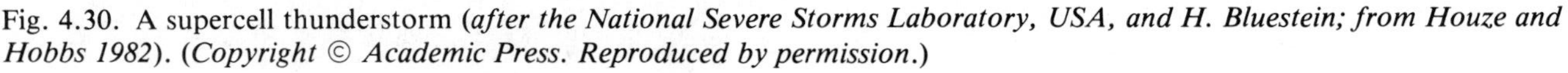

Fig. 4.30. A supercell thunderstorm (*after the National Severe Storms Laboratory, USA, and H. Bluestein; from Houze and Hobbs 1982*). (*Copyright © Academic Press. Reproduced by permission.*)

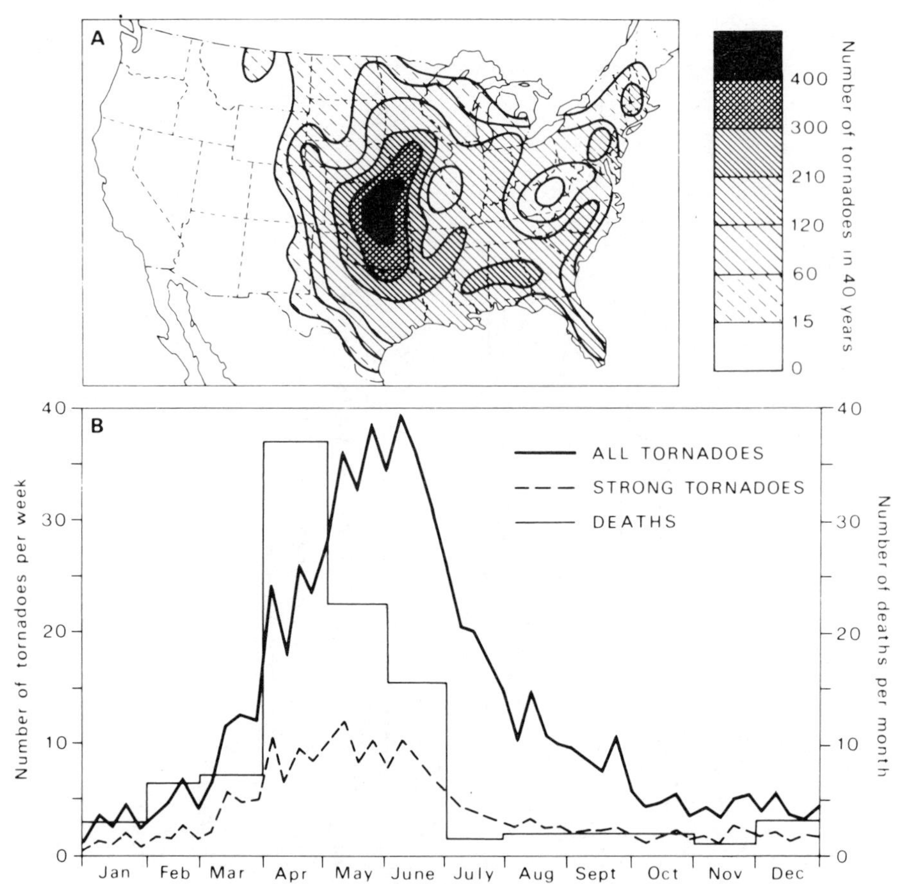

Fig. 4.31. Incidence of tornadoes in the United States. A Number of tornadoes reported over a 40-year period (*from Hindley 1977*). (*This first appeared in New Scientist, London, the weekly review of science and technology*). B Weekly average number of tornadoes (1950–82) and monthly averages of resulting deaths (1953–82) (*from The tornado by J. T. Snow*). (*Copyright © 1984 by Scientific American Inc. All rights reserved*).

moist tongue although tornadoes also occur in association with tropical cyclones (p. 292) and in other synoptic situations if the necessary vertical contrast is present in the temperature, moisture and wind fields.

The exact tornado mechanism is still not fully understood, because of the observational problems involved. Tornadoes generally develop in a meso-low on the periphery of severe rotating thunderstorm systems where horizontal convergence increases the vorticity and rising air is replenished by moist air from progressively lower levels as the vortex descends and intensifies (fig. 4.27). Such generating thunderstorms are identifiable, when seen in plan view on a radar display, by a 'hook echo' pattern representing spiral cloud bands about a small central eye. The pressures in the mesoscale thunderstorm low are only 2–5 mb less than in the surrounding environment. The tornado funnel has been observed to originate in the cloud base

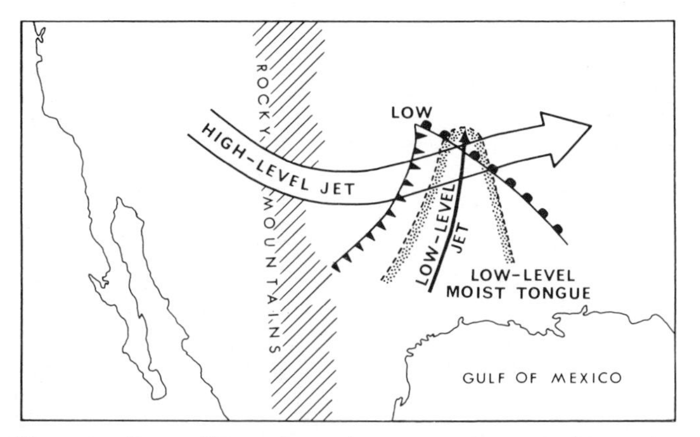

Fig. 4.32. The synoptic conditions favouring severe storms and tornadoes over the Great Plains.

and extend towards the surface and one idea is that convergence beneath the base of cumulonimbus clouds, aided by the interaction between cold precipitation downdraughts and neighbouring updraughts, may initiate rotation. Other observations suggest that the funnel forms simultaneously throughout a considerable depth of cloud, usually a towering cumulus. It appears that the upper portion of the tornado spire in this cloud may become linked to the main updraught of a neighbouring cumulonimbus, thereby causing rapid removal of air from the spire and allowing a violent pressure decrease at the surface. The pressure drop is estimated to exceed 200–250 mb in some cases and it is this which makes the funnel visible by causing air entering the vortex to reach saturation. The vortex is usually only a few hundred metres in diameter (pl. 25) and in an even more restricted band around the core the winds can attain speeds of 50–100 m s^{-1}. The fastest tornadoes often split into multiple vortices rotating anticlockwise with respect to the main tornado axis, each following a cycloidal path and the whole tornado system giving a complex pattern of destruction (fig. 4.33), with maximum wind speeds being experienced on the right-side boundary (in the northern hemisphere) where the translational and rotational speeds are combined. Destruction results not only from the high winds, for buildings near the path of the vortex may explode outwards owing to the pressure reduction outside. Tornadoes commonly occur in families and move in rather straight paths (typically between 10 and 100 km long and 100 m to 2 km wide) at velocities dictated by the low-level jet. Thirty-year averages indicate some 750 tornadoes per year in the United States, with 60% of these in April, May and June (fig. 4.31B). They cause 100 fatalities each year, on average, although most of the deaths and destruction result from a few enduring mature tornadoes, making up only 1·5% of the total reported. For example, the most severe recorded tornado travelled 200 km in 3 hours across Missouri, Illinois and Indiana on 18 March 1925, killing 689 people. These really intense tornadoes present problems as to their energy supply and it has

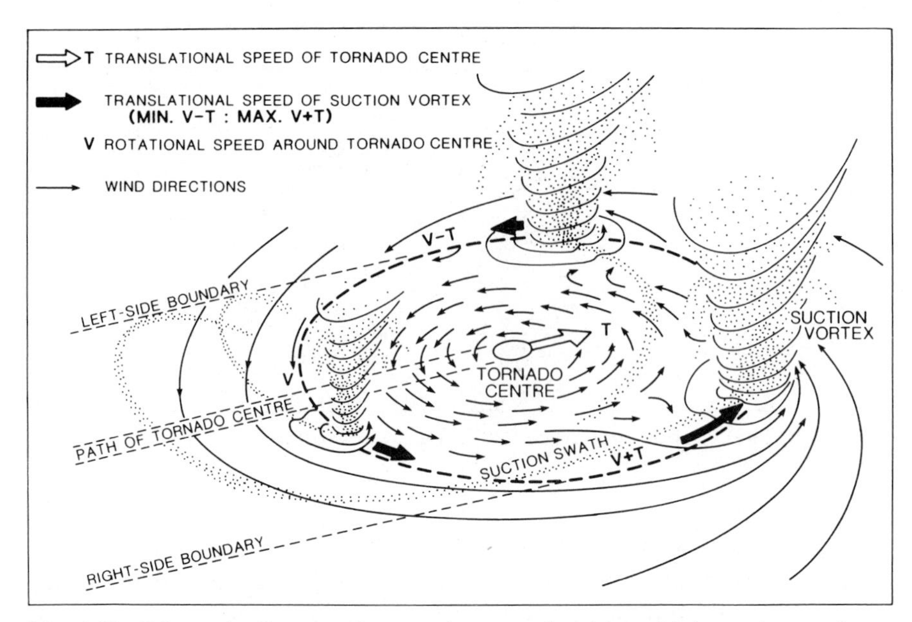

Fig. 4.33. Schematic diagram of a complex tornado with multiple suction vortices (*after Fujita, J. Atmos. Sci., 38, 1981, Fig. 15, p. 1251*).

recently been suggested that the release of heat energy by lightning, and other electrical discharges, may be a necessary additional source of energy.

Tornadoes are not unknown even in the British Isles. During 1960–82 there were 14 days per year with tornado occurrences. Most are minor outbreaks, but on 23 November 1981, 102 were reported during south-westerly flow ahead of a cold front. They are most common in autumn when cold air moves over relatively warm seas.

I Meteorological forecasts

National meteorological services perform a variety of activities in order to provide weather forecasts. The principal ones are data collection, the preparation of basic analyses and prognostic charts of atmosphere conditions for use by local weather offices, the preparation of short- and long-term forecasts for the public as well as special services for aviation, shipping, agricultural and other commercial and industrial users, and the issuance of severe weather warnings.

1 Data sources

The data required for forecasting and other services are provided by world-wide standard synoptic reports at 00, 06, 12 and 18 GMT (see App. 3), similar observations made hourly, particularly in support of national aviation requirements, upper-air soundings (at 00 and 12 GMT), satellite

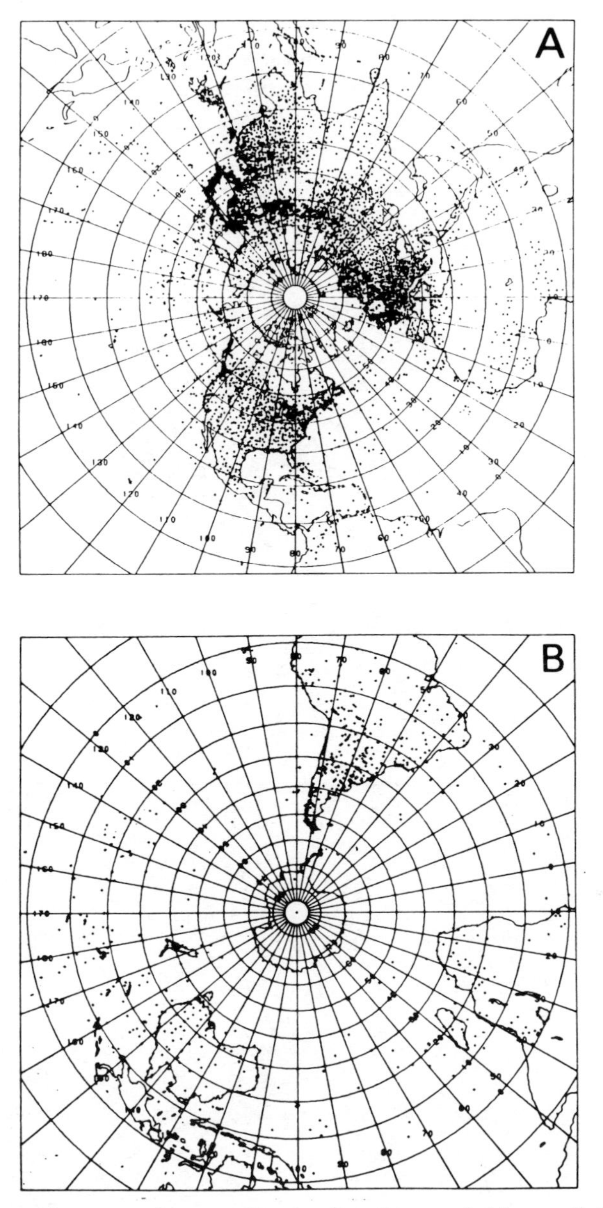

Fig. 4.34. Synoptic reports from surface land stations and ships available over the Global Telecommunications System at the National Meteorological Center, Washington, DC.

data and other specialized networks such as radar stations for severe weather. Under the World Weather Watch programme, synoptic reports are made at some 4000 land stations and by 7000 ships (fig. 4.34). There are about 700 stations making upper-air soundings (temperature, pressure, humidity and wind) (fig. 4.35). These data are transmitted in code via teletype and radio links to regional or national centres and into the high-speed Global Telecommunications System (GTS) connecting World Weather Centres in Melbourne, Moscow and Washington and eleven Regional Meteorological Centres for redistribution. Some 157 states and territories

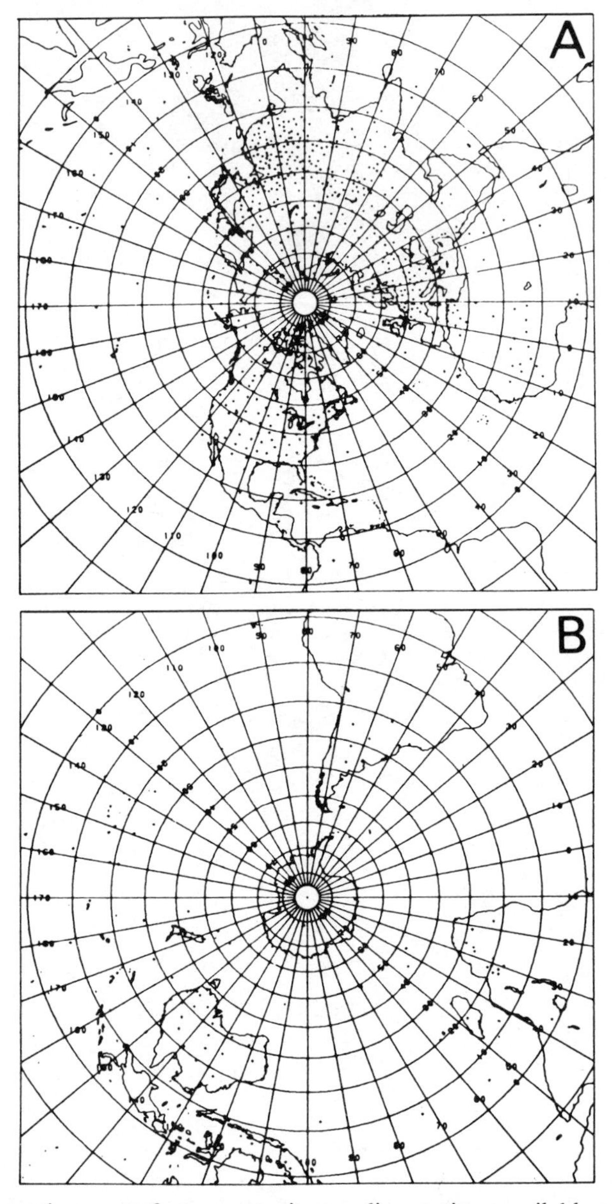

Fig. 4.35. Synoptic reports from upper-air sounding stations available over the Global Telecommunications System at the National Meteorological Center, Washington, DC (*from A. D. Hecht, Paleoclimate Models and Analysis, Wiley, New York, 1985*). (*Copyright © 1985 by John Wiley and Sons, Inc. Reproduced by permission*).

co-operate in this activity under the aegis of the World Meteorological Organization.

Meteorological information has been collected operationally by satellities of the United States and USSR since 1965 and, more recently, by the European Space Agency, India and Japan. There are two general categories of weather satellite polar orbiters providing global coverage twice per 24 hours in orbital strips over the poles (such as the United States NOAA and TIROS series (pls. 2 and 3), and the USSR's Meteor) and geosynchronous satellites (such as the Geostationary Operational Environmental Satellites (GOES) and Meteosat), giving repetitive (30-minute) coverage of almost one-third of the earth's surface in low middle latitudes (fig. 4.36). Information on the atmosphere is collected as digital data or direct readout visible and infrared images of cloud cover and sea-surface temperature, but also includes global temperature and moisture profiles through the atmosphere obtained from multi-channel infrared and microwave sensors which receive radiation emitted from particular levels in the atmosphere. Additionally, satellites have a data collection system (DCS) that relays data on numerous environmental variables from ground platforms or ocean buoys to processing centres; GOES can also transmit processed satellite images in facsimile and the NOAA polar orbiters have an automatic picture transmission (APT) system that is utilized at 900 stations worldwide.

2 *Forecasting*

Modern forecasting did not become possible until weather information could be rapidly collected, assembled and processed. The first development came in the middle of the last century with the invention of telegraphy, which permitted immediate analysis of weather data by the drawing of synoptic

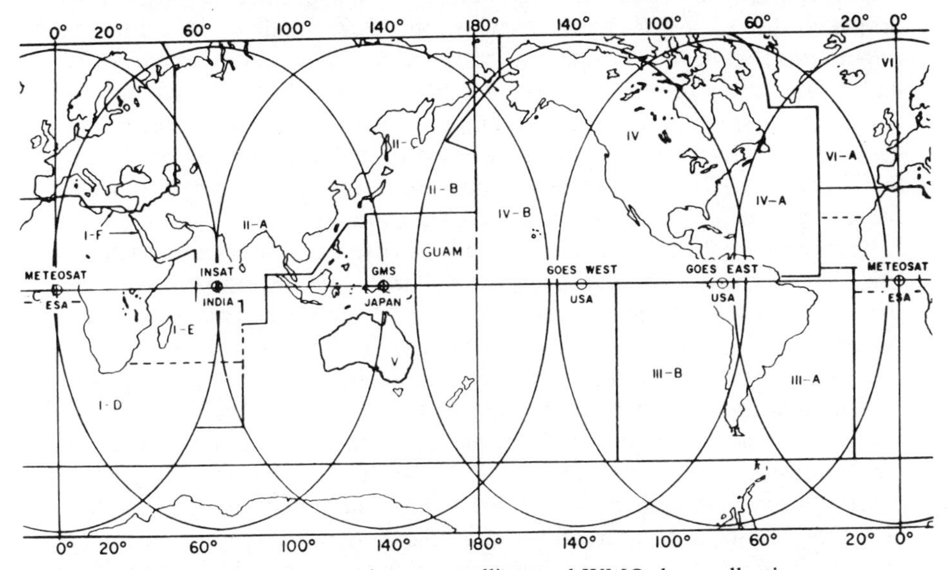

Fig. 4.36. Coverage of geostationary satellites and WMO data collection areas (rectangular areas and numbers) (*courtesy NOAA*).

charts. These were first displayed in Britain at the Great Exhibition of 1851. Sequences of weather change were correlated with barometric pressure patterns both in space and time by such workers as Fitzroy and Abercromby, but it was not until later that theoretical models of weather systems were devised – notably the Bjerknes' depression model described earlier.

Forecasts are usually referred to as short-range (1–2 days), medium (or extended) range (up to 5–7 days) and long-range (monthly or seasonal). The first two can for present purposes be considered together.

a Short-range forecasting Forecasting procedures developed up to the 1950s were based on synoptic principles but, since the 1960s, practices have been revolutionized by numerical forecasting methods and the adoption of 'nowcasting' techniques.

During the first half of the century, short-range forecasts were based on synoptic principles, empirical rules and extrapolation of pressure changes. The Bjerknes' model of cyclone development for middle latitudes and simple concepts of tropical weather (see ch. 6) served as the basic tools of the forecaster. The relationship between the development of surface lows and highs and the upper-air circulation was worked out during the 1940s and 1950s by C.-G. Rossby, R. C. Sutcliffe and others, providing the theoretical basis of synoptic forecasting. In this way, the position and intensities of low- and high-pressure cells and frontal systems were predicted.

Since 1955 in the United States – and 1965 in the United Kingdom – routine forecasts have been based on numerical models. These predict the evolution of physical processes in the atmosphere by determinations of the conservation of mass, energy and momentum. The basic principle is that the rise or fall of surface pressure is related to mass convergence or divergence, respectively, in the overlying air column. This prediction method was proposed by L. F. Richardson who, in 1922, made a laborious test calculation that gave very unsatisfactory results. The major reason for this lack of success is the fact that the *net* convergence or divergence in an air column is a small residual term compared with the large values of convergence and divergence at different levels in the atmosphere (see fig. 3.7). Small errors arising from observational limitations may, therefore, have a considerable effect on the correctness of the analysis.

Numerical methods developed in the 1950s use a less direct approach. The first developments assumed a one-level barotropic atmosphere with geostrophic winds and hence no convergence or divergence. The movement of systems could be predicted, but not changes in intensity. Despite the great simplifications involved in the barotropic model, it has been used for forecasting 500 mb contour patterns. The latest techniques employ multi-level baroclinic models and include frictional and other effects; hence the basic mechanisms of cyclogenesis are provided for. It is noteworthy that *fields* of continuous variables, such as pressure, wind and temperature, are handled and that fronts are regarded as secondary, derived features. The vast increase in the number of calculations which these models necessitate has required a new generation of larger, faster electronic computers to allow the preparation of a forecast map to keep sufficiently ahead of the weather changes!

Forecast practices in the major national centres are basically similar. However, the specific information relates to the National Meteorological Centre (NMC) in Washington, DC. The forecasts are essentially derived from twice-daily (00 and 12 GMT) prognoses of atmospheric circulation. Since most techniques are now largely automated, the analyses of synoptic fields are based on the previous 12-hour forecast maps as a first guess. Three different interpolation methods are used to obtain smoothed, gridded data on temperature, moisture, wind and geopotential height for the surface at standard pressure levels (850, 700, 500, 400, 300, 250, 200 and 100 mb) over the globe. The NMC currently has two basic prediction models: a spectral model with (6 or) 12 layers (from the boundary layer into the upper stratosphere), which is integrated for up to 10 days, and a regionally applicable nested grid model with finer horizontal resolution. It should be noted that typically the computer time required increases several-fold when the grid spacing is halved. The essential forecast products are MSL pressure, temperature and wind velocity for standard pressure levels, 1000–500 mb thickness, vertical motion and moisture content in the lower troposphere, and precipitation amounts.

Actual weather conditions are now commonly predicted using the Model Output Statistics (MOS) technique developed by the US National Weather Service. Rather than relating weather variables to the predicted pressure/height patterns and taking account of frontal models, for example, a series of regression equations are developed for specific locations between the variable of interest and up to 10 predictors calculated by the numerical models. Weather elements so predicted for numerous locations include daily maximum/minimum temperature, 12-hour probability of precipitation occurrences and precipitation amount, probability of frozen precipitation, thunderstorm occurrence, cloud cover and surface winds. These forecasts are distributed as facsimile maps and tables to weather offices for local use.

In the United States, there are also separate centres with responsibilities for forecasting tropical storms (the National Hurricane Center in Miami for the Atlantic area, with warning centres at San Francisco for the eastern North Pacific and Honolulu for the central North Pacific) and also the National Severe Storms Forecasting Center in Kansas City, Missouri. The latter depends greatly on satellite and radar data to issue severe weather alerts.

The British Meteorological Office's 1982 global model has a 1·5° latitude by 1·875° longitude horizontal grid and 15 levels between the surface and approximately 25 mb. All essential physical processes (diurnal radiation cycle, surface energy exchanges, cloud development and precipitation) are represented. Certain surface conditions are fixed, such as snow cover and sea ice extent, albedo, soil moisture and thermal capacity, while sea-surface temperatures are entered from daily observations. By repeated calculations for time steps of a few minutes, atmospheric motion, temperature, cloud cover, precipitation, etc., are forecast up to six days ahead. For a more limited area (80° W–40° E, 30°–80° N) a 'fine mesh' model (0·75° latitude, 0·9375° longitude grid) is used to forecast up to 36 hours ahead.

Errors in numerical forecasts arise from several sources. One of the most serious is the limited accuracy of the initial analyses due to data deficiencies. The coverage over the oceans is sparse and only a quarter of the possible ship reports may be received within 12 hours; even over land more than one-third of the synoptic reports may be delayed beyond 6 hours. However, satellite-derived information and aircraft reports can help fill some gaps for the upper air. Another limitation is imposed by the horizontal and vertical resolution of the models and the need to parameterize subgrid processes such as cumulus convection. The small-scale nature of the turbulent motion of the atmosphere means that some weather phenomena are basically unpredictable, for example, the specific locations of shower cells in an unstable air mass. Greater precision than the 'showers and bright periods' or 'scattered showers' of the forecast language is impossible with present techniques. The procedure for preparing a forecast is becoming much less subjective, although in complex weather situations the skill of the experienced forecaster still makes the technique almost as much an art as a science. Detailed regional or local predictions can only be made within the framework of the general forecast situation for the country and demand thorough knowledge of possible topographic or other local effects by the forecaster.

b 'Nowcasting' Severe weather is typically short-lived (< 2 hr) and, due to its mesoscale character (< 100 km), it affects local/regional areas necessitating site-specific forecasts. Included in this category are thunderstorms, gust fronts, tornadoes, high winds especially along coasts, over lakes and mountains, heavy snow and freezing precipitation. The development of radar networks, new instruments and high-speed communication links has provided a means of issuing warnings of such phenomena. Several countries have recently developed integrated satellite and radar systems to provide information on the horizontal and vertical extent of thunderstorms, for example. Such data are supplemented by networks of automatic weather stations (including buoys) that measure wind, temperature and humidity. In addition, for detailed boundary layer and lower troposphere data, there is now an array of vertical sounders – acoustic sounders (measuring wind speed and direction from echoes created by thermal eddies), specialized (Doppler) radar measuring winds in clear air by returns either from insects (3·5 cm wavelength radar) or from variations in the air's refractive index (10 cm wavelength radar). Nowcasting techniques use highly automated computers and image analysis systems to integrate data from a variety of sources rapidly. Interpretation of the data displays requires skilled personnel and/or extensive software to provide appropriate information. The prompt forecasting of wind shear and downburst hazards at airports is one example of the importance of nowcasting procedures.

Overall, the greatest benefits from improved forecasting can be expected in aviation and the electric power industry for forecasts less than 6 hours ahead, in transportation, construction and manufacturing for 12–24-hour forecasts and in agriculture for 2–5-day forecasts. In terms of economic losses, the last category could benefit the most from more reliable and more precise forecasts.

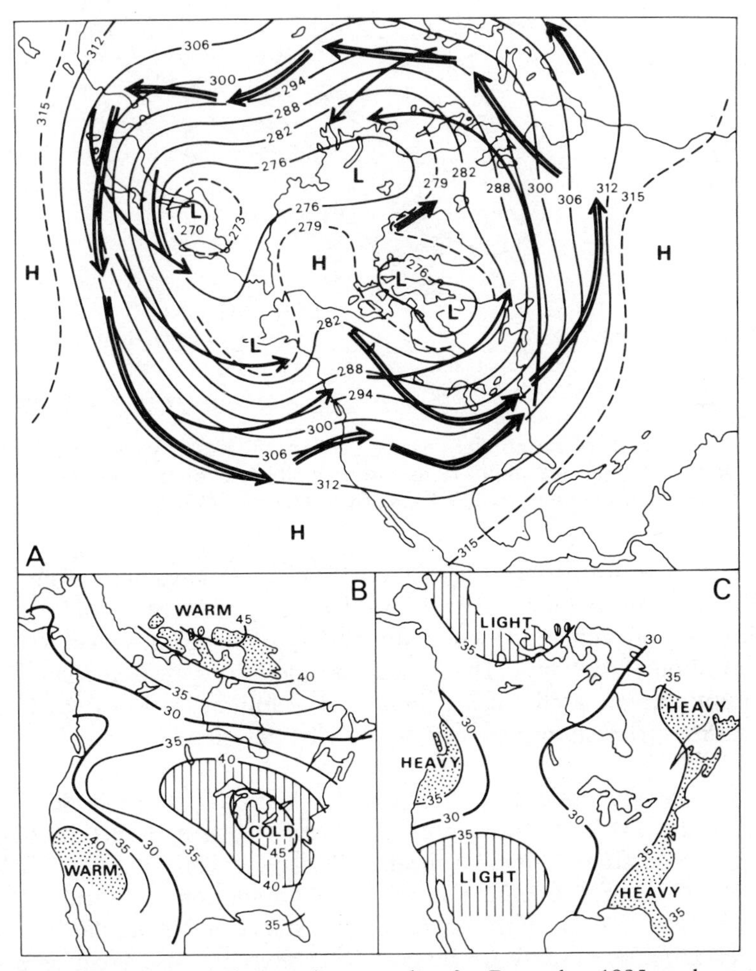

Fig. 4.37. Forecast of North American weather for December 1985 made one month ahead. A. Predicated 700-mb contours (gpdm). Solid arrows indicate main tracks of cyclones, open arrows of anticyclones, at sea-level. B and C. Forecast average temperature (B) and average precipitation (C) probabilities (in %). There are two classes of temperature, warm and cold, and similarly heavy and light for precipitation. Each of these classes is defined to occur 30% of the time in the long run; near-normal temperature or moderate precipitation occur 40% of the time. The 30% heavy lines indicate indifference (for any departure from average), but near-normal values are most likely in the unshaded areas (*from Monthly and Seasonal Weather Outlook, vol. 39, no. 23, November 28, 1985, Climate Analysis Center, NOAA, Washington, DC*).

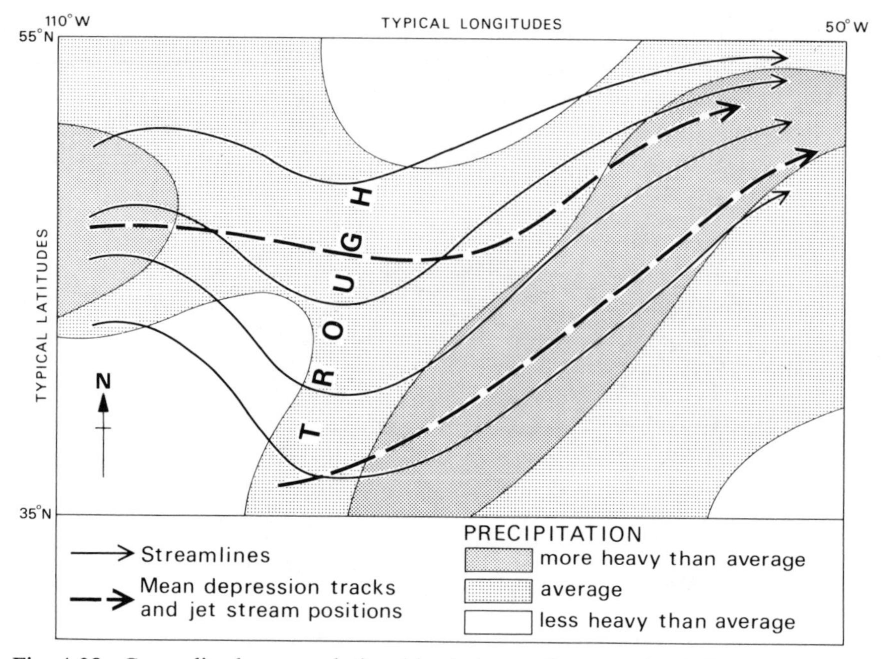

Fig. 4.38. Generalized mean relationships between the westerly airflow pattern, depression tracks, jet-stream positions and precipitation zones in mid-latitudes in the northern hemisphere (*after W. Klein; from Harman 1971*).

c Long-range forecasting The methods discussed above are unsuitable for predicting the probable trend of the weather for periods of a month or more, because they are concerned with individual synoptic disturbances with a life cycle of about 3 to 7 days. Theoretical considerations indicate that the limit of synoptic predictability using numerical techniques is less than 15 days. Two rather different approaches will now be described.

(1) Statistical methods The United States National Weather Service has issued 30-day forecasts twice monthly since 1948, using a method that has two principal steps. First, a mean 700 mb contour map is predicted (e.g. fig. 4.37A) for the coming month by a combination of several principles – a statistical analysis of the degree of persistence in the circulation pattern expected from one month to the next, extrapolation of current tendencies (such as blocking patterns) or the recognition of probable changes of regime in the large-scale atmospheric circulation, study of the possible effects of features such as snow cover and sea-surface temperature anomalies, and the examination of statistical records of the typical locations of troughs and ridges at that season. Second, the 700 mb map is interpreted in terms of surface weather conditions. This is partly based on analogues where the weather experienced during past cases with similar 700 mb patterns occurring in the same season serves as a guide to the conditions expected to recur. Next, the most probable anomalies of mean temperature and precipitation

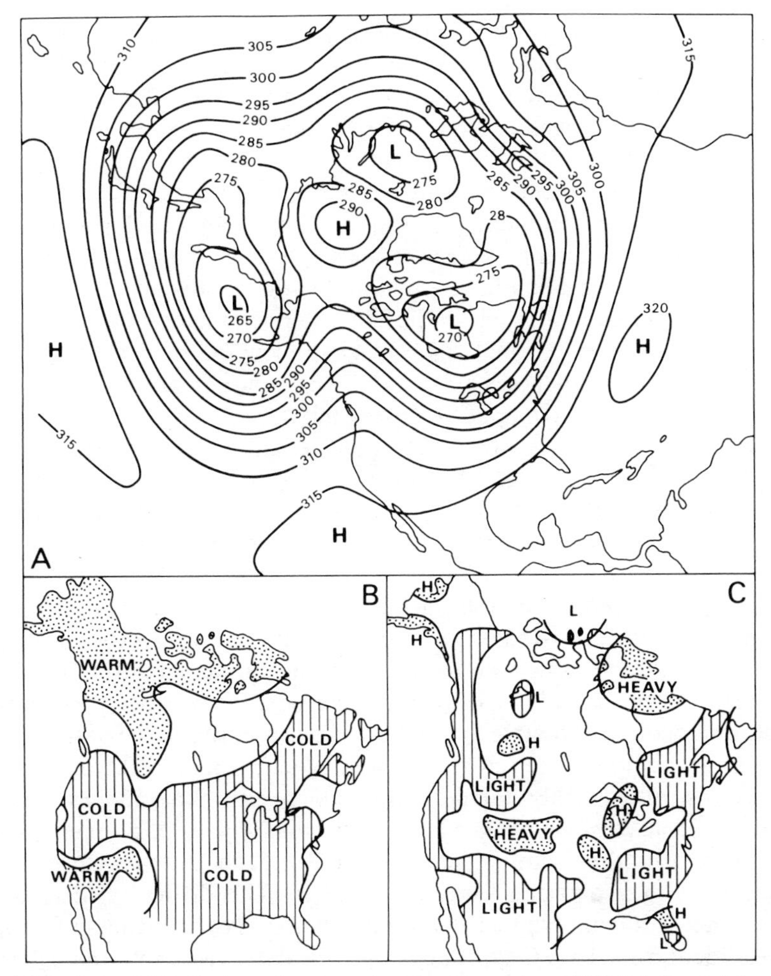

Fig. 4.39. Actual North American weather for December 1985 (cf. fig. 4.37). A Observed 700-mb contours (5 gpdm) corresponding to fig. 4.37A (from Climate Analysis Center, NOAA, Washington, DC.) B and C Observed temperature (B) and precipitation (C), corresponding to fig. 4.37B and C. (*From Monthly and Seasonal Weather Outlook, Vol. 40, Number 1, 1986. Climate Analysis Center, NOAA, Washington, DC*).

amount corresponding to the predicted contour map are derived from known relationships existing between them. For example, fig. 4.38 shows that more than average precipitation is probable ahead of the trough where there is maximum vorticity advection, especially in association with an active jet stream, whereas less than average precipitation is usual to the rear of the trough axis. With a low zonal index in the westerlies, long waves may persist in more or less stationary positions for periods of several weeks at a time, or even longer, and their mean positions are, therefore, reflected in surface patterns of temperature and precipitation (see also ch. 5, B.1).

The 30-day outlook for the United States is presented as maps of the probability of three simple classes each of temperature (near-normal, above/below normal) and precipitation (near-average, above/below the median) (fig. 4.37), together with tables for many cities.

The 90-day seasonal outlooks prepared by the Climate Analysis Center of the US National Weather Service are based primarily on persistence statistics that take account of trends and tendencies linking the circulation and weather patterns of the preceding several seasons. Overall, winter is more easily predictable and spring is the most variable season. Recurrent, large-scale abnormalities in climate, such as the El Niño phenomenon (see ch. 6, E.4), may prove useful in seasonal forecasting in view of their worldwide *teleconnections* (i.e. distant areas of the globe where the climatic anomalies show either the same, or almost opposite, tendencies).

Figure 4.39A illustrates the observed height field corresponding to fig. 4.37A for December 1985, showing that the pattern is well represented on the forecast chart. Figure 4.39B and C shows that in this case, as is usual, the temperature forecasts are more reliable than those for precipitation.

(2) Analogue methods An approach to long-term forecasting developed in Germany and Britain is based on the principle that sequences of weather events may tend to follow a similar course if the initial conditions are almost identical. The problem is then to find a period with weather conditions as closely analogous as possible to the present one and to use the past sequence of events as a guide to the future. The analogues are matched from a record of patterns of monthly temperature and pressure anomalies, and of sequences of *weather types*. The latter are actually types of pressure pattern or airflow over the country (see ch. 5, A.3). Each category tends to be associated with a particular type of weather. The difficulty in analogue prediction arises from the fact that no two patterns or sequences of weather are ever identical. There may, for example, be five reasonable analogues for a particular month, but examination of the succeeding weather sequences might show mild, rainy weather in two cases and cold spells in the other three. In the preparation of the forecast, therefore, many factors which can affect the weather trends, such as sea temperatures and the extent of snow cover, have to be taken into account. The problems facing the long-range forecasters need to be recognized before criticisms are levelled at them. The complexity of the atmosphere's behaviour makes tentative wording of their predictions necessary and occasional failures inevitable at present.

Summary

Ideal air masses are defined in terms of barotropic conditions, where isobars and isotherms are assumed to be parallel to each other and to the surface. The character of an air mass is determined by the nature of the source area, changes due to air mass movement, and its age. On a regional scale, energy exchanges and vertical mixing lead to a measure of equilibrium between surface conditions and those of the overlying air, particularly in quasi-stationary high-pressure systems. Air masses are conventionally identified in terms of temperature characteristics (arctic, polar, tropical) and source region (maritime, continental). Primary air masses originate in regions of semi-permanent anticyclonic subsidence over extensive surfaces of like properties. Cold-air masses originate either in winter continental anticyclones (Siberia and Canada), where snow cover promotes low temperatures and stable stratification, or over high-latitude sea ice. Some sources are seasonal, like Siberia, others are permanent, such as Antarctica. Warm-air masses originate either as shallow tropical continental sources in summer or as deep, moist layers over tropical oceans. Air-mass movement causes stability changes by thermodynamic processes (heating/cooling from below and moisture exchanges) and by dynamic processes (mixing, lifting/subsidence), producing secondary air masses (e.g. mP air). The age of an air mass determines the degree to which it has lost its identity as the result of mixing with other air masses and of vertical exchanges with the underlying surface.

Air-mass boundaries give rise to fronts or baroclinic zones a few hundred kilometres wide. The classical (Norwegian) theory of mid-latitude cyclones considers that fronts are a key feature of their formation and life cycle. Depressions tend to form along major frontal zones – the polar fronts of the North Atlantic and North Pacific regions and of the southern oceans. Less well defined arctic fronts lie poleward and there are other seasonal frontal zone as in the Mediterranean. The Inter-Tropical Convergence Zone, between air masses from the opposing subtropical anticyclones, is different in character from frontal zones of high latitudes (see ch. 6). Air masses and frontal zones move poleward/equatorward in summer/winter.

Newer cyclone theories regard fronts as rather incidental. Divergence of air in the upper troposphere is essential for large-scale uplift and low-level convergence. Surface cyclogenesis is therefore favoured on the eastern limb of an upper wave trough. Cyclones are basically steered by the quasi-stationary long (Rossby) waves in the hemispheric westerlies, the positions of which are strongly influenced by surface features (major mountain barriers and land/sea-surface temperature contrasts). Upper baroclinic zones are associated with jet streams at 300–200 mb which also follow the long-wave pattern.

continued

The idealized weather sequence in an eastward-moving frontal depression involves increasing cloudiness and precipitation with an approaching warm front; the degree of activity depends on whether the warm sector air is rising or not (ana- or kata-fronts, respectively). The following cold front is often marked by a narrow band of convective precipitation, but rain both ahead of the warm front and in the warm sector may also be organized into locally intense meso-scale cells and bands due to the 'conveyor belt' of air in the warm sector. Associated with this airflow organization, there is often a well developed squall line ahead of the cold front. In the central United States, especially, thunderstorm cells along such squall lines give severe thunder and hail conditions, sometimes with tornadoes. The up- and down-draughts in such cells set up clusters of developing and decaying storms.

Some low-pressure systems are essentially non-frontal. These include the lee depression encountered in the lee of mountain ranges; thermal lows due to summer heating; the polar air depression commonly formed in an outbreak of maritime Arctic air over oceans; and the upper cold low which is often a cutoff system in upper wave development or an occluded mid-latitude cyclone in the Arctic.

Analysis and forecasting of surface and upper-air weather maps is now highly automated in many national weather services. Short-range forecasting is based primarily on numerical prediction models. Automated integration of satellite and radar data is being widely developed for 'nowcasting' of severe weather events. Long-range forecasting is primarily statistical taking account of possible effects of surface conditions on the large-scale circulation structure. Analogue studies of past events may be helpful.

Notes

1. Resultant wind is the vector average of all wind directions and speeds.
2. This latter term is tending to be restricted to the tropical (hurricane) variety.
3. It is significant that some storms also occur downwind of other arid plateau regions in Mexico, the Iberian peninsula and West Africa.

5

Weather and climate in temperate latitudes

In the two preceding chapters the general structure of the circulation and air-mass characteristics in middle latitudes have been outlined and the behaviour and origin of extra-tropical depressions examined. The direct contribution of pressure systems to the daily and seasonal variability of weather in the westerly wind belt is quite apparent to inhabitants of the temperate lands. Nevertheless there are equally prominent contrasts of regional climate in mid-latitudes which reflect the interaction of geographical and meteorological factors. This chapter gives a selective synthesis of weather and climate in Europe and North America, drawing mainly on the principles already presented. The climatic conditions of the polar and subtropical margins of the westerly wind belt are examined in the final sections of the chapter. As far as possible different themes are used to illustrate some of the more significant aspects of the climate in each area.

A Europe

1 Pressure and wind conditions

The principal features of the mean North Atlantic pressure pattern are the Icelandic Low and the Azores High. These are present at all seasons (see fig. 3.24), although their location and relative intensity change considerably. The upper flow in this sector undergoes little seasonal change in pattern but the westerlies decrease in strength by over half from winter to summer. The other major pressure system influencing European climates is the Siberian winter anticyclone, the occurrence of which is intensified by the extensive winter snow cover and the marked continentality of Eurasia. Atlantic depressions frequently move towards the Norwegian or Mediterranean Seas in winter, but if they travel due east they occlude and fill long before they can penetrate into the heart of Siberia. Thus the Siberian high pressure is quasi-permanent at this season and when it extends westward severe conditions affect much of Europe. In summer, pressure is low over all of Asia and depressions from the Atlantic tend to follow a more zonal path. Although the depression tracks over Europe do not shift poleward in summer (as a result of the local *southward* displacement of the Atlantic Arctic Front), the depressions at this season are rather less intense and the diminished air-mass contrasts produce weaker fronts.

2 *Oceanicity and continentality*

Winter temperatures in north-west Europe are some 11°C (20°F) or more above the latitudinal average (see fig. 1.24), a fact usually attributed solely to the presence of the North Atlantic Drift. There is, however, a complex interaction between the ocean and atmosphere. The Drift, which originates from the Gulf Stream off Florida strengthened by the Antilles Current, is primarily a wind-driven current initiated by the prevailing south-westerlies. It flows at a velocity of 16 to 32 km (10–20 miles) per day and thus, from Florida, the water takes about eight or nine months to reach Ireland and about a year to Norway (see ch. 3, F.3). The south-westerly winds transport both sensible and latent heat acquired over the western Atlantic towards Europe, and although they continue to gain heat supplies over the north-eastern Atlantic this local warming arises in the first place through the drag effect of the winds on the warm surface waters. Warming of air masses over the north-eastern Atlantic is mainly of significance when polar or arctic air flows south-eastwards from Iceland. For example, the temperature in such airstreams in winter may rise by 9°C (17°F) between Iceland and northern Scotland. By contrast, maritime tropical air cools on average about 4°C (8°F) between the Azores and south-west England in winter and summer. One very evident effect of the North Atlantic Drift is the absence of ice around the Norwegian coastline. However, as far as the climate of north-western Europe is concerned the primary factor is the occurrence of prevailing *onshore* winds transferring heat into the area.

The influence of maritime air masses can extend deep into Europe because there are few major topographic barriers to airflow and because of the presence of the Mediterranean Sea. Hence the change to a more continental climatic regime is relatively gradual except in Scandinavia where the mountain spine produces a sharp contrast between western Norway and Sweden. There are numerous indices expressing this continentality, but most are based on the annual range of temperature. Gorczynski's continentality index (K) is:

$$K = 1{\cdot}7 \frac{A}{\sin\theta} - 20{\cdot}4$$

where A is the annual temperature range (°C) and θ is the latitude angle. K ranges between −12 at extreme oceanic stations and 100 at extreme continental stations. Some values in Europe are London 10, Berlin 21, Moscow 42. Figure 5.1 shows the variation of this index over Europe and fig. 5.2 plots its variation for Moscow between 1880 and 1980, showing its considerable oscillation with different large-scale atmospheric circulation regimes.

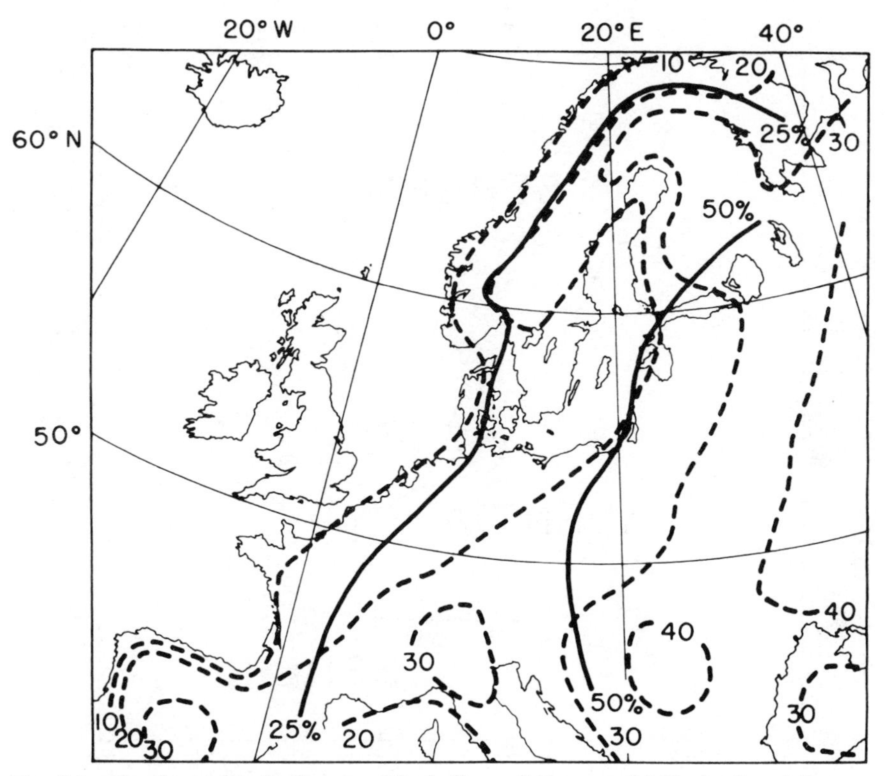

Fig. 5.1. Continentality in Europe. The indices of Gorczynski (dashed) and Berg (solid) are explained in the text (*partly after Blüthgen 1966*).

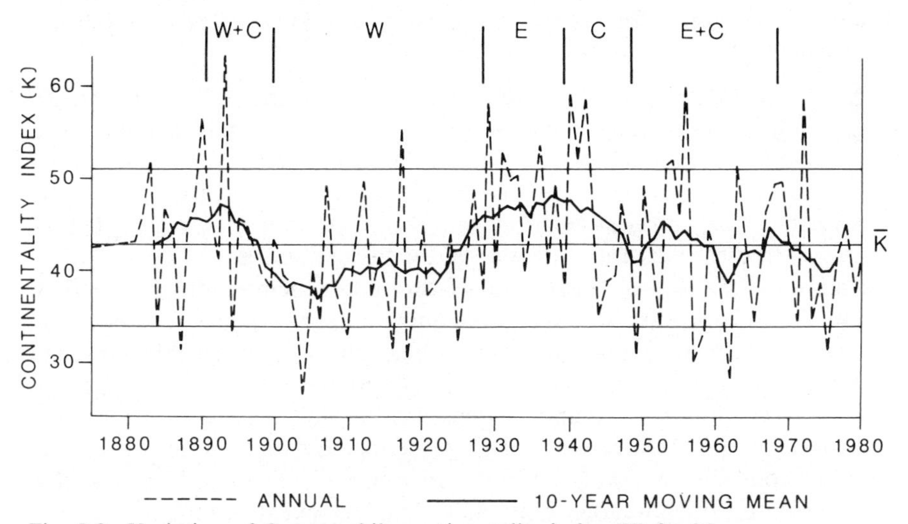

Fig. 5.2. Variation of Gorczynski's continentality index (K) for Moscow (1880–1980) associated with dominant westerly (W), easterly (E) and meridional (C) hemispheric circulation patterns. Annual figures, the mean value ($\bar{K}$), the 10-year moving mean and one standard deviation are shown (*from Poltaraus and Staviskiy 1986*).

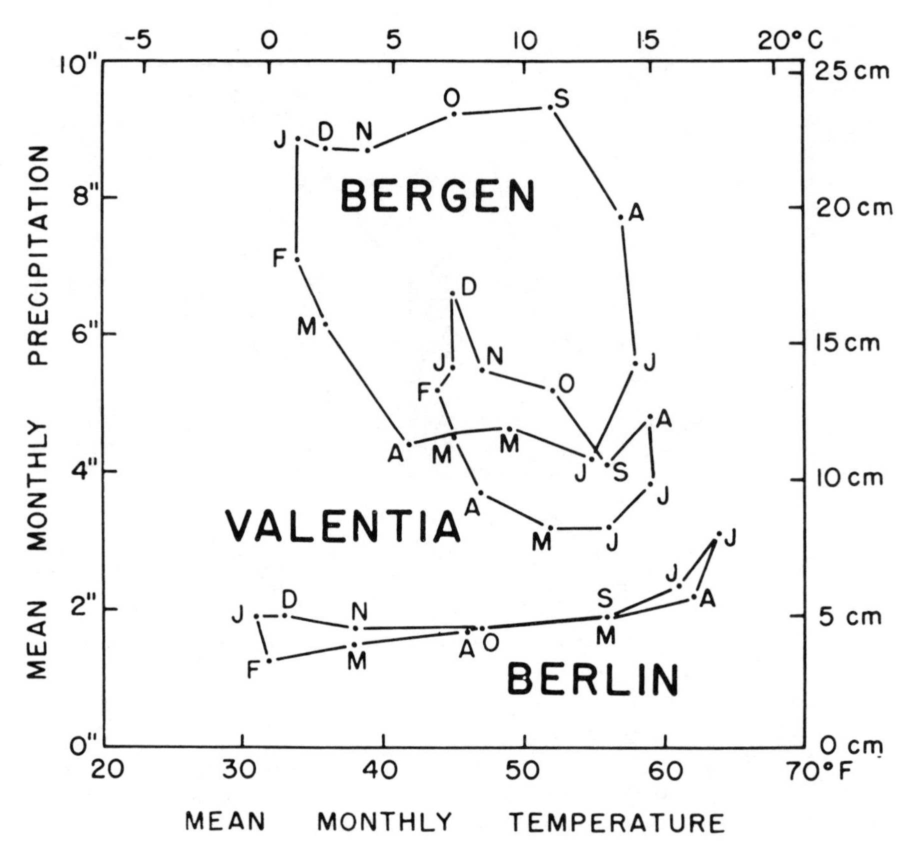

Fig. 5.3. Hythergraphs for Valentia (Eire), Bergen and Berlin. Mean temperature and precipitation values for each month are plotted.

A very different approach by Berg relates the frequency of continental air masses (*C*) to that of all air masses (*N*) as an index of continentality, i.e. $K = C/N$ (%). Figure 5.1 shows that non-continental air occurs at least half the time over Europe west of 15°E as well as over Sweden and most of Finland.

A further illustration of maritime and continental regimes is provided by fig. 5.3. *Hythergraphs* for Valentia (Eire), Bergen and Berlin demonstrate the seasonal changes of mean temperature and precipitation in different locations. Valentia has a winter rainfall maximum and equable temperatures as a result of oceanic situation, whereas Berlin has a considerable temperature range and a summer maximum of rainfall. Bergen receives larger rainfall totals due to orographic intensification and maximum in autumn and winter, its temperature range being intermediate to the other two. Such averages convey only a very general impression of climatic characteristics and therefore British weather patterns will now be examined in more detail.

3 *British airflow patterns and their climatic characteristics*

The daily weather maps for the British Isles sector (50°–60°N, 2°E–10°W)

Table 5.1 General weather characteristics and air masses associated with Lamb's 'Airflow Types' over the British Isles.

Type	
Westerly	Unsettled weather with variable wind directions as depressions cross the country. Mild and stormy in winter, generally cool and cloudy in summer (mP, mPw, mT).
North-westerly	Cool, changeable conditions. Strong winds and showers affect windward coasts especially, but the southern part of Britain may have dry, bright weather (mP, mA).
Northerly	Cold weather at all seasons, often associated with polar lows. Snow and sleet showers in winter, especially in the north and east (mA).
Easterly	Cold in the winter half-year, sometimes very severe weather in the south and east with snow or sleet. Warm in summer with dry weather in the west. Occasionally thundery (cA, cP).
Southerly	Warm and thundery in summer. In winter it may be associated with a low in the Atlantic giving mild, damp weather especially in the south-west or with a high over central Europe, in which case it is cold and dry (mT or cT, summer; mT or cP, winter).
Cyclonic	Rainy, unsettled conditions often accompanied by gales and thunderstorms. This type may refer either to the rapid passage of depressions across the country or to the persistence of a deep depression (mP, mPw, mT).
Anticyclonic	Warm and dry in summer, occasional thunderstorms (mT, cT). Cold and frosty in winter with fog, especially in autumn (cP).

from 1873 to the present day have been classified by H. H. Lamb according to the airflow direction or isobaric pattern. He identifies seven major categories: Westerly (W), North-westerly (NW), Northerly (N), Easterly (E) and Southerly (S) types – referring to the compass directions from which the airflow and weather systems are moving – and Cyclonic (C) or Anticyclonic (A) types when depressions or a high-pressure cell respectively dominate the weather map.

In theory, each category should produce a characteristic type of weather, depending on the season, and many writers use the term *weather type* to convey this idea. A few studies have been made of the *actual* weather conditions occurring in different localities with specific isobaric patterns – a field of study known as *synoptic climatology* – but the term *airflow type* seems preferable for Lamb's categories. The general weather conditions which are likely to be associated with a particular airflow type over the British Isles are summarized in table 5.1.

On an annual basis, the most frequent airflow type is Westerly; including cyclonic and anticyclonic subtypes, it has a 35% frequency in December–January and is almost as frequent in July–September (fig. 5.4). The minimum occurs in May (15% when Northerly and Easterly types reach their maxima (about 10% each). Pure Cyclonic patterns are most frequent

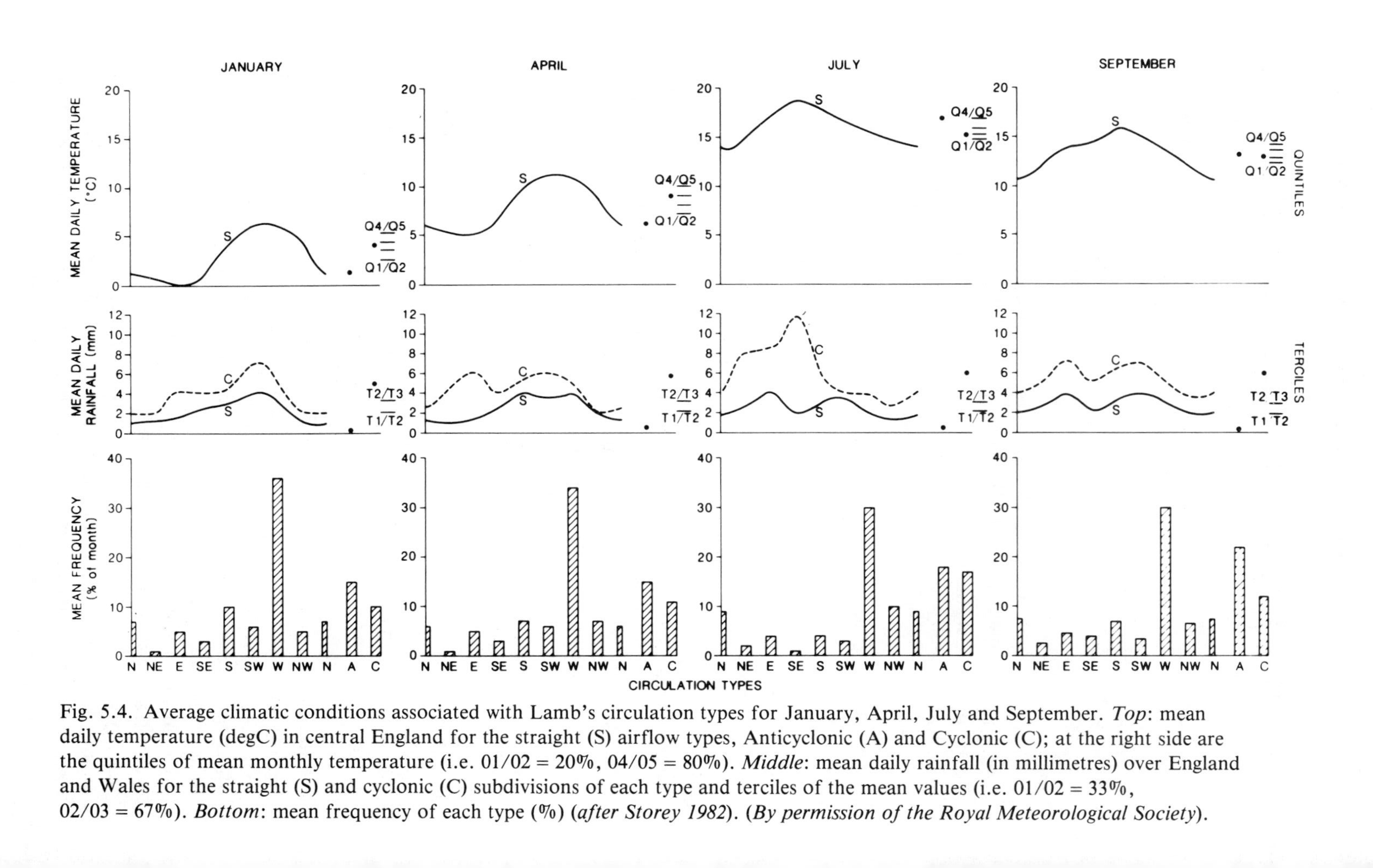

Fig. 5.4. Average climatic conditions associated with Lamb's circulation types for January, April, July and September. *Top*: mean daily temperature (degC) in central England for the straight (S) airflow types, Anticyclonic (A) and Cyclonic (C); at the right side are the quintiles of mean monthly temperature (i.e. 01/02 = 20%, 04/05 = 80%). *Middle*: mean daily rainfall (in millimetres) over England and Wales for the straight (S) and cyclonic (C) subdivisions of each type and terciles of the mean values (i.e. 01/02 = 33%, 02/03 = 67%). *Bottom*: mean frequency of each type (%) (*after Storey 1982*). (*By permission of the Royal Meteorological Society*).

(13–17%) in July–August and Anticyclonic patterns in June and September (20%); Cyclonic patterns have ⩾ 10% frequency in all months and Anticyclonic patterns ⩾ 13%. Figure 5.4 illustrates the mean daily temperature in central England and the mean daily precipitation over England and Wales for each type in the mid-season months.

The general properties of air masses have been examined in ch. 4, but certain aspects are of particular interest in respect of the British climate. The frequency of air masses in January, based on a study by Belasco for 1938–49, is illustrated for Kew in fig. 5.5. There is a clear predominance of polar maritime (mP and mPw) air which has a frequency of 30% or more in all months except March. The maximum frequency of mP air at Kew is 33% (with a further 10% mPw) in July. The proportion is even greater in western coastal districts with mP and mPw occurring in the Hebrides, for example, on at least 38% of days throughout the year.

North-westerly mP airstreams produce cool, showery weather at all seasons, especially on windward coasts. The air is unstable with cumuliform cloud, although inland in winter and at night the clouds disperse, giving low night temperatures. Over the sea, however, heating of the lower air continues by day and night in winter months so that showers and squalls can occur at any time, and these may affect windward coastal areas. The average daily mean temperatures with mP air (see tables 4.1 and 4.2) are within about ±1 °C of the seasonal means in winter and summer, depending on the precise track of the air. More extreme conditions occur with mA air, the temperature departures at Kew being approximately −4°C in summer and winter. The visibility in mA air is usually very good. The contribution of mP and mA air masses to the mean annual rainfall over a 5-year period at three stations in northern England is given in table 5.2, although it should be noted

Table 5.2 Percentage of the annual rainfall (1956–60) occurring with different synoptic situations (*after Shaw 1962 and R. P. Mathews unpublished*)

Station	*Synoptic categories*								
	Warm front	*Warm sector*	*Cold front*	*Occlusion*	*Polar low*	*mP*	*cP*	*Arctic*	*Thunderstorm*
Cwm Dyli (99 m or 324 ft)*	18	30	13	10	5	22	0·1	0·8	0·8
Squires Gate (10 m or 33 ft)†	23	16	14	15	7	22	0·2	0·7	3
Rotherham (21 m or 70 ft)‡	26	9	11	20	14	15	1·5	1·1	3

* Snowdonia.
† On the Lancashire coast (Blackpool).
‡ In the Don Valley, Yorkshire.

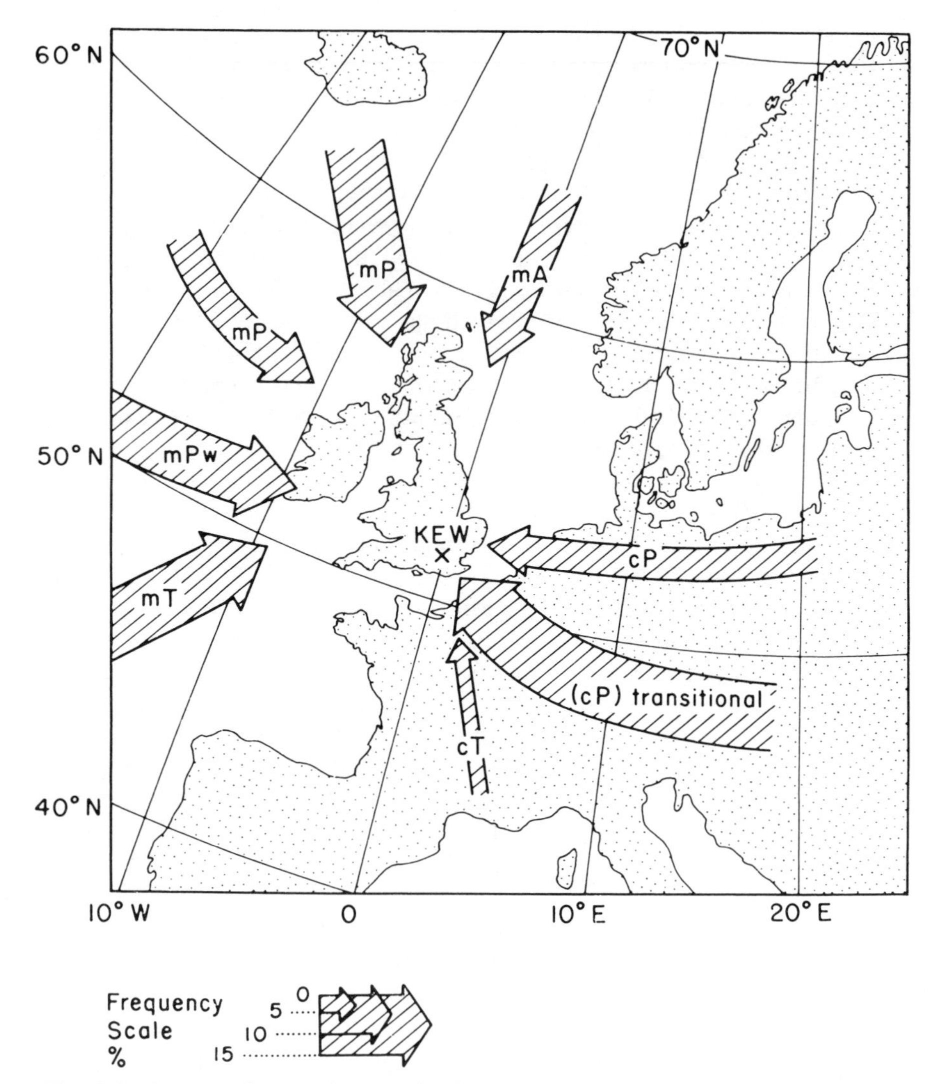

Fig. 5.5. Average air-mass frequencies for Kew (London) in January. Anticyclonic types are included according to their direction of origin (*based on Belasco 1952*).

that both air masses may also be involved in non-frontal polar lows. Over much of southern England, and in areas to the lee of high ground, northerly and north-westerly airstreams usually give clear sunny weather with few showers. There is some indication of this in table 5.2, for at Rotherham, in the lee of the Pennines, the percentage of the rainfall occurring with mP air is much lower than over Lancashire.

Maritime polar air which has followed a more southerly, cyclonic track over the Atlantic, approaching Britain from the south-west, or air which is moving northward ahead of a depression, is shown as mPw in fig. 5.2. This air has surface properties intermediate with mT air.

Maritime tropical air commonly forms the warm sector of depressions moving from between west and south towards Britain, but fig. 5.5 excludes cases of fronts and depressions (which have a frequency of about 10–12% throughout the year). Hence the characteristic weather conditions of mT air occur rather more often than the percentage frequency might suggest (in January 11% mT air and 4% Anticyclonic air originating from south-west of Britain). The weather is unseasonably mild and damp with mT air in winter (see table 4.1). There is generally a complete cover of stratus or stratocumulus cloud and drizzle or light rain (formed by coalescence) may occur, especially over high ground where low cloud produces hill fog. The clearance of cloud on nights with light winds readily cools the moist air to its dew point, forming mist and fog. Table 5.2 shows that a large proportion of the annual rainfall is associated with warm-front and warm-sector situations and therefore is largely attributable to convergence and frontal uplift within mT air. In summer the cloud cover with this air mass keeps temperatures closer to average than in winter (see tables 4.1 and 4.2); night temperatures tend to be high, but daytime maxima remain rather low.

True continental polar air only affects the British Isles between December and February and even then it is relatively infrequent. Mean daily temperatures are well below average and maxima rise to only a degree or so above freezing point. The air mass is basically very dry and stable but a track over the central part of the North Sea supplies sufficient heat and moisture to cause showers, often in the form of snow, over eastern England and Scotland. *In toto* this provides only a very small contribution to the annual precipitation, as table 5.2 shows, and on the west coast the weather is generally clear. A transitional cP–cT type of air mass reaches Britain from south-east Europe in all seasons, though less frequently in summer. Such airstreams are dry and stable.

Continental tropical air occurs on average about one day per month in summer, which accounts for the rarity of summer heat-waves since these south or south-east winds bring hot, settled weather. The lower layers are stable and the air is commonly hazy, but the upper layers tend to be unstable and occasionally intense surface heating may trigger off a thunderstorm. In winter such modified cT air sometimes reaches Britain from the Mediterranean bringing fine, hazy, mild weather.

4 Singularities and natural seasons

Most popular weather lore expresses the belief that each season has its own weather (for example, in England, 'February fill-dyke', 'April showers'), and ancient adages suggest that even the sequence of weather may be determined by the conditions established on a given date (e.g. 40 days of wet or fine weather following St Swithin's day, 15 July, in England). Some of these ideas are quite fallacious, but others contain more than a grain of truth if properly interpreted.

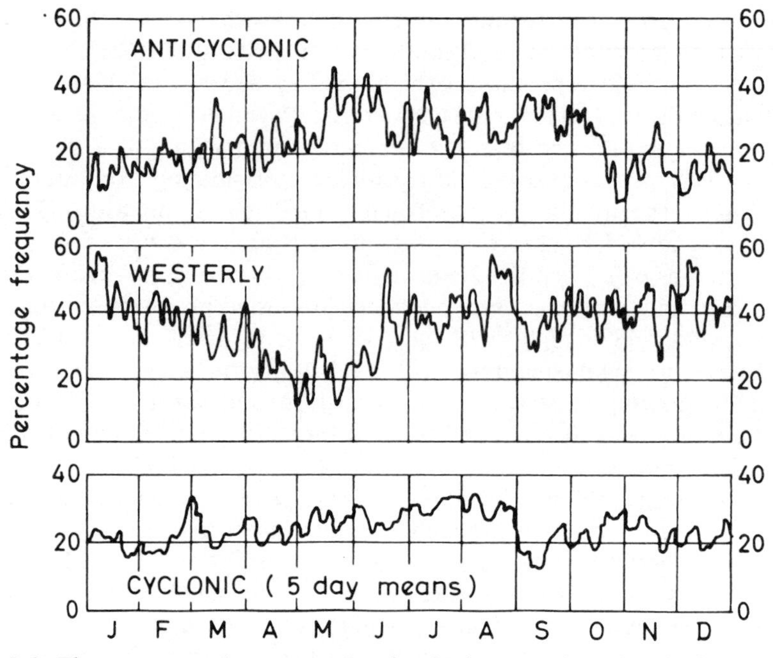

Fig. 5.6. The percentage frequency of anticyclonic, westerly and cyclonic conditions over Britain, 1898–1947 (*after Lamb 1950*).

The tendency for a certain type of weather to recur with reasonable regularity around the same date is termed a *singularity*. Many calendars of singularities have been compiled, particularly in Europe, but the early ones (which concentrated upon anomalies of temperature or rainfall) did not prove very reliable. Greater success has been achieved by studying singularities of circulation pattern and catalogues have been prepared for the British Isles by Lamb and for central Europe by Flohn and Hess. Lamb's results are based on calculations of the daily frequency of the airflow categories between 1898 and 1947, some examples of which are shown in fig. 5.6. A noticeable feature is the infrequency of the Westerly type in spring, the driest season of the year in the British Isles and also in northern France, northern Germany and in the countries bordering the North Sea. The catalogue of Flohn and Hess is based on a classification of large-scale patterns of airflow in the lower troposphere (*Grosswetterlage*) over central Europe originally proposed by F. Baur.

Some of the European singularities that occur most regularly are as follows:

(1) A sharp increase in the frequency of Westerly and North-westerly type over Britain takes place about the middle of June. These invasions of maritime air also affect central Europe and this period is sometimes referred to as the beginning of the European 'summer monsoon'.

(2) About the second week in September Europe and Britain are affected by a spell of Anticyclonic weather. This may be interrupted by Atlantic

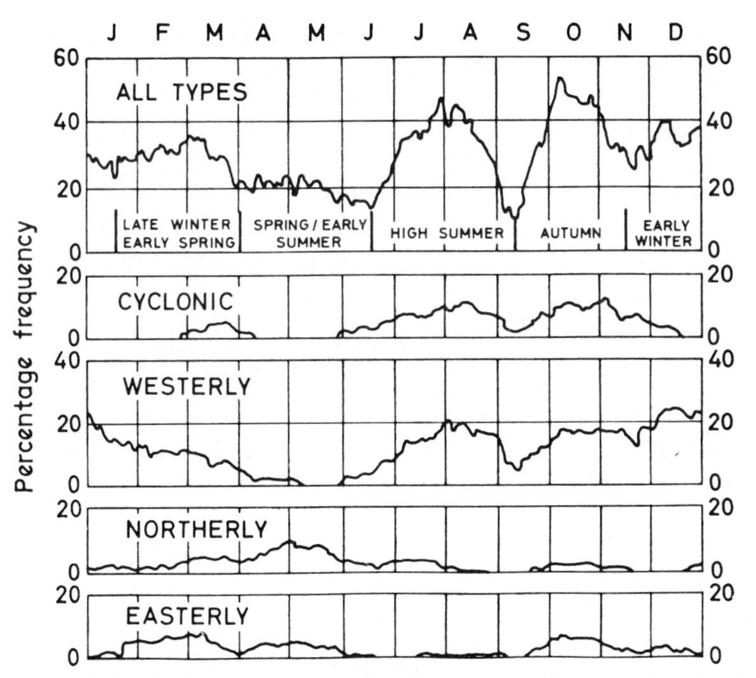

Fig. 5.7. The frequency of long spells (25 days or more) of a given airflow type over Britain, 1898–1947. The diagram showing all long spells also indicates a division of the year into 'natural seasons' (*after Lamb 1950*).

depressions giving stormy weather over Britain in late September, though anticyclonic conditions again affect central Europe at the end of the month and Britain during early October.

(3) A marked period of wet weather often affects western Europe and also the western half of the Mediterranean at the end of October, whereas the weather in eastern Europe generally remains fine.

(4) Anticyclonic conditions return to Britain and affect much of Europe about mid-November, giving rise to fog and frost.

(5) In early December Atlantic depressions push eastwards to give mild, wet weather over most of Europe.

In addition to these singularities, major seasonal trends are recognizable and for the British Isles Lamb identified five *natural seasons* on the basis of spells of a particular type lasting for 25 days or more during the period 1898–1947 (fig. 5.7). In as far as is possible to think in terms of a 'normal year', the seasons are:

(i) *Spring–early summer* (the beginning of April to mid-June). This is a period of variable weather conditions during which long spells are least likely. Northerly spells in the first half of May are the most significant feature, although there is a marked tendency for anticyclones to occur in late May–early June.

(ii) *High summer* (mid-June to early September). Long spells of various types may occur in different years. Westerly and North-westerly types are the most common and they may be combined with either Cyclonic or Anticyclonic types. Persistent sequences of Cyclonic type occur more frequently than Anticyclonic ones.
(iii) *Autumn* (the second week in September to mid-November). Long spells are again present in most years, Anticyclonic ones mainly in the first half, Cyclonic and other stormy ones generally in October–November.
(iv) *Early winter* (from about the third week in November to mid-January). Long spells are less frequent than in summer and autumn. They are usually of Westerly type giving mild, stormy weather.
(v) *Late winter and early spring* (from about the third week in January to the end of March). The long spells at this time of year can be of very different types, so that in some years it is mid-winter weather while in other years there is an early spring from about late February.

5 *Synoptic anomalies*

The mean climatic features of pressure, wind and the seasonal airflow regime provide an incomplete picture of climatic conditions. Some patterns of circulation occur irregularly and yet, because of their tendency to persist for weeks or even months, form an essential element of the climate.

Blocking patterns are an important example. It was noted in ch. 3, F.2 that the zonal circulation in middle latitudes sometimes breaks down into a cellular pattern. This is commonly associated with a split of the jet stream into two branches over higher and lower middle latitudes and the formation of a cut-off low (see ch. 4, G.4) south of a high-pressure cell. The latter is referred to as a *blocking anticyclone* since it prevents the normal eastward motion of depressions in the zonal flow. Figure 5.8 gives numerical values expressing the frequency of occurrence of blocking for part of the northern hemisphere with five major blocking centres shown (H). A major area of blocking is Scandinavia, particularly in spring. Depressions are diverted north-eastwards towards the Norwegian Sea or south-eastwards into southern Europe. This pattern, with easterly flow around the southern margins of the anticyclone, produces severe winter weather over much of northern Europe. In January–February 1947, for example, easterly flow across Britain as a result of blocking over Scandinavia led to extreme cold and frequent snowfall. Winds were almost continuously from the east between 22 January and 22 February and even daytime temperatures rose little above freezing point. Snow fell in some part of Britain every day from 22 January to 17 March 1947, and major snowstorms occurred as occluded Atlantic depressions moved slowly across the country. Other notably severe winter months – January 1881, February 1895 and January 1940 – were the result of similar pressure anomalies with pressure much above average to the north of the British Isles and below average to the south.

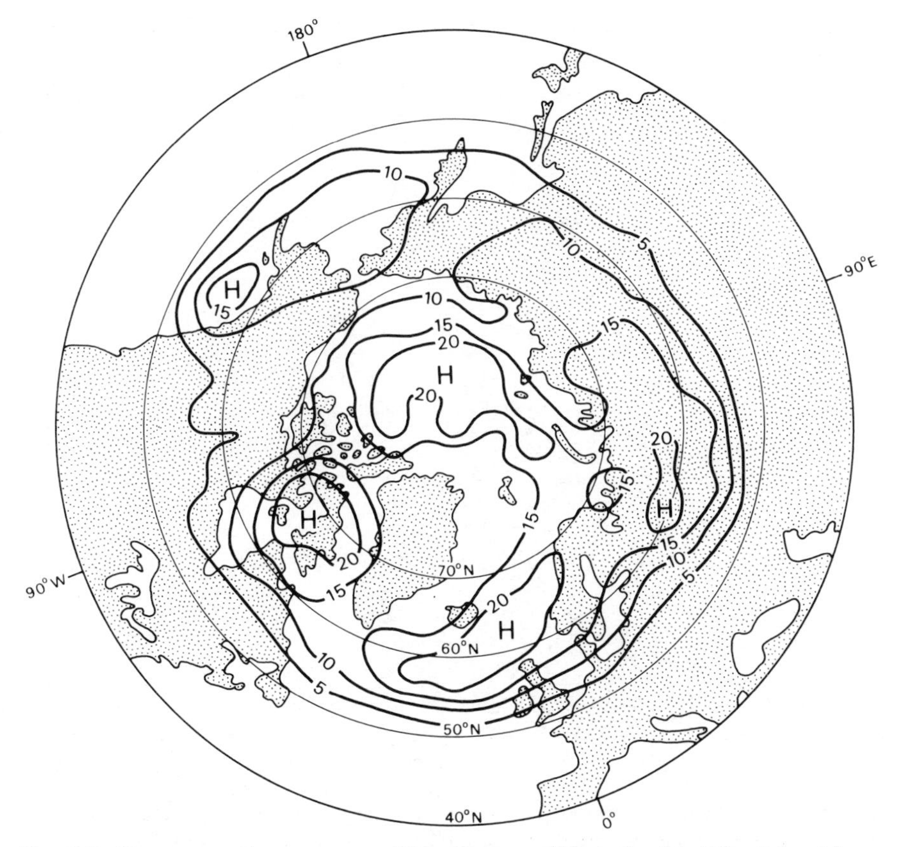

Fig. 5.8. Frequency of occurrence of blocking conditions for the 500 mb level for all seasons. Values were calculated as 5-day means for 381 × 381 km squares for the period 1946–78 (*from Knox and Hay 1985*). (*By permission of the Royal Meteorological Society*).

The average effects of a number of winter blocking situations over north-west Europe are shown in figs 5.9 and 5.10. Precipitation amounts are above normal mainly over Iceland and the western Mediterranean as depressions are steered around the blocking high following the path of the upper jet streams. Over most of Europe precipitation remains below average and this pattern is repeated with summer blocking. winter temperatures are above average over the north-eastern Atlantic and adjoining land areas, but below average over central and eastern Europe and the Mediterranean due to the northerly outbreaks of cP air (fig. 5.10). The negative temperature anomalies associated with cool northerly airflow in summer cover most of Europe, and only northern Scandinavia and the north-eastern Atlantic have above-average values.

Despite these generalizations the location of the block is of the utmost importance. For instance, in the summer of 1954 a blocking anticyclone across eastern Europe and Scandinavia allowed depressions to stagnate over the

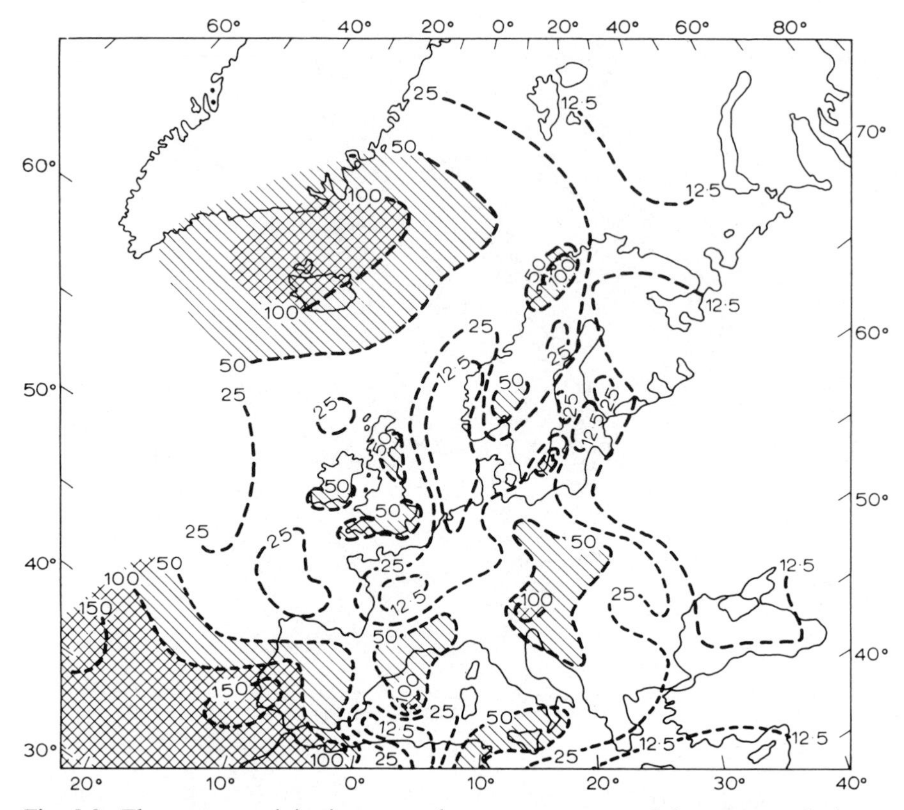

Fig. 5.9. The mean precipitation anomaly, as a percentage of the average, during anticyclonic blocking in winter over Scandinavia. Areas above normal are cross-hatched, areas recording precipitation between 50 and 100% of normal have oblique hatching (*after Rex 1950*).

British Isles giving a dull, wet August, whereas in 1955 the blocking was located over the North Sea and a fine, warm summer resulted. The 1975–6 drought in Britain and the continent was caused by persistent blocking over north-western Europe. Another less common location of blocking is Iceland. A notable example was the 1962–3 winter when persistent high pressure south-east of Iceland led to northerly and north-easterly airflow over Britain. Temperatures in central England at that time were the lowest since 1740, with a mean of 0°C for December 1962 to February 1963. Central Europe was affected by easterly airstreams and mean January temperatures there were 6°C below average.

6 Topographic effects

In various parts of Europe topography has a marked effect on the climate not only of the uplands themselves but also on that of adjacent areas. Apart from the more obvious effects upon temperatures, precipitation amounts and winds, the major mountain masses affect the movement of frontal

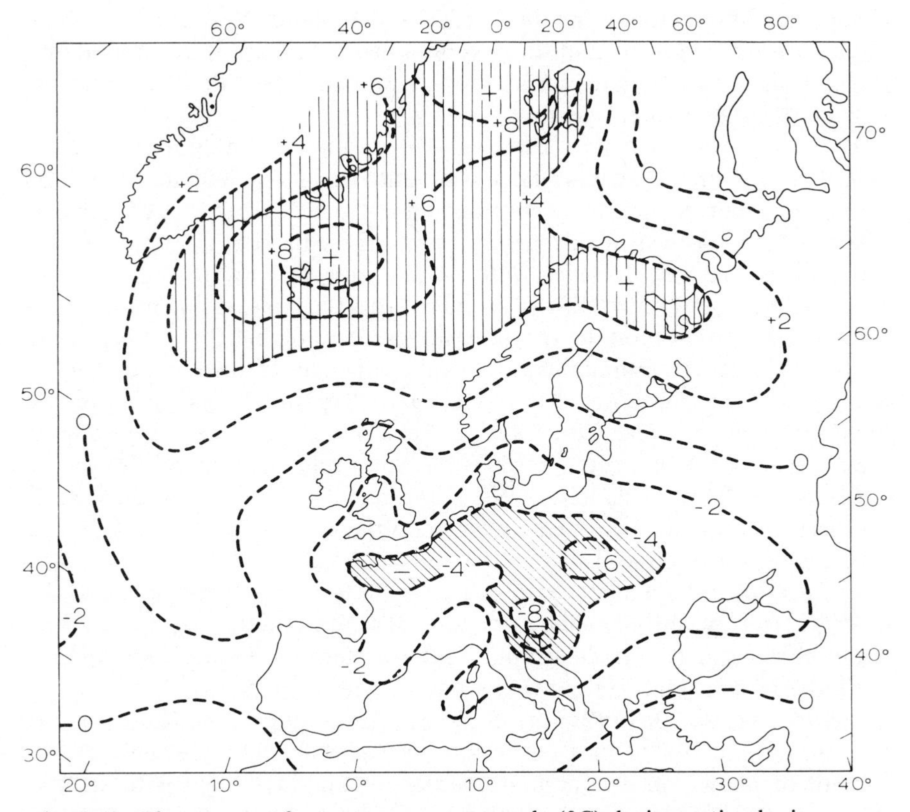

Fig. 5.10. The mean surface temperature anomaly (°C) during anticyclonic blocking in winter over Scandinavia. Areas more than 4°C above normal have vertical hatching, those more than 4°C below normal have oblique hatching (*after Rex 1950*).

systems too. Surface friction over mountain barriers tends to steepen the slope of cold fronts and decrease the slope of warm fronts so that the latter are slowed down and the former accelerated. The cyclogenetic effect of mountain barriers in producing lee depressions has already been discussed (see ch. 4, G.1).

The Scandinavian mountains are one of the most significant climatic barriers in Europe as a result of their orientation with regard to westerly airflow. Maritime air masses are forced to rise over the highland zone giving annual precipitation totals of over 250 cm (100 in) on the mountains of western Norway, whereas descent in their lee produces a sharp decrease in the amounts. The upper Gudbrandsdalen and Osterdalen in the lee of the Jotunheim and Dovre Mountains receive on average less than 50 cm (20 in), and similar low values are recorded in the Jämtland province of central Sweden around Östersund.

The mountains can equally function in the opposite sense. For example, Arctic air from the Barents Sea may move southwards in winter over the Gulf of Bothnia, usually when there is a depression over northern Russia, giving very low temperatures in Sweden and Finland. Western Norway is rarely affected, since the cold wave is contained to the east of the mountains. In consequence there is a sharp climatic gradient across the Scandinavian Highlands in the winter months.

The Alps provide a quite different illustration of topographic effects. Together with the Pyrenees and the mountains of the Balkans, the Alps effectively separate the Mediterranean climatic region from that of Europe. The penetration of warm air masses north of these barriers is comparatively rare and short-lived. However, with certain pressure patterns, air from the Mediterranean and north Italy is forced to cross the Alps, losing its moisture through precipitation on the southern slopes. Dry adiabatic warming on the northern side of the mountains can readily raise temperatures by 5°–6°C in the upper valleys of the Aar, Rhine and Inn. At Innsbruck there are approximately 50 days per year with föhn winds, with a maximum in spring, and these occurrences are often responsible for rapid melting of the snow, creating a risk of avalanches. When the airflow across the Alps has a northerly component föhns may occur in northern Italy, but their effects are generally less pronounced.

Features of upland climates in Britain will serve to illustrate some of the diverse effects of altitude. The mean annual rainfall on the west coasts near sea level is about 114 cm (45 in) but on the western mountains of Scotland, the Lake District and Wales averages exceed 380 cm (150 in) per year. The annual record is 653 cm (257 in) in 1954 at Sprinkling Tarn, Cumberland, and 145 cm (57 in) fell in a single month (October 1909) just east of the summit of Snowdon. The annual number of rain-days (days with at least 0·25 mm (0·01 in) of precipitation) increases from about 165 days in south-eastern England and the south coast to over 230 days in north-west Britain. There is little additional increase in the frequency of rainfall with height on the mountains of the north-west, so that the mean rainfall per rain-day rises sharply from 0·5 cm (0·2 in) near sea-level in the west and north-west to over 1·3 cm (0·5 in) in the Western Highlands, the Lake District and Snowdonia. This demonstrates that 'orographic rainfall' here is primarily due to an intensification of the normal precipitation processes and is not a special type. It is more appropriate, therefore, to recognize an orographic component which increases the amounts of rain associated with frontal depressions and unstable airstreams (see ch. 2, I.2).

Even quite low hills such as the Chilterns and South Downs cause a rise in rainfall, receiving about 12–13 cm (5 in) per year more than the surrounding lowlands. In South Wales, mean annual precipitation increases from 120 cm at the coast to 250 cm on the 500-m high Glamorgan Hills, 20 km inland. Studies using radar and a dense network of rain gauges indicate that orographic intensification is pronounced during strong low-level south-westerly airflows in frontal situations. Most of the enhancement of precipitation rate occurs in the lowest 1500 m. Figure 5.11 shows the mean enhancement according to wind direction over England and Wales, averaged for several days with fairly constant wind velocities of about 20 m s^{-1} and

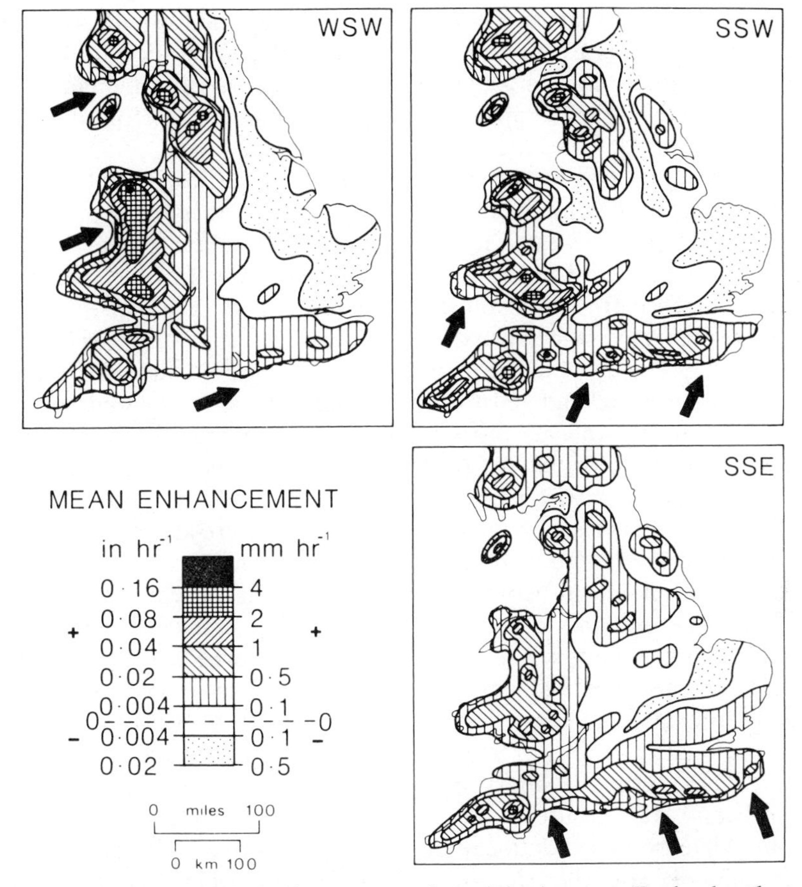

Fig. 5.11. Mean orographic enhancement of precipitation over England and Wales, averaged for several days of fairly constant wind direction of about 20 m s^{-1} and nearly saturated low-level airflow (*after Browning and Hill 1981*). (*By permission of the Royal Meteorological Society*).

nearly saturated low-level flow, attributable to a single frontal system on each day. Differences are apparent in Wales and southern England between winds from SSW and from WSW, whereas for SSE airflows the mountains of North Wales and the Pennines have little effect. Note also the areas of negative enhancement on the lee side of mountains.

The sheltering effects of the uplands produce low annual totals on the lee side (with respect to the prevailing winds). Thus, the lower Dee valley in the lee of the mountains of North Wales receives less than 75 cm (30 in) per year compared with over 250 cm (100 in) on Snowdonia.

The complexity of the various factors affecting rainfall in Britain is shown by the fact that a close correlation exists between annual totals in north-west Scotland, the Lake District and western Norway, which are directly affected by Atlantic depressions. At the same time there is an inverse relationship between annual amounts in the Western Highlands and lowland Aberdeenshire less than 240 km (150 miles) away. Annual precipitation in the latter area is more closely correlated with that in lowland eastern England. Essentially, the British Isles comprise two major climatic units for rainfall – firstly, an 'Atlantic' one with a winter season maximum, and, secondly, those central and eastern districts with 'continental' affinities in the form of a weak summer maximum in most years. Other areas (eastern Ireland, eastern Scotland, north-east England and most of midland England and the Welsh border counties) generally have a wet second half of the year.

The occurrence of snow is another measure of altitude effects. Near sea-level there are on average approximately 5 days per year with snow falling in south-west England, 15 days in the south-east and 35 days in northern Scotland. Between 60 and 300 m the frequency increases by about 1 day per 15 m of elevation and even more rapidly on higher ground. Approximate figures for northern Britain are 60 days at 600 m and 90 days at 900 m. The number of mornings with snow lying on the ground (more than half the ground covered) is closely related to mean temperature and hence altitude. Average figures range from about 5 days per year or less in much of southern England and Ireland, to between 30 and 50 days on the Pennines and over 100 days on Grampian Mountains. In the last area (on the Cairngorms) and on Ben Nevis there are several semi-permanent snowbeds at about 1160 m and it is estimated that the theoretical climatic snowline – above which there is *net* accumulation of snow – is at 1620 m (5300 ft) over Scotland.

The seasonal variability of lapse rates in mountain areas was mentioned in ch. 1, I. There also exist marked geographical variations even within the British Isles. One measure of these variations is the length of the 'growing season'. Meteorological data can be used to determine an index of growth opportunity by counting the number of days on which the mean daily temperature exceeds an arbitrary threshold value – commonly 6°C (43°F). Along the south-west coasts of England the 'growing season', as calculated on this basis, is nearly 365 days per year and in this area it decreases by about 9 days per 30 m of elevation, but in northern England and Scotland the decrease is only about 5 days per 30 m from between 250–270 days near sea-level. In continental climates the altitudinal decrease may be even more gradual; in central Europe and New England, for example, it is about 2 days per 30 m.

B North America

The North American continent spans nearly 60° of latitude and, not surprisingly, exhibits a wide range of climatic conditions. Unlike Europe, the west coast is backed by the Pacific coast ranges rising to over 2750 m, which lie across the path of depressions in the mid-latitude westerlies and prevent the extension of maritime influences inland. In the interior of the continent there are no significant obstructions to air movement, and the absence of any east–west barrier allows air masses from the Arctic or the Gulf of Mexico to sweep across the interior lowlands, causing wide extremes of weather and climate. Maritime influences in eastern North America are greatly limited by the fact that the prevailing winds are westerly, so that the temperature regime is continental. Nevertheless, the Gulf of Mexico is a major source of moisture supply for precipitation over the eastern half of the United States and, as a result, the precipitation regimes are different from those found in eastern Asia.

We will look first at the broad characteristics of the atmospheric circulation over the continent.

1 Pressure systems

The mean pressure pattern for the middle troposphere displays a prominent trough over eastern North America in both summer and winter (see fig. 3.19). One theory is that this is a lee trough caused by the effect of the western mountain ranges on the upper westerlies, but at least in winter the strong baroclinic zone along the east coast of the continent is undoubtedly a major contributory factor. The implications of this mean wave pattern are that cyclones tend to move south-eastward over the Midwest carrying continental polar air southward, while the cyclone paths are north-eastward along the Atlantic coast.

In individual months there may of course be considerable deviations from this average pattern, with important consequences for the weather in different parts of the continent, and, in fact, this relationship provides the basis for the monthly forecasts of the United States Weather Service. For example, if the trough is more pronounced than usual temperature may be much below average in the central, southern and eastern United States, whereas if the trough is weak the westerly flow is stronger with correspondingly less opportunity for cold outbreaks of polar air masses. Sometimes the trough is displaced to the western half of the continent causing a reversal of the usual weather pattern, since upper north-westerly airflow can bring cold, dry weather to the west while in the east there are very mild conditions associated with upper south-westerly flow. Precipitation amounts also depend on the depression tracks; if the upper trough is far to the west, depressions form ahead of it (see ch. 4, E) over the south central United States and move northeastwards towards the lower St Lawrence, giving more precipitation than usual in these areas and less along the Atlantic coast.

The major features of the surface pressure map in January (see fig. 3.24A) are the extension of the subtropical high over the south-western United States (called the Great Basin high) and the separate polar anticyclone of the Mackenzie district of Canada. Mean pressure is low off both the east and west coasts of higher middle latitudes, where oceanic heat sources indirectly give rise to the (mean) Icelandic and Aleutian lows. It is interesting to note that, on average, in December, of any region in the northern hemisphere for any month of the year, the Great Basin region has the most frequent occurrence of highs, whereas the Gulf of Alaska has the maximum frequency of lows. The Pacific coast as a whole has its most frequent cyclonic activity in winter as does the Great Lakes area, whereas over the Great Plains the maximum is in spring and early summer. Remarkably, the Great Basin in June has the most frequent cyclogenesis of any part of the northern hemisphere in any month of the year. Heating over this area in summer helps to maintain a shallow, quasi-permanent low-pressure cell, in marked contrast with the almost continuous subtropical high-pressure belt in the middle troposphere (fig. 3.19). Continental heating also indirectly assists in the splitting of the Icelandic low to create a secondary centre over north-eastern Canada. The west-coast summer circulation is dominated by the Pacific anticyclone while the south-eastern United States is affected by the Atlantic subtropical anticyclone cell.

Broadly, there are three prominent depression tracks across the continent in winter (see fig. 4.18). One group moves from the west along a more or less zonal path about 45°–50°N, whereas a second loops southward over the central United States and then turns north-eastward towards New England and the Gulf of St Lawrence. Some of these depressions originate over the Pacific, cross the western ranges as an upper trough and redevelop in the lee of the mountains. Alberta is a noted area for this process and also for primary cyclogenesis since the arctic frontal zone is over north-west Canada in winter. This frontal zone involves much-modified mA air from the Gulf of Alaska and cold dry cA (or cP) air. Depressions of the third group form along the main polar frontal zone, which in winter is off the east coast of the United States, and move north-eastward towards Newfoundland. Sometimes this frontal zone is present over the continent about 35°N with mT air from the Gulf and cP air from the north or modified mP air from the Pacific. Polar front depressions forming over Colorado move north-eastward towards the Great Lakes and others developing over Texas follow a more or less parallel path, further to the south and east, towards New England.

Between the Arctic and Polar Fronts a third frontal zone is distinguished by Canadian meteorologists. This *maritime* (arctic) frontal zone is present when mA and mP (or mPc and mPw) air masses interact along their common boundary. The three-front (i.e. four air-mass) model allows a detailed analysis to be made of the baroclinic structure of depressions over the North American continent using synoptic weather maps and cross-sections of the atmosphere. Figure 5.12 illustrates the three frontal zones and associated depressions on 29 May 1963. Along 95°W, from 60°N to 40°N, the following dew-point temperatures were reported in the four air masses: (–8°C) (17°F), (1°C) (33°F), 4°C (40°F) and 13°C (55°F).

In summer, the east-coast depressions are less frequent and the tracks across the continent are displaced northwards with the main ones moving over Hudson Bay and Labrador–Ungava, or along the line of the St Lawrence. These are associated mainly with a rather poorly-defined Maritime frontal zone. The Arctic Front is usually located along the north coast of Alaska, where there is a strong temperature gradient between the bare land and the cold Polar Sea and pack-ice. East from here the front is

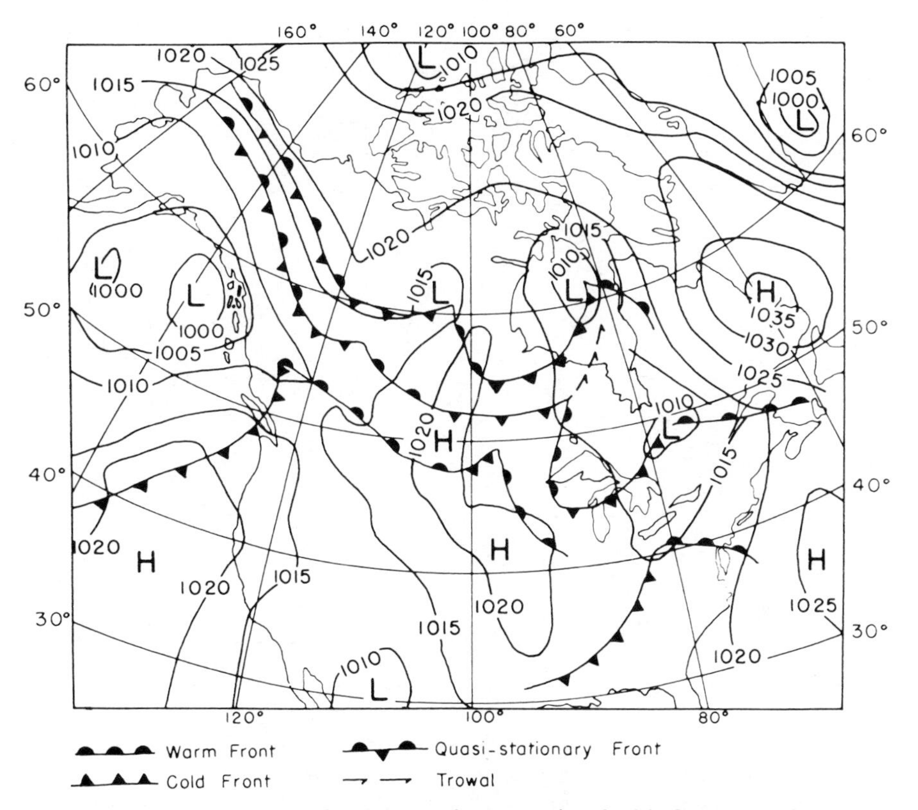

Fig. 5.12. A synoptic example of depressions associated with three-frontal zones on 29 May 1963 over North America (*based on charts of the Edmonton Analysis Office and the Daily Weather Report*).

very variable in location from day to day and year to year. Broadly, it occurs most often in the vicinity of northern Keewatin and Hudson Strait, although one study of air-mass temperatures and airstream confluence regions suggests that an arctic frontal zone occurs further south over Keewatin in July and that its mean position (fig. 5.13) is closely related to the boreal forest–tundra boundary. This relationship undoubtedly reflects the importance of arctic air-mass dominance for summer temperatures and consequently for tree-growth possibilities, but the precise nature of the interrelationships between atmospheric systems and vegetation boundaries requires more extensive investigation.

Several circulation singularities have been recognized in North America, as in Europe (ch. 5, A.4). Three which have received considerable attention in view of their prominence are: (*i*) the advent of spring in late March; (*ii*) the midsummer high-pressure jump at the end of June; (*iii*) the Indian summer in late September (and late October).

The arrival of spring is marked by different climatic responses in different parts of the continent. For example, there is a sharp decrease in March to April precipitation in California, due to the extension of the Pacific high,

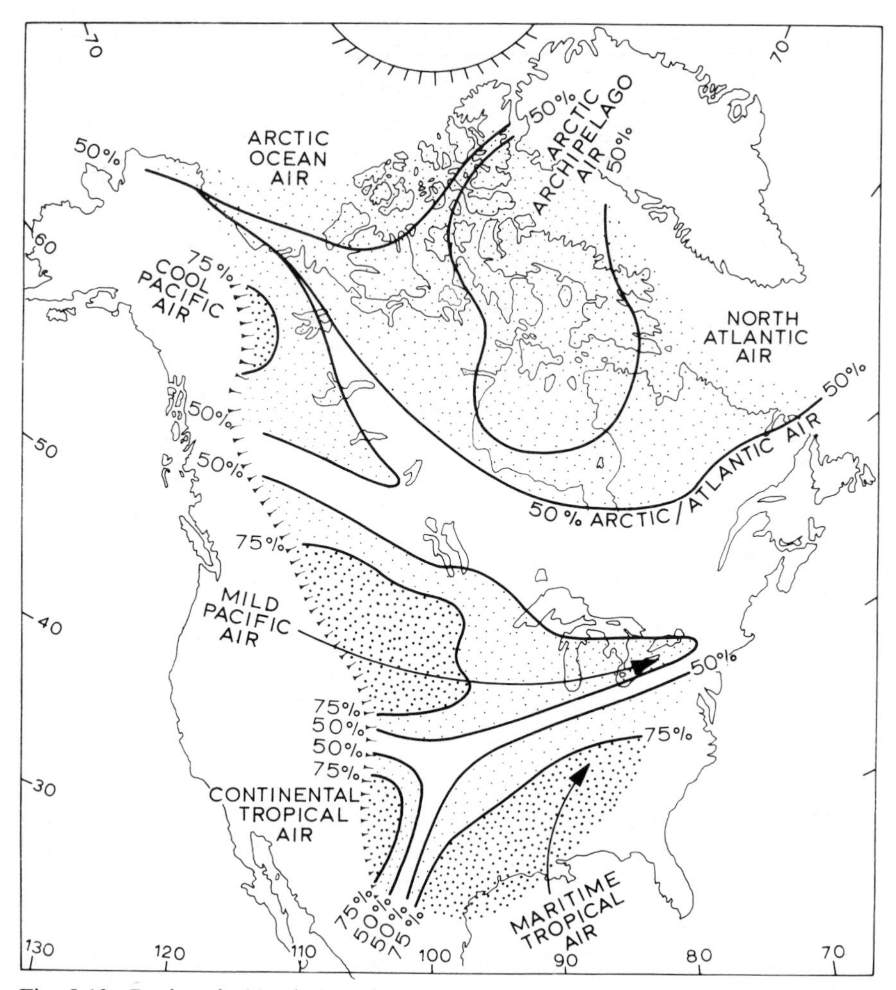

Fig. 5.13. Regions in North America east of the Rocky Mountains dominated by the various air-mass types in July for more than 50% and 75% of the time (*after Bryson 1966*). The 50% frequency lines correspond to mean frontal positions.

whereas precipitation intensity increases in the Midwest (fig. 5.14) as a result of more frequent cyclogenesis in Alberta and Colorado and a northward extension of maritime tropical air over the Midwest from the Gulf of Mexico. These changes are part of a hemispheric readjustment of the circulation since at the beginning of April the Aleutian low-pressure cell, which from September to March is located about 55°N, 165°W, splits into two with one centre in the Gulf of Alaska and the other over northern Manchuria. This represents a decrease in the zonal index (ch. 3, F.2).

In late June there is a rapid northward displacement of the Bermuda and North Pacific subtropical high-pressure cells. In North America this pushes the depression tracks northward also with the result that precipitation decreases from June to July over the northern Great Plains, parts of Idaho

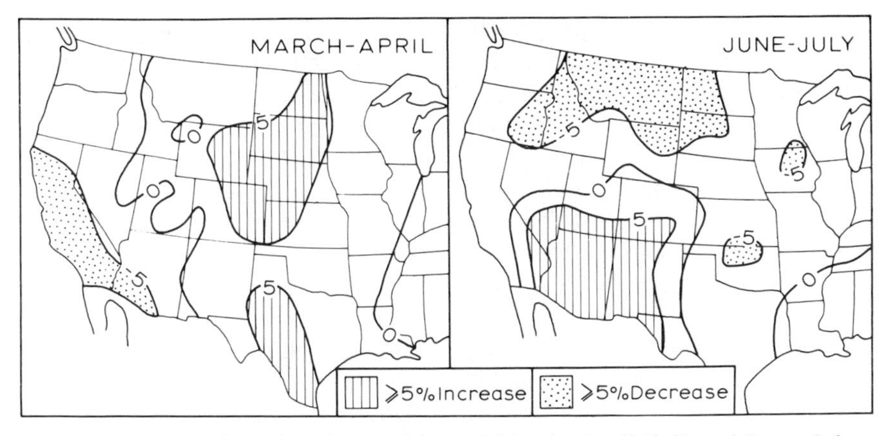

Fig. 5.14. The precipitation changes between March–April (*left*) and June–July (*right*), as a percentage of the mean annual total, for the central and western United States (*after Bryson and Lahey 1958*).

and eastern Oregon (fig. 5.14). Conversely, the south-westerly anticyclonic flow which affects Arizona in June is replaced by air from the Gulf of California and this causes the onset of the summer rains (see ch. 5, D.2). It has been suggested by Bryson and Lahey that these circulation changes at the end of June may be connected with the disappearance of snow cover from the arctic tundra. This leads to a sudden decrease of surface albedo from about 75 to 15% with consequent changes in the heat budget components and hence in the atmospheric circulation.

Frontal wave activity makes the first half of September a rainy period in the northern Midwest states of Iowa, Minnesota and Wisconsin, but after about the 20th of the month anticyclonic conditions return with warm air-flow from the dry south-west giving fine weather – the so-called Indian summer. Significantly, the hemispheric zonal index value rises in late September. This anticyclonic weather type has a second phase in the latter half of October, but at this time there are polar outbreaks. The weather is generally cold and dry, although if precipitation does occur there is a high probability of it being in the form of snow.

2 *The temperate west coast and cordillera*

The oceanic circulation of the North Pacific closely resembles that of the North Atlantic. The drift from the Kuro Shio current off Japan is propelled by the westerlies towards the western coast of North America and acts as a warm current between 40° and 60°N. Sea-surface temperatures are several degrees lower than in comparable latitudes off western Europe, however, due to the smaller volume of warm water involved. Also, in contrast to the Norwegian Sea, the shape of the Alaskan coastline prevents the extension of the drift to high latitudes (see fig. 3.38).

The Pacific coast ranges greatly restrict the inland extent of oceanic influences, and hence there is no extensive maritime temperate climate such as we have described for western Europe. The major climatic features duplicate those of the coastal mountains of Norway and those of New Zealand and southern Chile in the belt of Southern Westerlies. Topographic factors make the weather and climate of such areas very variable over short distances both vertically and horizontally, and, therefore, only a few salient characteristics are selected for consideration.

There is a regular pattern of rainy windward and drier lee slopes across the successive north-west to south-east ranges with a more general decrease towards the interior. The Coast Range in British Columbia has mean annual totals of precipitation exceeding 250 cm (100 in) with 500 cm (200 in) in the wettest places compared with 125 cm (50 in) or less on the summits of the Rockies, yet even on the leeward side of Vancouver Island the average figure at Victoria is only 70 cm (27 in). Analogous to the 'Westerlies-oceanic' regime of north-west Europe, there is a winter precipitation maximum along the littoral which also extends beyond the Cascades (in Washington) and the Coast Range (in British Columbia), but summers are drier due to the strong North Pacific anticyclone. The regime in the interior of British Columbia is transitional between that of the coastal region and the distinct summer maximum of central North America (fig. 5.15), although at Kamloops in the Thompson valley (annual average 25 cm or 10 in) there is a slight summer

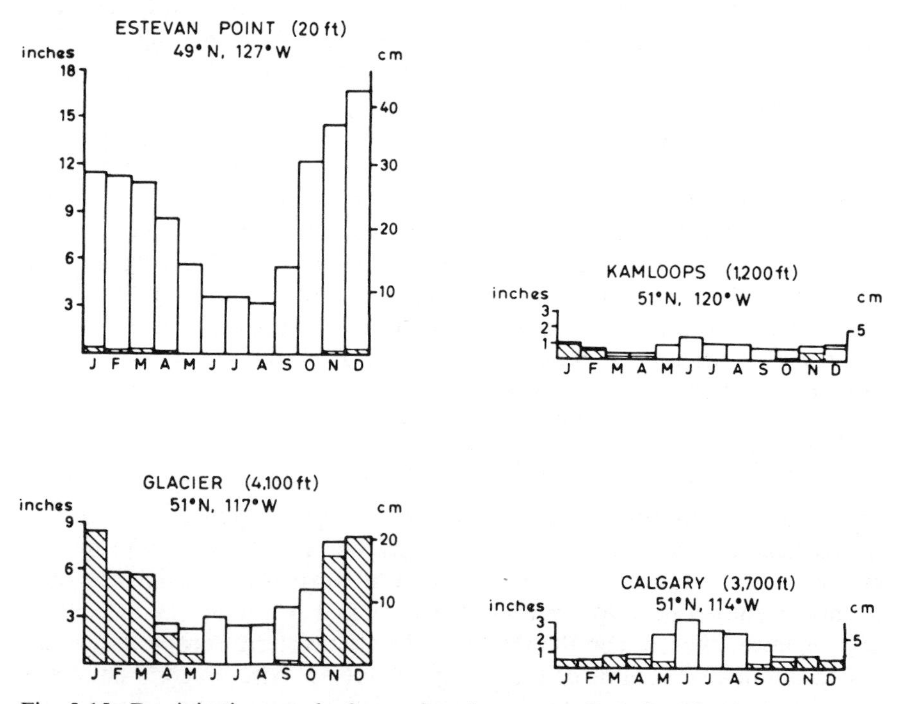

Fig. 5.15. Precipitation graphs for stations in western Canada. The shaded portions represent snowfall, expressed as water equivalent.

maximum associated with thunderstorm type of rainfall. In general, the sheltered interior valleys receive less than 50 cm (20 in) per year and in the driest years certain localities have recorded only 15 cm (6 in). Above 1000 m much of the precipitation falls as snow (fig. 5.15) and some of the greatest snow-depths in the world are reported from British Columbia, Washington and Oregon. For example, between 1000 and 1500 cm (400 and 600 in) falls on the Cascade Range at heights of about 1500 m (5000 ft) and even as far inland as the Selkirk Mountains the totals are considerable. The mean snowfall is 990 cm (390 in) at Glacier, British Columbia (elevation 1200 m) and this accounts for almost 70% of the annual precipitation (fig. 5.15). Near sea level on the outer coast, in contrast, very little precipitation falls as snow (for example, Estevan Point). It is estimated that the climatic snowline rises from about 1600 m (5250 ft) on the west side of Vancouver Island to 2900 m (9500 ft) in the eastern Coast Range. Inland its elevation increases from 2300 m (7550 ft) on the west slopes of the Columbia Mountains to 3100 m (10,170 ft) on the east side of the Rockies. This trend reflects the precipitation pattern referred to above.

Finally, mention must be made of the large diurnal variations which affect the cordilleran valleys. Strong diurnal rhythms of temperature (especially in summer) and wind direction are a feature of mountain climates and their effect is superimposed upon the general climatic characteristics of the area. Cold air drainage produces many remarkably low minima in the mountain valleys and basins. At Princeton, British Columbia (elevation 695 m), where the mean daily minimum in January is −14°C (7°F), there is on record an absolute low of−45°C (−49°F), for example. This leads in some cases to reversal of the normal lapse rate. Golden in the Rocky Mountain Trench has a January mean of −12°C (11°F) whereas 460 m (1500 ft) higher at Glacier (1248 m) it is −10°C (14°F).

3 Interior and eastern North America

Central North America has the typical climate of a continental interior in middle latitudes with hot summers and cold winters (see fig. 5.17), yet the weather in winter is subject to marked variability. This is determined by the steep temperature gradient between the Gulf of Mexico and the snow-covered northern plains; also by shifts of the upper wave patterns and jet stream. Cyclonic activity in winter is much more pronounced over central and eastern North America than in Asia, which is dominated by the Siberian anticyclone (see fig. 4.18), and consequently there is no climatic type with a winter minimum of precipitation in eastern North America.

The general temperature conditions in winter and summer are illustrated in fig. 5.16, showing the frequency with which hourly temperature readings exceed or fall below certain limits. The two chief features of all four maps are: (*a*) the dominance of the meridional temperature gradient, away from coasts, and (*b*) the continentality of the interior and east compared with the 'maritimeness' of the west coast. On the July maps additional influences are evident and these are referred to below.

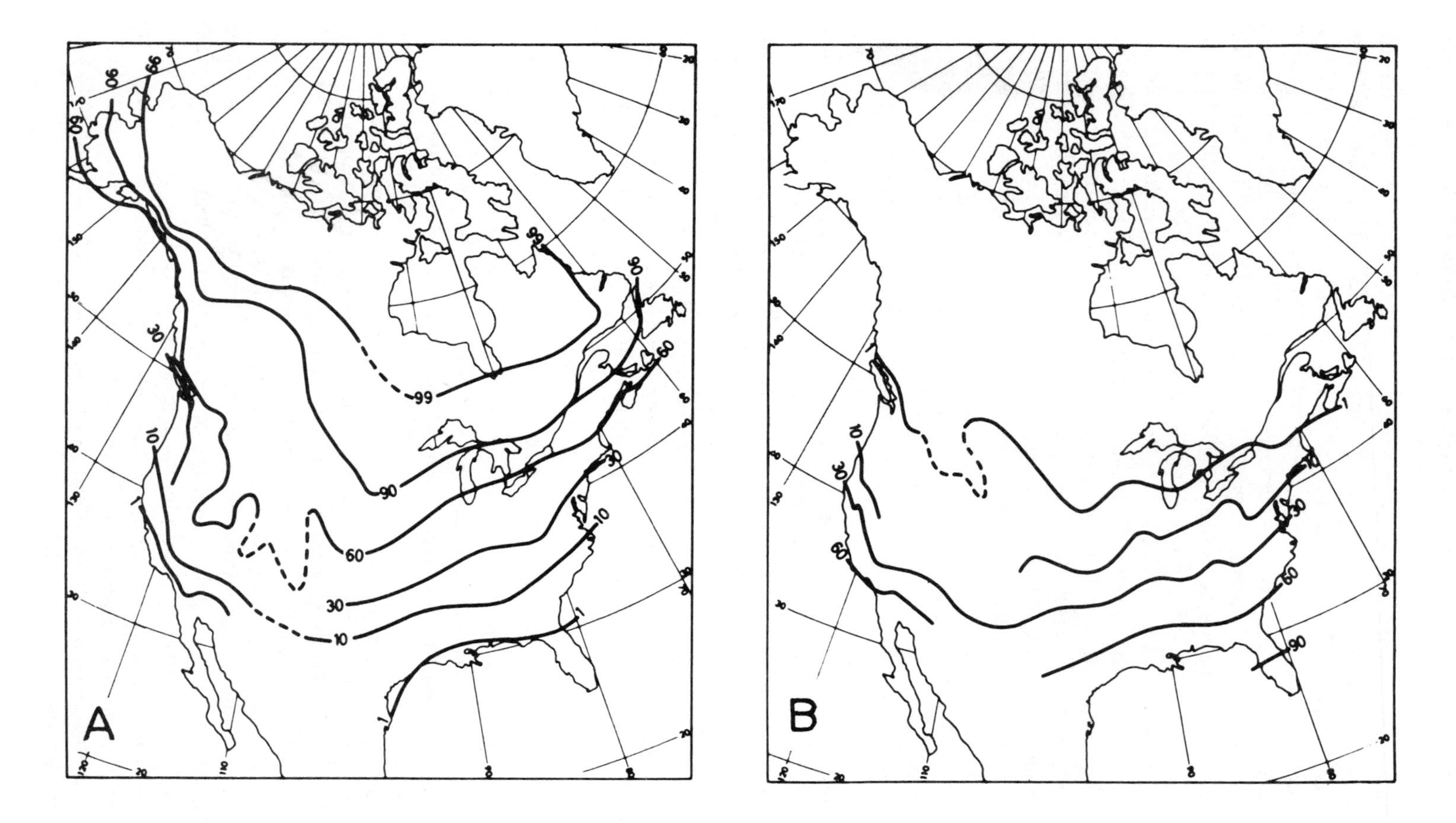
A
99
90
60
30
10
1
B
10
30
60
90
1

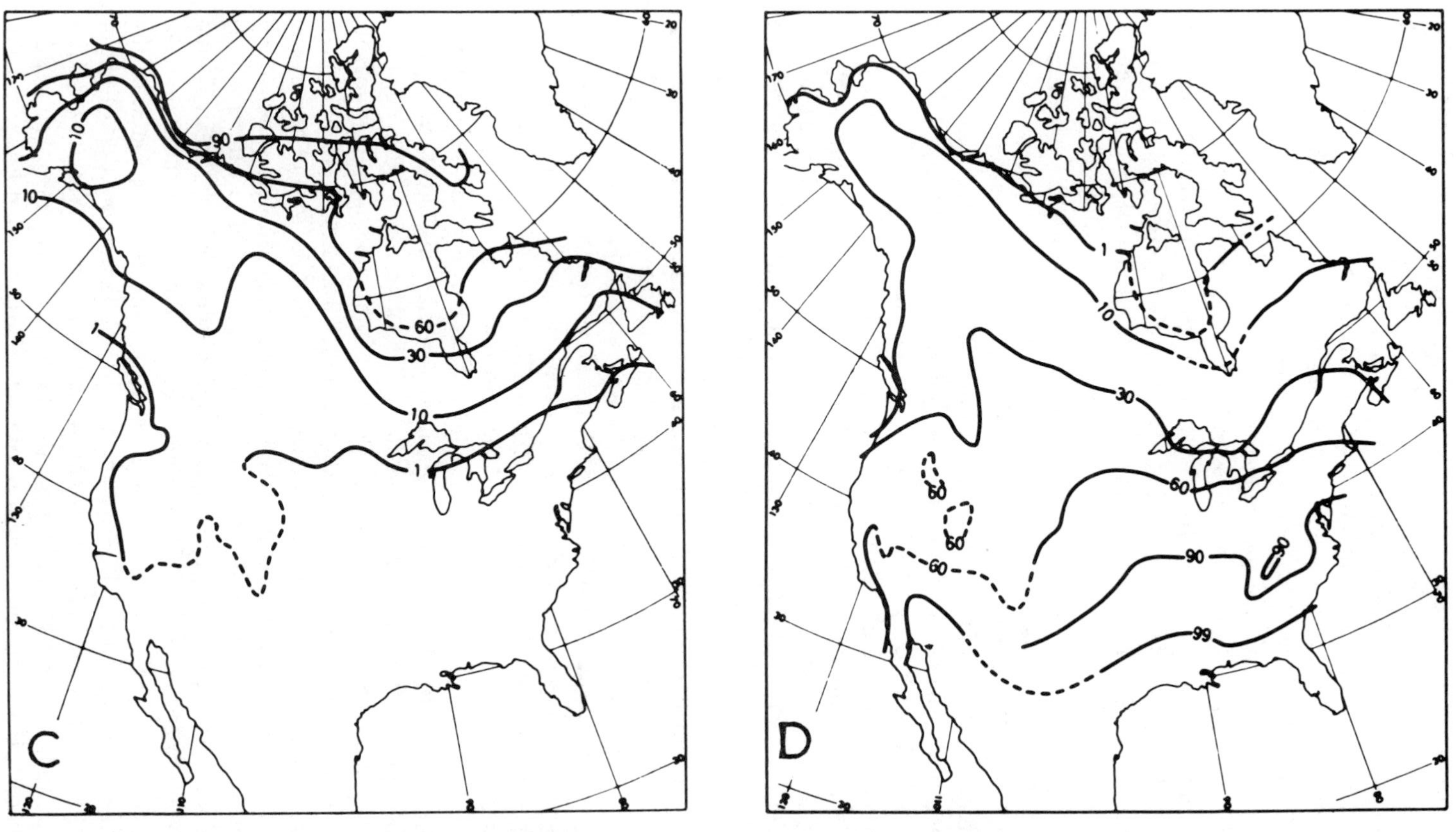

Fig. 5.16. The percentage frequency of hourly temperatures above or below certain limits for North America (*after Rayner 1961*).
A January temperatures <0°C. B January temperatures >10°C. C July temperatures <10°C. D July temperatures >21°C.

a Continental and oceanic influences The Labrador coast is fringed by the waters of a cold current, analogous to the Oya Shio off eastern Asia, but in both cases the prevailing westerlies greatly limit their climatic significance. The Labrador Current maintains drift ice off Labrador and Newfoundland until June and gives very low summer temperatures along the Labrador coast (fig. 5.16C). The lower incidence of freezing temperatures in this area in January is related to the movement of some depressions into the Davis Strait, carrying Atlantic air northwards. A major role of the Labrador Current is in the formation of fog. Advection fogs are very frequent between May and August off Newfoundland, where the Gulf Stream and Labrador Current meet. Warm, moist southerly airstreams are cooled rapidly over the cold waters of the Labrador Current and with steady, light winds such fogs may persist for several days, creating hazardous conditions for shipping. Southward-facing coasts are particularly affected and at Cape Race (Newfoundland), for example, there are on average 158 days per year with fog (visibility less than 1 km) at some time of day. The summer concentration is shown by the figures for Cape Race during individual months; May – 18 (days), June – 18, July – 24, August – 21 and September – 15.

Oceanic influence along the Atlantic coasts of the United States is very limited, and although there is some moderating effect on minimum temperatures at coastal stations this is scarcely evident on generalized maps such as fig. 5.16. More significant climatic effects are in fact found in the neighbourhood of Hudson Bay and the Great Lakes. Hudson Bay remains very cool in summer with water temperatures of about 7°–9°C and this depresses temperatures along its shore, especially in the east (fig. 5.12C and D). Mean July temperatures are 12°C (54°F) at Churchill (59°N) and 8°C (47°F) at Port Harrison (58°N), on the west and east shores respectively, compared for instance with 13°C (56°F) at Aklavik (68°N) on the Mackenzie delta. The influence of Hudson Bay is even more striking in early winter when the land is snow-covered. Westerly airstreams crossing the open water are warmed on average by 11°C (20°F) in November, and moisture added to the air leads to considerable snowfall in western Ungava (see the graph for Port Harrison, fig. 5.20). By the beginning of January the Bay is frozen over almost entirely and no effects are evident. The Great Lakes influence their surroundings in much the same way. Heavy winter snowfalls are a notable feature of the southern and eastern shores of the Great Lakes. In addition to contributing moisture to north-westerly streams of cold cA and cP air, the heat source of the open water in early winter produces a low-pressure trough which increases the snowfall as a result of convergence. Yet a further factor is frictional convergence and orographic uplift at the shoreline. Mean annual snowfall exceeds 250 cm (100 in) along much of the eastern shore of Lake Huron and Georgian Bay, the south-eastern shore of Lake Ontario, the north-eastern shore of Lake Superior and its southern shore east of about 90° 30′W. Extremes include 114 cm (45 in) in 1 day at Watertown, New York, and 894 cm (352 in) during the 1946–7 winter season at nearby Bennetts Bridge, both of which are close to the eastern end of Lake Ontario.

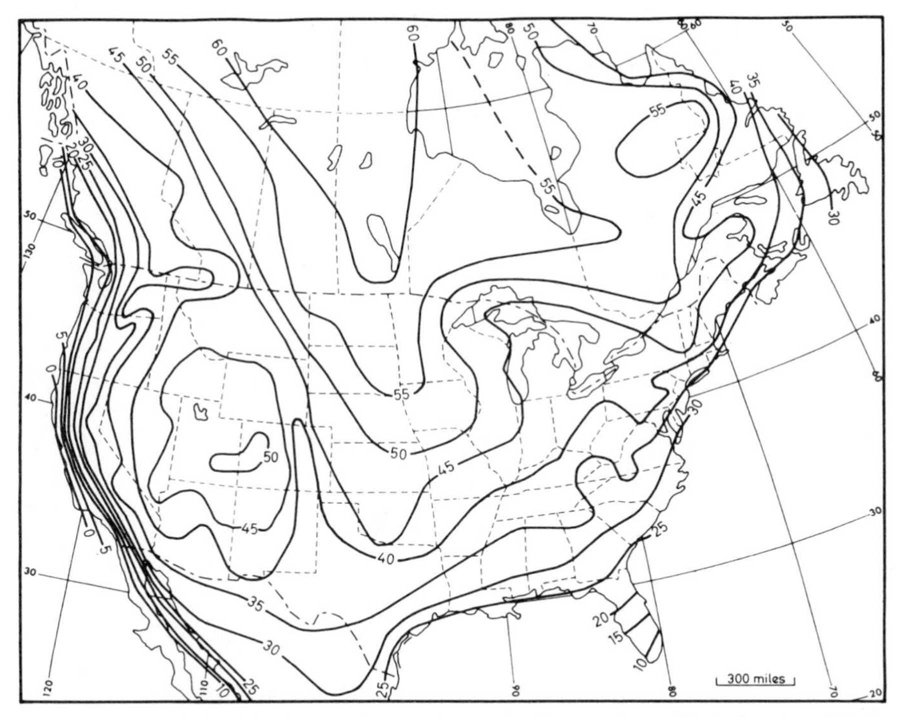

Fig. 5.17. Continentality in North America according to Conrad's index (*modified after Trewartha 1981*).

Transport in cities in these snow belts is quite frequently disrupted during winter snowstorms. The Great Lakes also provide an important tempering influence during winter months by raising average daily minimum temperatures at lakeshore stations by some 2°–4°C above those at inland locations. In mid-December the upper 60 m of Lake Erie has a uniform temperature of 5°C.

An indication of the seasonal range of temperature is provided by fig. 5.17, showing continentality (k) based on Conrad's formula:

$$k = \frac{1 \cdot 7\,A}{\sin\,(\phi + 10)} - 14$$

where A is the average annual temperature range in °C and ϕ is the latitude angle. The results in middle and high latitudes are similar to those obtained by Gorczynski's method (see ch. 5, A.2); with either of these empirical expressions it is only the relative magnitude of k that is of interest. The highest values form a tongue along the 100°W meridian with subsidiary areas on the 'Lake Plateau' of central Labrador–Ungava and on the high plateaux of Colorado and Utah. The 'maritimeness' of the Pacific coast, though of very limited inland extent, is pronounced, whereas on the east coast there is relatively high continentality. The map also illustrates the ameliorating effect of the Great Lakes.

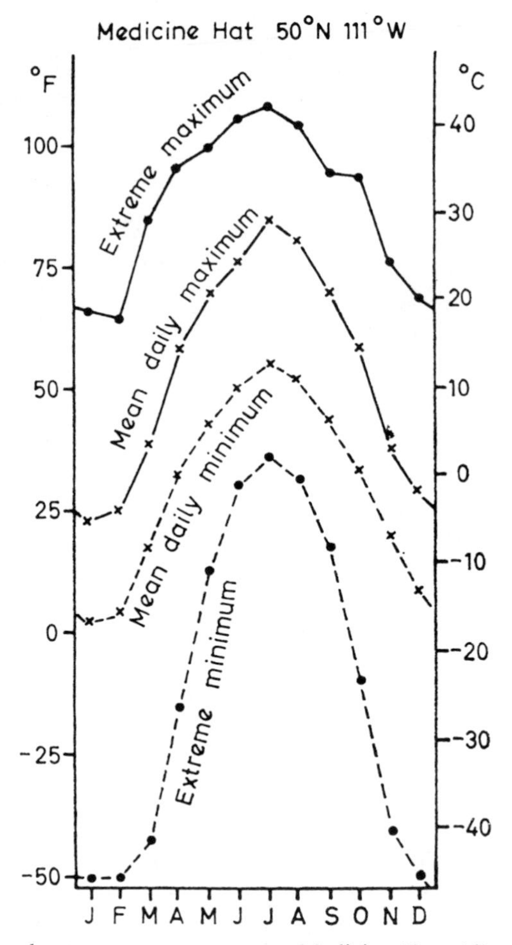

Fig. 5.18. Mean and extreme temperatures at Medicine Hat, Alberta.

b Warm and cold spells Two types of synoptic condition are of particular significance for temperatures in the interior of North America. One is the 'cold wave' caused by a northerly outbreak of cP air, which in winter regularly penetrates deep into the central and eastern United States and occasionally affects even the Gulf coast, injuring frost-sensitive crops. Cold waves arbitrarily defined as a temperature drop of at least 11°C (20°F) in 24 hours over most of the United States, and at least 9°C (16°F) in California, Florida and the Gulf Coast, to below a specified minimum depending on location and season. The winter criterion decreases from 0°C in California, Florida and the Gulf Coast to −18°C (0°F) over the northern Great Plains and the north-eastern States. The cold spells commonly occur with the build-up of a north–south anticyclone in the rear of a cold front. The polar air gives clear, dry weather with strong, cold winds, although if the winds follow snowfall, fine, powdery snow may be whipped up by the wind, creating blizzard conditions. These are quite common over the northern plains.

Another type of temperature fluctuation is associated with the *Chinook* winds in the lee of the Rockies (see ch. 3, C.2). The Chinook is particularly warm and dry as air of Pacific origin, after losing its moisture over the mountains, descends the eastern slopes and warms at the dry adiabatic lapse rate. The onset of the Chinook produces temperatures well above the seasonal normals so that snow is often thawed rapidly, and in fact the Indian word Chinook means snow-eater. Temperature rises of as much as 22°C (40°F) have been observed in 5 min and the occurrence of such warm spells is reflected by the high extreme maxima in winter months at Medicine Hat (fig. 5.18). In Canada the Chinook effect may be observed a considerable distance from the Rockies into south-west Saskatchewan, but in Colorado its influence is rarely felt more than about 50 km from the foothills. No adequate definition of a Chinook has yet been established, but using an arbitrary criterion of winter days with a maximum temperature of at least 4·4°C (40°F), R. W. Longley has shown that in the Lethbridge area Chinooks occur on 40% of days during December–February. However, since the phenomenon is the result of a particular type of airflow, it is evident that some wind characteristic should be included in future definitions.

Chinook conditions commonly develop in a Pacific airstream which is replacing a winter high-pressure cell over the high western plains. Sometimes the cold, stagnant cP air of the anticyclone is not dislodged by the descending Chinook and a marked inversion is formed, but on other occasions the boundary between the two air masses may reach ground level locally and, for example, the western suburbs of Calgary may record temperatures above 0°C while those to the east of the city remain below −15°C.

c Precipitation and the moisture balance Longitudinal influences are apparent in the distribution of annual precipitation, although this is in large measure a reflection of the topography. The 60 cm annual isohyet in the United States approximately follows the 100°W meridian (fig. 5.19), and westwards to the Rockies is an extensive dry belt in the rain shadow of the western mountain ranges. In the south-east totals exceed 125 cm, and 100 cm or more is received along the Atlantic coast as far as New Brunswick and Newfoundland.

The major sources of moisture for precipitation over North America are the Pacific Ocean and the Gulf of Mexico. The former need not concern us here since comparatively little of the precipitation falling over the interior appears to be derived from that source. The Gulf source is extremely important in providing moisture for precipitation over central and eastern North America, but the predominance of south-westerly airflow means that little precipitation falls over the western Great Plains (fig. 5.19). Over the south-eastern United States there is considerable evapotranspiration and this helps to maintain moderate annual totals northwards and eastwards from the Gulf by providing additional water vapour for the atmosphere. Along the east coast the Atlantic Ocean is an additional significant source of moisture for winter precipitation.

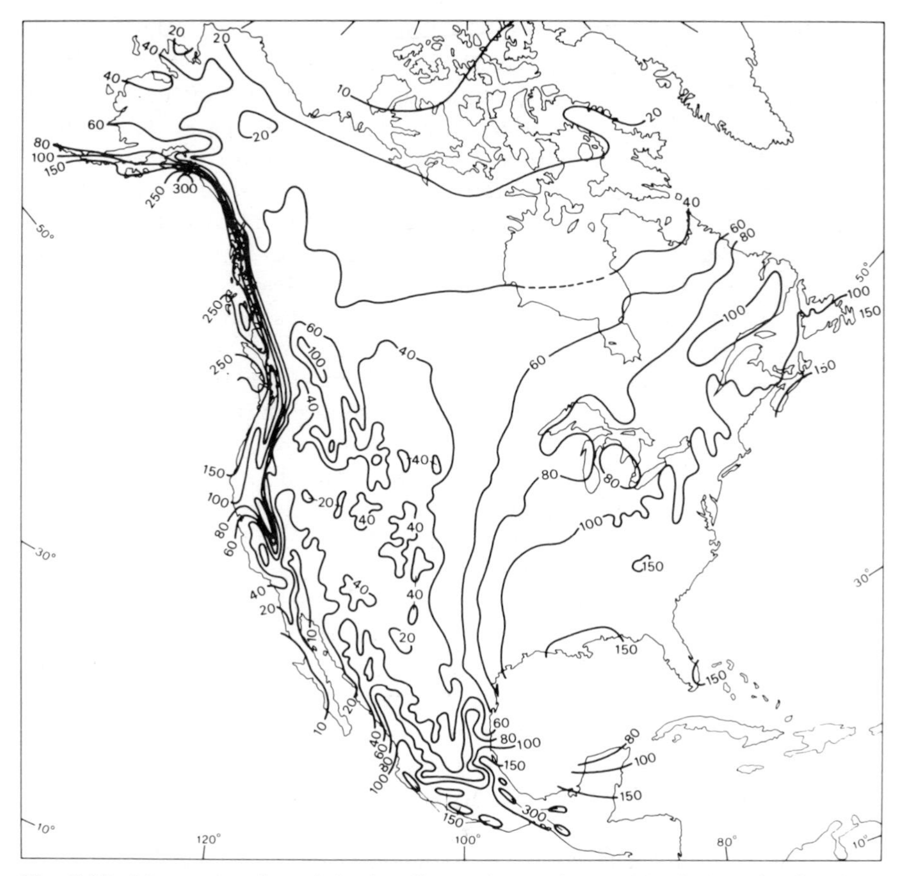

Fig. 5.19. Mean annual precipitation (in centimetres) over North America for the period 1931–60. Isohyets in the Arctic underestimate the true totals by 30–50% due to problems in recording snowfall accurately with precipitation gauges (*based on Hare and Hay, Court, and Mosiño Aleuán and Garcia*).

There are at least eight major types or seasonal precipitation regime in North America (fig. 5.20); the winter maximum of the west coast and the transition type of the intermontane region in middle latitudes have already been mentioned and the subtropical types are discussed in the next section. Four primarily mid-latitude regimes are distinguished east of the Rocky Mountains:

(i) A warm season maximum is found over much of the continental interior (e.g. Rapid City). In an extensive belt from New Mexico to the Prairie Provinces more than 40% of the annual precipitation falls in summer. In New Mexico, the rain occurs mainly with late summer thunderstorms, but May–June is the wettest time over the central and northern Great Plains due to more frequent cyclonic activity. Winters are quite dry over the Plains, but the mechanism of the occasional heavy snowfalls is of interest. They occur over the north-western Plains during easterly upslope flow, usually in a ridge of high pressure. Further north in

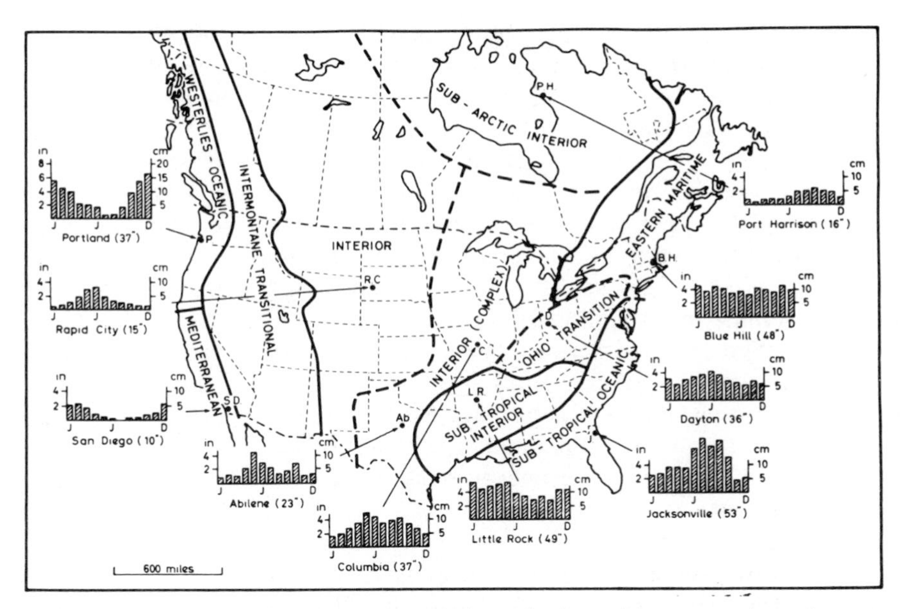

Fig. 5.20. Rainfall regimes in North America. The insert histograms show the mean monthly precipitation; January, June and December are indicated (*partly after Trewartha 1981*).

Canada the maximum is commonly in late summer or autumn when depression tracks are in higher middle latitudes. There is a local maximum in autumn on the eastern shores of Hudson Bay (e.g. Port Harrison) due to the effect of open water.

(ii) Eastward and southward of the first zone there is a double maximum in May and September. In the upper Mississippi region (e.g. Columbia) there is a secondary minimum, paradoxically in July–August when the air is especially warm and moist, and a similar profile occurs in northern Texas (e.g. Abilene). An upper-level ridge of high pressure over the Mississippi valley seems to be responsible for reduced thunderstorm rainfall in midsummer, and a tongue of subsiding dry air extends southwards from this ridge towards Texas. In September, renewed cyclonic activity associated with the seasonal southward shift of the polar front, at a time when mT air from the Gulf is still warm and moist, causes a resumption of rainfall. Subsequently, however, drier westerly airstreams affect the continental interior as the general airflow becomes more zonal.

The diurnal occurrence of precipitation in the central United States is rather unusual for a continental interior. Sixty per cent or more of the summer precipitation falls during nocturnal thunderstorms (2000–0800 hr True Solar Time) in central Kansas, parts of Nebraska, Oklahoma and Texas. Hypotheses suggest that the nocturnal thunderstorm rainfall which occurs, especially with extensive meso-scale convective systems (see p. 203) may be linked to a tendency for nocturnal convergence and rising air over the plains east of the Rocky Mountains. The terrain profile appears to play a role here, as a large-scale inversion layer forms at night over the mountains setting up a low-level jet east of the mountains just above the boundary layer. This southerly flow, at 500–1000 m above the surface, can supply the necessary low-level moisture influx and convergence for the storms (cf. fig. 4.32).

(iii) East of the upper Mississippi, in the Ohio valley and south of the lower Lakes, there is a transitional regime between that of the interior and the east-coast type. Precipitation is reasonably abundant in all seasons but the summer maximum is still in evidence (e.g. Dayton).

(iv) In eastern North America (New England, the Maritimes, Quebec and south-east Ontario) precipitation is fairly evenly distributed throughout the year (e.g. Blue Hill). In Nova Scotia and locally around Georgian Bay there is a winter maximum, due in the latter case to the influence of open water. In the Maritimes it is related to winter (and also autumn) storm tracks.

It is worth comparing the eastern regime with the summer maximum which is found over eastern Asia. There the Siberian anticyclone excludes cyclonic precipitation in winter and monsoonal influences are felt in the summer months.

The seasonal distribution of precipitation is of vital interest for agricultural purposes. Rain falling in summer, for instance, when evaporation losses are high is less effective than an equal amount in the cool season. Figure 5.21 illustrates the effect of different regimes in terms of the moisture balance, calculated according to Thornthwaite's method. At Halifax (Nova Scotia) there is sufficient moisture stored in the soil to maintain evaporation at its maximum rate (i.e. actual evaporation = potential evaporation), whereas at Berkeley (California) there is a computed moisture deficit of nearly 5 cm in August. This is a guide to the amount of irrigation water which may be required by crops, although in dry regimes the Thornthwaite method generally underestimates the real moisture deficit.

The mean monthly potential evaporation (*PE*) is calculated in the method developed by C. W. Thornthwaite from tables based on a complex equation relating *PE* to air temperature. This only gives a general guide to the true *PE* in view of the different factors which affect evaporation (see ch. 2, A), but the results are reasonably satisfactory in temperate latitudes. By determining the annual moisture surplus and the annual moisture deficit from graphs such as those in fig. 5.21, or from a monthly 'balance sheet', Thornthwaite obtained an index of aridity and humidity. The humidity index is $100 \times$ water surplus/*PE* and the aridity index is $100 \times$ water deficit/*PE*; there is generally a surplus in one season and a deficit in another. These may then be combined into a single moisture index (*Im*):

$$Im = \frac{100 \times \text{water surplus} - 60 \times \text{water deficit}}{PE}$$

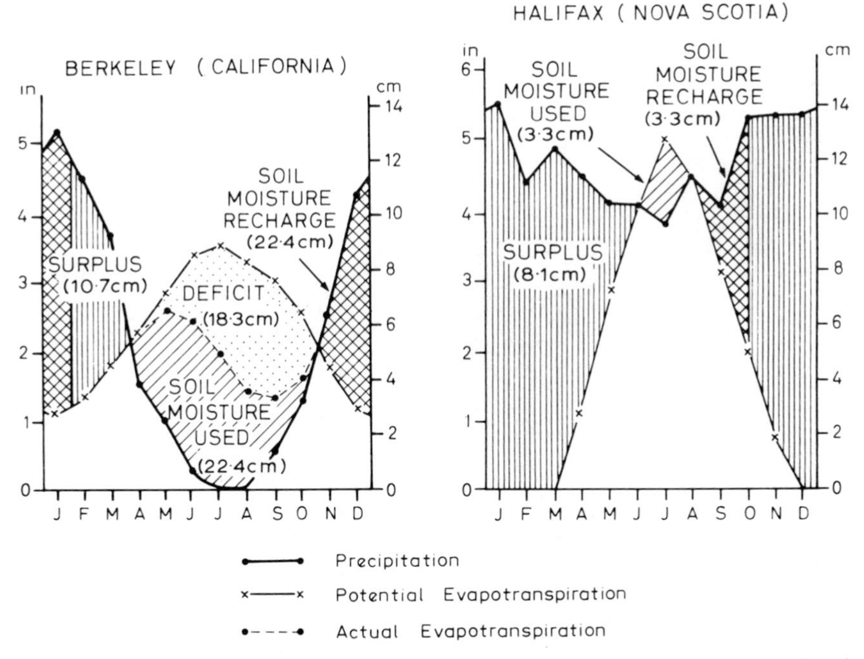

Fig. 5.21. The moisture balances at Berkeley, California and Halifax, Nova Scotia (*after Thornthwaite and Mather 1955*).

Here the deficit is given less weight as the surplus is held in the soil, whereas any deficit means that the actual evaporation rate falls below the potential value. The moisture index is used to define the following climatic types (see also App. 1.B):

Im	*climate*	*symbol*
>100	perhumid	A
20 to 100	humid	B (with 4 subdivisions)
0 to 20	moist subhumid	C_2
−20 to 0	dry subhumid	C_1
−40 to −20	semi-arid	D
<−40	arid	E

Figure 5.22 illustrates the distribution of these moisture regions over the United States. The zero line separating the moist climates of the east from the dry climates of the west (apart from the west coast) follows the 96th meridian almost exactly. The major humid areas are along the Appalachians, in the north-east and along the Pacific coast, while the most extensive arid area is the south-west. Some aspects of the precipitation climatology of this arid area are examined in ch. 5, D.2.

C The polar margins

The longitudinal differences in mid-latitude climates persist into the polar margins, giving rise to maritime and continental sub-types, modified by the

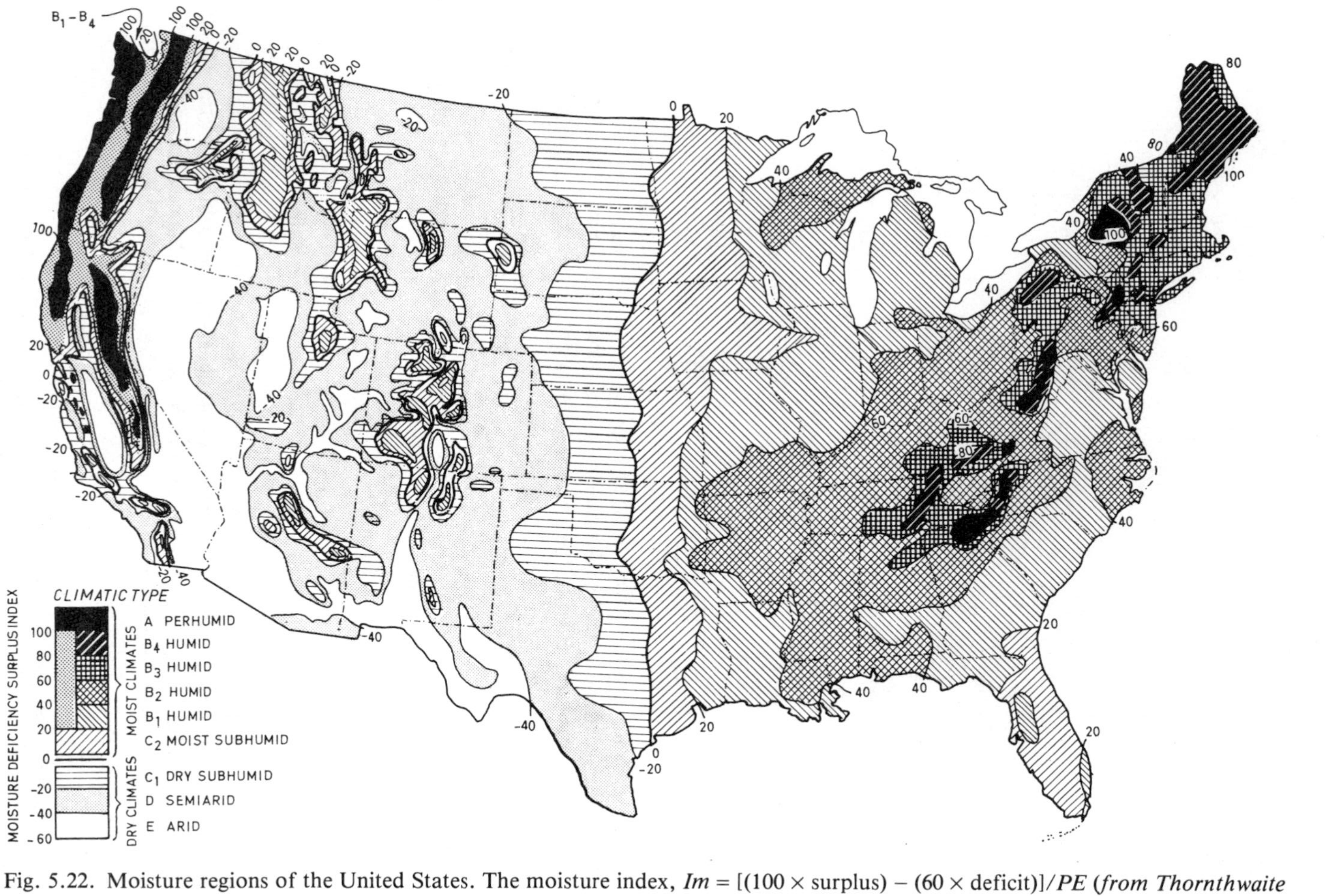

Fig. 5.22. Moisture regions of the United States. The moisture index, $Im = [(100 \times \text{surplus}) - (60 \times \text{deficit})]/PE$ (*from Thornthwaite and Mather 1955*).

extreme radiation conditions in winter and summer. For example, radiation receipts in summer along the arctic coast of Siberia compare favourably, by virtue of the long daylight, with those in lower middle latitudes. The maritime type is found in coastal Alaska, Iceland, northern Norway and adjoining parts of the USSR. Winters are cold and stormy, with very short days. Summers are cloudy but mild with mean temperatures about 10°C. For example, Vardø in north Norway (70°N, 31°E) has monthly mean temperatures of −6°C (21°F) in January and 9°C (48°F) in July, while Anchorage, Alaska (61°N, 150°W) records −11°C (12°F) and 14°C (57°F), respectively. Annual precipitation is generally between 60 and 125 cm (25 and 50 in), with a cool season maximum and about 6 months of snow cover.

The weather is mainly controlled by depressions which are weakly developed in summer. In winter the Alaskan area is north of the main depression tracks and occluded fronts and upper troughs (trowals) are prominent, whereas northern Norway is affected by frontal depressions moving into the Barents

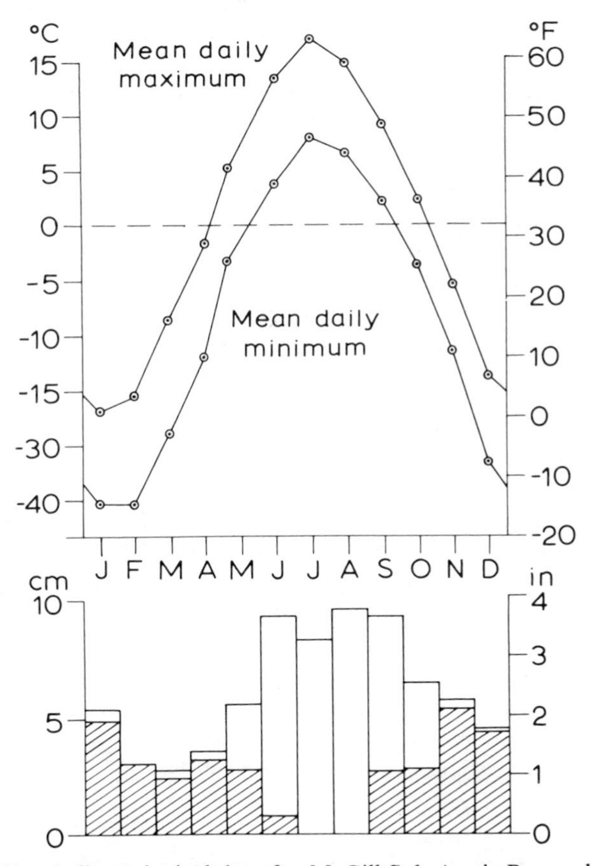

Fig. 5.23. Selected climatological data for McGill Sub-Arctic Research Laboratory, Schefferville, 1955–62 (*data from J. B. Shaw and D. G. Tout*). The shaded portions of the precipitation represent snowfall, expressed as water equivalent.

Sea. Iceland is similar to Alaska, though depressions often move slowly over the area and occlude, whereas others moving north-eastwards along the Denmark Strait bring mild, rainy weather.

The interior, cold-continental climates have much more severe winters although precipitation amounts are less. At Yellowknife (62°N, 114°W), for instance, the mean January temperature is only –28°C (–18°F). In these regions *permafrost* (permanently frozen ground) is widespread and often of great depth. In summer only the top 1–2 m of ground thaw and as the water cannot readily drain away this 'active layer' often remains waterlogged. Although frost may occur in any month, the long summer days usually give 3 months with mean temperatures above 10°C, and at many stations extreme maxima are 32°C (90°F) or more (fig. 5.16D). The Barren Grounds of Keewatin, however, are much cooler in summer due to the extensive areas of lake and muskeg, and only July has a mean daily temperature of 10°C. Labrador–Ungava to the east is rather similar, with very high cloud amounts and maximum precipitation in June–September (fig. 5.23). In winter conditions fluctuate between periods of very cold, dry, high-pressure weather and spells of dull, bleak, snowy weather as depressions move eastwards or occasionally northwards over the area. In spite of the very low mean temperatures in winter, there have been occasions when maxima have exceeded 4°C (40°F) during incursions of maritime Atlantic air. Such variability is not found in eastern Siberia which is intensely continental, apart from the Kamchatka peninsula, with the northern hemisphere's *cold pole* located in the remote north-east (see fig. 1.17A). Verkhoyansk and Oimyakon have a January mean of –50°C and both have recorded an absolute minimum of –67·7°C.

D The subtropical margins

1 The Mediterranean

The characteristic west coast climate of the subtropics is the Mediterranean type with hot, dry summers and mild, relatively wet winters. It is interposed between the temperate maritime type and the arid subtropical desert climate, but the Mediterranean regime is transitional in a special way for it is controlled by the westerlies in winter and by the subtropical anticyclone in summer. The seasonal change in position of the subtropical high and the associated subtropical westerly jet stream in the upper troposphere is evident in fig. 5.24. The type region is peculiarly distinctive, extending more than 3000 km into the Eurasian continent. Additionally, the configuration of seas and peninsulas produces great regional variety of weather and climate. The Californian region with similar conditions (fig. 5.20) is of very limited extent, and attention is therefore concentrated on the Mediterranean basin itself.

The winter season sets in quite suddenly in the Mediterranean as the summer eastward extension of the Azores high-pressure cell collapses. This

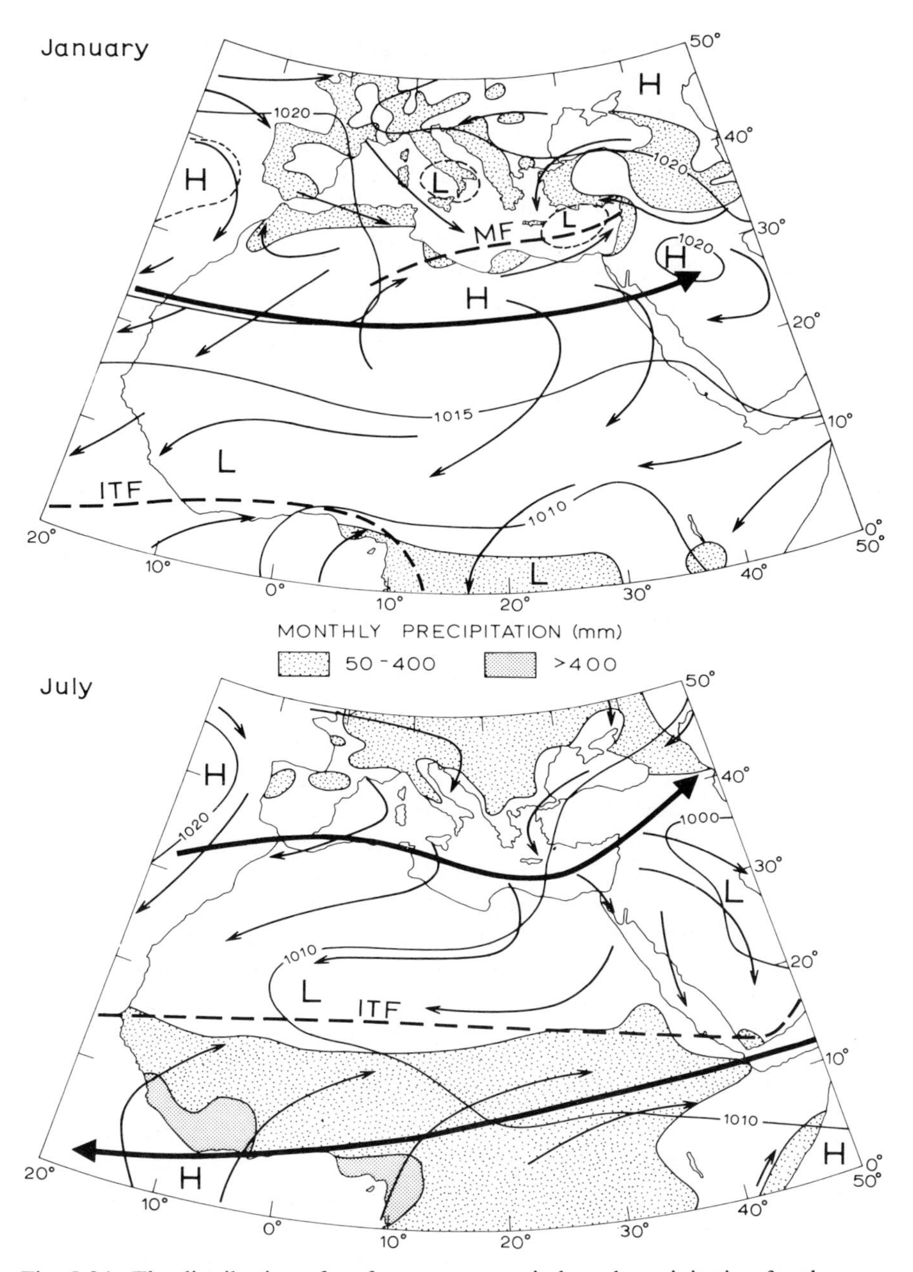

Fig. 5.24. The distribution of surface pressure, winds and precipitation for the Mediterranean and north Africa during January and July. The average positions of the Subtropical Westerly and Tropical Easterly Jet Streams, together with the Intertropical (ITF) and Mediterranean (MF) Fronts, are also shown. The ITF is now interpreted as Monsoon Trough (*partly after 'Weather in the Mediterranean' HMSO 1962; Crown Copyright Reserved*).

phenomenon can be observed on barographs throughout the region, but particularly in the western Mediterranean where a sudden drop in pressure occurs about 20 October and is accompanied by a marked increase in the probability of precipitation. The probability of receiving rain in any 5-day period increases dramatically from 50–70% in early October to 90% in late October. This change is associated with the first invasions by cold fronts, although thunder-shower rain has been common since August. The pronounced winter precipitation over the Mediterranean largely results from the relatively high sea-surface temperatures at that season; in January the sea temperature being about 2°C higher than the mean air temperature. Incursions of colder air into the region lead to convective instability along the cold front producing frontal and orographic rain. Incursions of arctic air are relatively infrequent (there being, on average, 6–9 invasions by cA and mA air each year), but the penetration by unstable mP air is much more common. It typically gives rise to deep cumulus development and is critical in the formation of Mediterranean depressions. The initiation and movement of these depressions (see fig. 5.25) is associated with a branch of the Polar Front Jet Stream located about 35°N. This jet develops during low index phases when the westerlies over the eastern Atlantic are distorted by a blocking anticyclone at about 20°W, leading to a deep stream of arctic air flowing southward over the British Isles and France.

Atlantic depressions entering the western Mediterranean as surface lows make up only 9% of those affecting the region (fig. 5.25); 17% form as baroclinic waves south of the Atlas Mountains (the so-called *Saharan depressions*, which are most important sources of rainfall in late winter and spring) and fully 74% develop in the western Mediterranean to the lee of the Alps and Pyrenees (see ch. 4, G.1). The combination of the lee effect and that of unstable surface air over the western Mediterranean explains the frequent formation of these *Genoa type depressions* when conditionally unstable mP air invades the region. These depressions are exceptional in that the instability of the local air in the warm sector gives unusually intense precipitation along the warm front, and the unstable mP air produces heavy showers and thunderstorm rainfall to the rear of the cold front, especially between 5° and 25°E. This warming of mP (or mA) air is so characteristic as to produce air designated as *Mediterranean* (see table 4.1). The mean boundary between this Mediterranean air mass and cT air flowing north-eastward from the Sahara is referred to as the Mediterranean front (fig. 5.24). There may be a temperature discontinuity as great as 12°–16°C (30°–40°F) across it in late winter. Sahara depressions and those from the western Mediterranean move eastward forming a belt of low pressure associated with this frontal zone and frequently draw cT northward ahead of the cold front as the warm, dust-laden *scirocco* (especially in spring and autumn when Saharan air may spread into Europe). The movement of Mediterranean depressions is greatly complicated both by relief effects and by their regeneration in the eastern Mediterranean by fresh cP air from Russia or south-east Europe. Although many depressions travel eastward over Asia, there is a strong tendency for

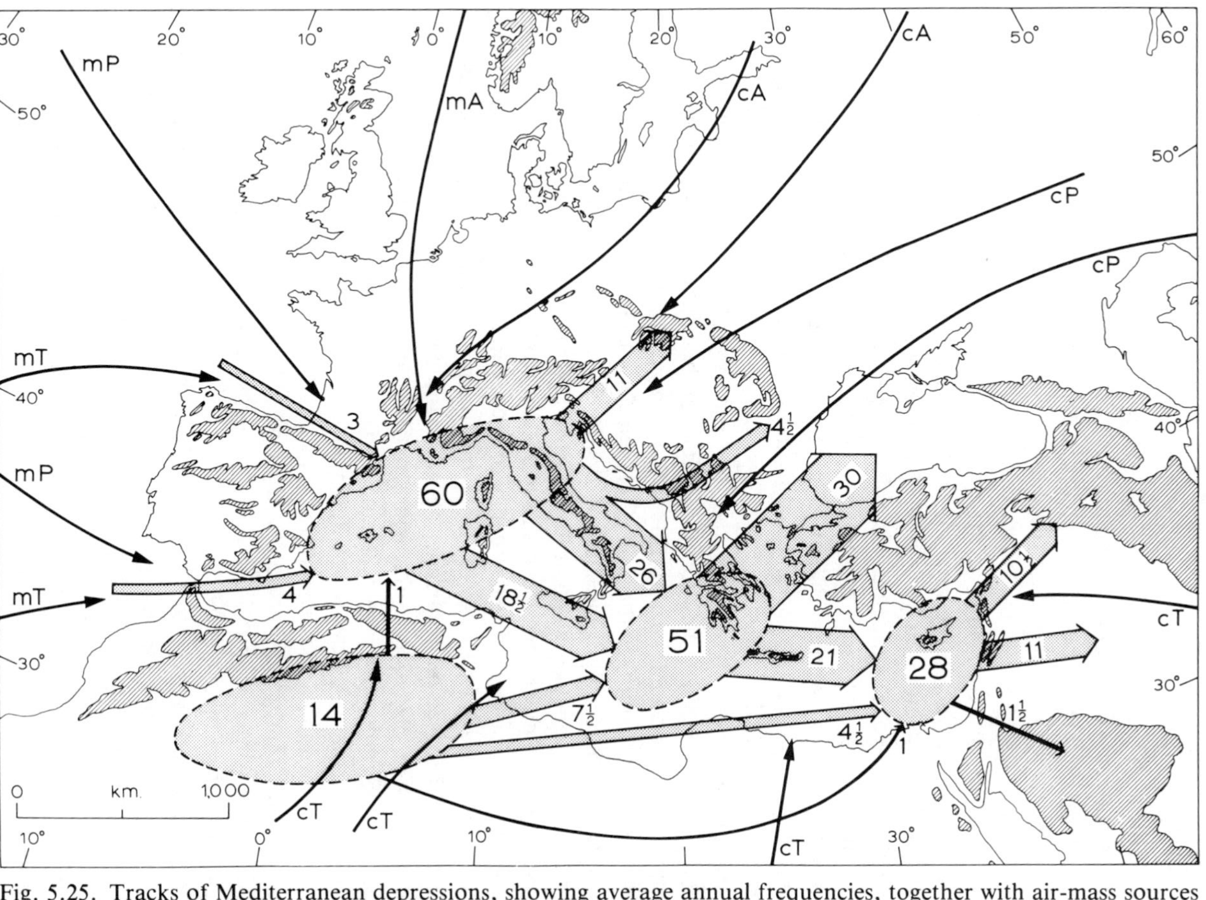

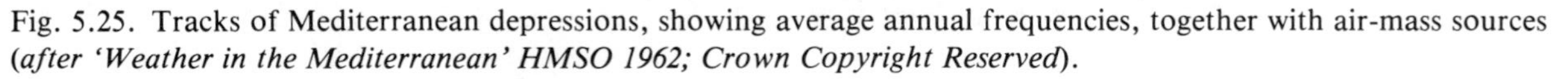
Fig. 5.25. Tracks of Mediterranean depressions, showing average annual frequencies, together with air-mass sources (*after 'Weather in the Mediterranean' HMSO 1962; Crown Copyright Reserved*).

low-pressure centres to move north-eastward over the Black Sea and Balkans, especially as spring advances. Winter weather in the Mediterranean presents considerable variation, however, particularly as the Subtropical Westerly Jet Stream is highly mobile and may occasionally even coalesce with the southerly-displaced Polar Front Jet Stream.

With high index zonal circulation over the Atlantic and Europe, depressions may pass far enough to the north so that their cold-sector air does not reach the Mediterranean, and then the weather there is generally settled and fine. Between October and April, anticyclones are the dominant circulation type for at least 25% of the time over the whole Mediterranean area and in the western basin for 48% of the time. This is reflected in the high mean pressure over the latter area in January (see fig. 5.24). Consequently, although the winter half-year is the rainy period, there are rather few rain-days. On average, rain falls on only 6 days per month during winter in northern Libya and south-east Spain, yet there are 12 rain-days per month in western Italy, the western Balkan peninsula and the Cyprus area. The higher frequencies (and totals) are related to the areas of cyclogenesis and to the windward side of peninsulas.

Regional winds are also related to the meteorological and topographic factors. The familiar cold, northerly winds of the Gulf of Lions (the *mistral*), which are associated with northly mP airflows, are best developed when a depression is forming in the Gulf of Genoa east of a high-pressure ridge from the Azores anticyclone. Katabatic and funnelling effects strengthen the flow in the Rhone valley and similar localities so that violent winds are sometimes recorded. The mistral may last for several days until the outbreak of polar or continental air ceases. The frequency of these winds depends on their definition. The average frequency of strong mistral in the south of France is shown in table 5.3 (based on occurrence at one or more stations from Perpignan to the Rhone in 1924–7). Similar winds may occur along the Catalan coast of Spain (the *tramontana*, fig. 5.26) and also in the northern Adriatic (the *bora*) and the northern Aegean Seas when polar air flows southward in the rear of an eastward-moving depression and is forced over the mountains (cf. ch. 3, C.2). In Spain, cold dry northerly winds occur in several different regions in winter. Figure 5.26 shows the *galerna* of the north coast and the *cierzo* of the Ebro valley.

The generally wet, windy and mild winter season in the Mediterranean is succeeded by a long indecisive spring lasting from March to May, with many false starts of summer weather.

Table 5.3 Number of days with strong mistral in the south of France (*after 'Weather in the Mediterranean', HMSO 1962*)

Speed	*J*	*F*	*M*	*A*	*M*	*J*	*J*	*A*	*S*	*O*	*N*	*D*	*Year*
⩾11 m s^{-1} (21 kt)	10	9	13	11	8	9	9	7	5	5	7	10	103
⩾17 m s^{-1} (33 kt)	4	4	6	5	3	2	0·6	1	0·6	0	0	4	30

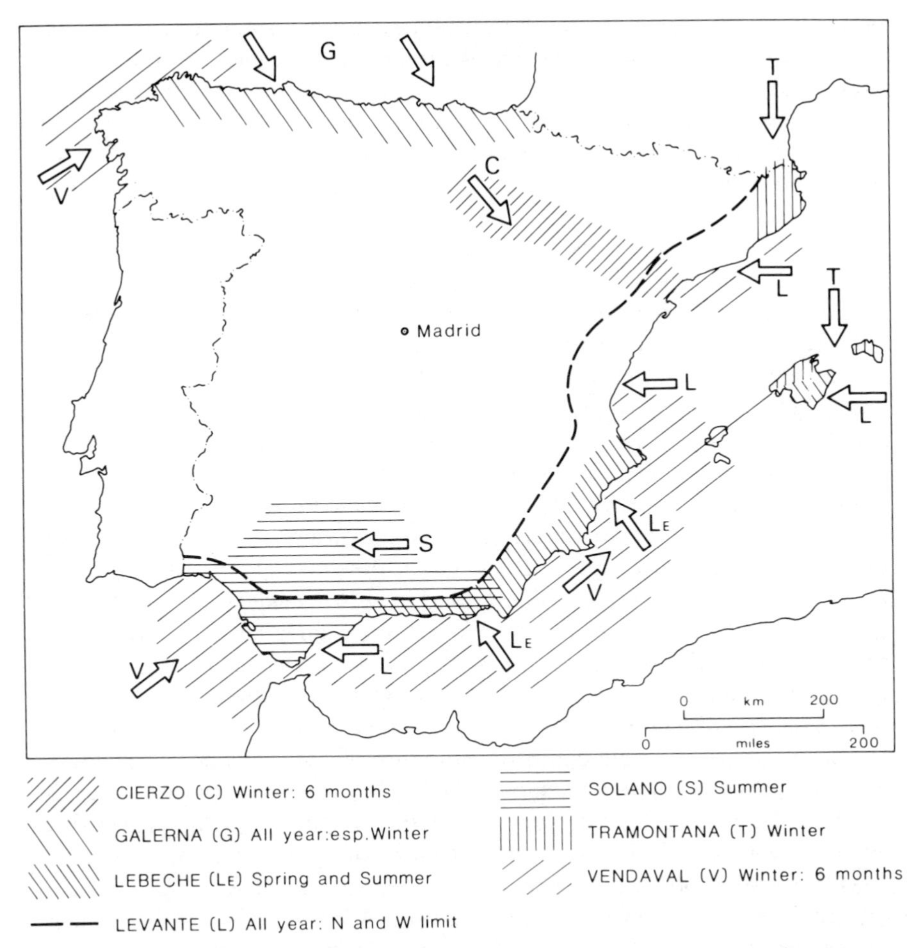

Fig. 5.26. Areas affected by the major regional winds in Spain as a function of season (*from Tout and Kemp 1985*). (*By permission of the Royal Meteorological Society*).

The spring period, like that of early autumn, is especially unpredictable. In March 1966 a trough moving across the eastern Mediterranean, preceded by a warm southerly *khamsin* and followed by a northerly airstream, brought up to 70 mm (2·8 in) of rain in only 4 hours to an area of the southern Negev desert. Although April is normally a dry month in the eastern Mediterranean, Cyprus having an average of only 3 days with 1 mm of rainfall or more, high rainfalls can occur as in April 1971 when four depressions affected the region. Two of these were Saharan depressions moving eastward beneath the zone of diffluence on the cold side of a westerly jet and the other two were intensified in the lee of Cyprus. The rather rapid collapse of the Eurasian high-pressure cell in April, together with the discontinuous northward and eastward extension of the Azores anticyclone, encourages the northward displacement of depressions, and, even if higher latitude air does penetrate south to the Mediterranean, the sea surface there is relatively cooler and the air is more stable than during the winter.

By mid-June the Mediterranean basin is dominated by the expanded Azores anticyclone to the west, while to the south the mean pressure field shows a low-pressure trough extending across the Sahara from southern Asia (fig. 5.24). The winds are predominantly northerly (e.g. the *etesians* of the Aegean) and represent an eastward continuation of the north-easterly trades. Locally, sea breezes reinforce these winds, but on the Levant coast they cause surface south-westerlies. The day-to-day weather of many parts of the north African coast is largely conditioned by land and sea breezes, involving air up to 1500 m (5000 ft) deep. Depressions are by no means absent in the summer months, but they are usually weak since the anticyclonic character of the large-scale circulation encourages subsidence, and air-mass contrasts are much reduced compared with winter (see table 4.2). Thermal lows form from time to time over Iberia and Anatolia, though thundery outbreaks are infrequent due to the low relative humidity.

The most important regional winds in summer are of continental tropical origin. There is a variety of local names for these usually hot, dry and dusty airstreams – *scirocco* (Algeria and the Levant), *lebeche* (south-east Spain) and *khamsin* (Egypt) – which move northwards ahead of eastward-moving depressions. In the Negev the onset of an easterly *khamsin* may cause the relative humidity to fall suddenly to less than 10% and temperatures to rise to as much as 48°C (118°F). In southern Spain the easterly *solano* brings hot, humid weather to Andalucia in the summer half-year, whereas the coastal *levante* – which has a long fetch over the Mediterranean – is moist and somewhat cooler (fig. 5.26). Such regional winds occur when the Azores high extends over western Europe with a low-pressure system to the south.

Many stations in the Mediterranean receive only a few millimetres of rainfall in at least one summer month, yet it is important to realize that the seasonal distribution does not conform to the pattern of simple winter maximum over the whole of the Mediterranean basin. Figure 5.27 shows that this is found in the eastern and central Mediterranean, whereas Spain, southern France, northern Italy and the northern Balkans have more complicated profiles with a maximum in autumn or peaks in both spring and autumn. This double maximum can be interpreted generally as a transition between the continental interior type with summer maximum and the Mediterranean type with winter maximum. A similar transition region occurs in the south-western United States (see fig. 5.20), but local topography in this intermontane zone introduces further irregularities into the regimes.

2 The semi-arid south-western United States

Both the mechanisms and patterns of the climates of areas dominated by the subtropical high-pressure cells are rather obscure at present. The inhospitable nature of these arid regions inhibits data collection, and yet the proper interpretation of infrequent meteorological events requires a close

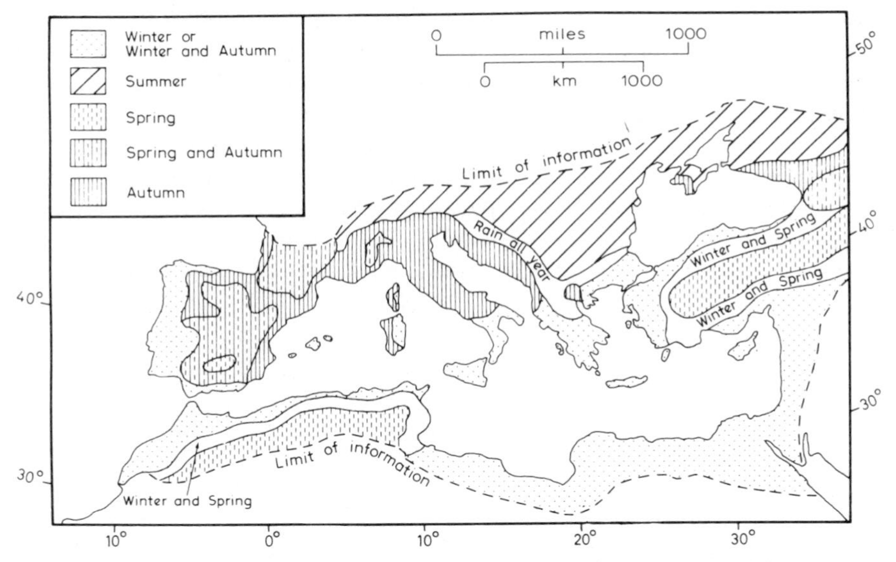

Fig. 5.27. Rainfall regimes in the Mediterranean area (*after Huttary 1950*).

network of stations maintaining continuous records over long periods. This difficulty is especially apparent in the interpretation of desert precipitation data, because much of the rain falls in very local storms irregularly scattered both in space and time. It is convenient to treat aspects of this climatic type here, in that most of the reliable data relate to the less arid regions marginal to the subtropical cell centres, and in particular to the south-western United States.

A series of observations at Tucson, Arizona, 730 m (2400 ft) between 1895 and 1957 showed a mean annual precipitation of 27·7 cm (10·91 in) falling on an average of about 45 days per year, with extreme annual figures of 61·4 cm (24·17 in) and 14·5 cm (5·72 in). Two moister periods in late November to March (receiving 30% of the mean annual precipitation) and late June to September (50%) are separated by more arid seasons from April to June (8%) and October to November (12%). The winter rains are generally prolonged and of low intensity (more than half the falls have an intensity of less than 0·5 cm (0·2 in) per hour), falling from altostratus clouds associated with the cold fronts of depressions which are forced to take southerly routes by strong blocking to the north. This occurs during phases of equatorial displacement of the Pacific subtropical high-pressure cell. The re-establishment of the cell in spring before the main period of intense surface heating and convectional showers is associated with the most persistent drought periods. Dry westerly to south-westerly flow from the eastern edge of the Pacific subtropical anticyclone is responsible for the low rainfall at this season. During one 29-year period in Tucson there were eight spells of more than 100 consecutive days of complete drought and twenty-four periods of more than 70 days. The dry conditions occasionally lead to dust storms. Yuma records 9 per year, on average, associated with winds averaging 10–15 m s^{-1}. They occur both with cyclonic systems in the cool season and with summer convective activity. Phoenix experiences 6–7 per year, mainly in summer, with visibility reduced below 1 km in nearly half of these events.

The period of summer precipitation (locally known as the summer 'monsoon') is quite sharply defined, beginning in the last week in June and lasting until the middle of September. Precipitation mainly occurs from convective cells initiated by surface heating, convergence, or, less commonly, orographic lifting when the atmosphere is destabilized by upper-level

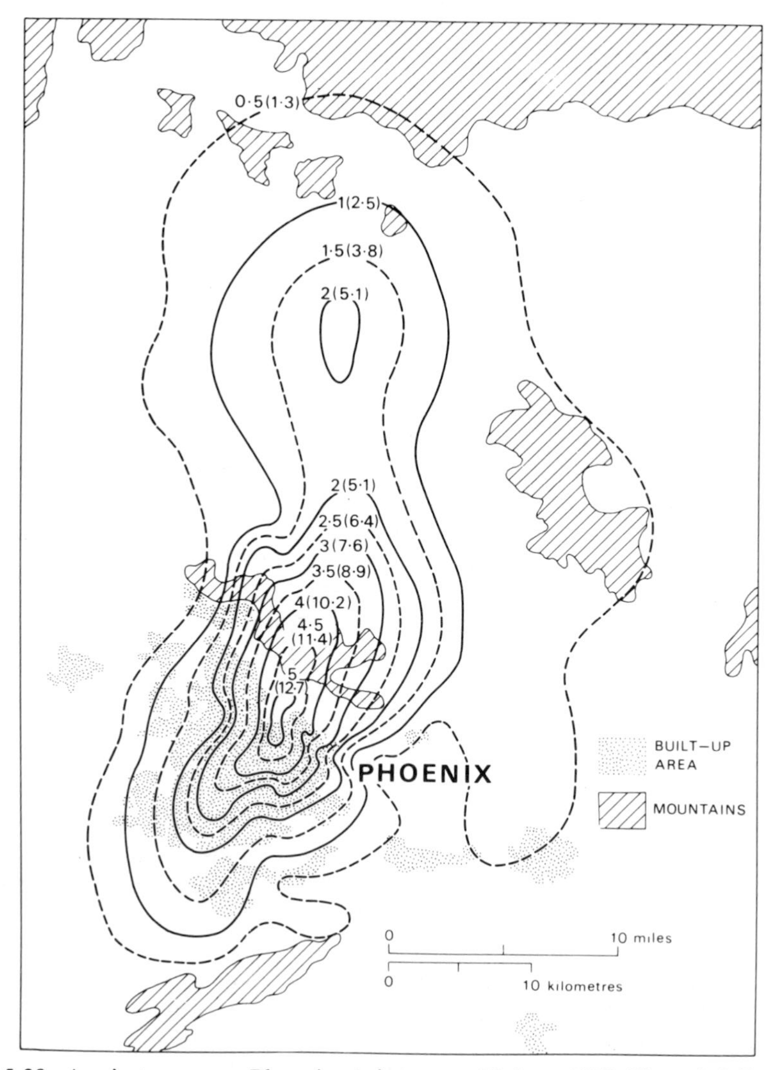

Fig. 5.28. A rainstorm over Phoenix, Arizona, on 22 June 1972. The rainfall between 0600 and 1200 hours is shown in inches and in centimetres (in parentheses) (*after Durrenberger and Ingram 1978*). (*Courtesy State Climatologist, Arizona*).

troughs in the westerlies. These summer convective storms form in meso-scale clusters, the individual storm cells covering together less than 3% of the surface area at any one time, and persisting for less than an hour on average. The storm clusters move across the country in the direction of the upper-air motion, and often seem to be controlled in movement by the existence of low-level jet streams at elevations of about 2500 m. The airflow associated with these storms is generally southerly along the southern and western margins of the Atlantic (or Bermudan) subtropical high, so that in contrast to the winter months the moisture is derived mainly from the Gulf of California, during 'surges' associated with the south-south-westerly low-level Sonoran jet (850–700 mb), and secondarily from the Gulf of Mexico during south-easterly flows. The southerly airflow regime at the surface and 700 mb (see figs. 3.19 and 3.24) over the Southwest often sets in abruptly around 1 July and it is therefore recognized as a singularity (see ch. 5, A.4, and fig. 5.14).

Precipitation from these cells is extremely local (pl. 13), and commonly concentrated in the mid-afternoon and evening. Intensities are much higher than in winter, half the summer rain falling at more than 0·4 in per hour. A particularly well documented storm occurred over Phoenix, Arizona, at 0600 hours on 22 June 1972. Moving north-east, it produced up to 127 mm (5 in) of rainfall in the following 6 hours including hailstones measuring 19 mm (0·75 in) in diameter (fig. 5.28). During the 29-year period about one quarter of the mean annual precipitation fell in storms giving 2·5 cm (1 in) or more per day. These intensities are much less than those associated with rain-storms in the humid tropics, but the sparsity of vegetation in the drier regions allows the rain to produce considerable surface erosion. Thus the highest measurements of surface erosion in the United States are from areas having 30–40 cm (12–15 in) of rain per year.

3 The interior and east coast of the United States

The climate of the subtropical south-eastern part of the United States has no exact counterpart in Asia, which is affected by the summer and winter monsoon systems. These are discussed in the next chapter and only the distinctive features of the North American subtropics are examined here. Seasonal wind changes are experienced in Florida, which is within the westerlies in winter and lies on the northern margin of the tropical easterlies in summer, but this is not comparable with the regime in southern and south-eastern Asia. Nevertheless, the summer season rainfall maximum (fig. 5.20 for Jacksonville) is a result of this change-over. In June the upper flow over the Florida peninsula changes from north-westerly to southerly as a trough moves westward and becomes established in the Gulf of Mexico. This deep, moist southerly airflow provides appropriate conditions for convection, and indeed Florida probably ranks as the area with the highest annual number of days with thunderstorms – 90 or more, on average, in the vicinity of Tampa. These often occur in late afternoon although two factors apart from diurnal heating are thought to be important. One is the effect of sea-breezes converging from both sides of the peninsula, and the other is the northward penetration of disturbances in the easterlies (see ch. 6). The latter may of course affect the area at any time of day. The westerlies resume control in September–October, although Florida remains under the easterlies during September when Caribbean tropical storms are most frequent and consequently the rainy season is prolonged. Tropical cyclones contribute an average of 10–15% of the annual rainfall near the Gulf Coast and in Florida.

The region of the Mississippi lowlands and the southern Appalachians to the west and north is not simply transitional to the 'Interior type' in terms at least of rainfall regime (fig. 5.20). The profile shows a winter–spring maximum and a secondary summer maximum. The cool season peak is related to westerly depressions moving north-eastward from the Gulf coast area, and it is significant that the wettest month is commonly March when the mean jet stream is farthest south. The summer rains are associated with convection in humid air from the Gulf, though this convection becomes less effective inland as a result of the subsidence created by the anticyclonic circulation in the middle troposphere referred to previously (see ch. 5, B.3.*c*).

Summary

Seasonal changes in the Icelandic Low and the Azores High, together with variations in cyclone activity, control the climate of western Europe. The eastward penetration of maritime influences related to these atmospheric processes and to the warm waters of the North Atlantic Drift is illustrated by mild winter temperatures, the seasonality of precipitation regimes and indices of continentality in western Europe. Topographic effects on precipitation, snowfall, length of growing seasons and local winds are particularly marked over the Scandinavian Mountains, the Scottish Highlands and the Alps. Weather types in the British Isles can be described in terms of seven basic airflow patterns, the frequency and effects of which vary considerably with season. Recurrent weather spells about a particular date (singularities), such as the tendency for anticyclonic weather in mid-September, have been recognized in Britain and major seasonal trends in occurrence of airflow regimes can be used to define five natural seasons. Abnormal weather conditions (synoptic anomalies) are associated particularly with blocking anticyclones which are especially prevalent over Scandinavia, and may give rise to cold dry winters and warm dry summers.

The climate of North America is similarly affected by pressure systems which generate air masses of varying seasonal frequency. In winter, the subtropical high-pressure cell extends north over the Great Basin with anticyclonic cP air to the north over Hudson Bay. Major depression belts occur at about 45°–50°N, from the central USA to the St Lawrence, and along the east coast of Newfoundland. The arctic front is located over north-west Canada, the polar front lies along the north-east coast of the

continued

United States, and between the two a maritime (arctic) front may occur over Canada. In summer the frontal zones move north, the arctic front lying along the north coast of Alaska, Hudson Bay and the St Lawrence being the main locations of depressions tracks. Three major North American singularities concern the advent of spring in early March, the midsummer northward displacement of the subtropical high-pressure cell, and the Indian summer of September–October. In western North America the Coast Ranges inhibit the eastward spread of precipitation which may vary greatly locally (e.g. in British Columbia), especially as regards snowfall. The strongly continental interior and east of the continent experiences some moderating effects of Hudson Bay and the Great Lakes in early winter, but with locally significant snow belts. The climate of the east coast is dominated by continental pressure influences. Cold spells are produced by winter outbreaks of high-latitude cA/cP air in the rear of cold fronts. Zonal westerly airflows give rise to Chinook winds in the lee of the Rockies. The major moisture sources of the Gulf of Mexico and the North Pacific produce regions of differing seasonal regime: the winter maximum of the west coast is separated by a transitional intermountain region from the interior with a general warm season maximum; the north-east has a relatively even seasonal distribution. Moisture gradients which strongly influence vegetation and soil types are predominantly east–west in central North America, in contrast to the isotherm pattern.

The polar margins have extensive areas of permanently frozen ground (permafrost) in the continental interiors, whereas the maritime regions of northern Europe and northern Canada–Alaska have cold stormy winters and cloudy milder summers influenced by the passage of depressions.

The subtropical margin of Europe is made up the Mediterranean region, lying between the belts dominated by the westerlies and the Saharan–Azores high pressure cells. The collapse of the Azores high-pressure cell in October allows depressions to move and form over the relatively warm Mediterranean Sea giving well marked orographic winds (e.g. *mistral*) and stormy rainy winters. Spring is an unpredictable season marked by the collapse of the Eurasian high-pressure cell to the north and the strengthening of the Saharan–Azores anticyclone. In summer, the latter gives dry, hot conditions with strong local southerly airstreams (e.g. *scirocco*). The simple winter rainfall maximum is most characteristic of the eastern and southern Mediterranean, whereas in the north and west autumn and spring rains become more important.

The semi-arid south-western United States comes under the complex influence of the Pacific and Bermuda high-pressure cells, having extreme rainfall variations with winter and summer maxima mainly due to depressions and local thunderstorms, respectively. The interior and east coast of the United States is dominated by westerlies in winter and southerly thundery airflows in summer.

6

Tropical weather and climate

Fifty per cent of the surface of the globe lies between latitudes 30°N and 30°S and over 75% of the world's population inhabits climatically tropical lands. Tropical climates are, therefore, of especial geographical interest.

The latitudinal limits of tropical climates vary greatly with longitude and season, and tropical weather conditions may reach well beyond the Tropics of Cancer and Capricorn. For example, the summer monsoon extends to 30°N in southern Asia, but only 20°N in West Africa, while in late summer and autumn tropical hurricanes may affect 'extra-tropical' areas of eastern Asia and eastern North America. Not only do the tropical margins extend seasonally polewards, but in the zone between the major subtropical high-pressure cells there is frequent interaction between temperate and tropical disturbances. Plate 22 illustrates a situation where there is such interaction between low and middle latitudes, whereas pl. 32, in contrast, shows distinct tropical and mid-latitude storms. In general, the tropical atmosphere is far from being a discrete entity and any meteorological or climatological boundaries must be arbitrary. There are, nevertheless, a number of distinctive features of tropical weather and these are discussed below.

A The assumed simplicity of tropical weather

The study of tropical weather has passed broadly through three stages. Firstly, for a long period, which only ended some years before the Second World War, tropical weather conditions, patterns and mechanisms were assumed to be much more simple and obvious than those in higher latitudes. This belief was partly due to the paucity of pre-war meteorological records, particularly over the vast tropical oceans, and partly due to certain theoretical and practical considerations. One reason was that temperature and hence air-mass contrasts seemed small compared with middle latitudes. Air masses were nevertheless identified on the basis of their moisture content, temperature and stability, although frontal activity was believed to be weak and weather systems correspondingly less evident. Obvious exceptions were tropical cyclones which were thought to result from special conditions of thermal convection. Another reason was the large extent of ocean surface which, it was assumed, would simplify the patterns of weather and climate.

Thus the following simple picture of *trade wind weather* evolved. Maritime tropical air masses, originating by subsidence in the subtropical high-pressure cells over the eastern halves of the oceans (see fig. 3.17), move steadily westwards and equatorwards in the easterly trade winds with quite constant speed and direction. Beneath the temperature inversion, formed between about 1 and 2 km altitude by subsidence, the air is moist with a layer of broken cumulus cloud. Conditions are invariably warm and dry, except where islands cause orographic clouds to form. Over the equatorial oceans the surface wind is light and variable (the doldrums) and the air is universally hot, humid and sultry (see ch. 3, E.1).

A further element of simplicity was thought to be provided by the insolation regime. The persistently high altitude of the sun in low latitudes and the equality of length between the days and nights means that seasonal variations of radiation are minimized. Hence the annual rhythm was thought to produce simple rainfall regimes with a single maximum following the summer solstice at the tropics, and a double peak at the equator in response to the passage of the overhead sun at the equinoxes. Diurnal patterns of land and sea breezes giving an afternoon build-up of convective activity and thundery showers were regarded as characteristic features of nearly all tropical climates.

Following the evolution of this simple picture of tropical weather processes, attempts were made between 1920 and 1940 to introduce mid-latitude frontal concepts. Little real progress was made, however, as a result of the apparent general absence of significant air-mass contrasts. Furthermore, the small surface pressure gradients of most tropical disturbances (other than the hurricane variety) tend to go unnoticed due to the large semi-diurnal pressure oscillation in the tropics. Pressure varies by about 2–3 mb with maxima around 1000 and 2200 hours and minima at 0400 and 1600 hours. It must also be recalled that the wind direction provides no guide to the pressure pattern in low latitudes. The small Coriolis force prevents the wind from attaining geostrophic balance and consequently the techniques used for analysing mid-latitude weather maps have to be abandoned.

Weather observations made during the Second World War revealed the inadequacy of the earlier views. It became evident that weather changes are frequent and complex with quite distinct types of weather systems in different tropical lands and with considerable climatic differences even over the ocean areas. It also became apparent that much smaller trigger mechanisms are necessary to produce disturbances in the flow of high-energy, tropical air masses than those associated with mid-latitude depressions, and yet, paradoxically, tropical cyclones are infrequent. The systems responsible for these contrasts are examined in the following sections.

B The intertropical convergence

The tendency for the trade wind systems of the two hemispheres to converge in the Equatorial (low-pressure) Trough has already been noted (see ch. 3. E).

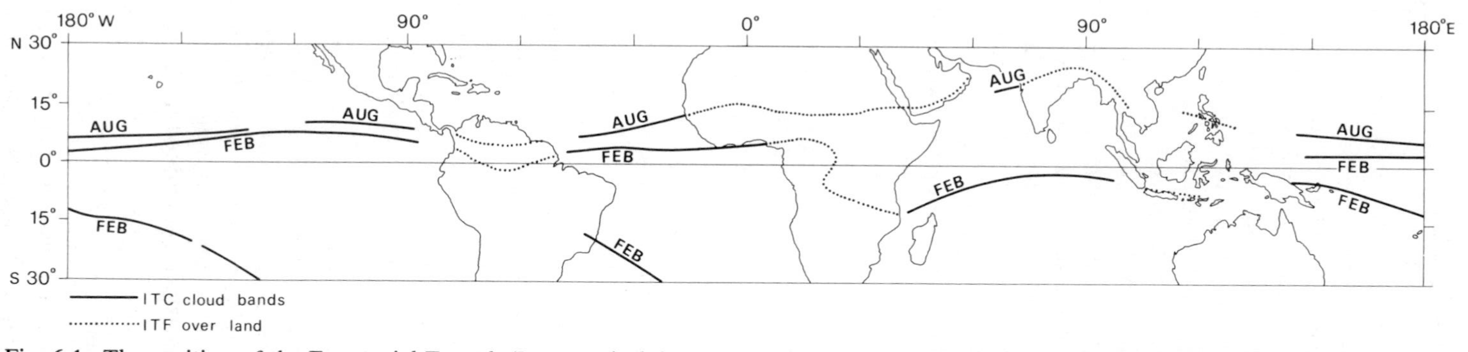

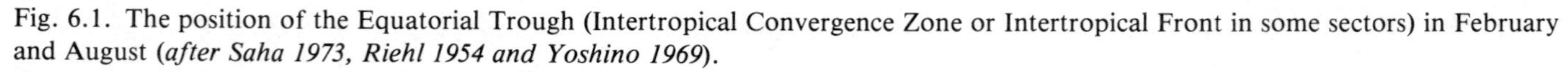

Fig. 6.1. The position of the Equatorial Trough (Intertropical Convergence Zone or Intertropical Front in some sectors) in February and August (*after Saha 1973, Riehl 1954 and Yoshino 1969*).

Views on the exact nature of this feature have been subject to continual revision. During the 1920s–1940s the frontal concepts developed in mid-latitudes were applied in the tropics and the streamline confluence of the North-East and South-East Trades was identified as the Intertropical Front (ITF). Over continental areas such as West Africa and southern Asia, where in summer hot, dry continental tropical air meets cooler, humid equatorial air, this term has some limited applicability (fig. 6.1). Sharp temperature and moisture gradients can occur, but the front is seldom a weather-producing mechanism of the mid-latitude type. Elsewhere in low latitudes true fronts (with a marked density contrast) are rare.

Recognition in the 1940s–1950s of the significance of wind field convergence in tropical weather-production led to the designation of the trade wind convergence as the Intertropical Convergence Zone (ITCZ). This feature is apparent on a *mean* streamline map, but areas of convergence grow and decay, either *in situ* or within disturbances moving westward (pl. 27), over periods of a few days. Moreover, convergence is infrequent even as a climatic feature in the doldrum zones (see fig. 3.27). Satellite photography has shown that over the oceans the position and intensity of the ITCZ varies greatly, even from day to day.

As climatic features the Equatorial Trough and ITC appear to move seasonally away from the equator (fig. 6.1) in association with the *Thermal Equator* (zone of seasonal maximum temperature). The location of the Thermal Equator is directly related to solar heating (figs. 1.15 and 1.17) and there is an obvious link between this and the Equatorial Trough in terms of thermal lows. However, if the ITC were to coincide with the Equatorial Trough then this zone of cloudiness would decrease incoming solar radiation, reducing the surface heating needed to maintain the low-pressure trough. In fact, this does not happen. Solar energy is available to heat the surface because the maximum surface wind convergence, uplift and cloud cover is commonly located several degrees equatorward of the trough. Figure 6.2 illustrates two patterns of Equatorial Trough and ITCZ. Cases involving convergence of two trade wind systems occur over the central North Atlantic in August and the eastern North Pacific in February, for example, while the situation where the Equatorial Trough is defined by easterlies on its poleward side and westerlies on its equatorward side is typical of West Africa in August and the western South Pacific in February. In this case the cloudiness maximum is distinct from the Equatorial Trough.

The dynamics of low-latitude atmosphere–ocean circulations are also involved. The convergence zone in the central equatorial Pacific moves seasonally between about 4°N in March–April and 8°N in September. This appears to be a response to the relative strengths of the north-east and south-east trades. The ratio of South Pacific/North Pacific trade wind strength exceeds 2 in September, but falls to 0·6 in April. Interestingly, the ratio varies in phase with the ratio of Antarctic–Arctic sea-ice areas; Antarctic ice is a maximum in September when Arctic ice is at its minimum. The convergence axis is often aligned close to the zone of maximum sea-surface

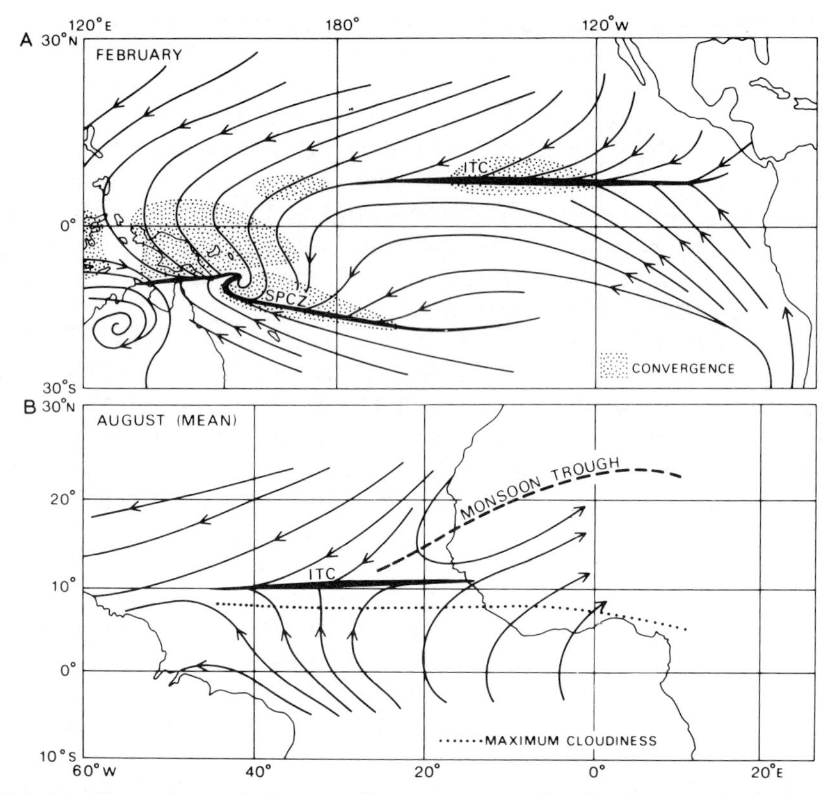

Fig. 6.2. Illustrations of (A) streamline convergence forming an Intertropical Convergence (ITC) and South Pacific Convergence Zone (SPCZ) in February (C. S. Ramage, personal communication, 1986), and (B) the contrasting patterns of monsoon trough over West Africa, streamline convergence over the central tropical North Atlantic, and axis of maximum cloudiness to the south for August (*from Sadler 1975a*).

temperatures, but is not anchored to it. Indeed, the SST maximum located within the Equatorial Countercurrent (see fig. 3.38) is a result of the interactions between the trade winds and horizontal and vertical motions in the ocean surface layer.

Aircraft studies have revealed the complex structure of the central Pacific ITCZ. When moderately strong trades provide horizontal moisture convergence, convective cloud bands form, but the convergent lifting may be insufficient for rainfall in the absence of upper-level divergence. Moreover, although the south-east trades cross the equator, the mean monthly resultant winds between 115° and 180°W have, throughout the year, a more southerly component north of the equator and a more northerly one south of it, giving a zone of divergence (due to the sign change in the Coriolis parameter) along the equator.

In the south-western sectors of the Pacific and Atlantic oceans, satellite cloudiness studies indicate the presence of two semi-permanent confluence zones (fig. 6.1). These do not occur in the eastern South Atlantic and South Pacific where there are cold ocean currents. The convergence shown in the western South Pacific in February is now recognized as an important discontinuity and zone of maximum cloudiness (see pl. 27) termed the South Pacific Convergence Zone (SPCZ); it extends from the eastern tip of Papua New Guinea to about 30°S, 120°W. At sea-level moist north-easterlies, west of the South Pacific subtropical anticyclone, converge with south-easterlies ahead of high-pressure systems moving eastward from Australia/New Zealand. The low-latitude section west of 180° longitude is part of the ITCZ system, whereas its south-eastern end is associated with wave disturbances and jet stream clouds on the South Pacific polar front. The link across the subtropics appears to reflect upper-level tropical mid-latitude transfers of moisture and energy, especially during subtropical storm situations. Hence, the SPCZ shows substantial short-term and interannual variability in its location and development.

C Tropical disturbances

It was not until the 1940s that detailed accounts were given of types of tropical disturbance other than the long-recognized tropical cyclone. However, our view of tropical weather systems has been radically revised following the advent of operational meteorological satellites in the 1960s. Special programmes of meterological measurements at the surface and in the upper air together with aircraft and ship observations have been carried out in the Pacific and Indian Oceans, the Caribbean and in the tropical eastern Atlantic.

It appears that five categories of weather system can be distinguished according to their space and time scales (fig. 6.3). The smallest, with a life span of a few hours, is the *individual cumulus*, 1–10 km in diameter, which is generated by dynamically-induced convergence in the Trade Wind boundary layer. In fair weather, cumulus clouds are generally aligned in 'cloud streets', more or less parallel to the wind direction (pl. 28), or form polygonal honeycomb-pattern cells, rather than scattered at random. This seems to be related to the boundary-layer structure and wind speed (see p. 87). There is little interaction between the air layers above and below the cloud base under these conditions, but in disturbed weather conditions up and downdraughts cause interaction between the two layers which intensifies the convection. Individual cumulus towers, associated with violent thunderstorms, develop particularly in the Intertropical Convergence Zone sometimes reaching to more than 20,000 m (65,000 ft) in height and having updraughts of 10–14 m s^{-1} (25–30 mph) (pl. 29). In this way the smallest scale of system can aid the development of larger disturbances. Convection is most active over sea surfaces with temperatures exceeding 26°C.

The second category of system develops through cumulus clouds becoming grouped into *meso-scale convective areas* (MCAs) up to 100 km across (fig. 6.3). In turn, several MCAs may comprise a *cloud cluster*, 100–1000 km

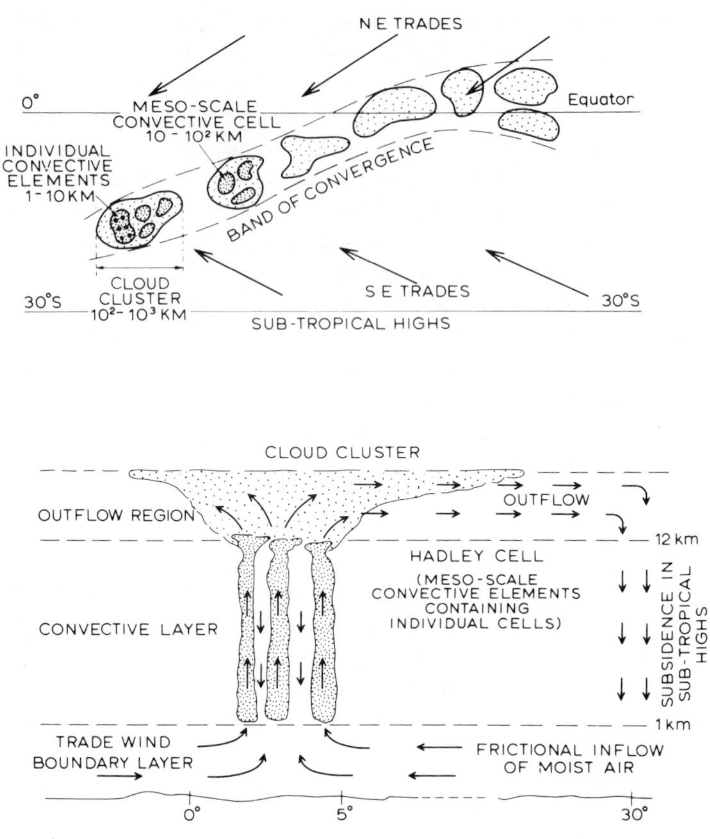

Fig. 6.3. The meso-scale and synoptic structure of the equatorial trough zone (ITCZ), showing a model of the spatial distribution (*above*) and of the vertical structure (*below*) of convective elements which form the cloud clusters (*from Mason 1970*).

in diameter. These subsynoptic scale systems are rather arbitrarily defined as they were initially identified as amorphous cloud areas on satellite images. They have been studied primarily from satellite photographs of the tropical Atlantic and Pacific oceans. Their definition is rather arbitrary, but they may extend over an area 2° square up to 12° square. It is important to note that the peak convective activity has passed when cloud cover is most extensive through the spreading of cirrus canopies. Clusters in the Atlantic, defined as more than 50% cloud cover extending over an area of 3° square, show maximum frequencies of 10–15 clusters per month near the ITC and also at 15°–20°N in the western Atlantic over zones of high sea-surface temperature. They consist of a cluster of meso-scale convective cells with the system having a deep layer of convergent airflow (see fig. 6.3). Some persist for only 1–2 days but others develop within synoptic-scale waves. Many aspects of their development and role remain to be determined.

The fourth category of tropical weather system includes the synoptic scale waves and cyclonic vortices (discussed more fully below) and the fifth group is represented by the planetary-scale waves. The planetary waves (with a wavelength from 10,000 to 40,000 km) need not concern us in detail. Two types occur in the equatorial stratosphere and another in the equatorial upper troposphere. While they may interact with lower tropospheric systems, they do not appear to be direct weather mechanisms. The synoptic scale systems which determine much of the 'disturbed weather' of the tropics are sufficiently important and varied as to be discussed under the headings of wave disturbances and cyclonic storms.

1 Wave disturbances

There are several types of wave travelling westward in the equatorial and tropical tropospheric easterlies; the differences between them probably result from regional and seasonal variations in the structure of the tropical atmosphere. Their wavelength is between about 2000–4000 km and they have a life span of 1–2 weeks, travelling some 6°–7° longitude per day.

The first wave type to be described in the Tropics was the *Easterly Wave* of the Caribbean area. This system is quite unlike a mid-latitude depression. There is a weak pressure trough which usually slopes eastward with height (fig. 6.4) and typically the main development of cumulonimbus cloud and thundery showers *behind* the trough line. This pattern is associated with horizontal and vertical motion in the easterlies. Behind the trough low-level air undergoes convergence, while ahead of it there is divergence (ch. 3, B.1). This follows from the equation for the conservation of potential vorticity (compare ch. 4, F), which assumes that the air travelling at a given level does not change its potential temperature (i.e. dry-adiabatic motion; see ch. 2, D):

$$\frac{f + \zeta}{\Delta p} = k$$

f = the Coriolis parameter, ζ = relative vorticity (cyclonic positive) and Δp = the depth of the tropospheric air column. Air overtaking the trough line is moving both polewards (f increasing) and towards a zone of cyclonic curvature (ζ increasing), so that if the left-hand side of the equation is to remain constant Δp must increase. This vertical expansion of the air column necessitates horizontal contraction (convergence). Conversely there is divergence in the air moving southward ahead of the trough and curving anticyclonically. The divergent zone is characterized by descending, drying air with only a shallow moist layer near the surface, while in the vicinity of the trough and behind it the moist layer may be 4500 m (15,000 ft) or more deep. When the easterly air flow is *slower* than the speed of the wave, the reverse pattern of low-level convergence ahead of the trough and divergence behind it is observed as a consequence of the potential vorticity equation. Often this is the case in the middle troposphere so that the pattern of vertical motion shown in fig. 6.4 is augmented.

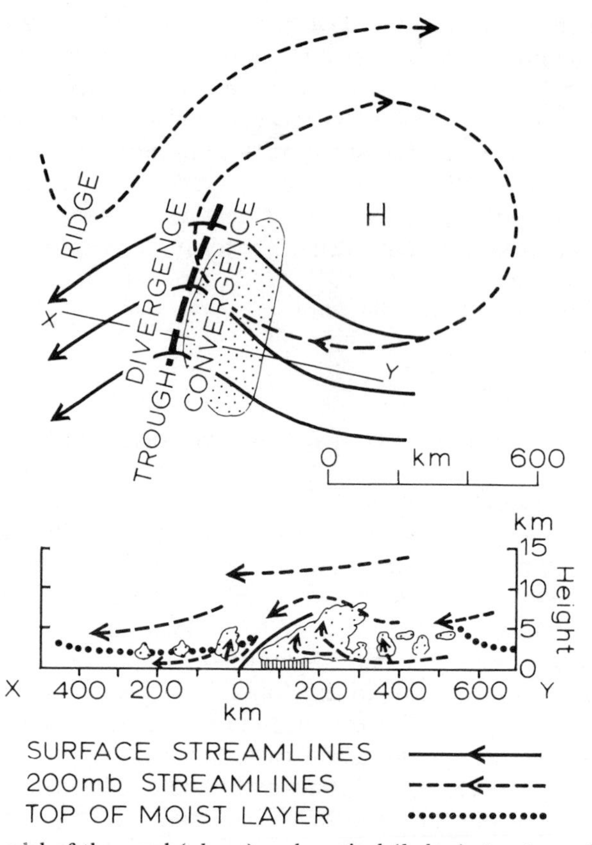

Fig. 6.4. A model of the areal (*above*) and vertical (*below*) structure of an easterly wave. Cloud is stippled and the precipitation area is shown in the vertical section. The streamlines symbols refer to the areal structure, and the arrows on the vertical section indicate the horizontal and vertical motions (*partly after Riehl and Malkus*).

The passage of such a transverse wave in the trades commonly produces the following weather sequence:

In the ridge ahead of the trough	fine weather, scattered cumulus cloud, some haze.
Close to the trough line	well-developed cumulus, occasional showers, improving visibility.
Behind the trough	veer of wind direction, heavy cumulus and cumulonimbus, moderate or heavy thundery showers and a decrease of temperature.

Satellite photography indicates that the simple easterly wave is rather less common than has been supposed. Many Atlantic disturbances show an

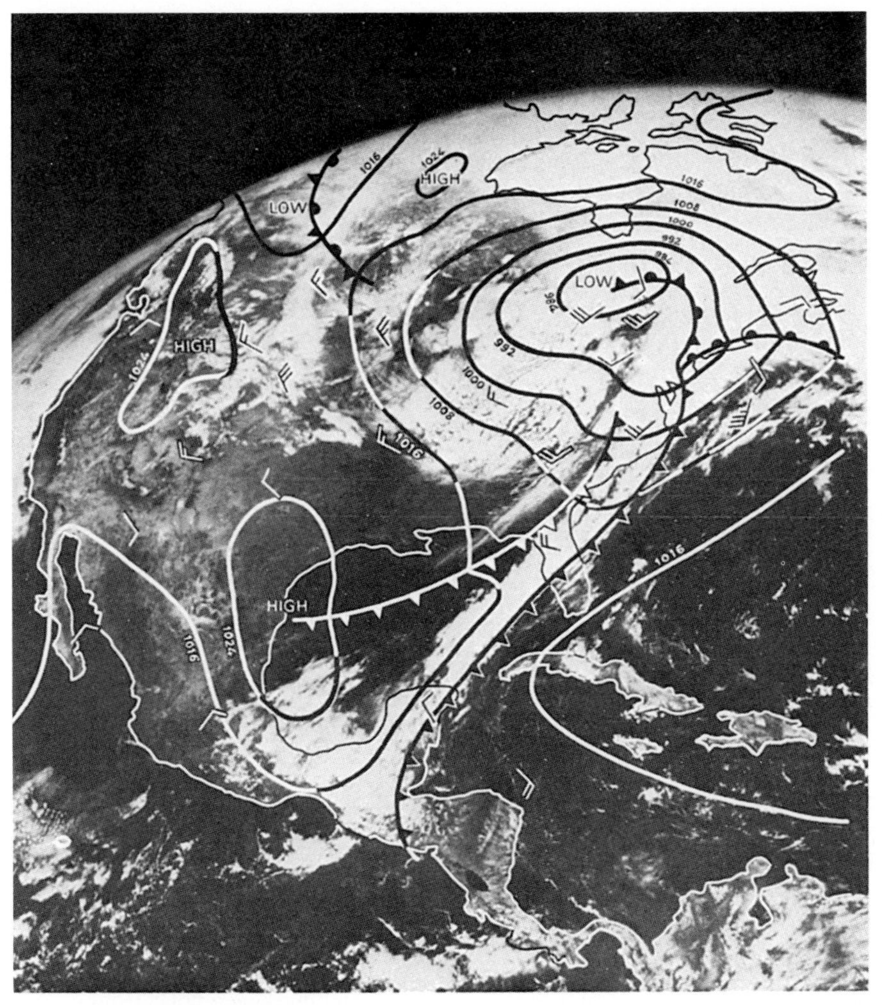

Plate 22. Satellite photograph of North and Central America taken from 37,000 km (23,000 miles) on an April day. A depression lies to the south-east of Hudson Bay and a belt of rain or drizzle marks the cold front from Yucatan to New England. This cloud band suggests high-level tropical/mid-latitude interaction. The Great Lakes are snowy and windy but elsewhere conditions are mostly clear with light winds (*NOAA photograph; Courtesy National Geographic Society*).

Plate 23. Infrared photograph from a NOAA satellite taken at 1502 hours GMT on 9 September 1983 showing a convective cloud band associated with a polar trough conveyor belt (low-level jet) wrapped around the leading edge of a cold pool lying to the north-west of a stratiform cloud band marking a polar front conveyor belt (upper-level jet) (*from Browning, 1985, fig. 21b, 314; see also fig. 4.14*). (*Permission from Dept of Electrical Engineering and Electronics, University of Dundee*). (*Reproduced by permission of the Controller of Her Majesty's Stationery Office*).

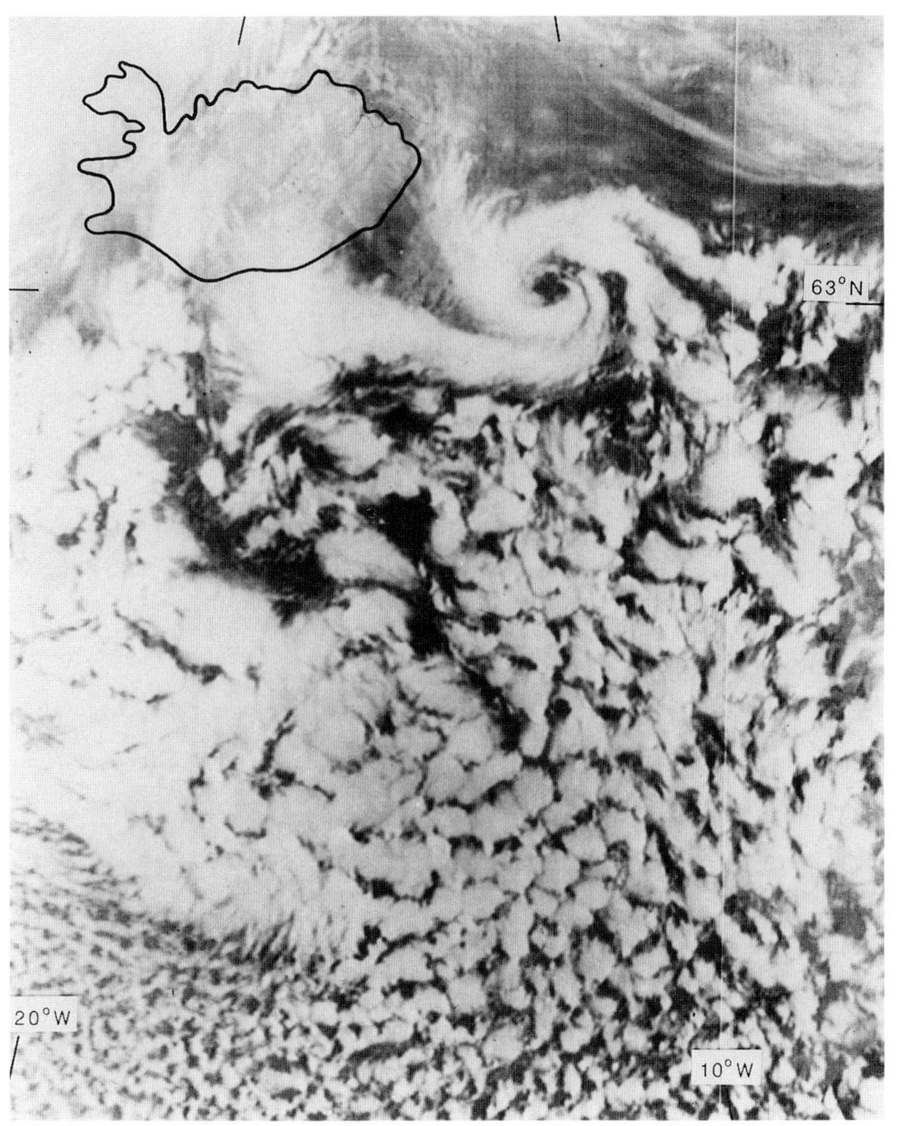

Plate 24. A polar low near Iceland, 14 January 1984, as seen on a visible band DMSP satellite image. This meso-scale low and the closed cellular cloud patterns to the south developed in a northerly airflow behind an occluded depression situated over the coast of Norway (*courtesy National Snow and Ice Data Center, University of Colorado, Boulder*).

Plate 25. View from 5 km (3 miles) distance of a tornado north-east of Tracy, Minnesota, on 13 June 1968 (*photograph courtesy of Eric Lantz and Associated Press*). A convectively unstable tongue of warm air extended north from Texas and by mid-afternoon its temperature had risen to 32°C (90°F), and severe thunderstorms had set in ahead of a cold front lying to its west. Surface pressure continued to drop in this belt, which was supported by a trough at 500 mb and surmounted by a jet of over 45 m s^{-1} (100 mph) at the tropopause extending from Oregon to eastern Canada. Thunderstorm activity reached a maximum at about 1800 hours as the unstable belt moved into Minnesota, and individual cells were shown by radar to have built up to over 15,240 m (50,000 ft). This combination of conditions was ideal for tornado inception and on that afternoon thirty-four funnels were sighted within 480 km (300 miles) of Minneapolis. The Tracy tornado appeared 13 km (8 miles) south-west of the town at 1900 hours and moved north-east at 13 m s^{-1} (35 mph) for 21 km (13 miles), cutting a 90 to 150 m (300 to 500 ft) wide belt of total destruction through Tracy, killing nine people, injuring 125 and causing $3 million worth of damage. Unlike most tornadoes, it did not lift off the ground on encountering the rough urban surface but 'dug in' for its whole course until it suddenly dissolved a few seconds after the photograph was taken (*description courtesy of the Director, National Severe Storms Forecast Center, Kansas City*).

Plate 26. Visible image taken by METEOSAT-2, geostationary some 35,900 km (22,300 miles) above the equator, at 1130 hours GMT on 25 September 1983. A broad ITCZ lies across West and East Africa exhibiting centres of varied convective activity including cumulus towers over Central and East Africa. The subtropical high-pressure belt is particularly well developed in the northern hemisphere, as are depressions in the belts of the northern and southern westerlies. A broad wedge of cloud extends west over the South Atlantic from Angola and Gabon in the tropical easterly flow and there is a narrow belt of low cloud in the zone of coastal upwelling off Namibia (*METEOSAT image supplied by the European Space Agency*).

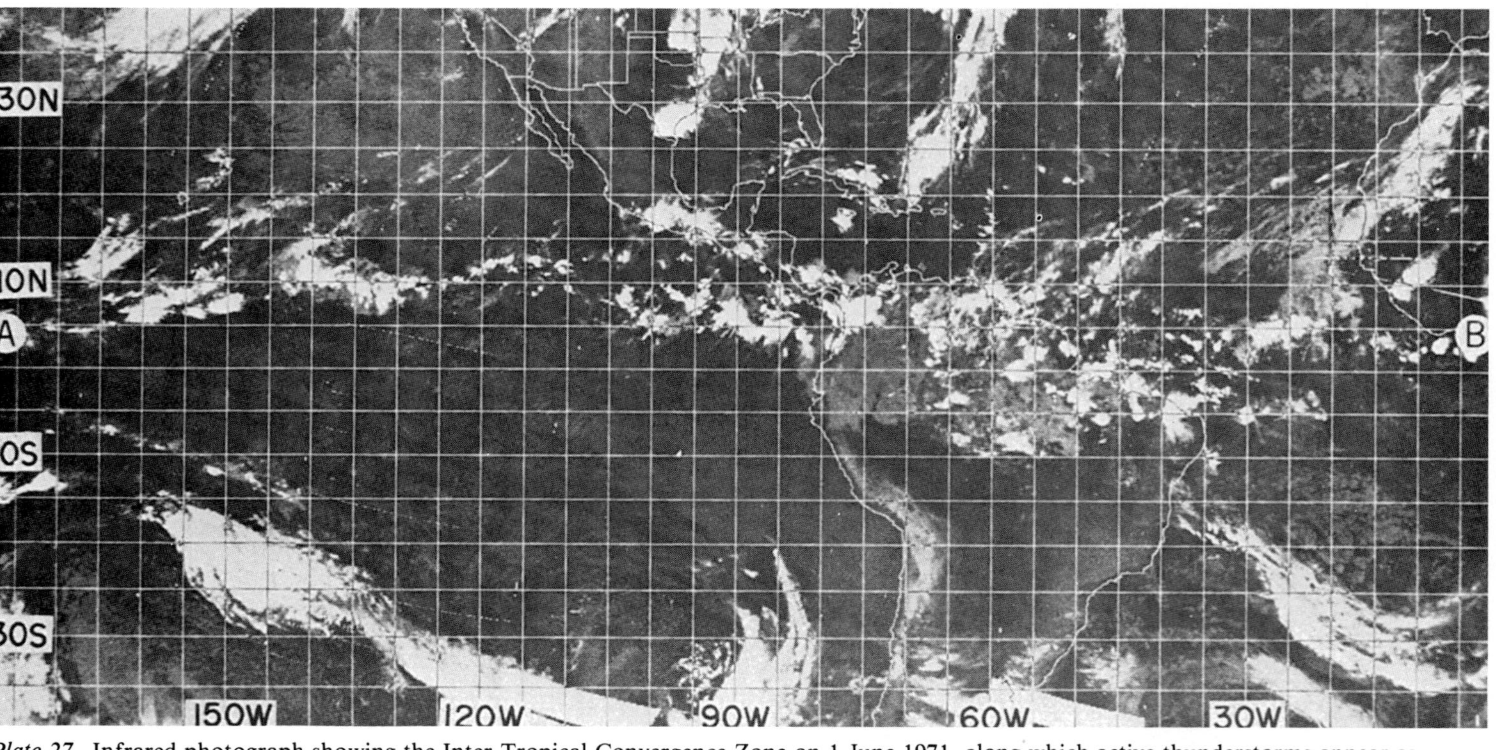

Plate 27. Infrared photograph showing the Inter-Tropical Convergence Zone on 1 June 1971, along which active thunderstorms appear as bright spots. There is also a cyclonic cloud band in the western South Pacific representing the South Pacific Convergence Zone (*NOAA-1 photograph, World Meteorological Organization 1973*).

Plate 28. View south over Florida from the Gemini V manned spacecraft at an elevation of 180 km (112 miles) on 22 August 1965, with Cape Kennedy launching site in the foreground. Cumulus clouds have formed over the warmer land, with a tendency to align in east–west 'streets', and are notably absent over Lake Okeechobee. In the south, thunder-head anvils can be seen (*NASA photograph*).

Plate 29. Cumulus towers with powerful thunderstorms along the ITCZ over Zaire photographed in April 1983 from the Space Shuttle at an elevation of 280 km (175 miles). The largest tower shows a double mushroom cap reaching to more than 15,240 m (50,000 ft) and the symmetrical form of the caps indicates a lack of pronounced airflow at high levels (*Courtesy NASA*).

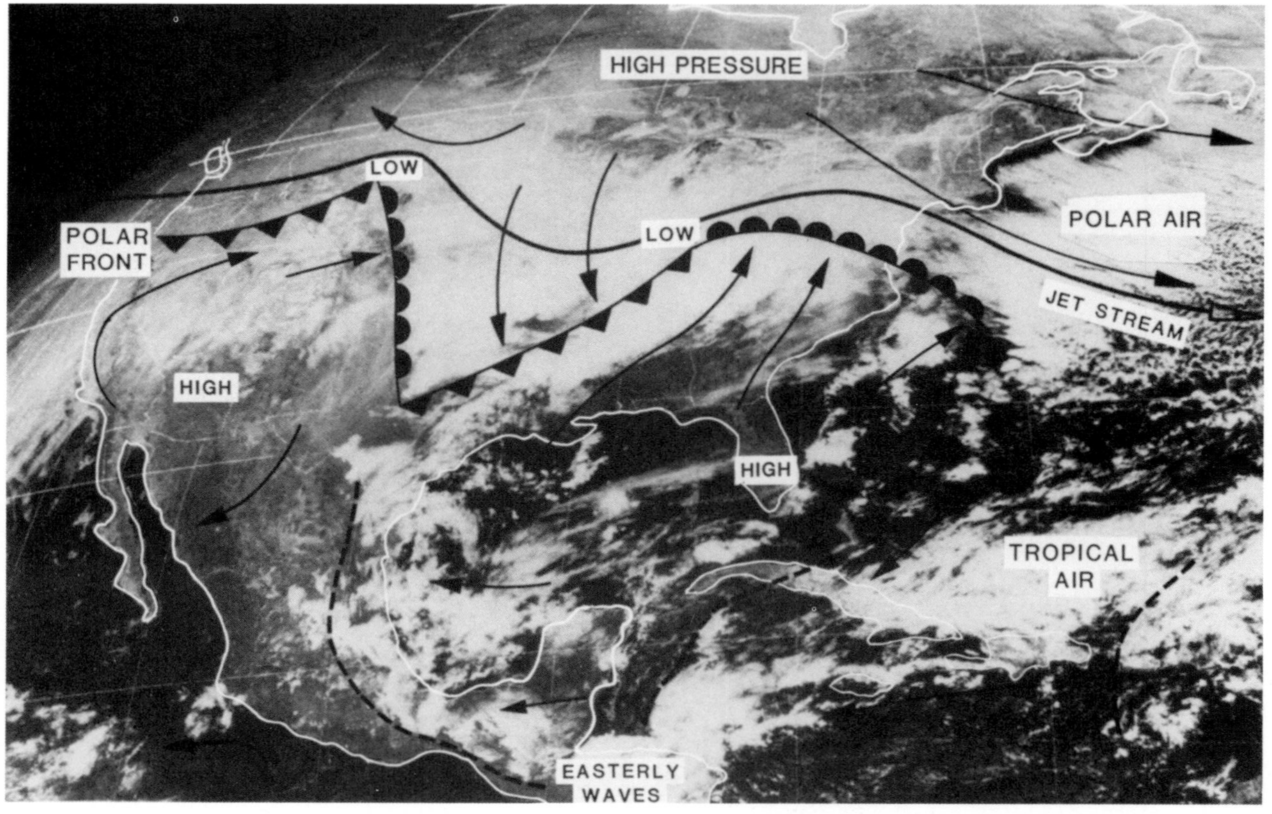

Plate 30. Satellite photograph of North America taken at 1700 hours GMT on 12 February 1979 from a GOES weather satellite located some 35,800 km (22,250 miles) above the equator. Two depressions are located below an upper jet stream located between polar and subtropical high-pressure cells. The more westerly depression is forming in the lee of the Rocky Mountains; cold polar air streaming off New England is becoming cloudy over the warmer sea; weak easterly waves (dashed) have developed equatorward of the subtropical high-pressure cells over the Caribbean and Central America (*courtesy NOAA*).

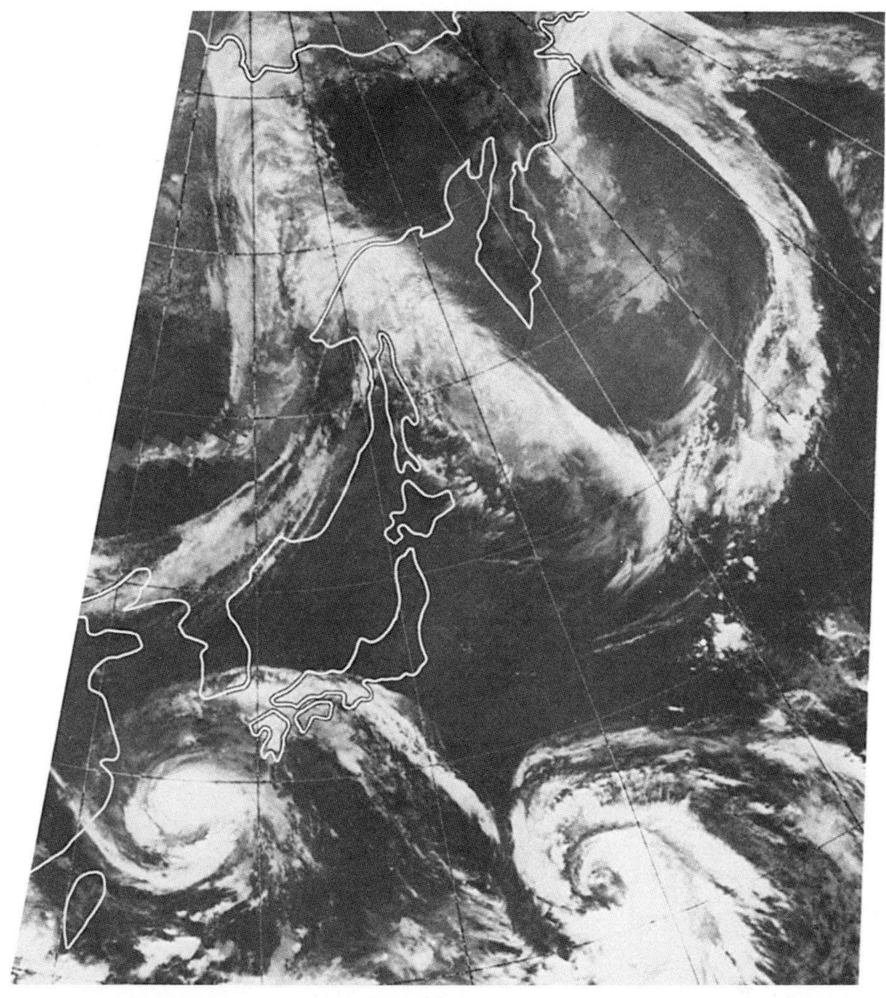

Plate 31. A satellite infrared mosaic of eastern Asia and the western North Pacific showing two mid-latitude depression systems and typhoons Wendy (28°N, 126°E) and Virginia (22°N, 147°E) on 29 July 1978, about 0900 local time (Tokyo). The typhoons had maximum winds of about 36 m s^{-1} (70 kt) and sea-level pressure minima of about 965 mb (Wendy) and 975 mb (Virginia). A subtropical high-pressure ridge about 35°N separates the tropical and mid-latitude storms minimizing any interaction (*Defense Meteorological Satellite Program imagery, National Snow and Ice Data Center, University of Colorado, Boulder*).

Plate 32. An air view looking south-eastwards towards the line of high cumulus towers marking the convergence zone near the Wake Island wave trough shown in fig. 6.8 (*from Malkus and Riehl 1964*).

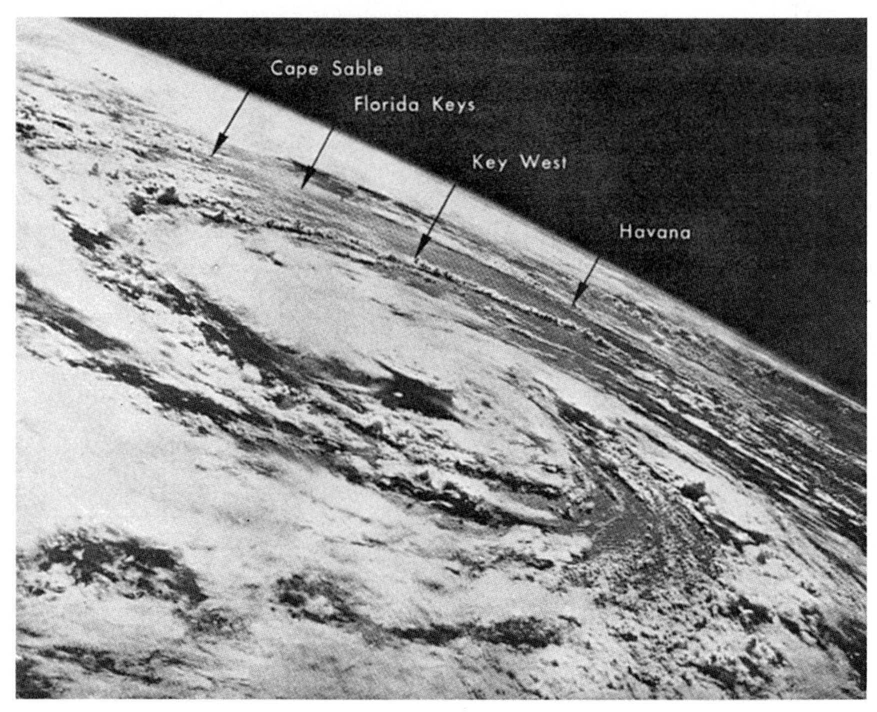

Plate 33. Hurricane 'Gladys' just west of Florida, photographed towards the south from Apollo 7 manned spacecraft at an elevation of 179 km (111 miles) on 17 October 1968. At this time, wind speeds up to 40 m s^{-1} (90 mph) were reported (*NASA photograph*).

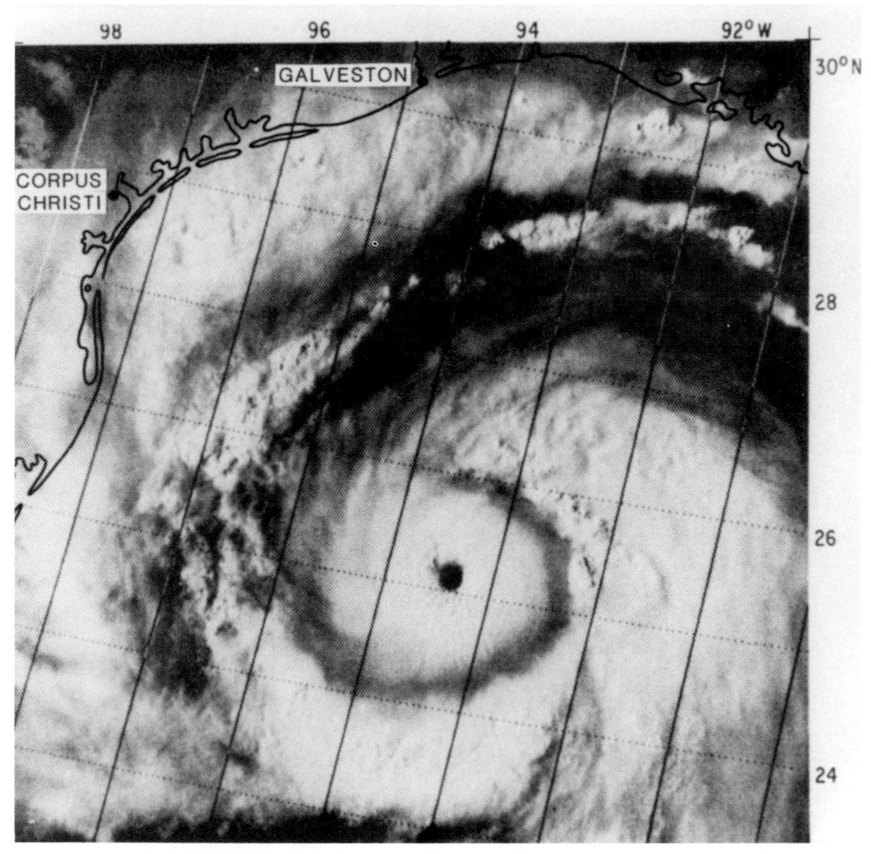

Plate 34. Visible satellite image of Hurricane Allen, located south of the Texas coast, taken at 2123 hours GMT on 8 August 1980 showing the clear eye (*photograph courtesy National Hurricane Center, Miami, Florida, USA*).

Plate 35. Visible satellite image showing five large tropical cloud clusters topped by cirrus shields situated between latitudes 5° and 10°N in the vicinity of West Africa, together with one squall-line cloud cluster at 15°N having a well-defined arc cloud squall line on its leading (south-west) edge. Taken by SMS-1 satellite at 1130 hours GMT on 5 September 1974 (courtesy NOAA).

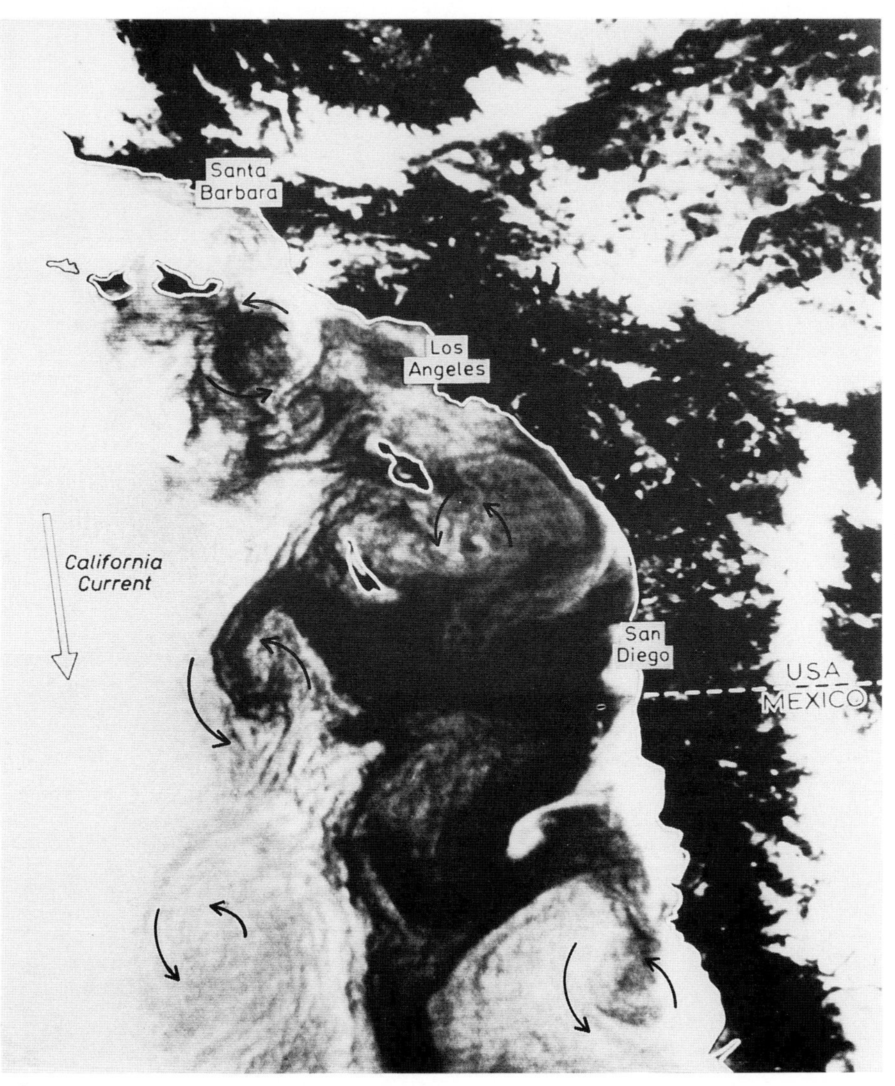

Plate 36. Infrared photograph showing large vortices of cold water (*light*) upwelling in the warmer surface coastal waters (*dark*) off southern California. The colder offshore California Current is clearly shown (*NASA photograph*).

Plate 37. Hydraulic jump in the lee (i.e. eastern side) of Elk Mountain, Wyoming, at 1345 hours on 1 February 1973, looking south-west from the north-eastern edge of Elk Mountain (*photograph courtesy Dr August H. Auer, Jr, University of Wyoming*).

Plate 38. View westwards over Edinburgh showing Arthur's Seat in the foreground and the Castle rising above urban smog in the middle distance. The smog thickens to the north (*right*) over the industrial suburb of Leith, merging with sea mist over the Firth of Forth at the far right. This photograph was taken some considerable time ago (*courtesy Aerofilms Ltd*).

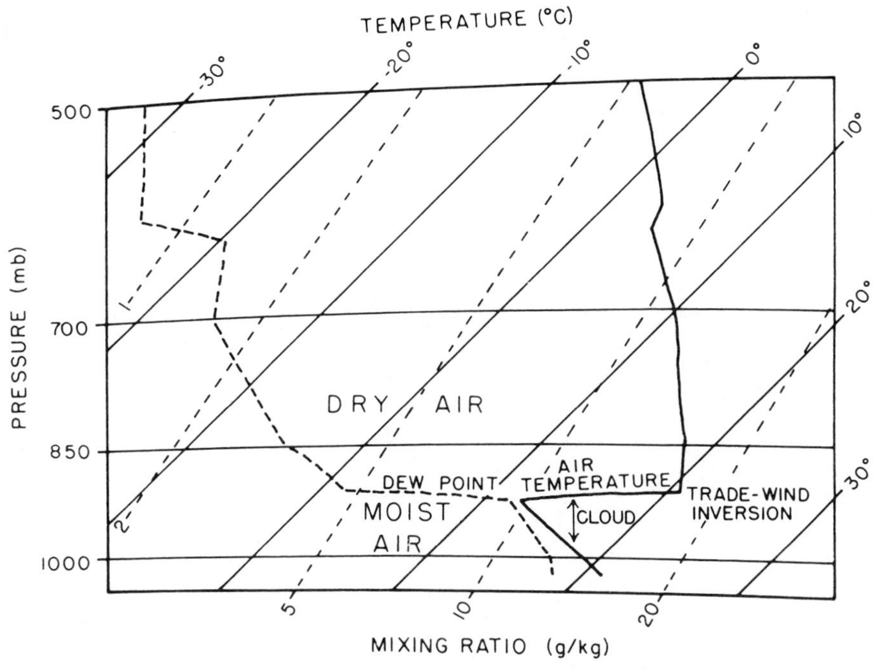

Fig. 6.5. The vertical structure of trade wind air at 30°N, 140°W at 0300 GMT on 10 July 1949. The mixing ratio is the saturation value (*based on Riehl 1954*).

'inverted V' wave-form in the low level wind field and associated cloud, or a 'comma' cloud related to a vortex. They are often apparently linked with a wave pattern on the ITC further south. West African disturbances that move out over the eastern tropical Atlantic usually exhibit low-level confluence and upper-level diffluence *ahead* of the trough, giving maximum precipitation rates in this same sector. Many disturbances in the easterlies have a closed cyclonic wind circulation at about the 600 mb level.

It is obviously difficult to trace the growth of wave disturbances over the oceans and in continental areas with sparse data coverage. However, some generalizations can be made. At least 8 out of 10 disturbances develop some 2°–4° latitude poleward of the equatorial trough. Convection is probably set off by convergence of moisture in the airflow, accentuated by friction, and then maintained by entrainment into the thermal convective plumes (fig. 6.3). About 100 tropical disturbances develop during the June–November hurricane season in the tropical Atlantic. More than half of those disturbances originate over Africa in association with the baroclinic zone between Saharan air and cooler, moist monsoon air. Many of them can be tracked westward into the eastern North Pacific. A quarter of the disturbances intensify into tropical depressions and 10% become 'named' storms.

Developments in the Atlantic are closely related to the structure of the trades. In the eastern sectors of the subtropical anticyclones active subsidence maintains a pronounced inversion at 450 to 600 m (1500 to 2000 ft) (fig. 6.5). Thus, the cool eastern tropical oceans are characterized by

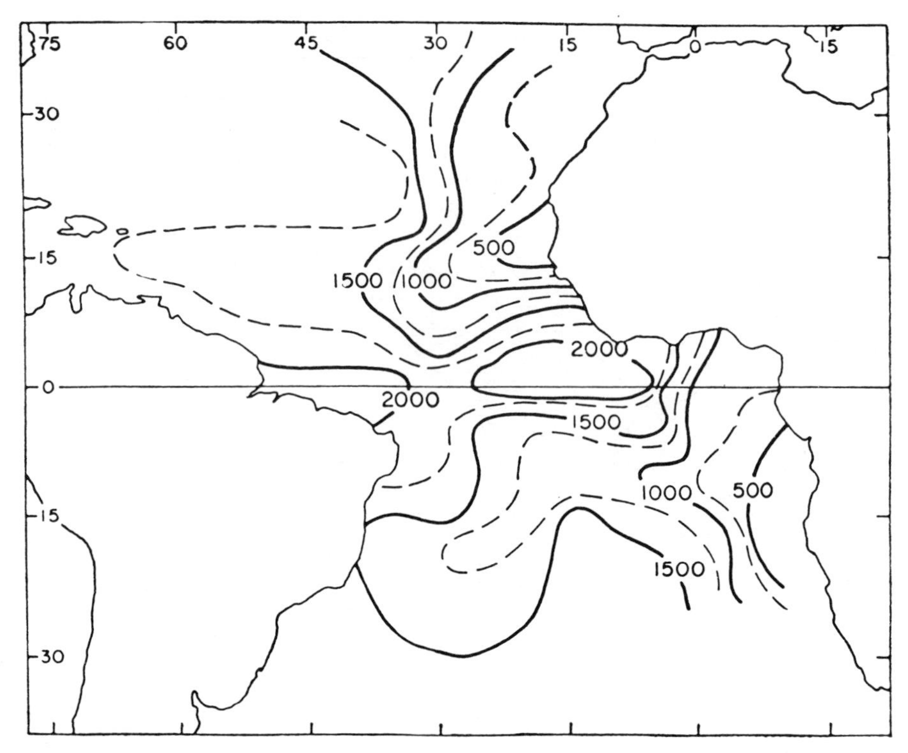

Fig. 6.6. The height (in metres) of the base of the trade wind inversion over the tropical Atlantic (*from Riehl 1954*).

extensive, but shallow, marine stratocumulus which gives little rainfall. Downstream the height of the inversion base rises (fig. 6.6) because the subsidence decreases away from the eastern part of the anticyclone and cumulus towers penetrate through the inversion from time to time, spreading moisture into the dry air above. Easterly waves tend to develop in the Caribbean when the trade wind inversion is weak or even absent during summer and autumn, whereas in winter and spring accentuated subsidence aloft inhibits their growth although disturbances may move westward above the inversion. Another factor which may initiate waves in the easterlies is the penetration of cold fronts into low latitudes. This is common in the sector between two subtropical high-pressure cells where the equatorward portion of the front tends to fracture, generating a westward-moving wave.

The influence of these features on regional climate is illustrated by the rainfall regime. For example, there is a late summer maximum at Martinique (fig. 6.7) in the Windward Islands (15°N) when subsidence is weak, although some of the autumn rainfall is associated with tropical storms. In many trade wind areas the rainfall occurs in a few rainstorms associated with some form of disturbance. Over a 10-year period Oahu (Hawaii) had an average of twenty-four rainstorms per year of which ten accounted for more than two-

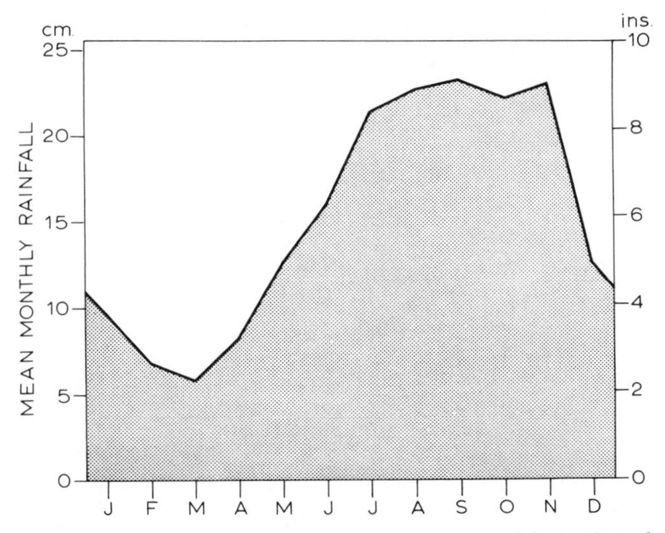

Fig. 6.7. Average monthly rainfall at Fort de France, Martinique (*based on* 'CLIMAT' *normals of the World Meteorological Organization for 1931–60*). The mean annual rainfall is 72·45 in (184 cm).

thirds of the annual precipitation. There is quite high variability of rainfall from year to year in such areas, since a small reduction in the frequency of disturbances can have a large effect on rainfall totals.

In the central equatorial Pacific the trade wind systems of the two hemispheres converge in the Equatorial Trough and wave disturbances may be generated if the trough is sufficiently removed from the equator (usually to the north) to provide a small Coriolis force to begin cyclone motion. These disturbances quite often become unstable forming a cyclonic vortex as they travel westwards towards the Philippines, but the winds do not necessarily attain hurricane strength. The synoptic chart for part of the north-west Pacific on 17 August 1957 (fig. 6.8), shows three developmental stages of tropical low-pressure systems. An incipient easterly wave has formed west of Hawaii which, however, filled and dissipated during the next 24 hours. A well-developed wave is evident near Wake Island, having spectacular cumulus towers extending up to over 9100 m (30,000 ft) along the convergence zone some 480 km (300 miles) east of it (pl. 31). This wave developed within 48 hours into a circular tropical storm with winds up to 20 m s^{-1} (46 mph), but not into a full hurricane. A strong, closed, north-west moving circulation is situated east of the Philippines. Equatorial waves may form on both sides of the equator in an easterly current located between about 5°N and S. In such cases divergence ahead of a trough in the northern hemisphere is paired with convergence behind a trough line located further to the west in the southern hemisphere. The reader may confirm that this should be so by applying the equation for the conservation of potential vorticity, remembering that both f and ζ operate in the reverse sense in the southern hemisphere.

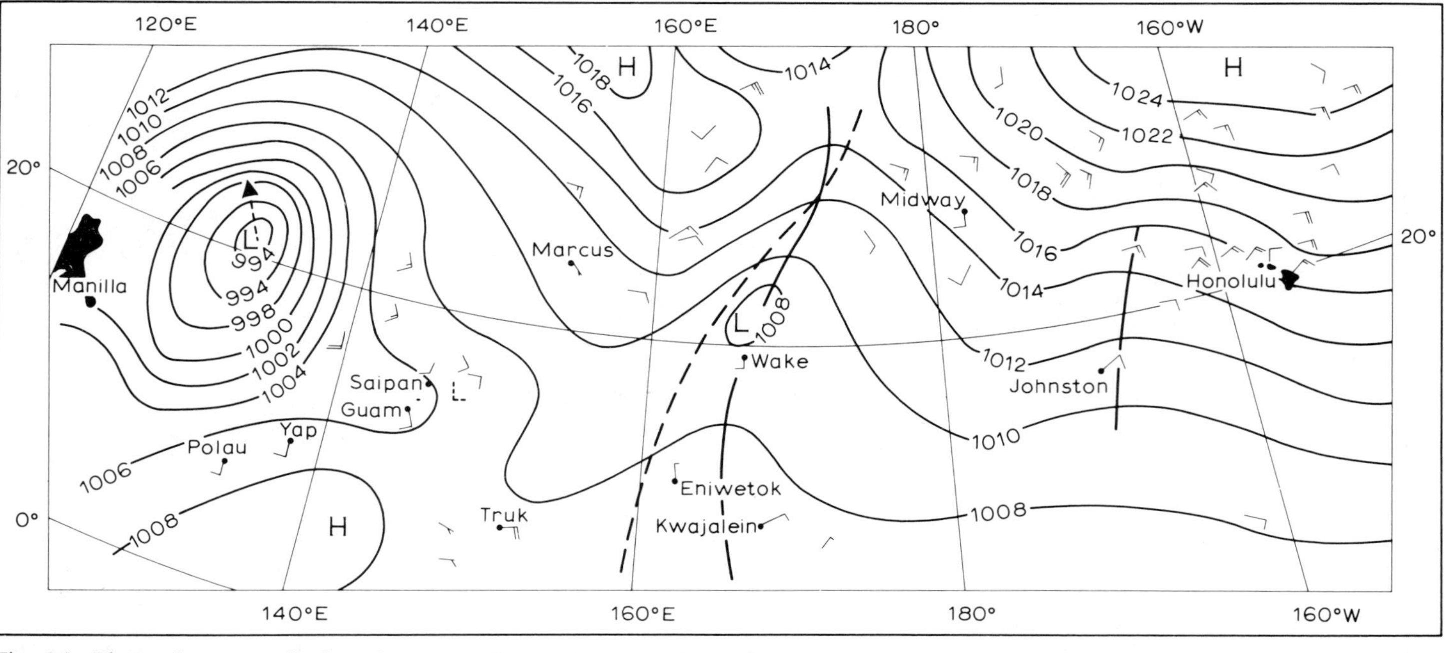

Fig. 6.8. The surface synoptic chart for part of the north-west Pacific on 17 August 1957. The movements of the central wave trough and of the closed circulation during the following 24 hours are shown by the dashed line and arrow, respectively. The dashed L just east of Saipan indicated the location in which another low-pressure system subsequently developed. Plate 31 shows the cloud formation along the convergence zone just east of Wake Island (*from Malkus and Riehl 1964*).

2 Cyclones

a Hurricanes The most notorious type of cyclone is the tropical hurricane (or typhoon). Some 80 or so cyclones each year are responsible, on average, for 20,000 fatalities, as well as causing immense damage to property and a serious shipping hazard, due to the combined effects of high winds, high seas, flooding from the heavy rainfall and coastal storm surges. As a result, considerable attention has been given to forecasting their development and movement so that their origin and structure are beginning to be understood. Naturally the catastrophic force of a hurricane makes it a very difficult phenomenon to investigate, but some assistance is now obtained from aircraft reconnaissance flights sent out during the 'hurricane season', from radar observations of cloud and precipitation structure, and from satellite photography (see pl. 33).

The typical hurricane system has a diameter of about 650 km (400 miles), less than half that of a mid-latitude depression, although typhoons in the Western Pacific are often much larger. The central pressure at sea-level is commonly 950 mb and exceptionally falls below 900 mb. Cyclonic-intensity storms are defined as having maximum sustained surface winds exceeding 33 m s^{-1} (74 mph) and in many storms they exceed 50 m s^{-1} (120 mph). The great vertical development of cumulonimbus clouds with tops at over 12,000 m (40,000 ft) reflects the immense convective activity concentrated in such systems. Radar and satellite studies show that the convective cells are normally organized in bands which spiral inward towards the centre.

The main tropical cyclone activity in both hemisphere is in late summer and autumn during times of maximum northward and southward displacements of the Equatorial Trough (table 6.1). A small number of storms affect both the western North Atlantic and North Pacific areas as early as May and

Table 6.1 Annual frequencies and usual seasonal occurrence of tropical cyclones (maximum sustained winds exceeding 25 m s^{-1}), 1958–77 (*after Gray 1979*). Area totals are rounded.

Location	*Annual frequency*	*Main occurrence*
Western North Pacific	26·3	July–October
Eastern North Pacific	13·4	August–September
Western North Atlantic	8·8	August–October
Northern Indian Ocean	6·4	May–June; October–November
Northern hemisphere total	54·6	
South-west Indian Ocean	8·4	January–March
South-east Indian Ocean	10·3	January–March
Western South Pacific	5·9	January–March
Southern hemisphere total	24·5	
Global total	79·1	

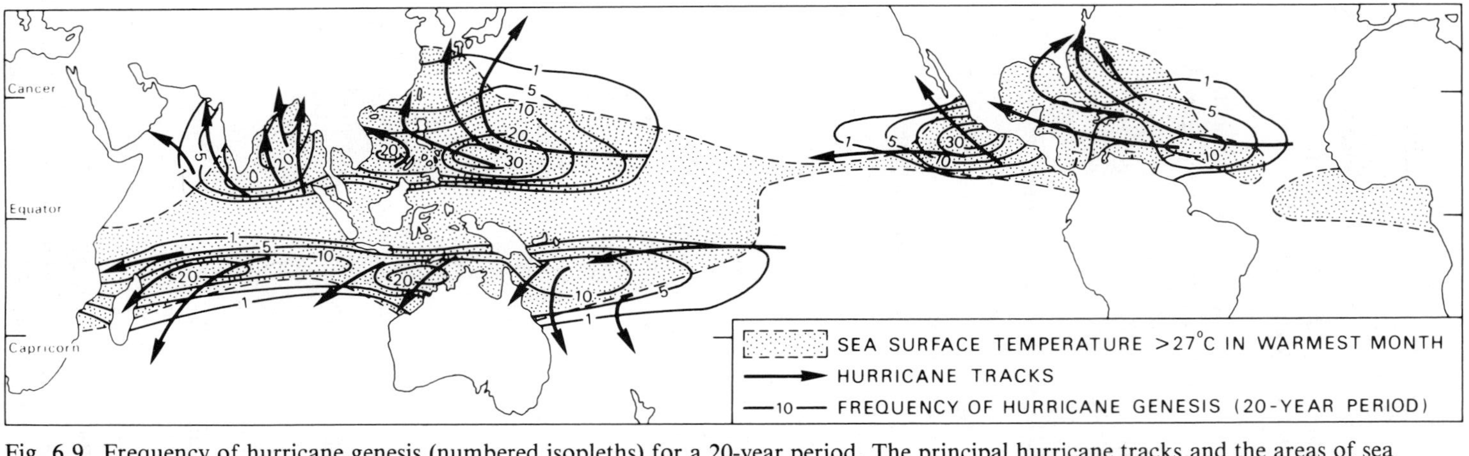

Fig. 6.9. Frequency of hurricane genesis (numbered isopleths) for a 20-year period. The principal hurricane tracks and the areas of sea surface having water temperatures greater than 27°C in the warmest month are also shown (*after Palmén 1948 and Gray 1979*).

as late as December, and have occurred in every month in the latter area. In the Bay of Bengal, there is also a secondary early summer maximum. The annual frequency of cyclones shown in table 6.1 are only approximate since in some cases it may be uncertain as to whether or not the winds actually exceeded hurricane force. In addition, storms in the more remote parts of the South Pacific and Indian oceans frequently escaped detection prior to the use of weather satellites.

A number of conditions are necessary, even if not always sufficient, for hurricane formation. One requirement as shown by fig. 6.9 is an extensive ocean area with a surface temperature greater than 27°C (80°F). Cyclones rarely form near the equator, where the Coriolis parameter is close to zero, or in zones of strong vertical wind shear (i.e. beneath a jet stream), as both factors inhibit the development of an organized vortex. There is also a definite connection between the seasonal position of the Equatorial Trough and zones of hurricane formation, which is borne out by the fact that no hurricanes occur in the South Atlantic (where the trough never lies south of 5°S) or in the south-east Pacific (where the trough remains north of the equator). On the other hand, recent satellite photographs over the north-east Pacific show an unexpected number of cyclonic vortices in summer, many of which move westwards near the trough line about 10°–15°N. About 60% of tropical cyclones seem to originate 5°–10° latitude poleward of the Equatorial Trough in the doldrum sectors, where the trough is at least 5° latitude from the equator. The development regions of hurricanes mainly lie over the western sections of the Atlantic, Pacific and Indian Oceans where the subtropical high-pressure cells do not cause subsidence and stability and the upper flow is divergent.

Early theories of hurricane development held that convection cells generated a sudden and massive release of latent heat to provide energy for the storm. Although convection cells were regarded as an integral part of the hurricane system, their scale was thought to be too small for them to account for the growth of a storm hundreds of kilometres in diameter. Recent research, however, is modifying this picture considerably. Energy is apparently transferred from the cumulus-scale to the large-scale circulation of the storm through the organization of the clouds into spiral bands (fig. 6.10 and pl. 33) although the nature of the process is still being investigated. There is now ample evidence to show that hurricanes form from pre-existing disturbances, but while many of these disturbances develop as closed low-pressure cells few attain full hurricane intensity. The key to this problem is high level outflow (fig. 6.11). This does not require an upper tropospheric anticyclone, but can occur on the eastern limb of an upper trough in the westerlies. This outflow in turn allows the development of very low pressure and high wind speeds near the surface. A distinctive feature of the hurricane is the warm vortex, since other tropical depressions and incipient storms have a cold core area of shower activity. The warm core develops through the action of 100–200 cumulonimbus towers releasing latent heat of condensation; about 15% of the area of cloud bands is giving

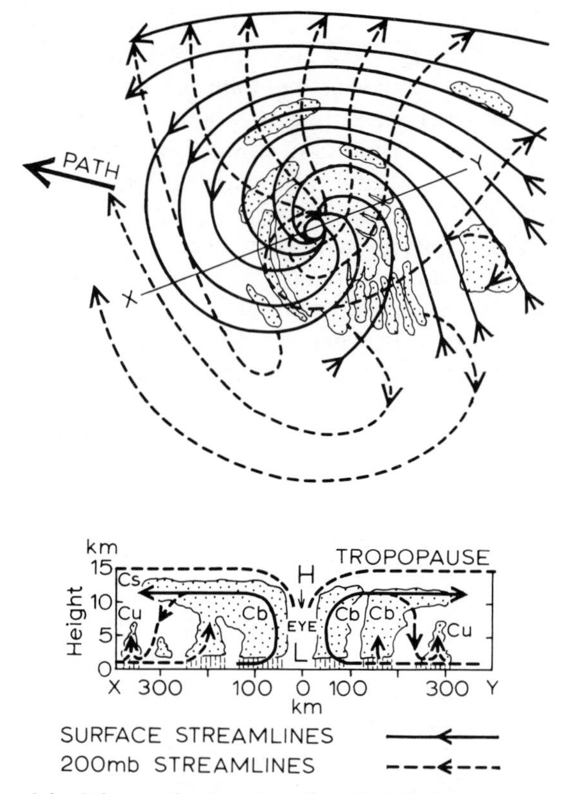

Fig. 6.10. A model of the areal (*above*) and vertical (*below*) structure of a hurricane. Cloud is stippled and areas of precipitation are shown in the vertical section. The streamlines symbols refer to the upper diagram.

rain at any one time. Observations show that although these 'hot towers' form less than 1% of the storm area within a radius of about 400 km (230 miles), their effect is sufficient to change the environment. The warm core is vital to hurricane growth because it intensifies the upper anticyclone, leading to a 'feedback' effect by stimulating the low-level influx of heat and moisture which further intensifies convective activity, latent heat release and therefore the upper-level high pressure. This enhancement of a storm system by cumulus convection is termed Conditional Instability of the Second Kind, or CISK (cf. the basic parcel instability described on p. 79). The thermally-direct circulation converts the heat increment into potential energy and a small fraction of this – about 3% – is transformed into kinetic energy. The remainder is exported by the anticyclonic circulation at about the 12 km (200 mb) level.

In the *eye*, or innermost region of the storm (fig. 6.10), adiabatic warming of descending air accentuates the high temperatures, although since high temperatures are also observed in the eye-wall cloud masses, subsiding air can only be one contributory factor. Without this sinking air in the eye, the

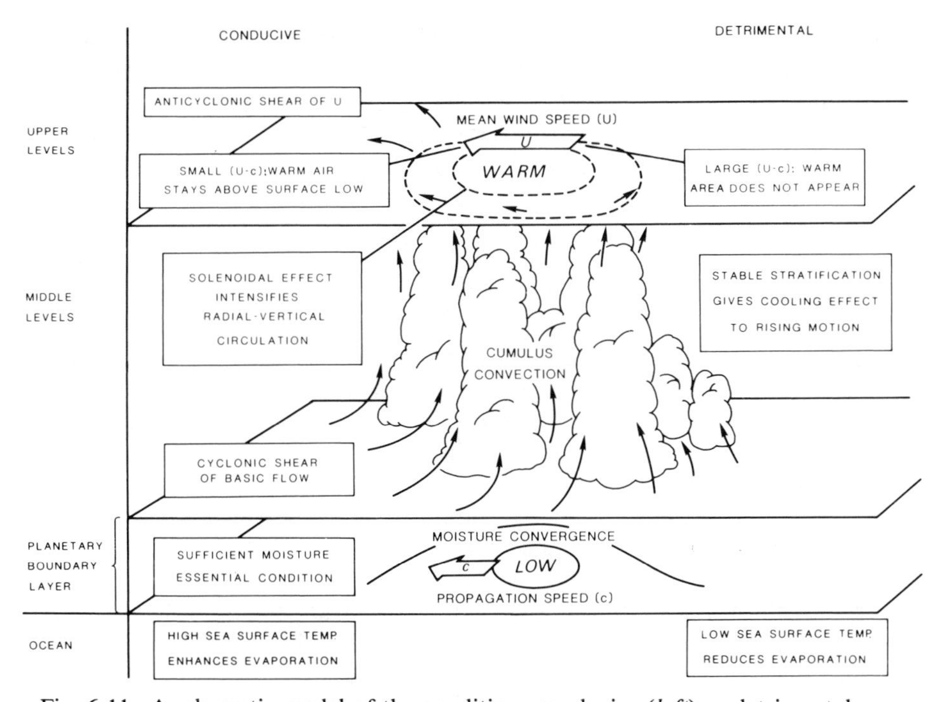

Fig. 6.11. A schematic model of the conditions conducive (*left*) or detrimental (*right*) to the growth of a tropical storm in an easterly wave; *U* is the mean upper-level wind speed and *c* is the rate of propagation of the system. The warm vortex creates a thermal gradient that intensifies both the radial motion around it and the ascending air currents, termed the solenoidal effect (*from Kurihara 1985*). (*Copyright © Academic Press. Reproduced by permission*).

central pressure could not fall below about 1000 mb. The eye has a diameter of some 30–50 km (20–30 miles), within which the air is virtually calm and the cloud cover may be broken. The mechanics of the eye's inception are still largely unknown. If the rotating air conserved absolute angular momentum, wind speeds would become infinite at the centre and clearly this is not the case. The strong winds surrounding the eye are more or less in cyclostrophic balance, with the small radial distance providing a large centripetal acceleration (see p. 119). The air rises when the pressure gradient can no longer force it further inward. It is possible that the cumulonimbus anvils play a vital role in the complex link between the horizontal and vertical circulations around the eye by redistributing angular momentum in such a way as to set up a concentration of rotation near the centre.

The supply of heat and moisture combined with low frictional drag at the sea surface, the release of latent heat through condensation and the removal of the air aloft are essential conditions for the maintenance of hurricane intensity. As soon as one of these ingredients diminishes the storm decays. This can occur quite rapidly if the track (determined by the general upper tropospheric flow) takes the vortex over a cool sea surface or over land. In the latter case the increased friction causes greater cross-isobar air motion, temporally increasing the convergence and ascent. At this stage, increased vertical wind shear in thunderstorm cells may generate tornadoes, especially in the north-east quadrant of the storm (in the northern hemisphere). However, the most important effect of a land track is that cutting off of the moisture supply removes one of the major sources of heat. Rapid decay also occurs when cold air is drawn into the circulation or when the upper-level divergence pattern moves away from the storm.

Hurricanes usually move at some 16–24 kmph (10–15 mph), controlled primarily by the rate of movement of the upper warm core. Commonly they recurve poleward around the western margins of the subtropical high-pressure cells, entering the circulation of the westerlies, where they die out or degenerate into extra-tropical depressions (pl. 20). Some of these systems retain an intense circulation and the high winds and waves can still wreak havoc. This is not uncommon along the Atlantic coast of the United States and occasionally eastern Canada. Similarly, in the western North Pacific, recurved typhoons are a major element in the climate of Japan (see ch. 6, D.4) and may occur in any month. There is an average frequency of twelve typhoons per year over southern Japan and neighbouring sea areas.

To sum up. The hurricane develops from an initial disturbance which, under favourable environmental conditions, grows first into a tropical depression and then a tropical storm (with wind speeds of 17–33 m s^{-1} or 39–73 mph). The tropical storm stage may persist 4–5 days, whereas the hurricane stage usually lasts for only 2–3 days (4–5 days in the western Pacific). The main energy source is latent heat derived from condensed water vapour, and for this reason hurricanes are generated and continue to gather strength only within the confines of warm oceans. The cold-cored tropical storm is transformed into a warm-cored hurricane in association with the release of latent heat in cumulonimbus towers, and this establishes or intensifies an upper tropospheric anticyclonic cell. Thus high-level outflow maintains the ascent and low-level inflow in order to provide a continual generation of potential energy (from latent heat) and the transformation of this into kinetic energy. The inner eye which forms by sinking air is an essential element in the life cycle.

b Other tropical disturbances Not all cyclonic systems in the tropics are of the intense hurricane variety. There are two other major types of cyclonic vortex. One is the *monsoon depression* which affects southern Asia during the summer. This disturbance is somewhat unusual in that the flow is westerly at low levels and easterly in the upper troposphere (see fig. 6.26). It is more fully described in ch. 6, D.4.

The second type of system is usually relatively weak near the surface, but well-developed in the middle troposphere. In the eastern North Pacific and Indian Ocean such lows are referred to as *subtropical cyclones*. Some develop from the cutting-off in low latitudes of a cold upper-level wave in the westerlies (compare ch. 4, G.4). They possess a broad eye of some 150 km

(100 miles) in radius with little cloud, surrounded by a belt of cloud and precipitation about 300 km (200 miles) wide. In late winter and spring a few such storms make a major contribution to the rainfall of the Hawaiian Islands. These cyclones are very persistent and tend eventually to be re-absorbed by a trough in the upper westerlies. Other subtropical cyclones occur over the Arabian Sea in summer and make a major contribution to summer ('monsoon') rains in north-western India. These systems show upward motion mainly in the upper troposphere. Their development may be linked to air export at upper levels of cyclonic vorticity from the persistent heat low over Southern Asia, especially Arabia.

An infrequent and distinctly different weather system, known as a Temporal, occurs along the Pacific coasts of Central America in autumn and early summer. Its main feature is an extensive layer of altostratus, fed by individual convective cells, producing sustained moderate rainfall. These systems originate in the ITCZ over the eastern tropical North Pacific Ocean and are maintained by large-scale lower tropospheric convergence, localized convection and orographic uplift.

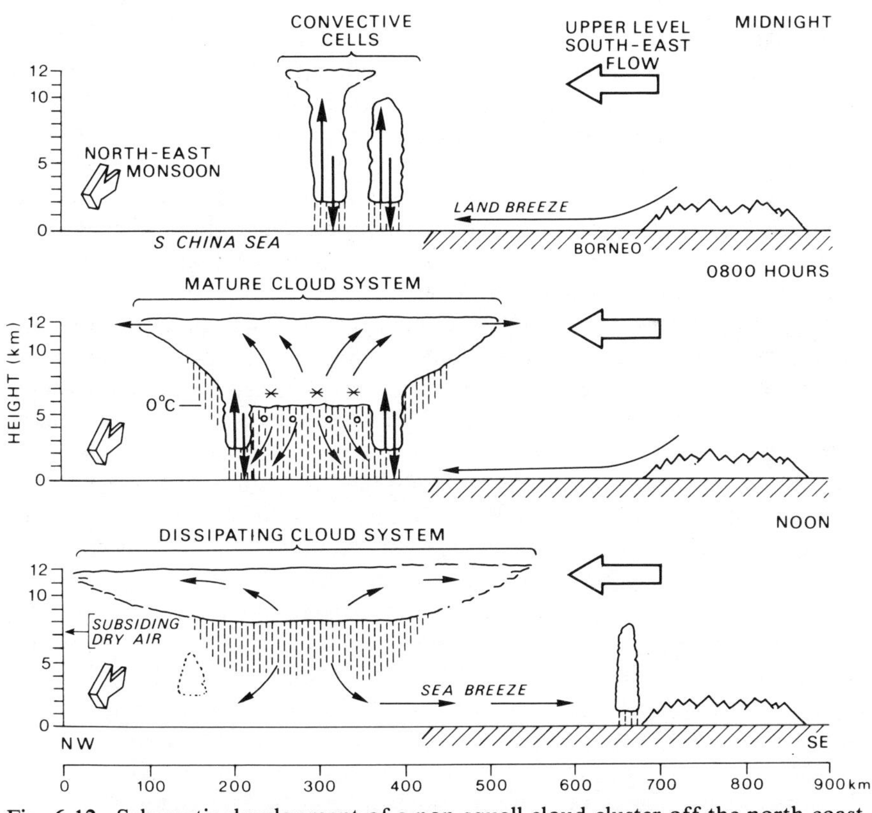

Fig. 6.12. Schematic development of a non-squall cloud cluster off the north coast of Borneo: large arrows indicate the major circulation; small arrows, the local circulation; vertical shading, the zones of rain; stars, ice crystals; and circles, melted raindrops (*after Houze et al. 1981*).

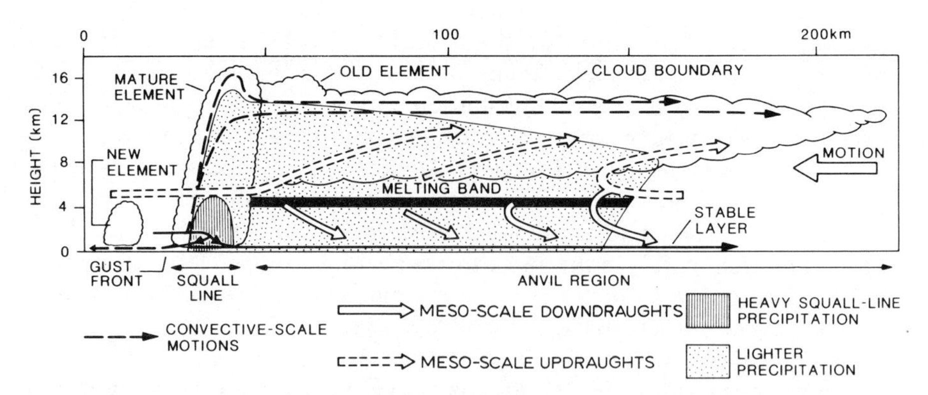

Fig. 6.13. Cross-section of a tropical squall-line cloud cluster showing locations of precipitation and ice particle melting. Dashed arrows show the air motion generated by the squall-line convection and the broad arrows the meso-scale circulation (*after Houze; from Houze and Hobbs 1982*).

3 *Tropical cloud clusters*

These systems fall into two categories: non-squall and squall line. The former contain one or more meso-scale precipitation areas. They occur diurnally, for example, off the north coast of Borneo in winter where they are initiated by convergence of a nocturnal land breeze and the north-east monsoon flow (fig. 6.12). By morning (0800 LST), cumulonimbus cells give precipitation. The cells are linked by an upper-level cloud shield which persists when the convection dies out around noon as a sea-breeze system replaces the nocturnal convergent flow.

Squall-line systems in the tropics (fig. 6.13) form the leading edge of a line of cumulonimbus cells. The squall line and gust front move forward in the low-level flow and by the formation of new cells. These mature and eventually dissipate to the rear of the main line. The process is analogous to that of mid-latitude squall lines (see fig. 4.23) but the tropical cells are weaker. Squall-line systems cross Malaya from the west during the south-west monsoon season. They appear to be initiated by the convergent effects of land breezes in the Malacca Straits. These particular linear systems known as *sumatras*, give heavy rain and often thunder.

In west Africa, systems known as *disturbance lines* are an important feature of the climate in the summer half-year, when low-level south-westerly monsoon air is overrun by dry, warm Saharan air. The meridional air mass contrast helps set up the lower-tropospheric African easterly jet shown in fig. 6.28 which illustrates the zonation of climate and associated weather systems. Disturbance lines tend to form when there is divergence in the upper troposphere north of the Tropical Easterly Jet (see fig. 6.29 also). They are several hundred kilometres long and travel westward associated with upper easterlies at about 50 km h^{-1} (30 mph) giving squalls and thundery showers before dissipating over cold-water areas of the North Atlantic.

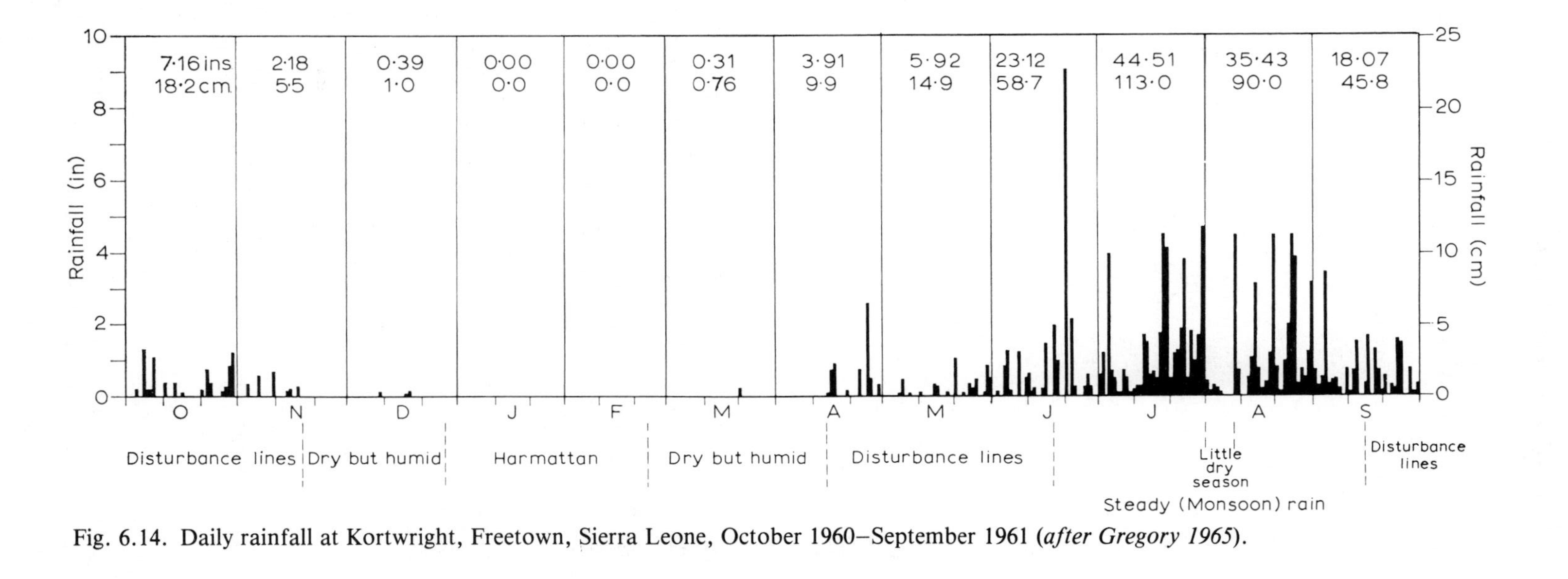

Fig. 6.14. Daily rainfall at Kortwright, Freetown, Sierra Leone, October 1960–September 1961 (*after Gregory 1965*).

Spring and autumn rainfall in west Africa is derived in large part from these disturbances. Figure 6.14 for Kortright (Freetown, Sierra Leone) illustrates the daily rainfall amounts in 1960–1 associated with disturbance lines at 8°N. The summer monsoon rains make up the greater part of the total here, but their contribution diminishes northward.

D The Asian monsoon

The name *monsoon* is derived from the Arabic word (*mausim*) which means season, so explaining its application to large-scale seasonal reversals of the wind regime. The Asiatic seasonal wind reversal is notable for its immense extent and the penetration of its influence beyond tropical latitudes. For example, the surface circulation over China reflects this seasonal change:

	January	*July*
North China	60% of winds from W, NW and N	57% of winds from SE, S and SW
South-east China	88% of winds from N, NE and E	56% of winds from SE, S and SW

However, such seasonal wind shifts at the surface are quite widespread and occur in many regions which would not traditionally be considered as monsoonal (fig. 6.15). Although there is a rough accordance between these traditional regions and those experiencing over 60% frequency of the prevailing octant, it is obvious that a variety of unconnected mechanisms can produce significant seasonal wind shifts. Nor is it possible to establish a simple relationship between seasonality of rainfall (fig. 6.16) and seasonal

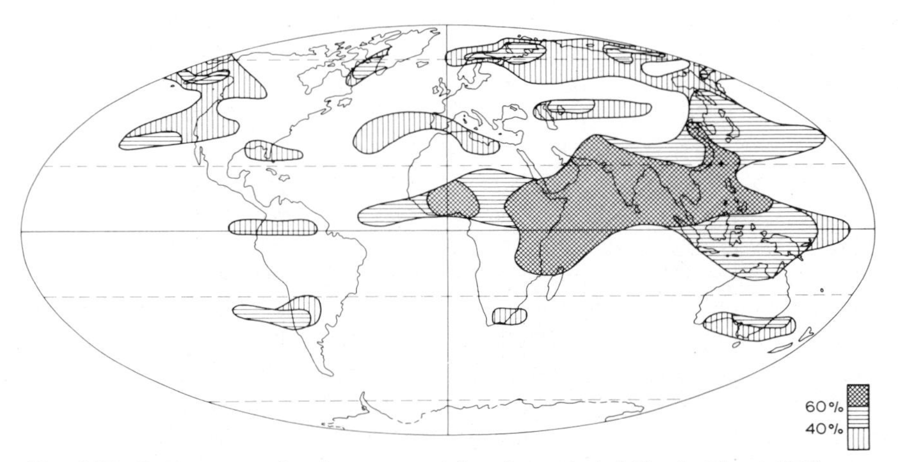

Fig. 6.15. Regions experiencing a seasonal surface wind shift of at least 120°, showing the frequency of the prevailing octant (*after Khromov*).

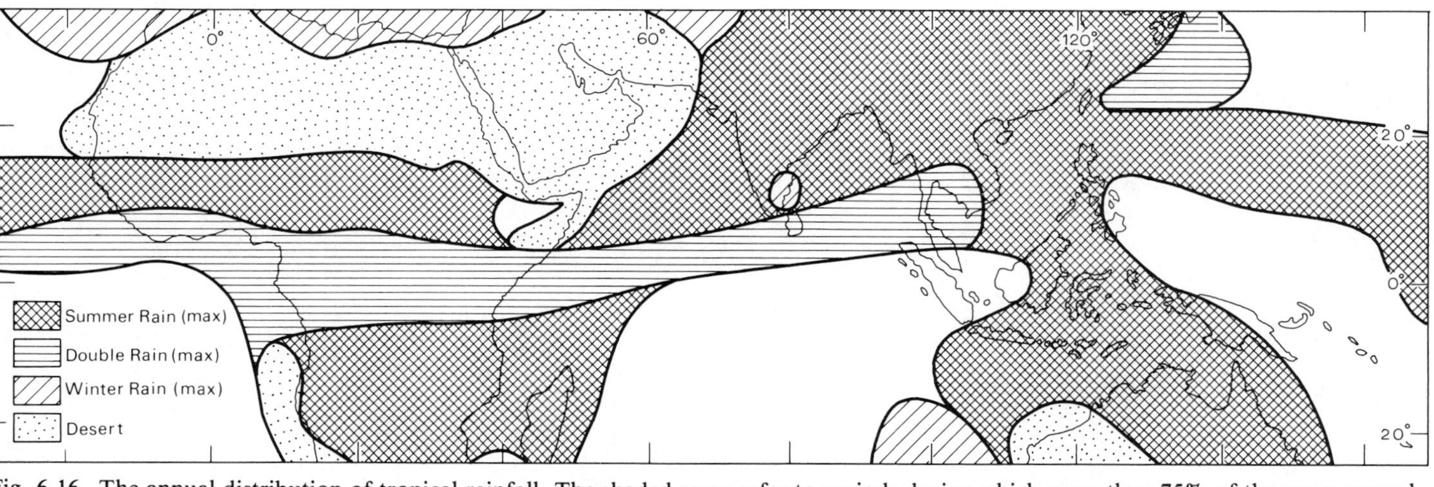

Fig. 6.16. The annual distribution of tropical rainfall. The shaded areas refer to periods during which more than 75% of the mean annual rainfall occurs. Areas with less than 250 mm y^{-1} (10 in y^{-1}) are classed as deserts, and the unshaded areas are those needing at least 7 months to accumulate 75% of the annual rainfall and are thus considered to exhibit no seasonal maximum (*after Ramage 1971*).

wind shift. Areas traditionally designated as 'monsoonal' include some of the tropical and near-tropical regions experiencing a summer rainfall maximum and most of those having a double rainfall maximum. It is clear that a combination of criteria (for example, from figs. 6.15 and 6.16) is necessary to approach an adequate definition of monsoonal areas.

In summer the Equatorial Trough and the subtropical anticyclones are everywhere displaced northwards in response to the changing pattern of solar heating of the earth, and in southern Asia this movement is magnified by the effects of the land mass. However, the attractive simplicity of the traditional explanation, which envisages a monsoonal 'sea breeze' directed towards a summer thermal low pressure over the continent, unfortunately provides an inadequate basis for understanding the workings of the system. The Asiatic monsoon regime is a consequence of the interaction of both planetary and regional factors, both at the surface and in the upper troposphere. It is convenient to look at each season in turn.

1 Winter

Near the surface this is the season of the outblowing 'winter monsoon', but aloft westerly airflow dominates. This, as we have seen, reflects the hemispheric pressure distribution. A shallow layer of cold, high-pressure air is centred over the continental interior, but this has disappeared even at 700 mb (see fig. 3.19) where there is a trough over eastern Asia and zonal circulation over the continent. The upper westerlies split into two currents to the north and south of the high Tibetan (Qinghai-Xizang) plateau (fig. 6.17), to reunite again off the east coast of China (fig. 6.18). The plateau, which exceeds 4000 m (13,000 ft) over a vast area, is a tropospheric cold source in winter, particularly over its western part, although the strength of this source depends on the extent and duration of snow cover (snow-free ground acts as a heat source for the atmosphere in *all* months). Below 600 mb, the tropospheric heat sink gives rise to a shallow, cold plateau anticyclone, which is best developed in December and January. The two jet-stream branches have been attributed to the disruptive effect of the topographic barrier on the airflow, but this is limited to altitudes below about 4 km. In fact, the northern jet is highly mobile and may be located far from the Tibetan plateau. Two currents are also found to occur farther west where there is no obstacle to the flow. The branch over northern India corresponds with a strong latitudinal thermal gradient (from November to April) and it is probable that this factor, combined with the thermal effect of the barrier to the north, is responsible for the anchoring of the southerly jet. This southern branch is the stronger, with an average speed of more than 40 m s^{-1} (90 mph) at 200 mb, compared with about 20–25 m s^{-1} (45–55 mph) in the northern one. Where the two unite over north China and south Japan the average velocity exceeds 66 m s^{-1} (148 mph) (fig. 6.19).

Air subsiding beneath this upper westerly current gives dry outblowing northerly winds from the subtropical anticyclone over north-western India

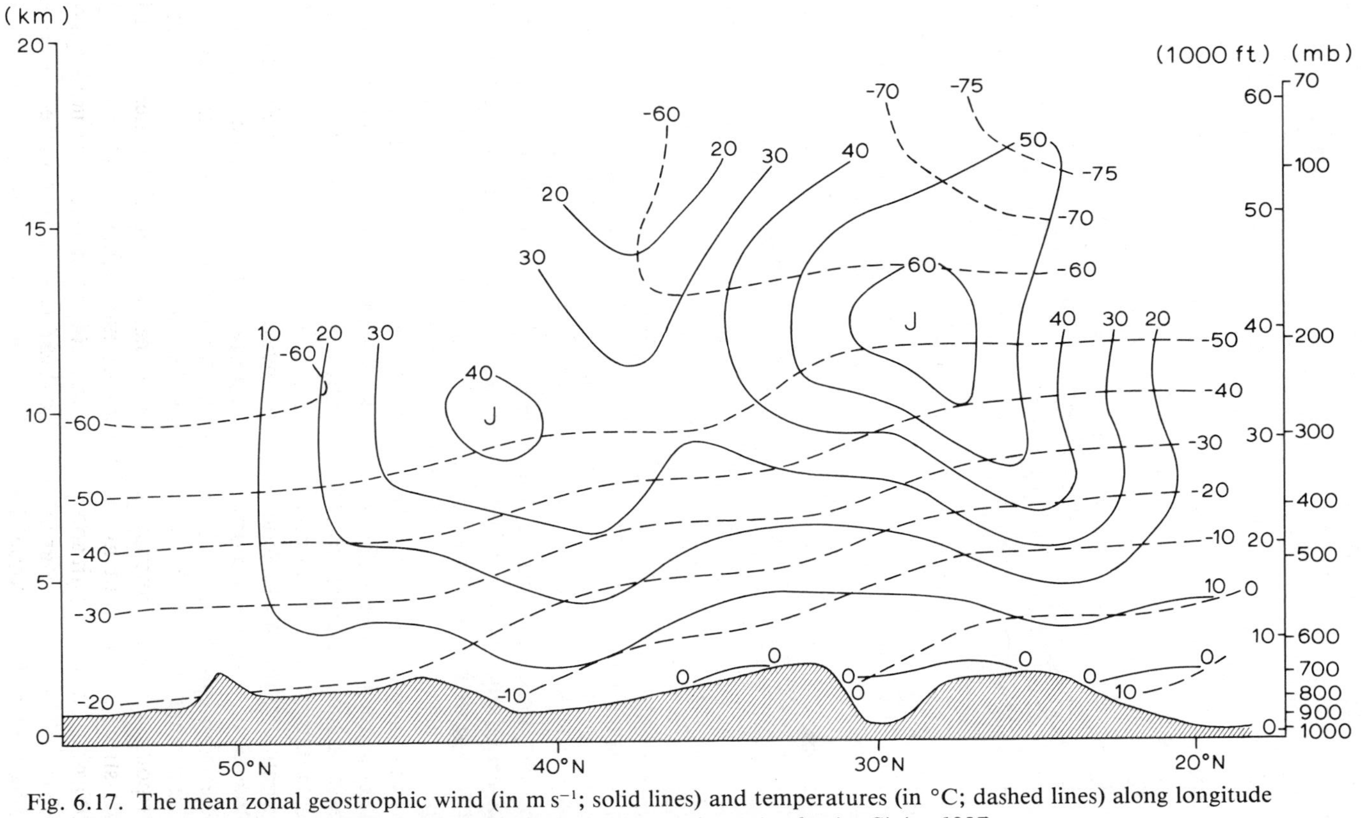

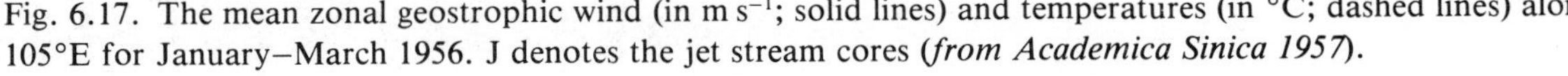
Fig. 6.17. The mean zonal geostrophic wind (in m s^{-1}; solid lines) and temperatures (in °C; dashed lines) along longitude 105°E for January–March 1956. J denotes the jet stream cores (*from Academica Sinica 1957*).

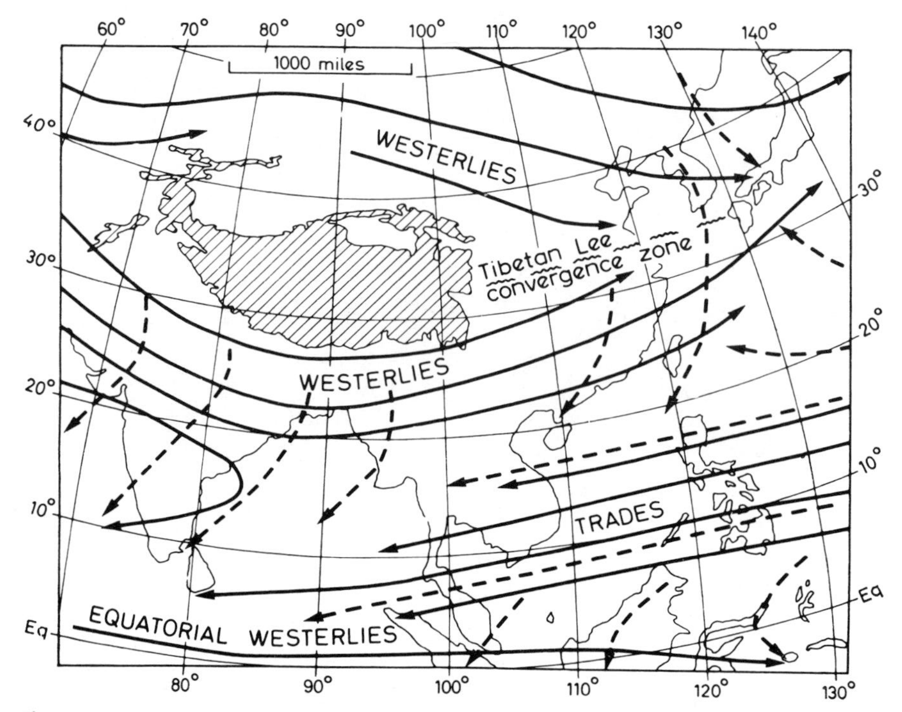

Fig. 6.18. The characteristic air circulation over southern and eastern Asia in winter (*after Thompson 1951; Flohn 1968, Frost and Stephenson 1965; and others*). Solid lines indicate airflow at about 3000 m (10,000 ft), and dashed lines that at about 600 m (2000 ft). The names refer to the wind systems aloft.

and Pakistan. The surface wind direction is north-westerly over most of northern India, becoming north-easterly over Burma and Bangladesh and easterly over peninsular India. Equally important is the steering of winter depressions over northern India by the upper jet. The lows, which are not usually frontal, appear to penetrate across the Middle East from the Mediterranean and are important sources of rainfall for northern India and Pakistan (e.g. Kalat, fig. 6.20), especially as it falls when evaporation is at a minimum.

Some of these westerly depressions continue eastwards, redeveloping in the zone of jet-stream confluence about 30°N, 105°E over China, beyond the area of subsidence in the immediate lee of Tibet (fig. 6.18), and it is significant that the mean axis of the winter jet stream over China shows a close correlation with the distribution of winter rainfall (fig. 6.21). Other depressions affecting central and north China travel within the westerlies north of Tibet or are initiated by outbreaks of fresh cP air. In the rear of these depressions there are invasions of very cold air (e.g. the *buran* blizzards of Mongolia and Manchuria). The effect of such cold waves, comparable with the *northers* in the central and southern United States, is greatly to reduce mean temperatures. Winter mean temperatures in less-protected southern

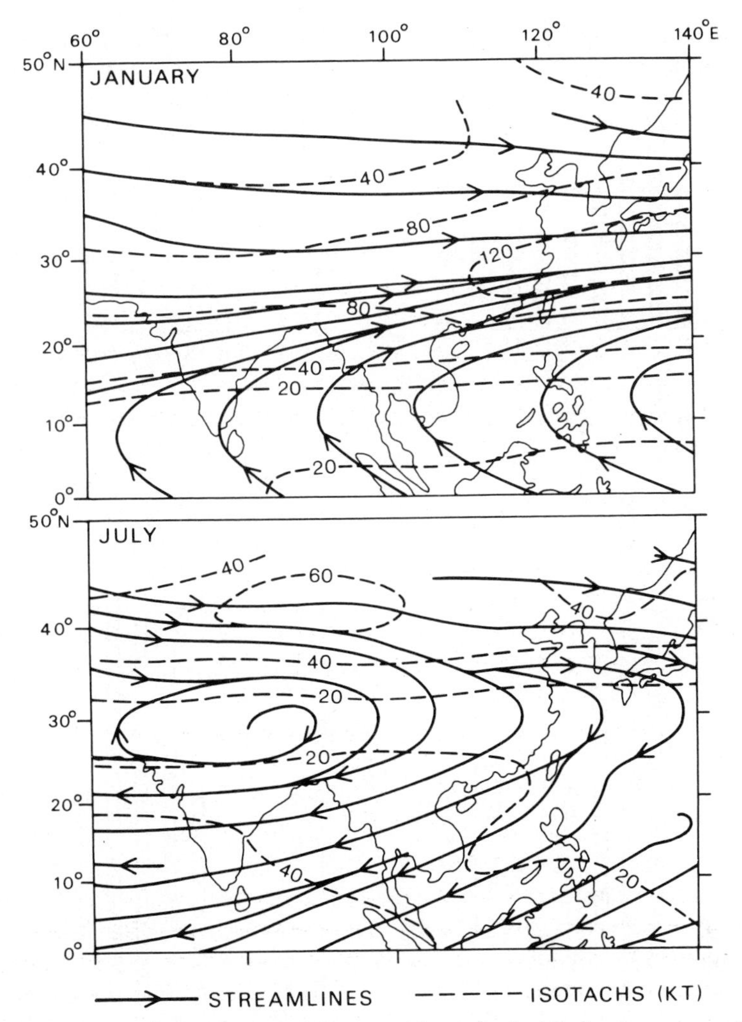

Fig. 6.19. Mean 200 mb streamlines (→) and isotachs (---) in knots over south-east Asia (A) January (B) July, based on aircraft reports and sounding data (*from Sadler 1975b*). (*Courtesy Dr J. C. Sadler, University of Hawaii*).

China are considerably below those at equivalent latitudes in India and, for example, temperatures at Calcutta and Hong Kong (both approximately 22½°N) are, respectively, 19°C (67°F) and 16°C (60°F) in January and 22°C (71°F) and 15°C (59°F) in February.

2 Spring

The key to change during this transition season is again found in the pattern of the upper airflow. In March the upper westerlies begin their seasonal migration northwards, but whereas the northerly jet strengthens and begins

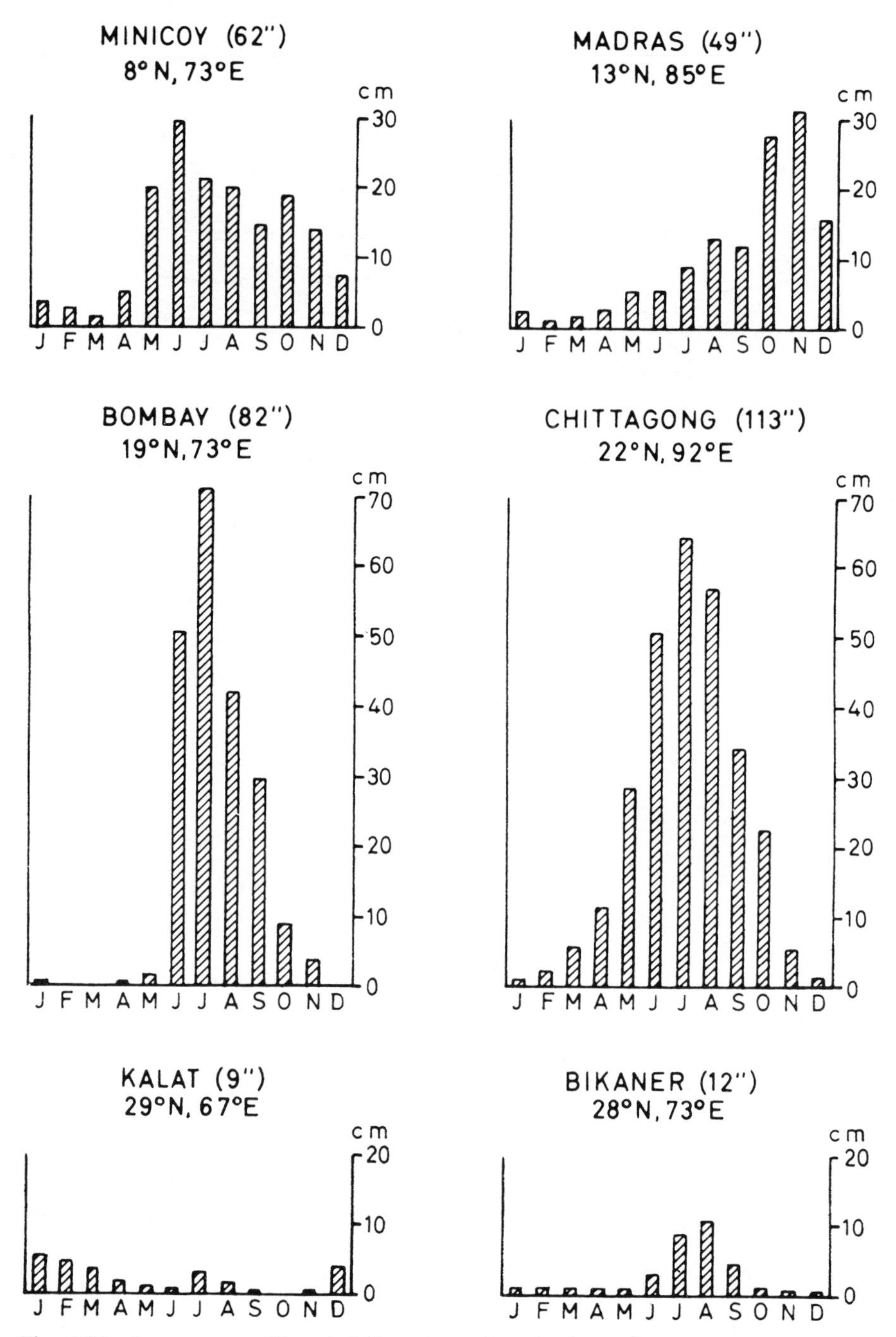

Fig. 6.20. Average monthly rainfall at six stations in the Indian region (*based on* 'CLIMAT' *normals of the World Meteorological Organization for 1931–60*). The annual total in inches is given after the station name (see App. 2 for the metric conversion).

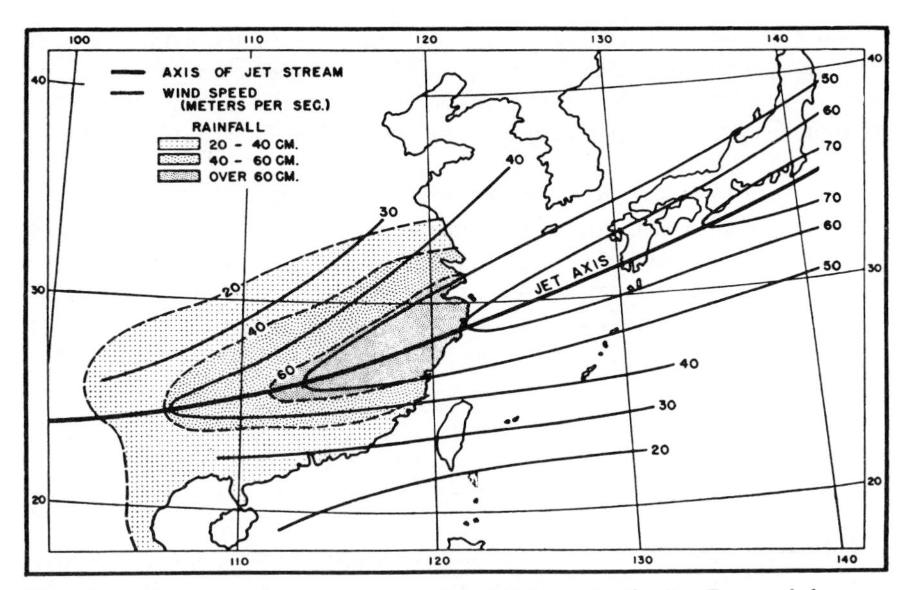

Fig. 6.21. The mean winter jet-stream axis at 12 km over the Far East and the mean winter precipitation over China in cm (*after Mohri and Yeh; from Trewartha 1958*).

to extend across central China and into Japan, the southerly branch remains positioned south of Tibet, although weakening in intensity.

The weather over northern India becomes hot, dry and squally in response to the greater solar radiation heating. Mean temperatures at Delhi rise from 23°C (74°F) in March to 33°C (92°F) in May. The thermal low-pressure cell (see ch. 4, G.2) reaches its maximum intensity at this time, but although onshore coastal winds develop the onset of the monsoon is still a month away and other mechanisms cause only limited precipitation. Some precipitation occurs in the north with 'westerly disturbances', particularly towards the Ganges delta where the low-level inflow of warm, humid air is overrun by dry, potentially cold air, triggering squall lines known as *nor'westers*. In the north-west, where less moisture is available the convection generates violent squalls and dust-storms termed *andhis*. The mechanism of these storms is not yet fully known, though high-level divergence in the waves of the subtropical westerly jet stream appears to be essential. The early onset of summer rains in Bengal, Bangladesh, Assam and Burma (e.g. Chittagong, fig. 6.20) is favoured by an orographically produced trough in the westerlies at 300 mb, which is located at about 85°–90°E in May. Low-level convergence of maritime air from the Bay of Bengal, combined with the upper-level divergence ahead of the 300 mb trough, generates thunder squalls. Tropical disturbances in the Bay of Bengal are another source of these early rains. Rain also falls during this season over Sri Lanka and south India (e.g. Minicoy, fig. 6.20) in response to the northward movement of the Equatorial Trough.

China has no equivalent of India's hot, pre-monsoon season. The low-level, north-easterly winter monsoon (reinforced by subsiding air from the upper westerlies) persists in north China and even in the south it only begins to be replaced by maritime tropical air in April–May. Thus, at Canton mean temperatures rise from only 17°C (63°F) in March to 27°C (80°F) in May, some 6°C (12°F) less than the mean values over northern India.

Westerly depressions are most frequent over China in spring. They form more readily over central Asia at this season as the continental anticyclone begins to weaken; also many develop in the jet-stream confluence zone in the lee of the plateau. The average number crossing China per month during 1921–31 was as follows:

J	*F*	*M*	*A*	*M*	*J*	*J*	*A*	*S*	*O*	*N*	*D*	*Year*
7	8	9	11	10	8	5	3	3	6	7	7	86

Hence spring is wetter than winter and, over most of central and southern China, the 3 months March–May contribute a quarter to a third of the annual rainfall.

3 Early summer

Generally during the last week in May the southern branch of the high-level jet begins to break down, becoming intermittent and then gradually shifting northward over the Tibetan plateau. At 500 mb and below, however, the plateau exerts a blocking effect on the flow and the jet axis there jumps from the south to north side of the plateau from May to June. Over India, the Equatorial Trough pushes northwards with each weakening of the upper westerlies south of Tibet, but the final *burst* of the monsoon, with the arrival of the humid, low-level south-westerlies, is not accomplished until the upper-air circulation has switched to its summer pattern (fig. 6.19 and 6.22). The low-level changes are related to the establishment of a high-level easterly jet stream over southern Asia about 15°N. One theory suggests that this takes place in June as the col between the subtropical anticyclone cells of the west Pacific and the Arabian Sea at the 300 mb level is displaced north-westwards from a position about 15°N, 95°E in May towards central India. The north-westward movement of the monsoon (fig. 6.23) is apparently related to the extension over India of the upper tropospheric easterlies.

The reorganization of the upper airflow has widespread effects in southern Asia. It is directly linked with the *Mai-yu* rains of China (which reach a peak about 10–15 June), the onset of the south-west Indian monsoon and the northerly retreat of the upper westerlies over the whole of the Middle East.

It must nevertheless be emphasized that it is still uncertain how far these changes are caused by events in the upper air or indeed whether the onset of the monsoon initiates a readjustment in the upper-air circulation. The presence of the Tibetan plateau is certainly of importance even if there is no

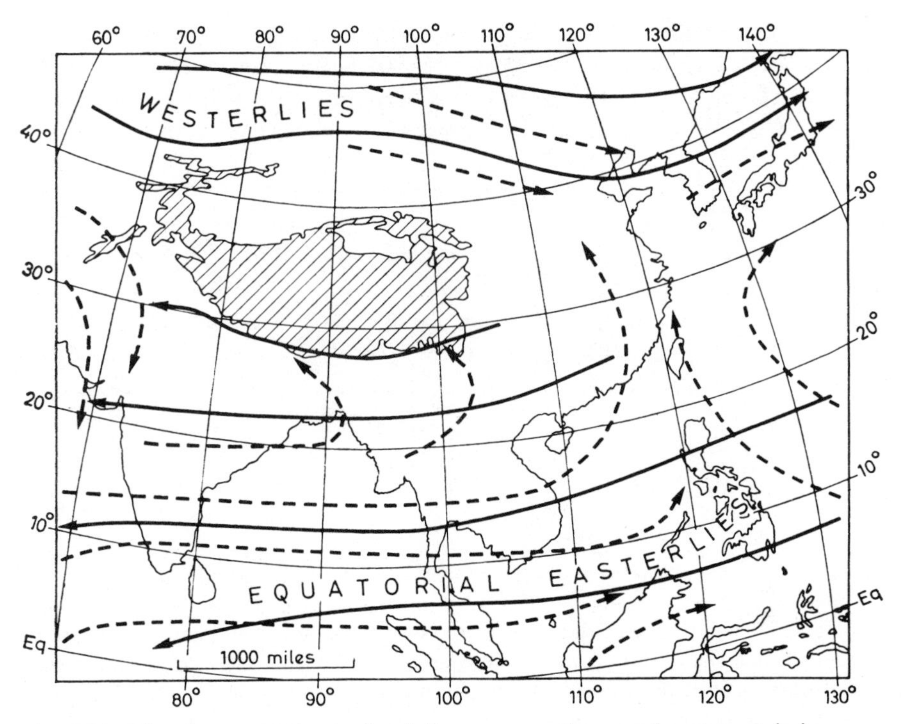

Fig. 6.22. The characteristic air circulation over southern and eastern Asia in summer (*after Thompson 1951; Flohn 1968; Frost and Stephenson 1965; and others*). Solid lines indicate air flow at about 6000 m (20,000 ft) and dashed lines at about 600 m (2000 ft). Note that the low-level flow is very uniform between about 600 and 3000 m.

significant barrier effect on the upper airflow. The plateau surface is strongly heated in spring and early summer (R_n is about 180 W m^{-2} in May) and nearly all of this is transferred via sensible heat to the atmosphere. This results in the formation of a shallow heat low on the plateau overlain, about 450 mb, by a warm anticyclone (see fig. 3.14). The plateau atmospheric boundary layer now extends over an area about twice the size of the plateau surface itself. Easterly airflow on the southern side of the upper anticyclone undoubtedly assists in the northward shift of the subtropical westerly jet stream. At the same time, the pre-monsoonal convective activity over the south-eastern rim of the plateau provides a further heat source, by latent heat release, for the upper anticyclone. The seasonal wind reversals over and around the Tibetan plateau have led Chinese meteorologists to distinguish a 'Plateau Monsoon' system, distinct from that over India.

Over China, the zonal westerlies retreat northwards in May–June and the westerly flow becomes concentrated north of the Tibetan plateau. The equatorial westerlies spread across south-east Asia from the Indian Ocean, giving a warm, humid air mass at least 3000 m deep, but the summer monsoon over southern China is apparently influenced less by the westerly

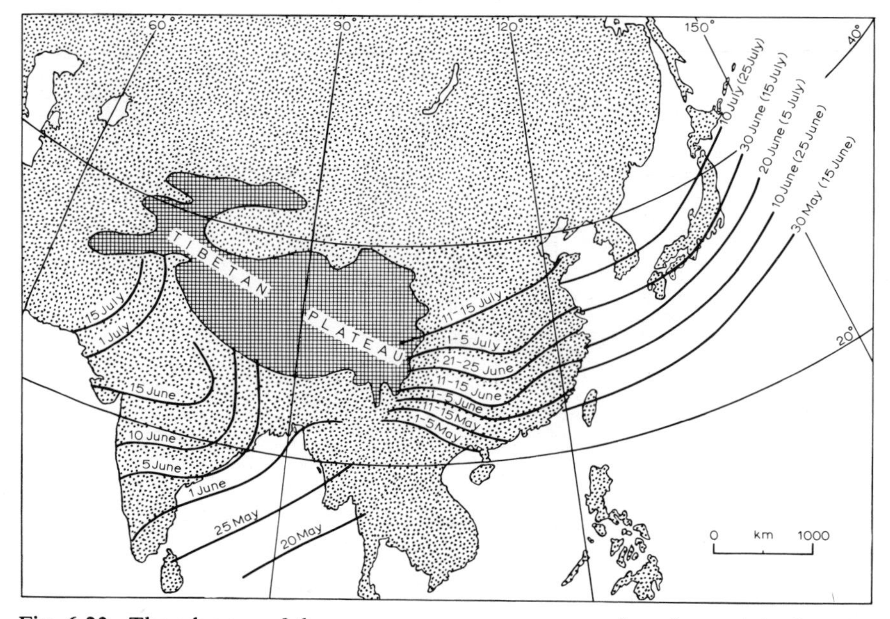

Fig. 6.23. The advance of the summer monsoon over south and east Asia, based on: for India, the beginning of the rainy season (*after Chatterjee*); for China, the northward shift of the 5-day mean wet-bulb temperature of 24°C (*after Tu and Hwang*); for Japan, the onset of the 'bai-u rains' (*after Takahashi and, in parentheses, after Kurashima 1968*).

flow over India than by southerly airflow over Indonesia near 100°E. Also contrary to earlier views, the Pacific is only a moisture source when tropical south-easterlies extend westwards to affect the east coast.

The Mai-yu 'front' involves both the monsoon trough and the East Asian–West Pacific Polar Front, with weak disturbances moving eastward along the Yangtze valley and occasional cold fronts from the north-west. Its location shifts northward in three stages, from south of the Yangtze River in early May to north of it by the end of the month and into northern China in mid-July (fig. 6.23), where it remains until late September.

4 Summer

By mid-July monsoon air covers most of southern and south-eastern Asia (fig. 6.22) and in India the Equatorial Trough is located about 25°N. North of the Tibetan plateau there is a rather weak upper westerly current with a (subtropical) high-pressure cell over the plateau. The south-west monsoon in southern Asia is overlain by strong upper easterlies (fig. 6.19) with a pronounced jet at 150 mb (about 15 km or 50,000 ft) which extends westwards across South Arabia and Africa (fig. 6.24). No easterly jets have so far been observed over the tropical Atlantic or Pacific. The jet is related to a steep lateral temperature gradient with the upper air getting progressively colder to the south.

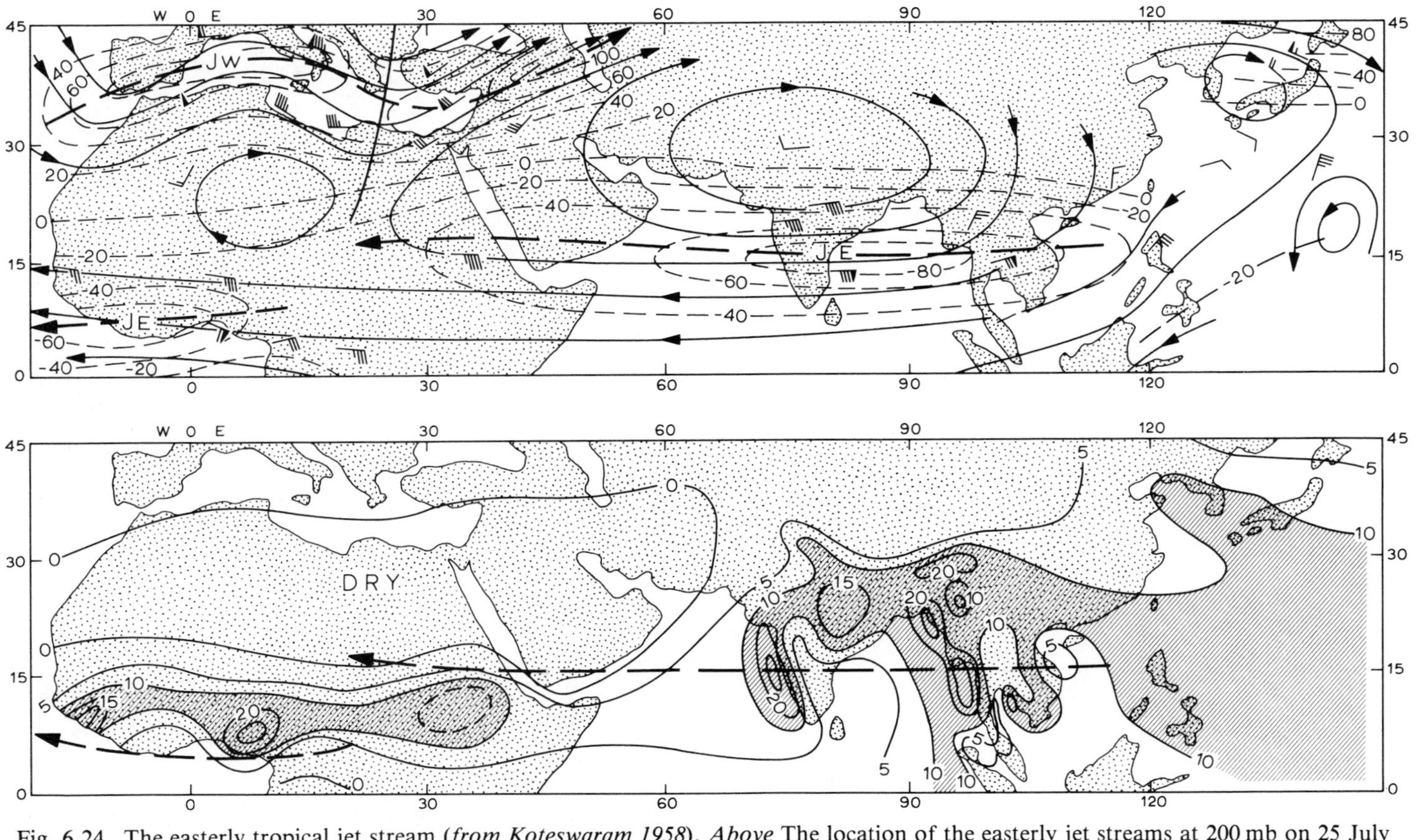

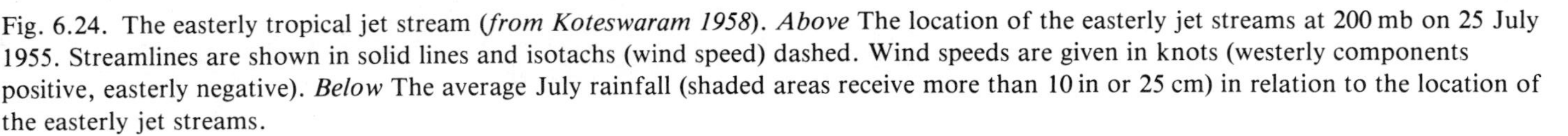
Fig. 6.24. The easterly tropical jet stream (*from Koteswaram 1958*). *Above* The location of the easterly jet streams at 200 mb on 25 July 1955. Streamlines are shown in solid lines and isotachs (wind speed) dashed. Wind speeds are given in knots (westerly components positive, easterly negative). *Below* The average July rainfall (shaded areas receive more than 10 in or 25 cm) in relation to the location of the easterly jet streams.

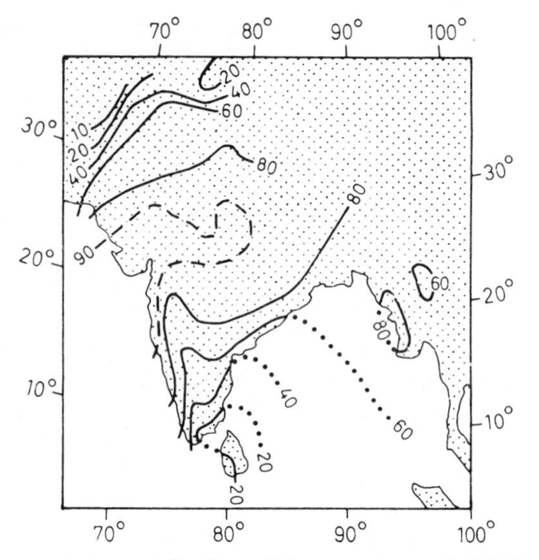

Fig. 6.25. The percentage contribution of the monsoon rainfall (June to September) to the annual total (*after Rao and Ramamoorthy 1960, in Indian Meteorological Department 1960; and Ananthakrishnan and Rajagopalachari 1964, in Hutchings 1964*).

An important characteristic of the tropical easterly jet is the location of the main belt of summer rainfall on the right (i.e. north) side of the axis upstream of the wind maximum and on the left side downstream, except for areas where the orographic effect is predominant (fig. 6.24). The mean jet maximum is located about 15°N, 50°–80°E.

The monsoon current does not give rise to a simple pattern of weather over India, despite the fact that much of the country receives 80% or more of its annual precipitation during the monsoon season (fig. 6.25). In the north-west a thin wedge of monsoon air is overlain by subsiding continental air. The inversion prevents convection and consequently little or no rain falls in the summer months in the arid north-west of the subcontinent (e.g. Bikaner and Kalat, fig. 6.20). This is similar to the Sahel zone in West Africa, discussed below.

Around the head of the Bay of Bengal and along the Ganges valley the main weather mechanisms in summer are the 'monsoon depressions' (see p. 292) which usually move westwards or north-westwards across India, steered by the upper easterlies. On average, they occur about twice a month, apparently when an upper trough becomes superimposed over a surface disturbance in the Bay of Bengal. Monsoon depressions are some 1000–1250 km across, with a cyclonic circulation up to about 8 km, and a typical life time of 2–5 days. They produce average daily rainfalls of 10–20 cm, occurring mainly as convective rains in the south-west quadrant of the depression (fig. 6.26). The main rain areas typically lie south of the Equatorial Trough (in the south-west quadrant of the monsoon depressions,

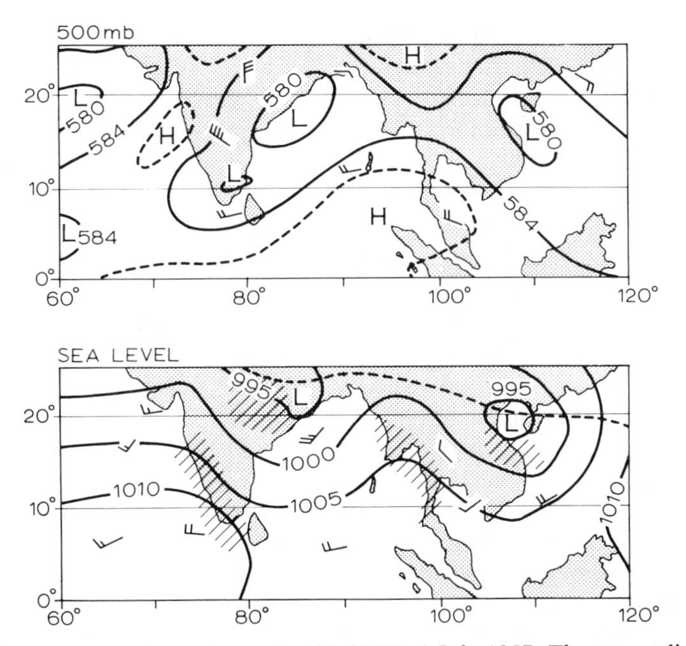

Fig. 6.26. Monsoon depressions of 1200 GMT, 4 July 1957. The upper diagram shows the height (in tens of metres) of the 500 mb surface, the lower one the sea-level isobars. The broken line in the lower diagram represents the Equatorial Trough, and precipitation areas are shown by the oblique shading (*based on the IGY charts of the Deutscher Wetterdienst*).

resembling an inverted mid-latitude depression), and also tend to occur on windward coasts and mountains of India, Burma and Malaya. Without such disturbances the distribution of monsoon rains would be controlled to a much larger degree by orography.

It has recently been discovered that part of the south-west monsoonal flow occurs in the form of a 15–45 m s^{-1} (30–100 mph) jet stream at a level of only 1000–1500 m (3250–5000 ft). This jet flows north-westward from Madagascar and crosses the equator from the south over east Africa where its core is often marked by a streak of cloud (similar to that in pl. 16) and where it may bring excessive local rainfall. It is then deflected by the East African plateau to the north-east across the Arabian Sea towards the west coast of the Indian peninsula. The south-westerly current over the Indian Ocean is quite dry near the equator, apart from a shallow moist layer at the surface. Moisture is acquired over the Arabian Sea, although even there an inversion indicates the presence of dry upper air originating perhaps over Arabia or East Africa and from subsidence due to the convergent upper easterlies. Convective instability is only released as the air slows down and converges at the coast and is forced over the Western Ghats. At Mangalore (13°N) there are on average 25 rain-days per month in June, 28 in July and 25 in August. The monthly rainfall averages are respectively 98 cm (38·6 in), 106 cm (41·7 in) and 58 cm (22·8 in), accounting for 75% of the annual total. In the lee of the Ghats amounts are much reduced and there are semi-arid areas receiving less than 64 cm (25 in) per year.

In southern India, excluding the south-east, there is a marked tendency for less rainfall when the Equatorial Trough is farthest north. Figure 6.20 shows a maximum at Minicoy in June with a secondary peak in October as the Equatorial Trough and its associated disturbances withdraw southward. This double peak occurs in much of interior peninsular India south of about 20°N and in western Sri Lanka, although autumn is the wettest period.

It is important to realize that the monsoon rains are highly variable from year to year, emphasizing the role played by disturbances in generating rainfall within the favourable environment created by the moist south-westerlies. *Breaks* occur in the monsoon rains when, during low-index periods, mid-latitude westerlies accompanied by the jet push southwards, weakening the Tibetan anticyclone or displacing it north-eastwards. The monsoon trough also shifts northward diminishing the rains over most of India. Westerly troughs travel along the southern edge of the Himalayas, giving heavy rain on the mountain slopes but little rain elsewhere. In part this may be due to the eastward extension across central India of the subtropical high over Arabia.

The strong surface heat source over the Tibetan plateau, which is most effective during the day, gives rise to a 50–85% frequency of cumulonimbus clouds over central and eastern Tibet in July. Late afternoon rain or hail showers are generally accompanied by thunder, but half or more of the precipitation falls at night, accounting for 70–80% of the total in south-central and south-eastern Tibet. This may be related to large-scale plateau-induced local wind systems. However, the central and eastern plateau also has a frequency maximum of shear lines and associated weak lows at 500 mb during May through September. These plateau systems are shallow (2–2.5 km) and only 400–1000 km in diameter, but they are associated with cloud clusters on satellite imagery.

To the east over China, the surface airflow is south-westerly and the upper winds are weak with only a diffuse easterly current over southern China. According to traditional views the monsoon current reaches northern China by July. The annual rainfall regime shows a distinct summer maximum with, for example, 64% of the annual total occurring at Tientsin (39°N) in July and August. Nevertheless, much of the rain falls during thunderstorms associated with shallow lows and the existence of the ITCZ in this region is doubtful (see fig. 6.1). The southerly winds, referred to above, which predominate over northern China in summer are not necessarily linked with the monsoon current farther south. Indeed this idea is the result of incorrect interpretation of streamline maps (of instantaneous airflow direction) as ones showing air trajectories (or the actual paths followed by air parcels). The depiction of the monsoon over China on fig. 6.21 is, in fact, based on a wet-bulb temperature value of 24°C. Cyclonic activity in northern China is attributable to the West Pacific Polar Front, forming between cP air and much-modified mT air.

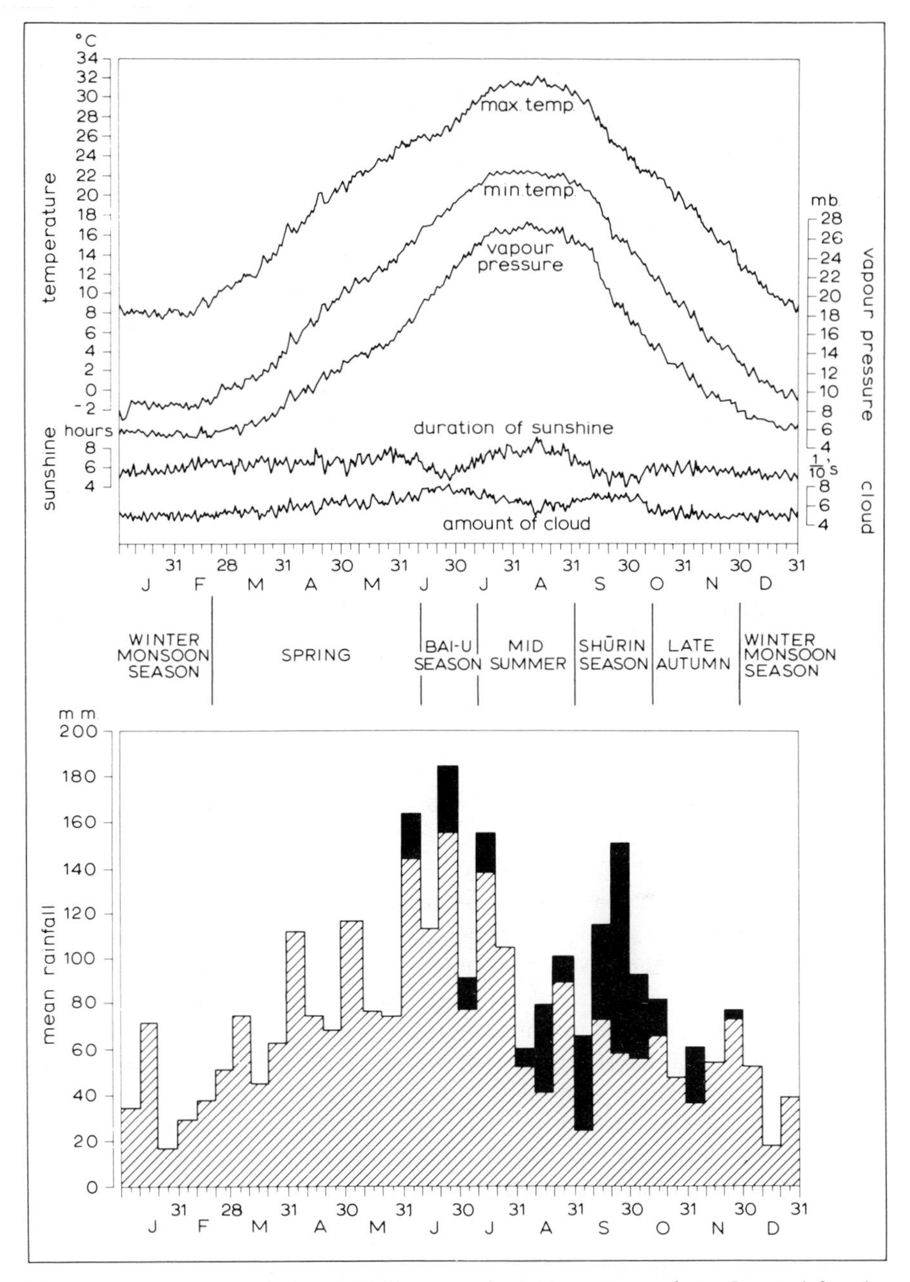

Fig. 6.27. Seasonal variation of daily normals at Nagoya, southern Japan (*above*), suggesting six natural seasons (*from Maejima 1967*). *Below* are average 10-day precipitation amounts for a station in southern Japan, indicating in black the proportion of rainfall produced by typhoon circulations. The latter reaches a maximum during the Shurin season (*after Saito 1959; from Trewartha 1981*).

In central and southern China the three summer months account for about 40–50% of the annual average precipitation, with another 30% or so being received in spring. In south-east China there is a rainfall singularity in the first half of July; a secondary minimum in the profile seems to result from the westward extension of the Pacific subtropical anticyclone over the coast of China.

A similar pattern of rainfall maxima occurs over southern and central Japan (fig. 6.27), comprising two of the six natural seasons which have been recognized there. The main rains occur during the *Bai-u* season of the south-east monsoon resulting from waves, convergence zones and closed circulations moving mainly in the tropical airstream round the Pacific subtropical anticyclone, but partly originating in a south-westerly stream which is the extension of the monsoon circulation of south Asia. The south-east circulation is displaced westwards from Japan by a zonal expansion of the subtropical anticyclone during late July and August giving a period of more settled sunny weather. The secondary precipitation maximum of the *Shurin* season during September and early October coincides with an eastward contraction of the Pacific subtropical anticyclone allowing low-pressure systems and typhoons from the Pacific to swing north towards Japan. Although much of the Shurin rainfall is believed to be of typhoon origin (fig. 6.27) some of it is undoubtedly associated with the southern sides of depressions, moving along the southward-migrating Pacific Polar Front to the north, because there is a marked tendency for the autumn rains to begin first in the north of Japan and to spread southwards.

5 Autumn

Autumn sees the southward retreat of the Equatorial Trough and the break-up of the summer circulation systems. By October the easterly trades of the Pacific affect the Bay of Bengal at the 500 mb level and generate disturbances at their confluence with the equatorial westerlies. This is the major season for Bay of Bengal cyclones and it is these disturbances, rather than the onshore northeasterly monsoon, which cause the October/November maximum of rainfall in south-east India (e.g. Madras, fig. 6.20).

During October the westerly jet re-establishes itself south of the Tibetan plateau, often within a few days, and cool season conditions are restored over most of southern and eastern Asia.

E The African monsoon

The annual climatic regime over West Africa shows important similarities to that over southern Asia, but also differences caused particularly by the geography of land–sea distribution and topography. In winter, the area is overlain by easterly trades whereas, in June, the northward shift of the Equatorial or Monsoon Trough (see fig. 6.2B) allows moist unstable and relatively cool south-westerly airflow to spread northward from the Gulf of

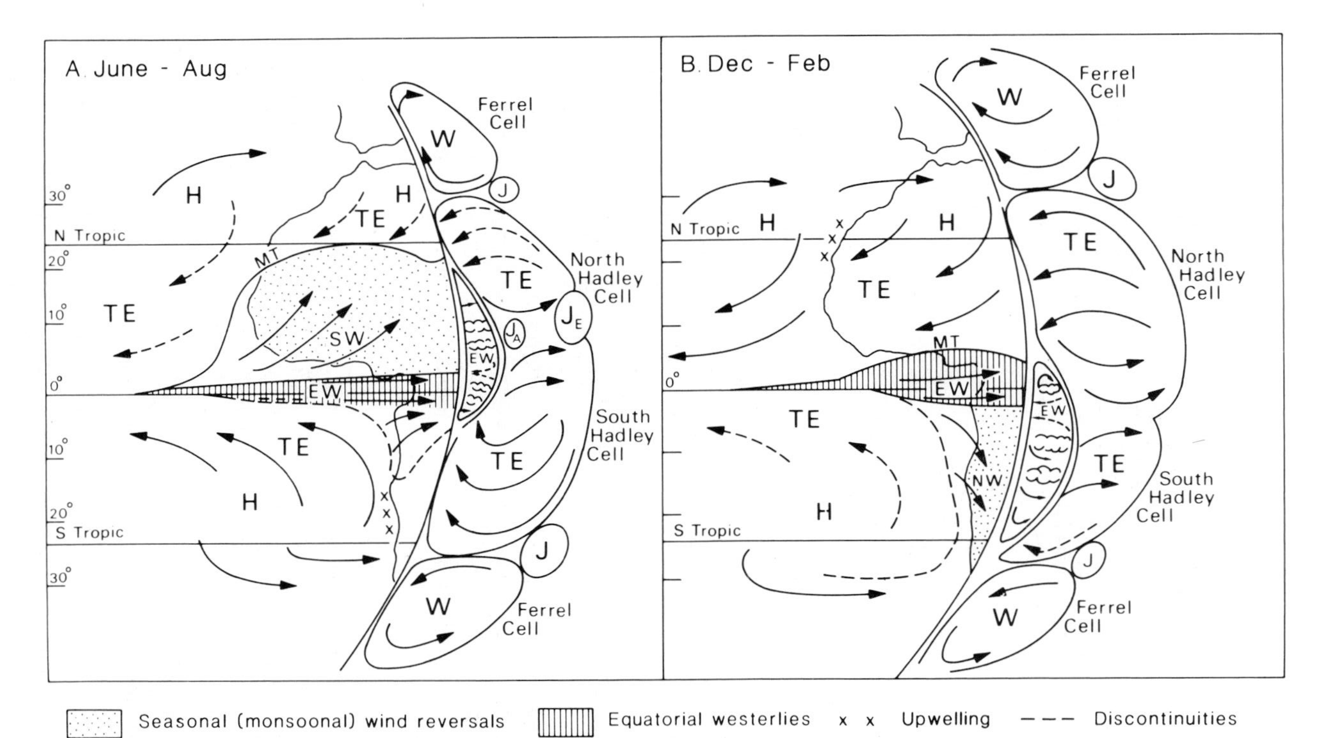

Fig. 6.28. The major circulation in Africa in (A) June–August and (B) December–February: H, subtropical high-pressure cells; EW, equatorial westerlies (moist, unstable but containing the Congo high-pressure ridge); NW, the north-westerlies (summer extension of EW in the southern hemisphere); TE, tropical easterlies (trades); SW, south-westerly monsoonal flow in the northern hemisphere; W, extratropical westerlies; J, subtropical westerly jet stream; J_A, the (easterly) African jet streams; and MT, monsoon trough (*from Rossignol-Strick 1985*). (*By permission Elsevier Science Publishers B.V., Amsterdam*).

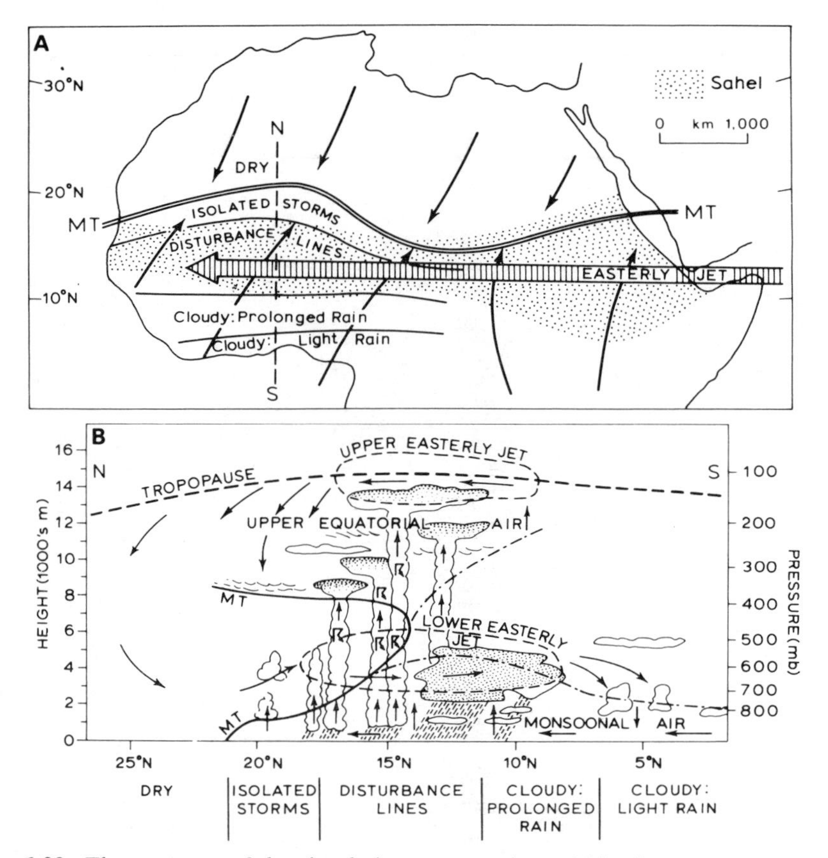

Fig. 6.29. The structure of the circulation over northern Africa in August: (A) surface airflow and Easterly Tropical Jet (*reproduction from the Geographical Magazine, London*); (B) vertical structure and resulting precipitation zones over West Africa; R = thunderstorm activity, MT = monsoon trough (*from Musk 1983* (*Copyright © Academic Press. Reproduced by permission*).

Guinea. At upper levels, there are tropical easterlies strengthening into the high-level Tropical Easterly Jet Stream with a regional band of wind maximum in the middle troposphere termed the African Easterly Jet Stream (fig. 6.28). At the most northerly advance of the trough in August, four climatic zones can be identified over West Africa (fig. 6.29):

1 A coastal belt of cloud and light rain related to frictional convergence within the monsoon flow, overlain here by subsiding easterlies.

2 A quasi-stationary zone of disturbances associated with deep stratiform cloud yielding prolonged light rains. Low-level convergence south of the jet axes, apparently associated with easterly wave disturbances from east central Africa, causes instability in the monsoon air.

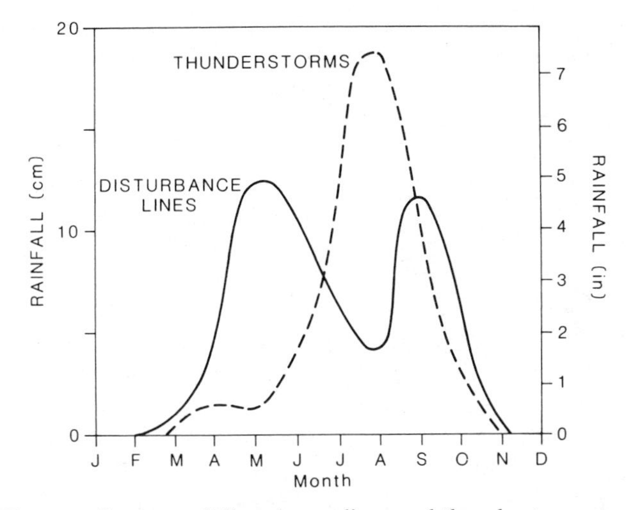

Fig. 6.30. The contributions of disturbance lines and thunderstorms to the average monthly precipitation at Minna, Nigeria (9.5°N) (*after Omotosho 1985*). (*By permission of the Royal Meteorological Society*).

3 A broad zone underlying the upper easterly jet streams which help activate disturbance lines (see p. 294) and thunderstorms. North-south lines of deep cumulonimbus cells move westward steered by the upper jets. The southern wetter part of this zone is termed the Soudan, the northern part the Sahel, but popular usage assigns the name Sahel to the whole belt.

4 Just south of the Monsoon Trough, the shallow tongue of humid air is overlain by drier subsiding air. There are only scattered showers and occasional thunderstorms.

Monsoon rains in the coastal zone of Nigeria (4°N) contribute 28% of the annual total (about 200 cm), thunderstorms 51% and disturbance lines 21%. At 10°N, 52% of the total (about 100 cm) is due to disturbance lines, 40% to thunderstorms and only 9% to the monsoon. Over most of the country, rainfall from disturbance lines has a double frequency maximum, thunderstorms a single one in summer (see fig. 6.30 for Minna, 9.5°N). In the northern parts of Nigeria and Ghana, rain falls in summer months, mostly from isolated storms or disturbance lines. The high variability of these rains from year to year characterizes the drought-prone Sahel environment.

The summer rainfall in the northern Soudano-Sahelian belts is determined partly by the northward penetration of the Monsoon Trough which may range up to 500–600 km beyond its average position (fig. 6.31) and by the strength of the easterly jet streams. The latter affects the frequency of disturbance lines.

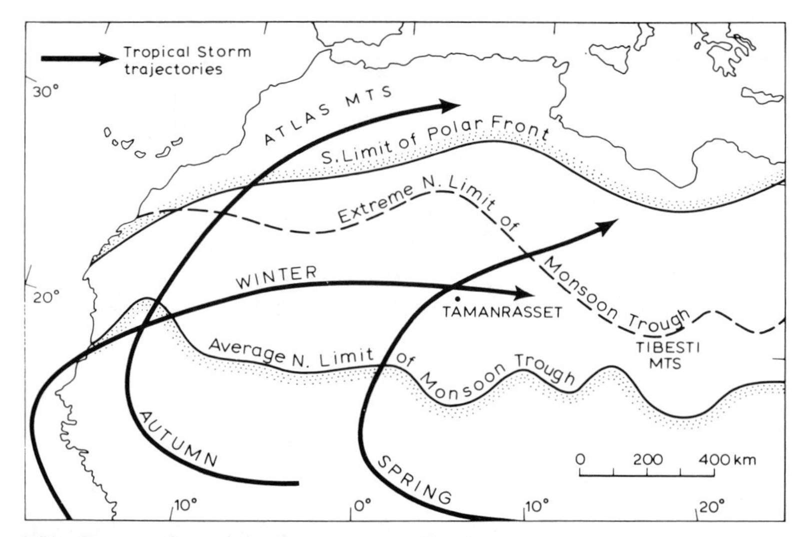

Fig. 6.31. Extent of precipitation systems affecting western and central North Africa and typical tracks of Soudano-Saharan depressions (*after Dubief and Yacono, from Barry 1981*).

F Other sources of climatic variations in the tropics

The major systems of tropical weather and climate have now been discussed, yet various other elements help to create contrasts in tropical weather both in space and time.

1 Diurnal variations

Diurnal weather variations are particularly evident at coastal locations in the trade wind belt and in the Indonesia–Malesian Archipelago. Land and sea breeze regimes (see ch. 3, C.3) are well developed, as the heating of tropical air over land can be up to five times that over adjacent water surfaces. The sea breeze normally sets in between 0800 and 1100, reaching a maximum velocity of 6–15 m s^{-1} (13–33 mph) about 1300 to 1600 and subsiding around 2000. It may be up to 1000–1200 m (about 3000–4000 ft) in height with a maximum velocity at an elevation of 200–400 m (650–1300 ft) and normally penetrates some 20–60 km (about 12–40 miles) inland.

In northern Australia, sea-breeze phenomena apparently extend up to 200 km inland from the Gulf of Carpenteria by late evening. During the August–November dry season, this may create suitable conditions for the bore-like 'Morning Glory' – a linear cloud roll and squall line which propagates, usually from the north-east, on the inversion created by the maritime air and nocturnal cooling. Sea breezes are usually associated with a heavy build-up of cumulus cloud and afternoon downpours. On large islands under calm conditions the sea breezes converge towards the centre so

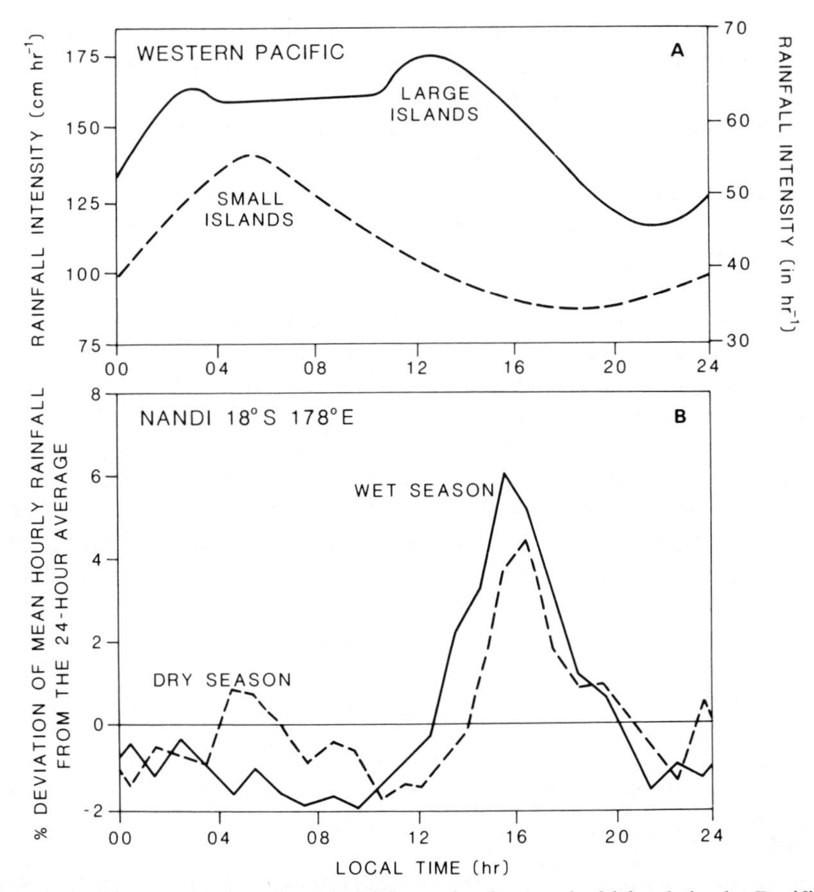

Fig. 6.32. Diurnal variation of rainfall intensity for tropical islands in the Pacific. A. Large and small islands in the western Pacific (*after Gray and Jacobson 1977*). B. Wet and dry seasons for Nandi (Fiji) in the south-west Pacific (percentage deviation from the daily average) (*after Finkelstein 1964, in Hutchings 1964*).

that an afternoon maximum of rainfall is observed. Under steady trade winds the pattern is displaced downwind so that descending air may be located over the centre of the island. A typical case of afternoon maximum is illustrated by fig. 6.32B for Nandi (Viti Levu, Fiji) in the south-west Pacific. The station has a lee exposure in both wet and dry seasons. This rainfall pattern is commonly believed to be widespread in the tropics, but over the open sea and on small islands a night-time maximum (often with a peak near dawn) seems to occur and even large islands can display this nocturnal regime when there is little synoptic activity. Figure 6.32A illustrates this nocturnal pattern at four small island locations in the western Pacific. Even large islands may show this effect, as well as the afternoon maximum associated with sea-breeze convergence and convection. There are several theories concerning the nocturnal rainfall peak. Recent studies point to a radiative effect, involving more effective nocturnal cooling of cloud-free areas around the meso-scale cloud systems. This favours subsidence which, in turn, enhances low-level convergence into the cloud systems and strengthens the ascending air currents. Strong cooling of cloud tops, relative to their surroundings, may also produce localized destabilization and encourage droplet growth by mixing of droplets at different temperatures (see ch. 2, G). This effect would be at a maximum near dawn. Another factor is that the sea–air temperature difference, and consequently the oceanic heat supply to the atmosphere, is largest about 0300–0600 hours. Yet a further hypothesis suggests that the semi-diurnal pressure oscillation encourages convergence and therefore convective activity in the early morning and evening, but divergence and suppression of convection around midday.

The Malayan peninsula displays very varied diurnal rainfall regimes in summer. The effects of land and sea breezes, anabatic and katabatic winds and topography greatly complicate the rainfall pattern by their interactions with the low-level south-westerly monsoon current. For example, there is a nocturnal maximum in the Malacca Straits region associated with the convection set off by the convergence of land breezes from Malaya and Sumatra (cf. p. 294), whereas on the east coast of Malaya the maximum occurs in the late afternoon-early evening when sea breezes extend about 30 km inland against the monsoon south-westerlies, and convective cloud develops in the deeper sea breeze current over the coast strip. On the interior mountains the summer rains have an afternoon maximum due to the unhindered convection process.

2 Topographic effects

Relief and surface configuration have a marked effect on rainfall amounts in tropical regions where hot, humid air masses are frequent. On the south-western slope of Mount Cameroon, Debundscha (9 m elevation) receives 1116 cm yr^{-1} on average (1960–80) from the south-westerly monsoon. In the Hawaiian Islands the mean annual total exceeds 760 cm (300 in) on the mountains, with one of the world's largest mean annual totals of 1199 cm (472 in) at 1569 elevation on Mt Waialeale (Kauai), but land on the lee side suffers correspondingly accentuated sheltering effects with less than 50 cm (20 in) over wide areas. On Hawaii itself the maximum falls on the eastern slopes at about 900 m, whereas the 4200 m summits of Mauna Loa and Mauna Kea, which rise above the trade wind inversion, receive only 25–50 cm (10–20 in). On the Hawaiian island of Oahu, the maximum precipitation occurs on the western slopes, just leeward of the 850 m (2785 ft) summit with respect to the easterly trade winds. The following measurements in the Koolau Mountains, Oahu, show that the orographic factor is pronounced during summer when precipitation is associated with the easterlies, but in winter when precipitation is from cyclonic disturbances it is more evenly distributed:

Location	*Elevation*		*Source of rainfall*		
	metres	*feet*	*Trade winds*	*Cyclonic disturbances*	
			28 May–3 Sept. 1957	*2–28 Jan. 1957*	*5–6 Mar. 1957*
Summit	850	(2785)	71·3 cm (28·09 in)	49·9 cm (19·63 in)	32·9 cm (12·94 in)
760 m (2500 ft) west of summit	625	(2050)	121·0 cm (47·64 in)	54·4 cm (21·41 in)	37·0 cm (14·56 in)
7600 m (25,000 ft) west of summit	350	(1150)	39·9 cm (12·97 in)	46·7 cm (18·38 in)	33·4 cm (13·16 in)

(After Mink 1960)

The Khasi Hills in Assam are an exceptional instance of the combined effect of relief and surface configuration. Part of the monsoon current from the head of the Bay of Bengal (fig. 6.22) is channelled by the topography towards the high ground and the sharp ascent, which follows the convergence of the airstream in the funnel-shaped lowland to the south, results in some of the heaviest annual rainfall totals recorded anywhere. Mawsyuran (1400 m elevation), 16 km west of the more famous station of Cherrapunji, has a mean annual total (1941–69) of 1221 cm (481 in) and can claim to be the wettest spot in the world. Cherrapunji (1340 m) during the same period averaged 1102 cm; extremes recorded there include 569 cm (224 in) in July and 2440 cm (959 in) in 1974 (see fig. 2.24). However, throughout the monsoon area, topography plays a secondary role in determining rainfall distribution to the synoptic activity and large-scale dynamics.

Really high relief produces major changes in the main weather characteristics and is best treated as a special climatic type. The Kenya plateau, situated on the equator, has an average elevation of about 1500 m, above which rise the three volcanic peaks of Mt Kilimanjaro (5800 m), Mt Kenya (5200 m) and Ruwenzori (5200 m) nourishing permanent glaciers above 4270 m. Annual precipitation on the summit of Mt Kenya is about 114 cm (44 in), similar to amounts on the plateau to the south, but on the southern slopes between 2100 and 3000 m, and on the eastern slopes between about 1400 and 2400 m totals exceed 250 cm (100 in). Kabete (at an elevation of 1800 m near Nairobi) exhibits many of the features of tropical mountain climates, having a small annual temperature range (mean monthly temperatures are 19°C (67°F) for February and 16°C (60°F) for July), a high diurnal temperature range (averaging 9·5°C (17°F) in July and 13°C (24°F) in February) and a large average cloud cover (mean 7–8/10ths).

3 Cool ocean currents

Between the western coasts of the continents and the eastern rims of the subtropical high-pressure cells the ocean surface is relatively cold (see fig. 3.38). This is the result of the importation of water from higher latitudes by the dominant currents and the slow upwelling (sometimes at the rate of about 1 m in 24 hours) of water from intermediate depths due to the Ekman effect (ch. 3, F.3) and to the coastal divergence (ch. 3, B.1). This concentration of cold water gently cools the local air to dew point. As a result, dry warm air degenerates into a relatively cool, clammy, foggy atmosphere with a comparatively low temperature and little range along the west coast of North America off California (pl. 36), off South America between latitudes 4°S and 31°S, and off south-west Africa (8°S and 32°S). Thus Callao, on the Peruvian coast, has a mean annual temperature of 19·4°C (67°F), whereas Bahia (at the same latitude on the Brazilian coast) has a corresponding figure of 25°C (77°F).

The temperature effect of offshore cold currents is not limited to coastal stations as it is carried inland during the day at all times of the year by a pronounced sea-breeze effect (ch. 3, C.3). Along the west coasts of South America and south-west Africa the sheltering effect from the dynamically stable easterly trades aloft provided by the nearby Andes and Namib Escarpment, respectively, allows incursions of shallow tongues of cold air to roll in from the south-west. These tongues of air are capped by strong inversions at between 600 and 1500 m (2000 and 5000 ft) reinforcing the regionally low trade-wind inversions (see fig. 6.6) and thereby precluding the development of strong convective cells, except where there is orographically forced ascent. Thus, although the cool maritime air perpetually bathes the lower western slopes of the Andes in mist and low stratus cloud and Swakopmund (South-West Africa) has an average of 150 foggy days a year, little rain falls on the coastal lowlands. Lima (Peru) has a total mean annual precipitation of only 4·6 cm (1·8 in), although it receives frequent drizzle during the winter months of June to September, and Swakopmund in Namibia has a mean annual rainfall of 1·6 cm (0·65 in). Heavier rain occurs on the rare instances when large-scale pressure changes cause a cessation of the diurnal sea breeze or when modified air from the South Atlantic or South Indian Ocean is able to cross the continents at a time when the normal dynamic stability of the trade-wind is disturbed. In south-west Africa the inversion is most likely to break down during either October or April allowing convectional storms to form, and Swakopmund recorded some 5·1 cm (2 in) of rain on a single day in 1934. Under normal conditions, however, the occurrence of precipitation is mainly limited to the higher seaward mountain slopes. From central Colombia to northern Peru the diurnal tide of cold air rolls inland for some 60 km (38 miles), rising up the seaward slopes of the Western Cordillera and overflowing into the longitudinal Andean valleys like water over a weir (fig. 6.33). Plate 37 illustrates a similar overflow, giving rise to a downstream hydraulic jump where the flow decelerates, in the lee of mountains in

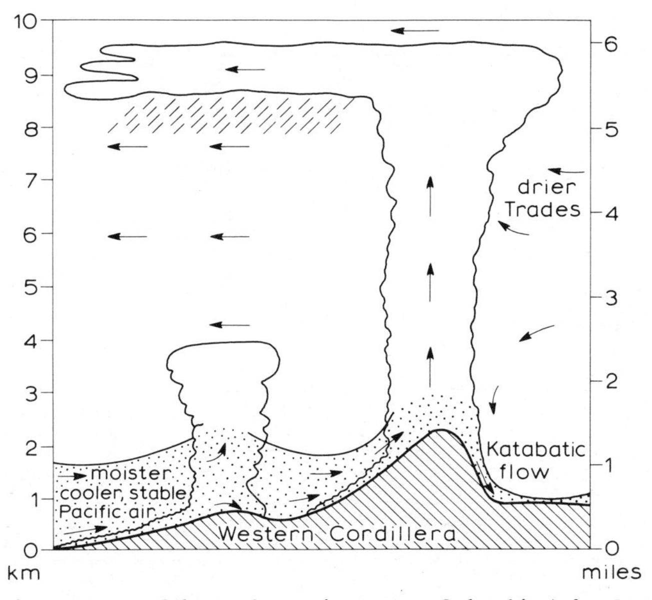

Fig. 6.33. The structure of the sea breeze in western Colombia (*after Lopez; from Fairbridge 1967*).

Wyoming. On the west-facing slopes of the Andes of Colombia, air ascending or banked-up against the mountains, may under suitable conditions trigger off convectional instability in the overlying trades and produce thunderstorms. In south-west Africa, however, the 'tide' flows inland for some 130 km (80 miles) and rises up the 1800 m (6000 ft) Namib Escarpment without producing much rain because convectional instability is not produced and the adiabatic cooling of the air is more than offset by the radiational heating from the warm ground.

4 El Niño–Southern Oscillation (ENSO) events

Each year, usually starting in December, a weak southward flow of warm water to about 6°S along the coast of Ecuador replaces the northward Peru Current and its associated cold upwelling (fig. 3.38). The phenomenon is known as El Niño (the child), after the Christ child. At irregular intervals, of 2 to 10 years, this warm water becomes much more extensive and the coastal upwelling ceases entirely, with catastrophic consequences. The absence of the cold water and its nutrients leads to massive mortality of fish, and the birdlife which feeds on them, causing economic disaster for the fishing and guano industries of Ecuador, Peru and northern Chile. Major El Niños occurred in 1925–6, 1939–41, 1957–8, 1972–3 and 1982–3, but in between are less severe events giving an average recurrence interval of three years.

This phenomenon forms part of large-scale 1–4°C warming of the ocean surface in the eastern equatorial Pacific. It is associated with a decrease in

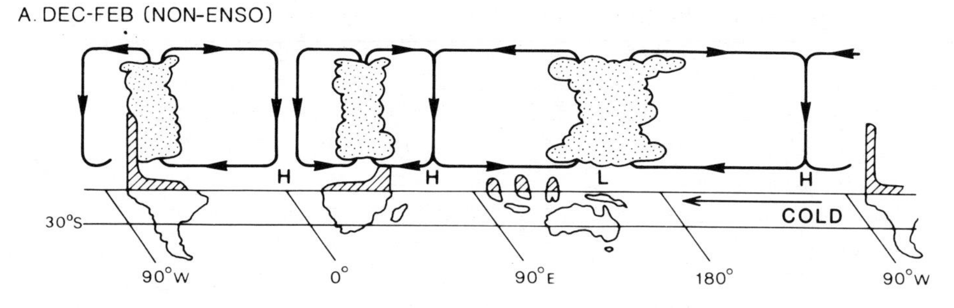

B. DEC-FEB 1982-3 (ENSO)

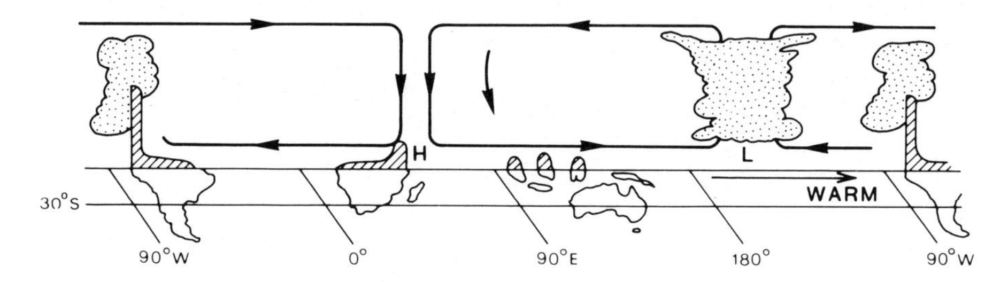

C. DEC-FEB SEA SURFACE TEMPERATURES (NON-ENSO)

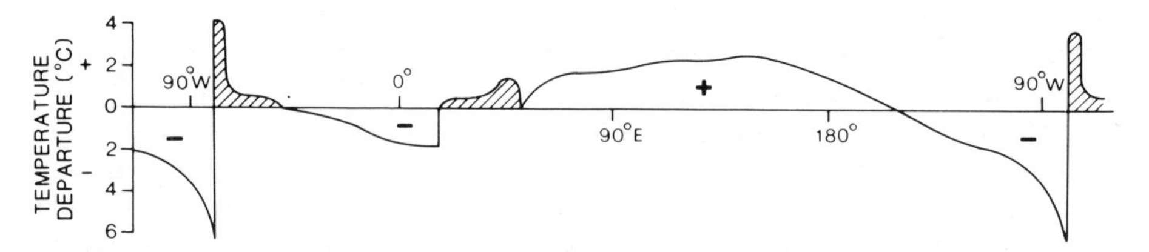

D. STRONG EASTERLY WIND (NON-ENSO) E. WEAK EASTERLY WIND (ENSO)

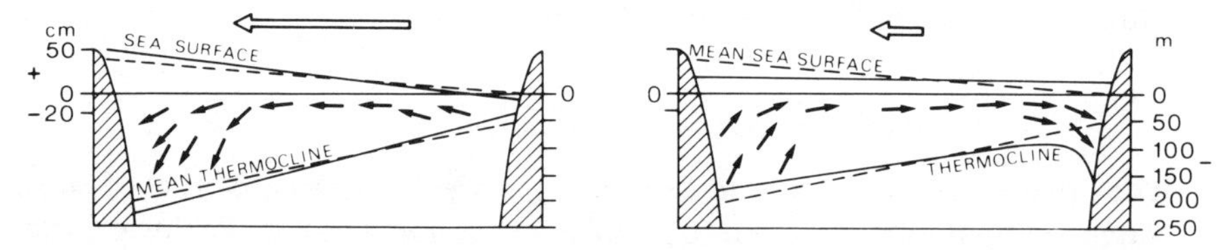

Fig. 6.34. Schematic cross-sections of the Walker Circulation along the equator based on computations of Y. Tourre (1984). A Mean December–February regime (non-ENSO); rising air and heavy rains occur over the Amazon basin, central Africa and Indonesia–western Pacific. B December–February 1982–3 ENSO pattern; the ascending Pacific branch is shifted east of the Date Line and suppressed convection occurs elsewhere due to subsidence. C Departure of sea-surface temperature from its equatorial zonal mean, corresponding to non-ENSO case (A). D Strong trades cause sea-level to rise and the thermocline to deepen in the western Pacific for case (A). E Winds relax, sea-level rises in the eastern Pacific as water mass moves back eastward and the thermocline deepens off South America during ENSO events (*based on Wyrtki 1982; by permission World Meteorological Organization 1985*).

the normal atmospheric pressure gradient between the subtropical high-pressure cell in the eastern South Pacific and low pressure over the Indonesia region (fig. 6.34). Fluctuations in this pressure gradient were first described in the 1920s by Sir Gilbert Walker (after whom the large-scale east–west vertical 'Walker circulation' cells are named, see p. 153. They are now referred to as the Southern Oscillation. The theory outlined in fig. 6.34 proposes that the normal (non-ENSO) trades push water offshore, allowing coastal upwelling, and raise sea-level in the western South Pacific. As the South Pacific high pressure weakens and the south-east trades slacken, this weakens the wind-driven upwelling allowing the ITC and its associated rainfall to extend southward to Peru. In fact, strong trades did not precede the pronounced 1982–3 ENSO. Equatorial westerlies in the western Pacific that develop during the low phase of the Southern Oscillation (perhaps in association with cyclone pairs north and south of the equator) create an eastward flow of warm water as large-scale, internal oceanic (Kelvin) waves. This depresses the thermocline off South America (fig. 6.34E) preventing cold water from reaching the surface.

These closely related oceanic and atmospheric phenomena, together referred to as ENSO events, appear to give rise to anomalous weather regimes not only in the southern hemisphere, but also further afield. In the northern hemisphere winter, ENSO events with equatorial heating anomalies are associated with a strong trough and ridge teleconnection pattern, known as the Pacific–North American (PNA) pattern (fig. 6.35), which may bring cloud and rain to the south-west United States and north-west Mexico.

5 Disturbances within continental subtropical high-pressure belts

A major source of climatic variation in the subtropical margins is the infrequent occurrence of low-pressure disturbances which disrupt the continental subtropical high-pressure cells. The two most significant of these are the Saharan and Australian cells, which strengthen in winter and are weakened by low-level thermal activity in summer.

The dominance of high-pressure conditions in the Sahara is marked by the low average precipitation figures for this region. Over most of the central Sahara the mean annual precipitation is less than 25 mm, except for the high plateaus of the Ahaggar and Tibesti, which receive more than 100 mm. Parts of western Algeria have gone at least 2 years without more than 0·1 mm of rain in any 24-hour period, and most of south-west Egypt as much as 5 years. However, 24-hour storm rainfalls approaching 50 mm (more than 75 mm over the high plateaus) may be expected in scattered localities and, during a 35-year period of record, excessive short-period rainfall intensities occurred in the vicinity of west-facing slopes in Algeria, such as at Tamanrasset (46 mm in 63 minutes) (Fig. 6.36), El Golea (8·7 mm in 3 minutes) and Beni Abbes (38·5 mm in 25 minutes). During the summer, rainfall variability is introduced into the southern Sahara by the variable northward penetration

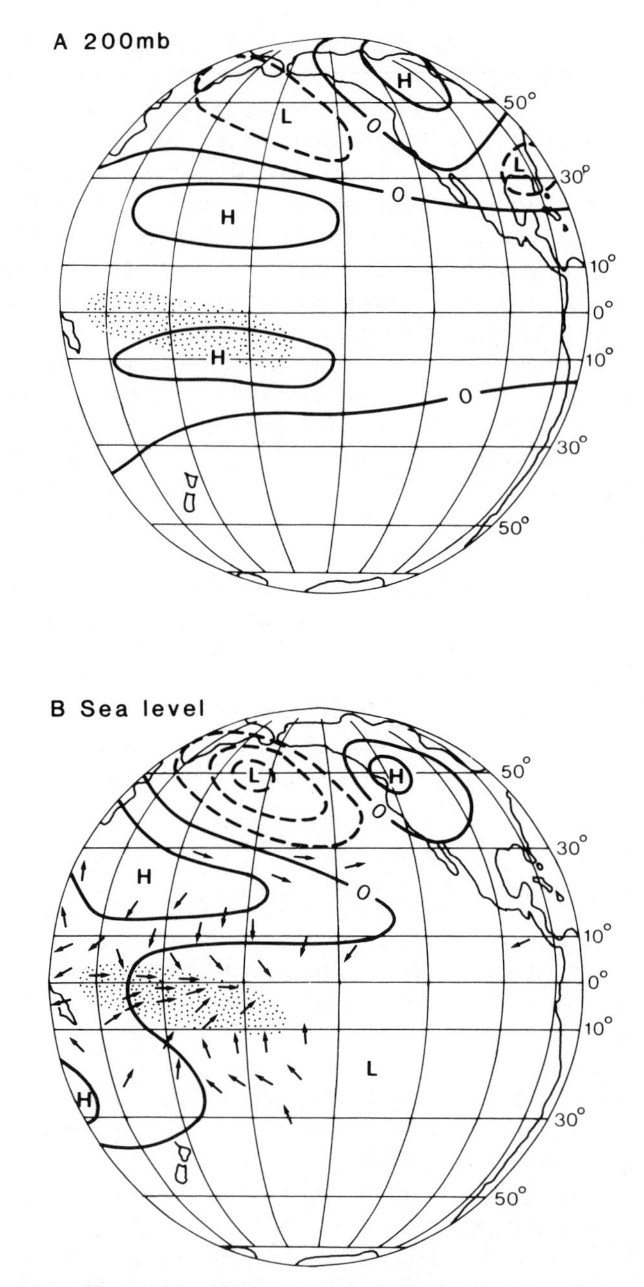

Fig. 6.35. Schematic Pacific–North America (PNA) circulation pattern in the upper troposphere during an ENSO event in December–February. The shading indicates a region of enhanced rainfall associated with anomalous westerly surface wind convergence in the equatorial western Pacific (*after Shukla and Wallace 1983*).

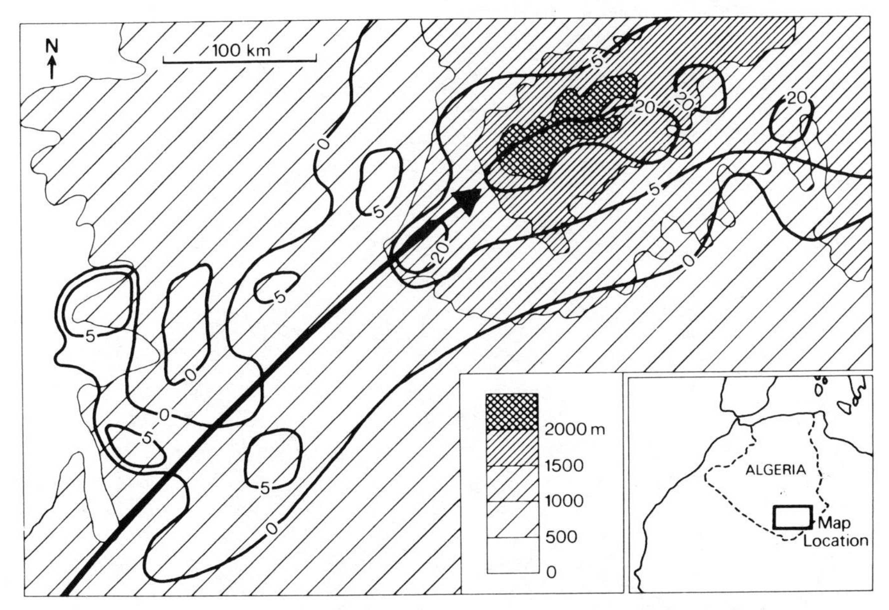

Fig. 6.36. Track of a storm and the associated 3-hour rainfall (mm) during September 1950 around Tamanrasset in the vicinity of the Ahaggar Mountains, southern Algeria (*partly after Goudie and Wilkinson 1977*).

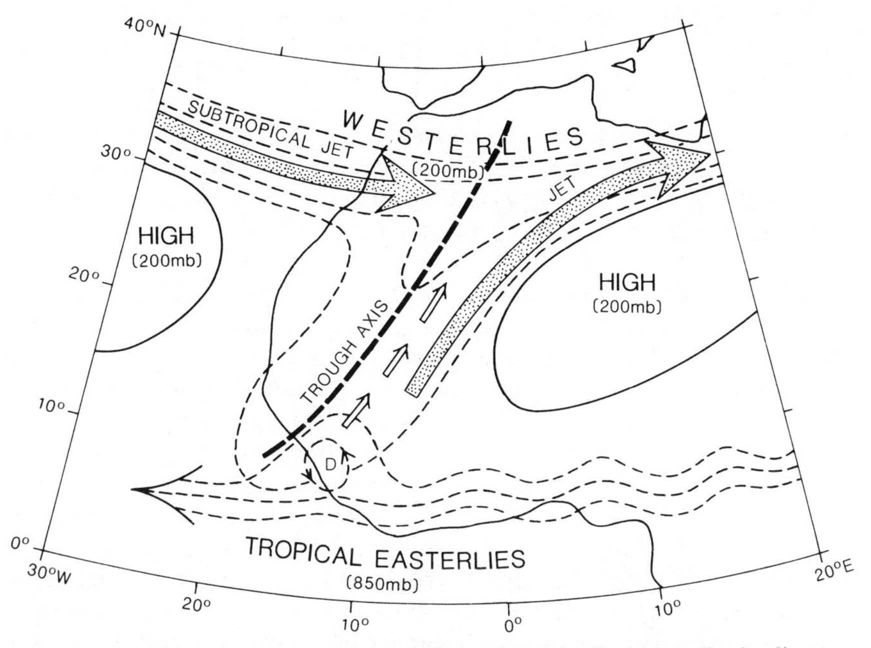

Fig. 6.37. Interaction between the westerlies and the tropical easterlies leading to the production of Saharan depressions (D) which move north-eastward along a trough axis (*after Nicholson and Flohn 1980*). (*Copyright © 1980/1982 by D. Reidel Publishing Company. Reprinted by permission*).

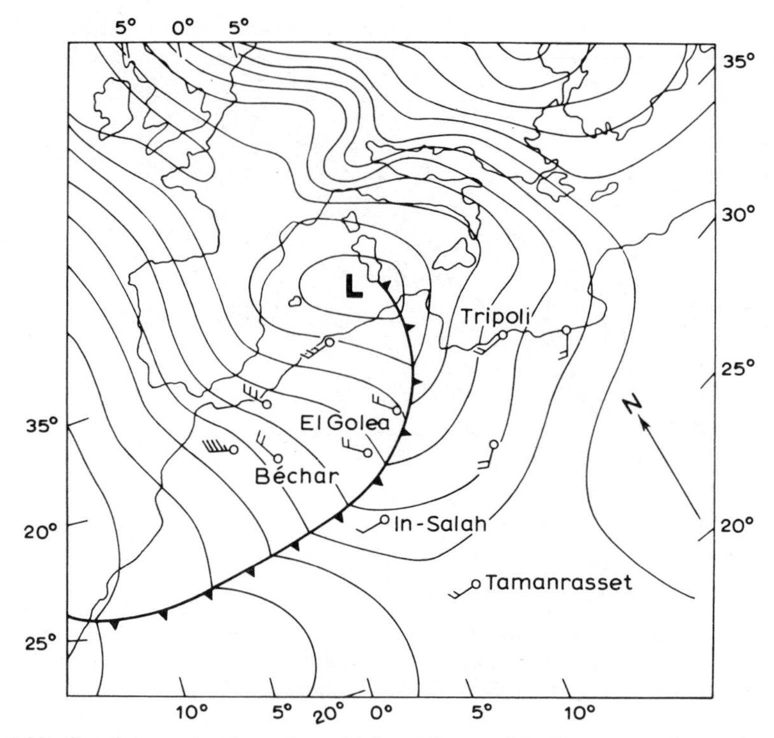

Fig. 6.38. Southern extension of a cold front from a Mediterranean depression into the Sahara at 1200 hrs on 17 January 1972 (*from Breed et al. 1979*).

of the Monsoon Trough (see figs. 6.2B and 6.31) which allows on occasion tongues of moist south-westerly air to penetrate far north and produce short-lived low-pressure centres. Study of these Saharan depressions has permitted a clearer picture to emerge of the region previously designated as the 'temperate front' (see chs. 3, D.1 and 4, E). In the upper troposphere at about 200 mb (12 km), the westerlies overlie the poleward flanks of the sub-tropical high-pressure belt. Occasionally the individual high-pressure cells contract away from one another as meanders develop in the westerlies between them which may extend equatorward so as to interact with the low-level underlying tropical easterlies (fig. 6.37). This interaction may lead to the development of lows which then move north-east along the meander trough associated with rain and thunder. By the time they reach the central Sahara, they are frequently 'rained out' and give rise to duststorms, but they may be reactivated further north by the entrainment of moist Mediterranean air. The conditions of subtropical cell separation and the interaction of westerly and easterly circulations are most likely to occur around the equinoxes (fig. 6.31), or sometimes in winter if the otherwise dominant Azores high-pressure cell contracts westwards. A less extreme case of the equatorward extension of the westerlies is the occasional penetration of cold fronts south from the Mediterranean bringing heavy rain to restricted desert

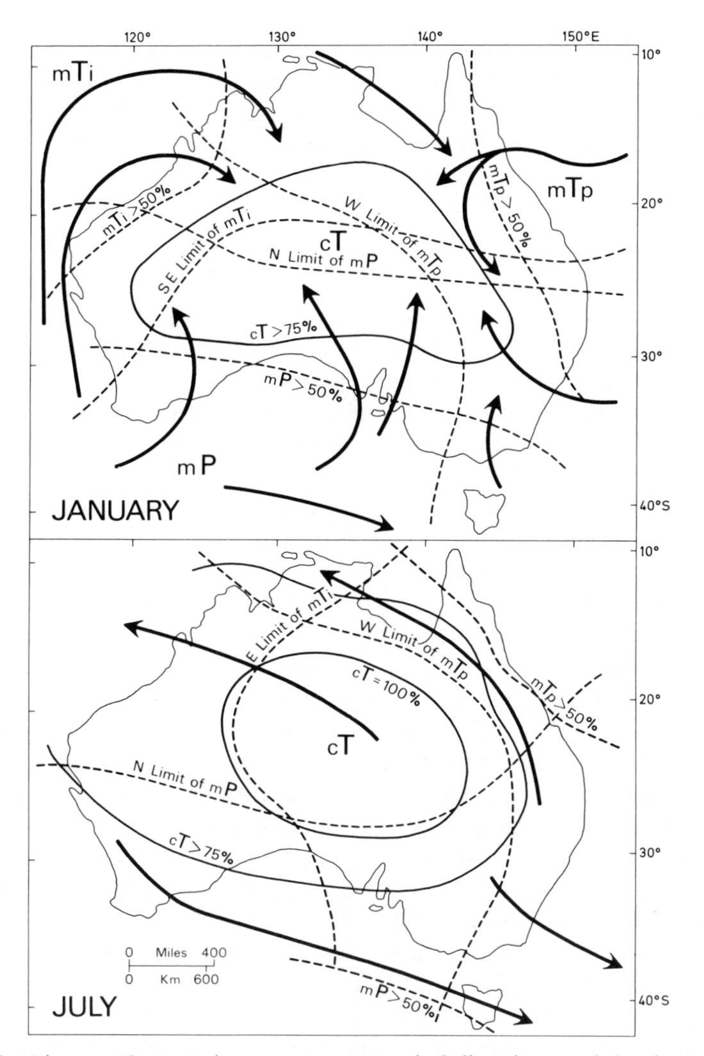

Fig. 6.39. Air-mass frequencies, source areas, wind directions and dominance of the cT high-pressure cell over Australia in summer (*above*) and winter (*below*) (*after Gentilli 1971*).

areas (fig. 6.38). In December 1976 such a depression produced up to 40 mm of rain during 2 days in southern Mauretania.

The continental high-pressure cell over central Australia similarly inhibits average rainfall amounts, which total less than 250 mm per year over fully 37% of the whole country. Indeed, one station exceeded 230 mm in only 1 out of 42 consecutive years. Detailed work on the variations in pressure of this cell has suggested that the apparently stable high-pressure cell is in reality caused by the local intensification of a constant progression of eastward-travelling anticyclones. Thus their seasonal and shorter-term variation in intensity allows the periodic inflow of the surrounding maritime tropical air masses from the Pacific (mTp) and Indian (mTi) Oceans and maritime polar air from the south. Occasional heavy rainfalls in the continental interior may result from northward incursions of troughs bringing mP air in January and eastward extension of mT air from the Indian Ocean in July (Fig. 6.39).

Summary

The tropical atmosphere differs significantly from that in middle latitudes. Temperature gradients are generally weak and weather systems are mainly produced by airstream convergence triggering convection in the moist surface layer. Strong longitudinal differences in climate exist as a result of the zones of subsidence (ascent) on the eastern (western) margins of the subtropical high-pressure cells. In the eastern oceans, there is typically a strong trade wind inversion at about 1 km with dry subsiding air above, giving fine weather. Downstream this stable lid is gradually raised by convective clouds as the trades flow westward. Cloud masses are frequently organized into amorphous 'clusters' on a subsynoptic scale. The trade wind systems of the two hemispheres converge, but not in a spatially or temporally continuous manner. This Intertropical Convergence Zone also shifts poleward over the land sectors in summer, associated with the monsoon regimes of Asia, Africa and Australia.

Wave disturbances in the tropical easterlies are variable in character. The 'classical' easterly wave has maximum cloud build-up and precipitation behind (east of) the trough line. This distribution follows from the conservation of potential vorticity by the air. About 10% of wave disturbances later intensify to become tropical storms or cyclones. This development requires a warm sea-surface and low-level convergence to maintain the sensible and latent heat supply and upper-level divergence to maintain ascent. Cumulonimbus 'hot towers' nevertheless account for a small fraction of the spiral cloud bands. Tropical cyclones are most numerous in the western oceans of the northern hemisphere in the summer–autumn seasons.

The monsoon seasonal wind-reversal of southern Asia is the product of global and regional influences. The orographic barrier of the Himalaya and Tibetan Plateau plays an important role. In winter, the subtropical westerly jet stream is anchored south of the mountains. Subsidence occurs over northern India giving north-easterly surface (trade) winds. Occasional depressions from the Mediterranean penetrate to north-western India–Pakistan. The circulation reversal in summer is triggered by the development of an upper anticyclone over the elevated Tibet Plateau with upper easterly flow over India. This change is accompanied by the northward extension of low-level south-westerlies in the Indian Ocean which appear

continued

first in southern India and along the Burma coast and then extend north-westward. Rainfall is concentrated in spells associated with 'monsoon depressions' which travel westward steered by the upper easterlies.

Variability in tropical climates also occurs through diurnal effects, such as land–sea breezes, local topographic and coastal effects on airflow, and the penetration of extratropical weather systems and airflow into lower latitudes. At irregular intervals, the equatorial Pacific sector experiences an ENSO event with major repercussions both regionally and globally.

7

Small-scale climates

Meteorological phenomena encompass a wide range of space and time scales, from the instantaneous gusts of wind that swirl-up leaves and litter to the global-scale wind systems that shape the annual planetary climate. The weather systems discussed in chs. 4 and 5 are conventionally designated as synoptic-scale systems, whereas tornadoes and thunderstorms (with a spatial scale of 1–50 km and a time scale of a few hours) are referred to as *meso-scale systems*. Other wind systems of comparable scale to the latter, like mountain and valley winds and land–sea breezes, can give rise to distinctive *local climates* (see ch. 3, C). Small-scale turbulence, with wind eddies of a few metres dimension and lasting only a few seconds, represent the domain of *micrometeorology*. Atmospheric airflow disturbances in the boundary layer near the earth's surface can be classified into four main types, depending on their altitudinal and horizontal scales – i.e. synoptic and meso-scale, thermal, topographic and surface friction on 'flat' surfaces of varying roughness (fig. 7.1). Their life span and the kinetic energy which they involve are illustrated in fig. 7.2, in comparison with those for a range of human activities. For our purposes, we can consider such phenomena in relation to the climatic processes within a crop canopy, a forest stand or a cluster of city buildings.

A Surface-energy budgets

We will first review the processes of energy exchange between the atmosphere and an unvegetated surface. In ch. 1, F, it was noted that the surface energy budget equation is usually written:

$$R_n = H + LE + G$$

where R_n, the net all-wavelength radiation

$= [S(1 - a)] + L_n$.

S = incoming short-wave radiation,

a = fractional albedo of the surface, and

L_n = the net outgoing long-wave radiation.

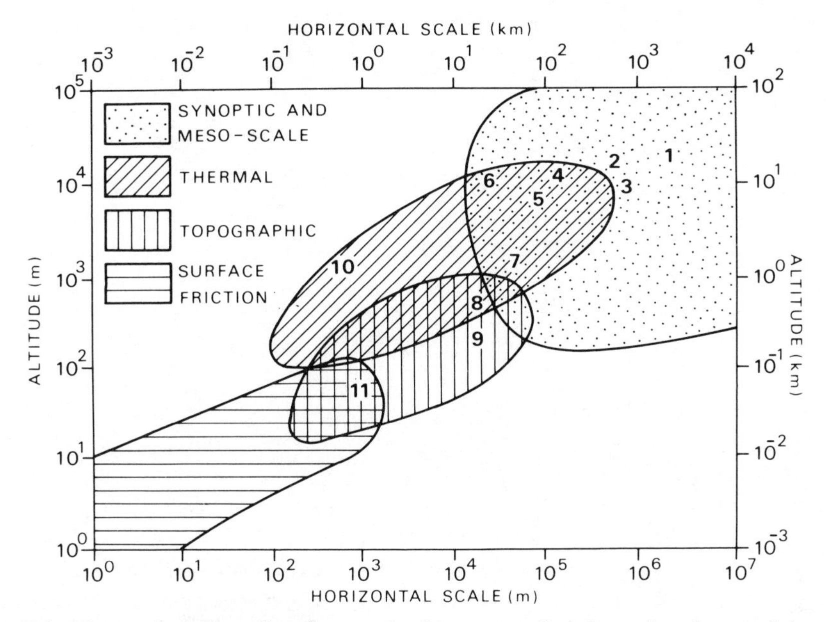

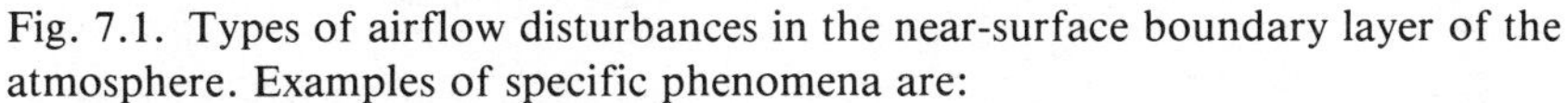

Fig. 7.1. Types of airflow disturbances in the near-surface boundary layer of the atmosphere. Examples of specific phenomena are:
1 Depression (altitude 12 km; horizontal scale 1500–2000 km);
2 Hurricane (15 km; 700 km);
3 Easterly wave (10 km; 800 km);
4 Thunderstorm (13 km; 100 km);
5 Mountain lee wave (7 km; 75 km);
6 ITCZ convective cell (15 km; 10–100 km);
7 Sea breeze (1 km; 50 km);
8 Urban heat island (300 m+; 20 km);
9 Urban friction hump (of buildings);
10 Small cumulus cell (2000 m; 100–1000 m);
11 Forest (60 m; 1500 m) (*from Anderson 1971*).

R_n is usually positive by day since the absorbed solar radiation exceeds the net outgoing long-wave radiation; at night, when $S = 0$, R_n is determined by the negative magnitude of L_n.

The surface energy flux terms are

G = ground heat flux
H = turbulent sensible heat flux to the atmosphere
LE = turbulent latent heat flux to the atmosphere (E = evaporation; L = latent heat of vaporization).

Positive values denote a flux *away* from the surface interface. By day, G, H, and LE are usually offset by the supply of conductive heat from the soil (G) and turbulent heat from the air (H) (see fig. 7.3A). Occasionally, condensation may contribute heat to the surface.

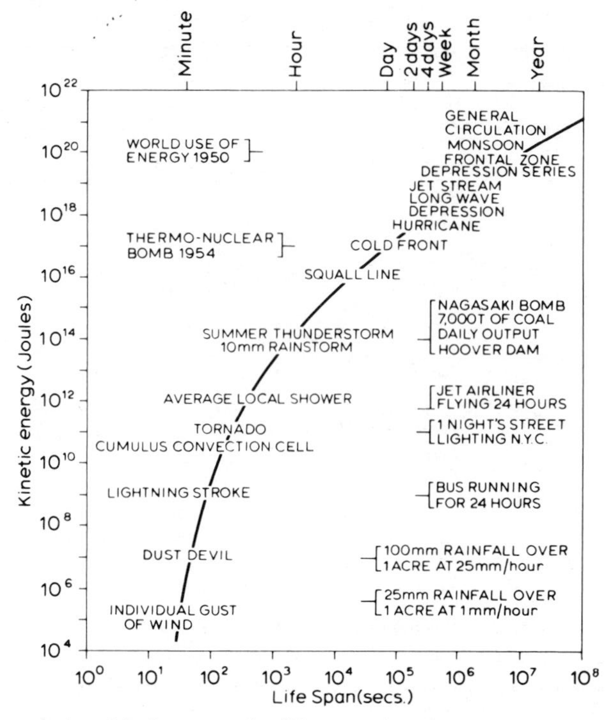

Fig. 7.2. The relationship between the life span (in seconds) of a range of meteorological phenomena and their kinetic energy (in joules). The kinetic energy of a number of human activities is also shown (*after Koppány 1975*).

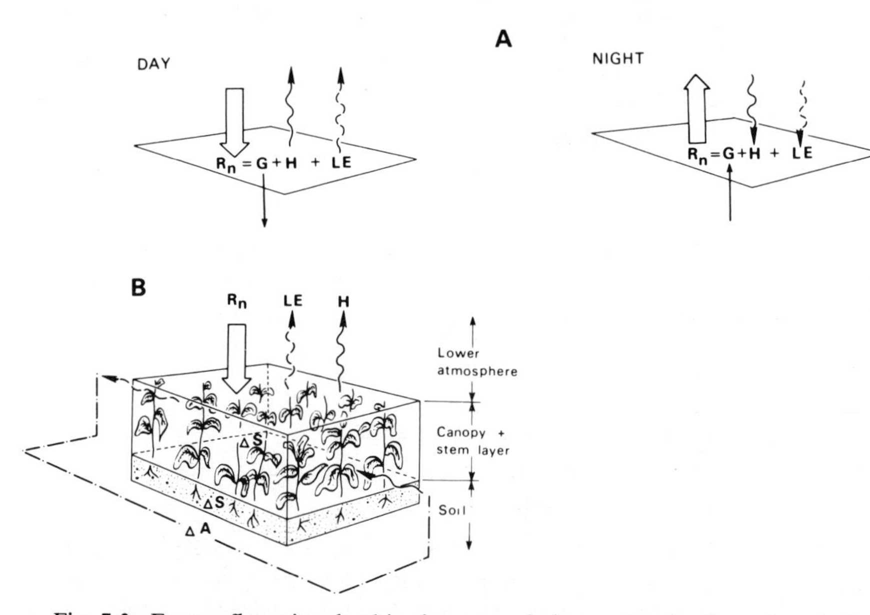

Fig. 7.3. Energy flows involved in the energy balance of a simple surface during day and night (A) and a vegetated surface (B) (*after Oke 1978*).

Commonly, there is a small residual heat storage (ΔS) in the soil in spring/summer and a return of heat to the surface in autumn/winter. Where a vegetation canopy is present there may also be a small additional biochemical heat storage, due to photosynthesis, as well as physical heat storage by leaves and stems (fig. 7.3B).

An additional energy component to be considered in areas of mixed canopy cover (forest/grassland, desert/oasis), and in water bodies, is the horizontal transfer (*advection*) of heat by wind and currents (ΔA; fig. 7.3B). The atmosphere transports both sensible and latent heat.

B Non-vegetated natural surfaces

The energy exchanges of dry desert surfaces are relatively simple and straightforward. Figure 7.4 illustrates the instantaneous noon and evening fluxes on a granite surface in July in California and the resulting large temperature range. Surface properties modify the heat penetration, as shown in fig. 7.5 from mid-August measurements in the Sahara. The maximum surface temperatures reached on bare dark-coloured basalt and light-coloured sandstone were almost identical, but the greater thermal conductivity of the former ($3 \cdot 1$ W m^{-1} K^{-1} for basalt, versus $2 \cdot 4$ W m^{-1} K^{-1} for sandstone) gives a greater diurnal range and deeper penetration of the diurnal temperature wave, to about 1 m in the basalt. In sand (fig. 7.4C), the temperature wave is negligible at 30 cm due to the low conductivity of intergranular air. Note that the surface range of temperature is several times that in the air. Sand also has an albedo of about $0 \cdot 35$ compared with $0 \cdot 2$ for a rock surface.

A representative diurnal pattern of energy exchange over desert surfaces is shown in fig. 7.6. The 2 m air temperature varies between 17° and 29°C, although the surface of the dry lake bed reaches 57°C at midday. R_n reaches a maximum about 1300 hours. At this time most of the heat is transferred to the air by turbulent convection, while in the early morning the heating goes

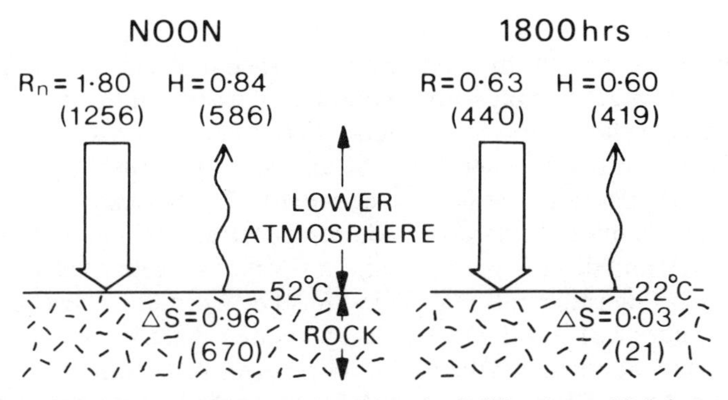

Fig. 7.4. Energy balance of a granite surface in California on 18 July at noon (solar altitude 70°) and 1800 hours (solar altitude 10°). Figures in cal cm^{-2} and in W m^{-2} in parentheses (*based on data from Miller 1965*).

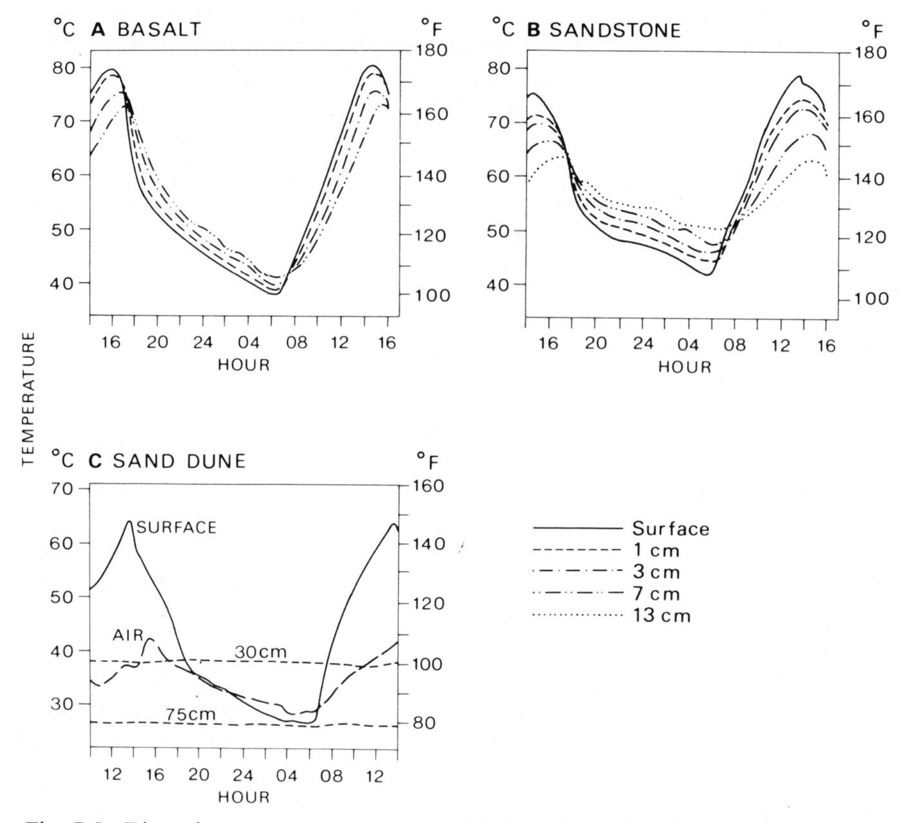

Fig. 7.5. Diurnal temperatures near, at and below the surface in the Tibesti region, central Sahara, in mid-August 1961. A At the surface and at 1 cm, 3 cm and 7 cm below the surface of a basalt. B At the surface and at 1 cm, 3 cm, 7 cm and 13 cm below the surface of a light-coloured sandstone. C In the surface air layer, at the surface and at 30 cm and 75 cm below the surface of a sand dune (*after Peel 1974*).

into the ground. At night, this soil heat is returned to the surface, offsetting radiational cooling. Over a 24-hour period, about 90% of the net radiation goes into sensible heat, 10% into ground flux.

For a water body, the energy fluxes are very differently apportioned. Figure 7.7 illustrates the seasonal regime for Lake Mead, Arizona, in 1952–3. The incoming short-wave radiation penetrates to about 10 m depth (see ch. 1, D.5) and there is an important horizontal advective term (ΔA) due to the changing density stratification in the lake. Warmer water rises to the surface in winter (ΔA positive) whereas in summer there is a large loss as a result of turbulent mixing of the water. There is a strong annual cycle in the flux into and out of the water body (G), whereas the evaporative loss in excess of 200 cm per year occurs at all seasons. Wind effects in autumn cause LE to exceed the net radiation term.

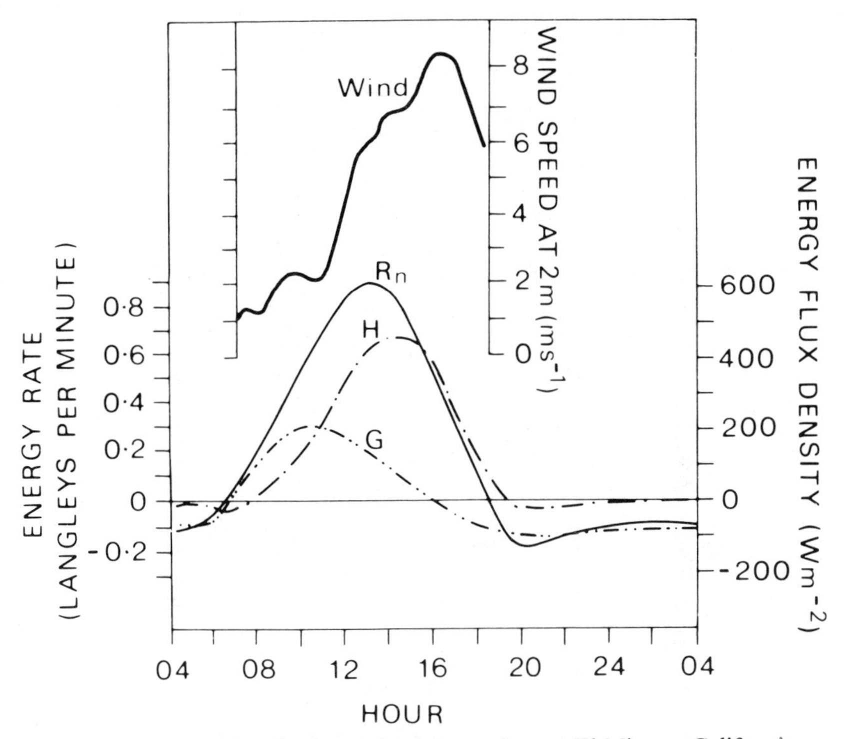

Fig. 7.6. Energy flows involved at a dry-lake surface at El Mirage, California (35°N), on 10–11 June 1950. Wind speed due to surface turbulence was measured at a height of 2 m (*after Vehrencamp 1953 and Oke 1978*).

C Vegetated surfaces

From the viewpoint of energetics, as well as of climate within the plant canopy, it is useful to consider short crops and forests separately.

1 Short green crops

Short green crops, up to a metre or so high, supplied with sufficient water and exposed to similar solar radiation conditions, all have a similar receipt of net radiation (R_n). This is largely because of a the small range of albedos, 20–30% for short green crops compared with 9–18% for forests. Canopy structure appears to be the primary reason for this difference.

General figures for rates of energy dispersal at noon on a June day in a 20 cm high stand of grass in the higher mid-latitudes are:

	W m^{-2}
Net radiation at the top of the crop	550
Physical heat storage in leaves	6
Biochemical heat storage (i.e. growth processes)	22
Received at soil surface	200

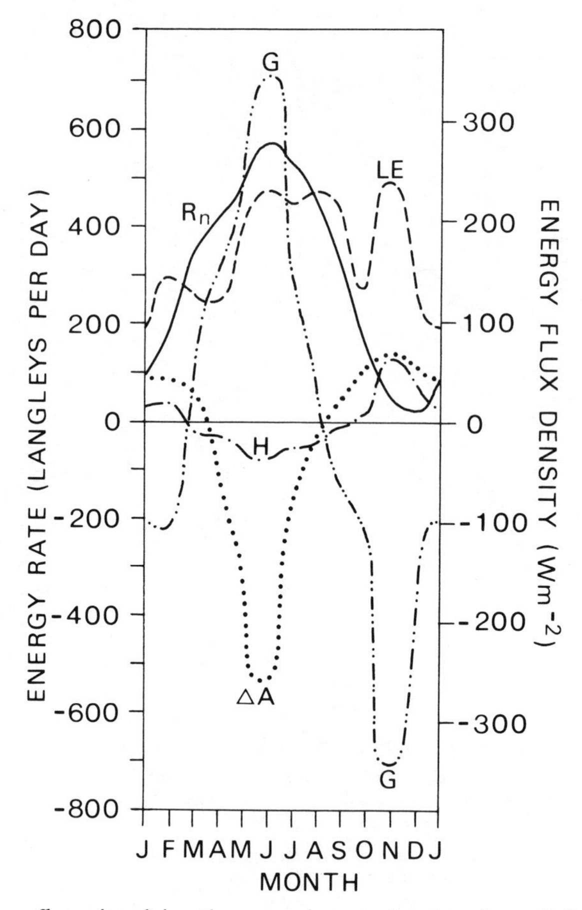

Fig. 7.7. Energy flows involving the upper layers of water. Annual figures for Lake Mead, Arizona (36·1°N) during 1952–3 (*after Sellers 1965*).

Figure 7.8A and B shows the diurnal and annual energy balances of a field of short grass near Copenhagen (55·7°N). For an average 24-hour period in June about 58% of the incoming radiation is involved in evapotranspiration, whereas in December the small amount of net outgoing radiation (i.e. negative R_n) is composed of 55% heat supplied by the soil and 45% sensible heat transfer from the air to the grass.

It is possible to generalize regarding the micro-climates of short growing crops (fig. 7.9; Oke 1978):

Temperature – In early afternoon there is a temperature maximum just below the vegetation crown where the maximum energy absorption is occurring; the temperature is lower near the soil surface where heat flows into the soil. At night the crop cools mainly by long-wave emission and by some continued transpiration, producing a temperature minimum at about two-thirds

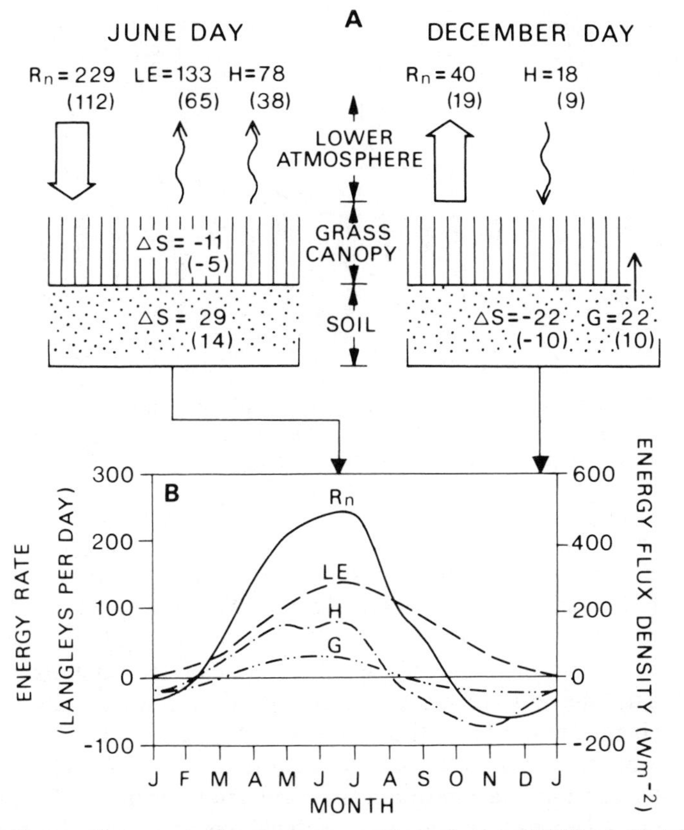

Fig. 7.8. Energy fluxes over short grass near Copenhagen (56°N). A Totals for a day in June (17 hours daylight; maximum solar altitude 58°) and December (7 hours daylight; maximum solar altitude 11°). Units are cal cm^{-2} per day and in W m^{-2} in parentheses. B Seasonal curves of net radiation (R_n), latent heat (LE), sensible heat (H) and ground-heat flux (G). (*Data from Miller 1965; and after Sellers 1965.*)

the height of the crop. Under calm conditions, a temperature inversion may form just above the crop.

Wind speed – This is a minimum in the upper crop canopy where the foliage is most dense. Below there is a slight increase and a marked increase above.

Water vapour – The maximum diurnal evapotranspiration rate and supply of water vapour occurs at about two-thirds the crop height where the canopy is most dense.

Carbon dioxide – During the day CO_2 is absorbed by the photosynthesis of growing plants and emitted at night due to respiration. This maximum sink and source of CO_2 is at about two-thirds the crop height.

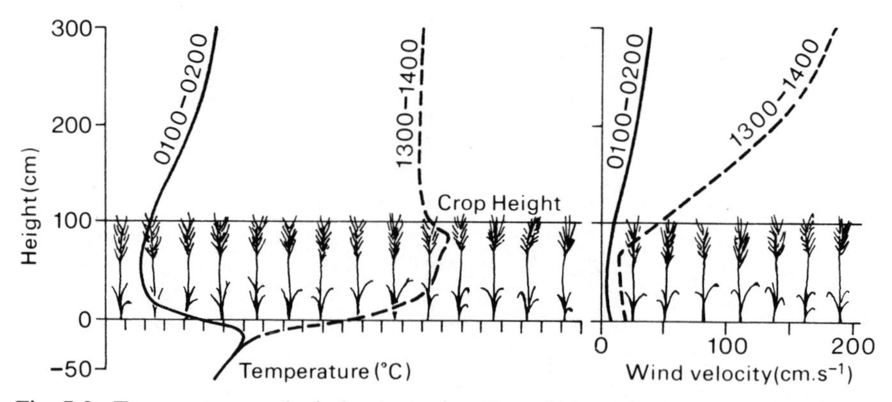

Fig. 7.9. Temperature and wind-velocity profiles within and above a metre-high stand of barley at Rothamsted, southern England, on 23 July 1963 at 0100–0200 hours and 1300–1400 hours (*after Long et al. 1964*).

Finally, it is instructive to look at the conditions accompanying the growth of irrigated crops. Figure 7.10A and B shows the energy relationships in a 1 m high stand of irrigated Sudan Grass at Tempe, Arizona, on 20 July 1962. The air temperature varied between 25°C (77°F) and 45°C (113°F). During the day evapotranspiration in the dry air is near its potential and LE (anomalously high due to a local temperature inversion) exceeds R_n, the deficiency being made up by a transfer of sensible heat from the air (negative *H*). Evaporation continues during the night due to a quite high wind speed (7 m s^{-1}) sustained by the continued heat flow from the air. The evapotranspiration thus gives comparatively low diurnal temperatures within irrigated desert crops.

2 Forests

The vertical structure of a forest, which depends on the vegetational species, the ecological associations, the age of the stand and other botanical considerations, largely determines the forest microclimate. Much of the climatic influence of a forest can be explained in terms of the geometry of the forest, including morphological characteristics, size, coverage and stratification. Morphological characteristics include amount of branching (bifurcation), the periodicity of growth (i.e. evergreen or deciduous), together with the size, density and texture of the leaves. Tree size is obviously important. In temperate forests the sizes may be closely similar, whereas in tropical forests a great variety of sizes may be present locally. Crown coverage is important in terms of the physical obstruction presented by the canopy to radiation exchange and air movements.

An obvious example of the microclimatic effects of different spatial forest organizations can be gained by comparing the features of tropical rain forests and temperate forests. In tropical forests the average height of the taller trees is of the order of 46–55 m (150–180 ft), individuals rising to over

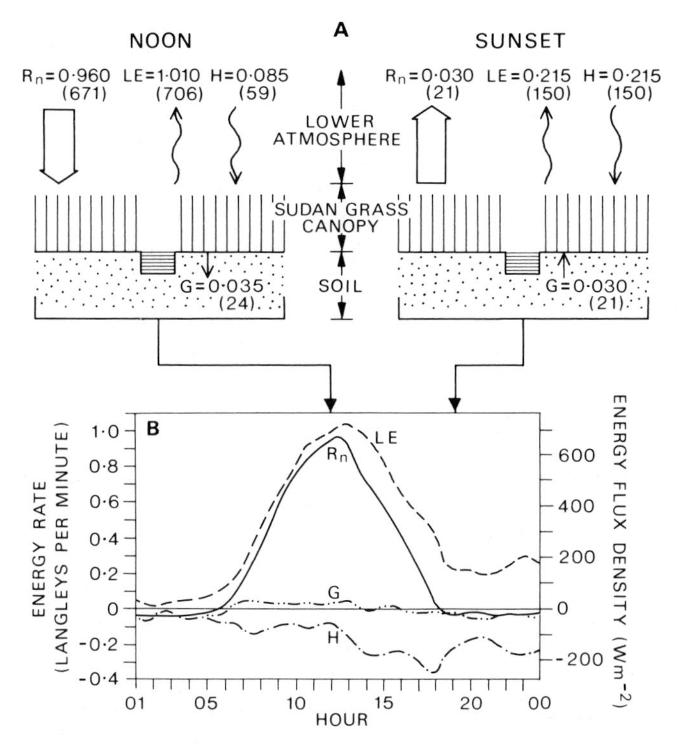

Fig. 7.10. Energy flows involved in the energy diurnal balance of irrigated Sudan Grass at Tempe, Arizona, on 20 July 1962. Figures given in cal cm^{-2} min^{-1} and in W m^{-2} in parentheses (*after Sellers 1965*).

60 m (200 ft). The dominant height of temperate forests is up to 30 m, so that neither temperate nor most tropical forests can compare in height with the western American redwoods (*Sequoia sempervirens*). Tropical forests commonly possess a great variety of species, there being seldom less than 40 per hectare (100 hectares = 1 km^2) and sometimes over 100; comparing with less than twenty-five (occasionally only one) tree species with a trunk diameter greater than 10 cm (or 4 in) in Europe and North America. Many British woodlands, for example, have almost continuous canopy stratification from low shrubs to the tops of 36 m beeches, whereas tropical forests are strongly stratified with dense undergrowth, unbranching trunks, and commonly two upper strata of foliage. This stratification, the second or lower of which is usually the more dense, results in rather more complex microclimates in tropical forests than in temperate stands.

It is convenient to describe the climatic effects of forest stands in terms of their modification of energy transfers and of the airflow, their modification of the humidity environment, and their modification of the thermal environment.

a Modification of energy transfers Forest canopies significantly change the pattern of incoming and outgoing radiation. The short-wave reflectivity of

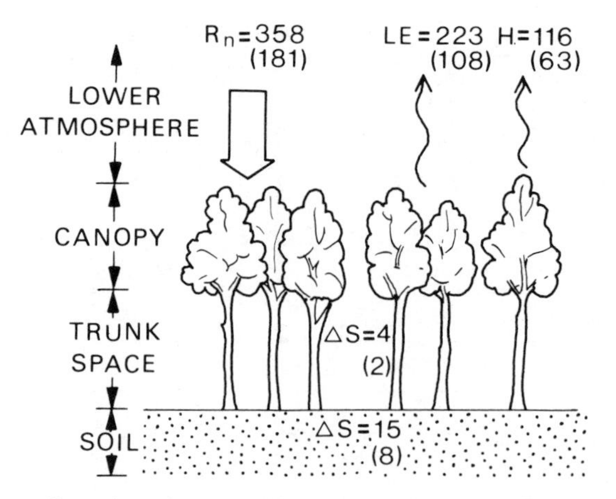

Fig. 7.11. Energy flows in a 30-year-old oak stand in the Tellerman Experimental Forest, Voronezh District, USSR, on an average summer day (June to August). Figures given in cal cm^{-2} day^{-1} and in W m^{-2} in parentheses (*after Sukachev and Dylis 1968*).

forests depends somewhat on the characteristics of the trees and their density. Coniferous forest have albedos of about 8–14% and values for deciduous range between 12 and 18%, increasing as the canopy becomes more open. Values for semi-arid savannas and scrub woodland are much higher.

Besides reflecting energy, the forest canopy traps energy (fig. 7.11), and it has been calculated that for dense red beeches (*Fagus sylvatica*) 80% of the incoming radiation is intercepted by the tree-tops and less than 5% reaches the forest floor. The greatest trapping occurs in sunny conditions, for when the sky is overcast the incoming diffuse radiation has greater possibility of penetration laterally to the trunk space (fig. 7.12). Visible light, however, does not give an altogether accurate picture of total energy penetration for more ultraviolet than infrared radiation is absorbed in the crowns. For example, only 7·6% of short-wave radiation (less than 0·5 μm) reached a forest floor in Nigeria, as against 45·3% greater than 0·6 μm. As far as light penetration is concerned there are obviously great variations depending on type of tree, tree spacing, time of year, age, crown density and height. About 50–75% of the outside light intensity may penetrate to the floor of a birch–beech forest, 20–40% for pine, 10–25% for spruce and fir; but for tropical Congo forests the figure may be as low as 0·1% and one of 0·01% has been recorded for a dense elm stand in Germany. One of the most important effects of this is to reduce the length of daylight. For deciduous trees, more than 70% of the light may penetrate when they are leafless. Tree age is also important in that this controls both crown cover and height. Figure 7.12 shows this rather complicated effect for spruce in the Thuringian Forest, Germany. For a Scots pine (*Pinus sylvestris*) forest in Germany 50% of the outside light intensity was recorded at 1·3 years, only 7% at 20 years and 35% at 130 years.

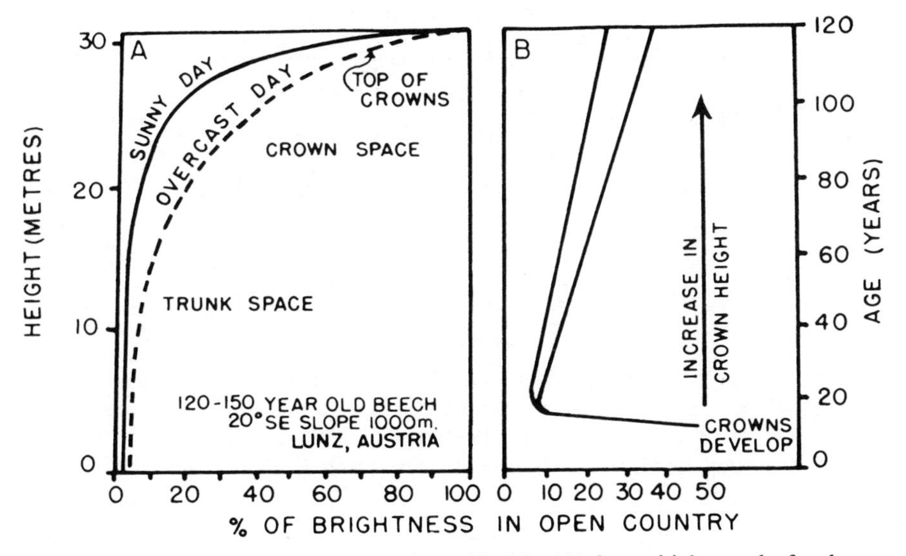

Fig. 7.12. Amount of light as a function of height (A) for a thick stand of red beeches (*Fagus sylvatica*) in Austria, and as a function of age (B) for a Thuringian spruce forest (*after Geiger 1965*).

b Modification of airflow Forests impede both the lateral and vertical movement of air, but it is more convenient to treat the latter in connection with thermal modifications. In general, air movement within forests is slight compared with that in the open, and quite large variations of outside wind velocity have little effect inside woods (fig. 7.13). Measurements for European forests show that 30 m (100 ft) of penetration reduces wind velocities to 60–80%, 60 m (200 ft) to 50% and 120 m (400 ft) to only 7%. A speed of 2·2 m s^{-1} (4·9 mph) outside a Brazilian evergreen forest was found to be reduced to 0·5 m s^{-1} (1·1 mph) at about 100 m (300 ft) within, and to be negligible at 1000 m. In the same location external storm winds of 28 m s^{-1} (62·7 mph) were reduced to 2 m s^{-1} (4·2 mph) some 11 km (7 miles) deep in the forest. Where there is a complex vertical structuring of the forest wind velocities become more complex. Thus whereas in the crowns (23 m; 75 ft) of a Panama rain forest the wind velocity was 75% of that outside, it was only 20% in the undergrowth (2 m; 6·5 ft). Other influences include the density of the stand and the season. For dense pine stands in Idaho simultaneous recordings showed the wind velocity to be 0·6 m s^{-1} (1·3 mph) in a cut area, 0·4 m s^{-1} (0·9 mph) half cut, and 0·1 m s^{-1} (0·2 mph) in the uncut stand. The effect of season on wind velocities in deciduous forests is shown by fig. 7.13. Observations in a Tennessee mixed-oak forest showed January forest-wind velocities to be 12% of those in the open whereas those in August had dropped to 2%.

Knowledge of the effect of forest barriers on winds has been utilized in the construction of wind breaks to protect crops and soil, and, for example, the cypress breaks of the southern Rhone valley and the Lombardy poplars

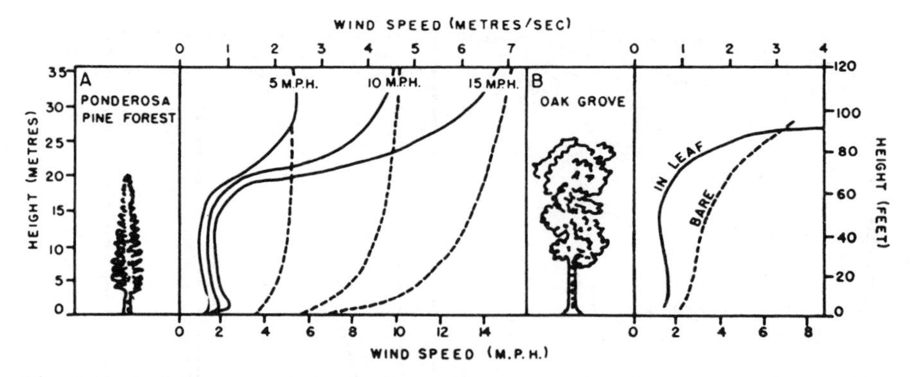

Fig. 7.13. Influence on wind velocity profiles exercised by (A) a dense 45-year-old ponderosa pine (*Pinus ponderosa*) stands in the Shasta Experimental Forest, California (after Fons, and Kittredge 1948), the dashed lines indicating the wind profile over open country; and (B) an oak grove (*after R. Geiger and H. Amann, and Geiger 1965*).

(*Populus nigra*) of the Netherlands form distinctive features of the landscape. It has been found that the denser the obstruction the greater the shelter immediately behind it, although the downwind extent of its effect is reduced by lee turbulence set up by the barrier. The maximum protection is given by the filtering mechanism produced by a break of about 40% penetrability (fig. 7.14). An obstruction begins to have an effect about eighteen times its own height upwind, and the downwind effect can be increased by the *back coupling* of more than one belt (fig. 7.14A).

There are also some less obvious microclimatic effects of forest barriers. One of the most important is that the reduction of horizontal air movement in forest clearings increases the frost hazard on winter nights. Another aspect is the removal of dust and fog droplets from the air by the filtering action of forests. Measurements 1·5 km upwind, on the lee side and 1·5 km downwind of a kilometre-wide German forest gave dust counts (particles per litre) of 9000, less than 2000 and more than 4000, respectively. Measurements in association with a 2 m high and 13 m thick shelter belt on the south-east Hokkaido coast, Japan, in July 1952 showed this filtering effect on advection fog rolling in from the sea, in that 20 m downwind of the obstruction the humidity was only 0·1 g m^{-3} (mean wind velocity 2·55 m s^{-1}), compared with 0·3 g m^{-3} (mean wind velocity 3·4 m s^{-1}) a similar distance upwind. In extreme cases so much fog can be filtered from laterally moving air that *negative interception* can occur, where there is a higher precipitation catch within a forest than outside. The winter rainfall catch outside a eucalyptus forest near Melbourne, Australia was 50 cm (19·7 in), whereas inside the forest it was 60 cm (23·6 in).

c Modification of the humidity environment The humidity conditions within forest stands contrast strikingly with those in the open. Evaporation from the forest floor is usually much less because of the decreased direct

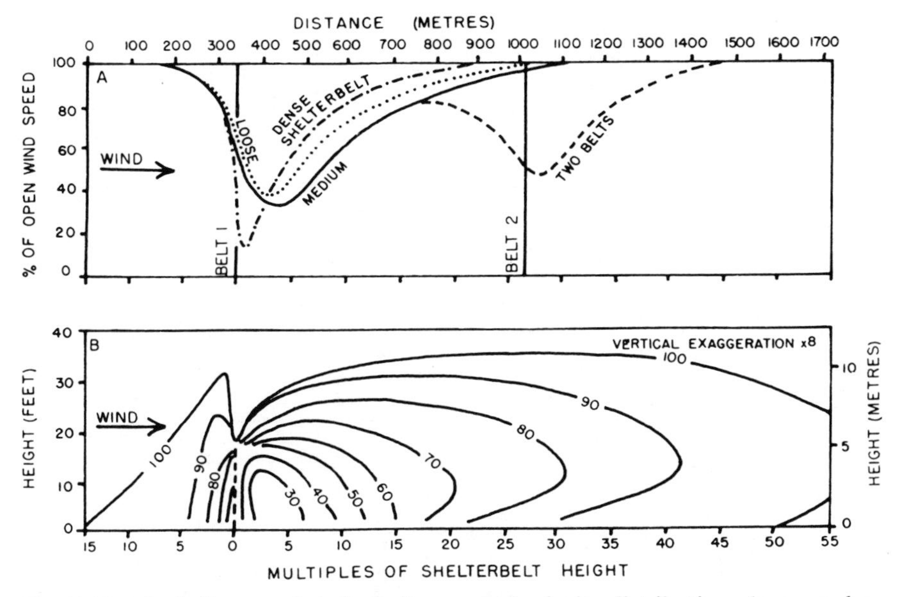

Fig. 7.14. The influence of shelterbelts on wind-velocity distributions (expressed as percentages of the velocity in the open). A The effects of one shelter belt of three different densities, and of two back-coupled medium-dense shelter belts (*after W. Nägeli, and Geiger 1965*). B The detailed effects of one half-solid shelter belt (*after Bates and Stoeckeler, and Kittredge 1948*).

sunlight, lower wind velocities, lower maximum temperatures, and generally higher forest air humidity. Evaporation from the bare floors of pine forests is 70% of that in the open for Arizona in summer and only 42% for the Mediterranean region, although such measurements have little real significance in that water losses from vegetated surfaces are controlled by the plant evapotranspiration.

Unlike many cultivated crops, forest trees exhibit a wide range of physiological resistance to transpiration processes and, as a consequence, the proportions of forest energy flows involved in evapotranspiration (LE) are varied, as is the inversely linked contribution by sensible heat exchange (H). Figure 7.15 compares diurnal energy flows during July in respect of a pine forest in eastern England and a fir forest in British Columbia. In the former area only $0 \cdot 33\, R_n$ is employed for LE due to the high resistance of the pines to transpiration, whereas $0 \cdot 66\, R_n$ is similarly employed in the British Columbia fir forest, especially during the afternoon. Like short green crops, however, only a very small proportion of R_n is used ultimately for tree growth, an average figure being about $1 \cdot 3$ W m^{-2} (1 kcal cm^{-2} yr^{-1}), of which some 60% produces wood tissue and 40% forest litter.

During daylight leaves transpire water through open pores, or *stomata*, so that this loss is controlled by the length of day, the leaf temperature (modified by evaporational cooling), the leaf surface area, the tree species and its age, as well as by the meteorological factors of available radiant

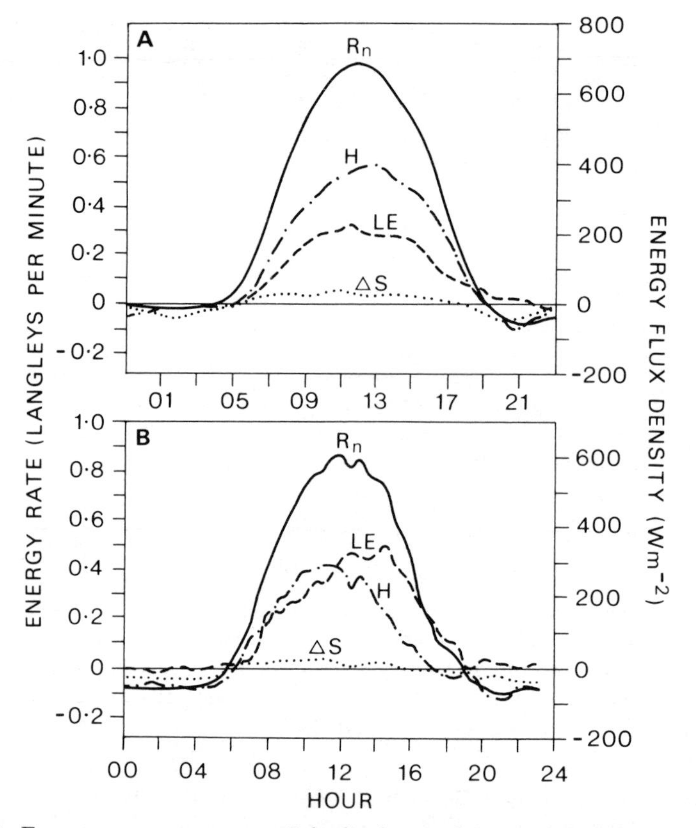

Fig. 7.15. Energy components on a July day in two forest stands. A Scots and Corsican pine at Thetford, England (52°N), on 7 July 1971. Cloud cover was present during the period 0000–0500 hours (*data from Gay and Stewart 1974; after Oke 1978*). B Douglas fir stand at Haney, British Columbia (49°N), on 10 July 1970. Cloud cover was present during the period 1100–2000 hours (*data from McNaughton and Black 1973; after Oke 1978*).

energy, atmospheric vapour pressure and wind speed (see ch. 2, A). Total evaporation figures are therefore extremely varied; also the evaporation of water intercepted by the vegetation surfaces enters into the totals, in addition to direct transpiration. Calculations made for a catchment covered with Norway spruce (*Picea abies*) in the Harz Mountains of Germany showed an estimated annual evapotranspiration of 34 cm and additional interception losses of 24 cm.

The humidity of forest stands is very much linked to the amount of evapotranspiration and increases with the density of vegetation present (fig. 7.16A). The increase of forest humidity over that outside averages some 3–10% of relative humidity and is especially marked in summer (fig. 7.16B). Mean annual relative humidity forest excesses for Germany and Switzerland are for beech 9·4%, Norway spruce (*Picea abies*) 8·6%, larch 7·9% and Scots pine (*Pinus sylvestris*) 3·9%. However, humidity comparisons in these

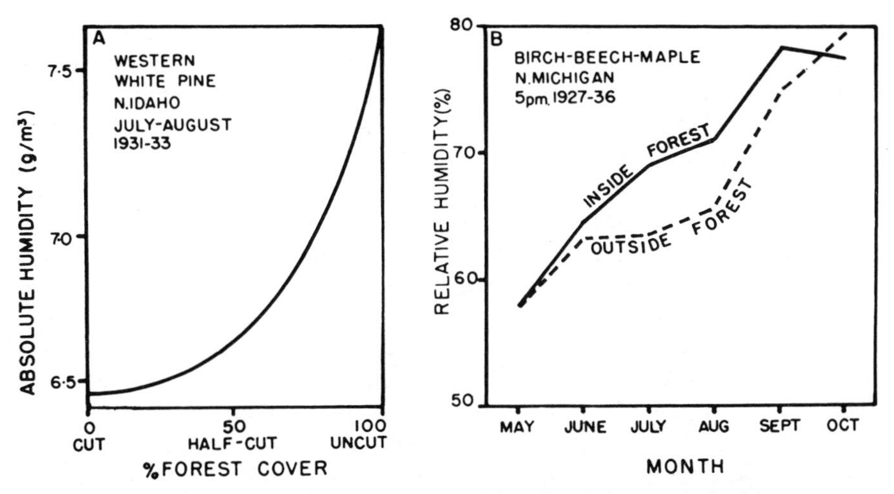

Fig. 7.16. The effects of (A) percentage of white pine (*Pinus monticola*) cover on summer absolute humidity in Idaho (*after Kittredge 1948*); and (B) season on relative humidity in a Michigan birch–beech–maple forest (*after US Dept. Agriculture Yearbook 1941*).

terms are rather unsatisfactory in that forest temperatures differ strikingly from those in the open. Forest vapour pressures were found to be higher within an oak stand in Tennessee than outside for every month except December. Tropical forests exhibit almost complete night saturation irrespective of elevation in the trunk space, whereas during the day humidity is inversely related to elevation.

The influence of forest structures on precipitation is still very much an unresolved problem. This is partly due to the difficulties of comparing rain-gauge catches in the open with those near forests, within clearings or beneath trees. For example, on the windward side of a forest the dominance of low-level upcurrents decreases the amount of precipitation actually caught in the rain gauge, whereas the reverse occurs where there are lee side downdraughts. In small clearings the low wind velocities cause little turbulence around the opening of the gauge and catches are generally greater than outside the forest, although the actual precipitation amounts may be identical. On the other hand, it is sometimes found that the larger the clearing the more prevalent are downdraughts and consequently the precipitation catch increases. In a 25 m high pine and beech forest in Germany, catches in clearings of 12 m diameter were only 87% of that upwind of the forest, but this catch rose to 105% in clearings of 38 m diameter. An analysis of precipitation records for Letzlinger Heath (Germany) before and after afforestation suggested a mean annual increase of 6%, with the greatest excesses occurring during drier years. It is generally agreed, however, that forests have little effect on cyclonic rain, but that they may have a marginal orographic effect in increasing lifting and turbulence, which is of the order of 1–3% in temperate regions.

A far more important obstructional influence of forests on precipitation is in terms of the direct interception of rainfall by the canopy. This obviously varies with crown coverage, with season, and with the rainfall intensity. Measurements in German beech forests indicate that, on average, they intercept 43% of precipitation in summer and 23% in winter. Pine forests may intercept up to 94% of low-intensity precipitation but as little as 15% of high intensities, the average for temperate pines being about 30%. The intercepted precipitation either evaporates on the canopy, runs down the trunk, or drips to the ground. Assessment of the total precipitation reaching the ground (the *throughfall*) requires very detailed measurements of the stem flow and the contribution of drips from the canopy. Canopy evaporation is not necessarily a total loss of moisture from the woodland, since the solar energy used in the evaporating process is not available to remove soil moisture or transpiration water, but the vegetation does not derive the benefit of the cycling of the water through it via the soil. Evaporation from the canopy is very much a function of net radiation receipts (20% of the total precipitation evaporates from the canopy of Brazilian evergreen forests), and of the type of species. Some Mediterranean oak forests yield virtually no stem-flow and their 35% interception almost all evaporates from the canopy.

Recent investigations of the water balance of forests provide some evidence that evergreen forests may permit 10–50%, more evapotranspiration than grass in the same climatic conditions. Grass normally reflects 10–15% more solar radiation than coniferous tree species and hence less energy is available for evaporation. In addition, trees have a greater surface roughness, which increases turbulent air motion and, therefore, the evaporation efficiency. Evergreens also allow transpiration to occur throughout the year. Nevertheless many more detailed and careful studies are required to check these results and to test the various hypotheses.

d Modification of the thermal environment From what has been said it is apparent that forest vegetation has an important effect on micro-scale temperature conditions. Shelter from the sun, blanketing at night, heat loss by evapotranspiration, reduction of wind speed and the impeding of vertical air movement all influence the temperature environment. The most obvious effect of canopy blanketing is that inside the forest daily maximum temperatures are lower and minima are higher (fig. 7.17A). This is particularly apparent during periods of high summer evapotranspiration which depress daily maximum temperatures and cause mean monthly temperatures in tropical and temperate forests to fall well below those outside. In temperate forests at sea-level the mean annual temperature may be about 0·6°C (1°F) less than that in surrounding open country, the mean monthly differences may reach 2·2°C (4°F) in summer but not exceed 0·1°C in winter, and on hot summer days the difference can be more than 2·8°C (5°F). Mean monthly temperatures and temperature ranges for temperate beech, spruce and pine forests are given in fig. 7.17B and C, which also show that when trees

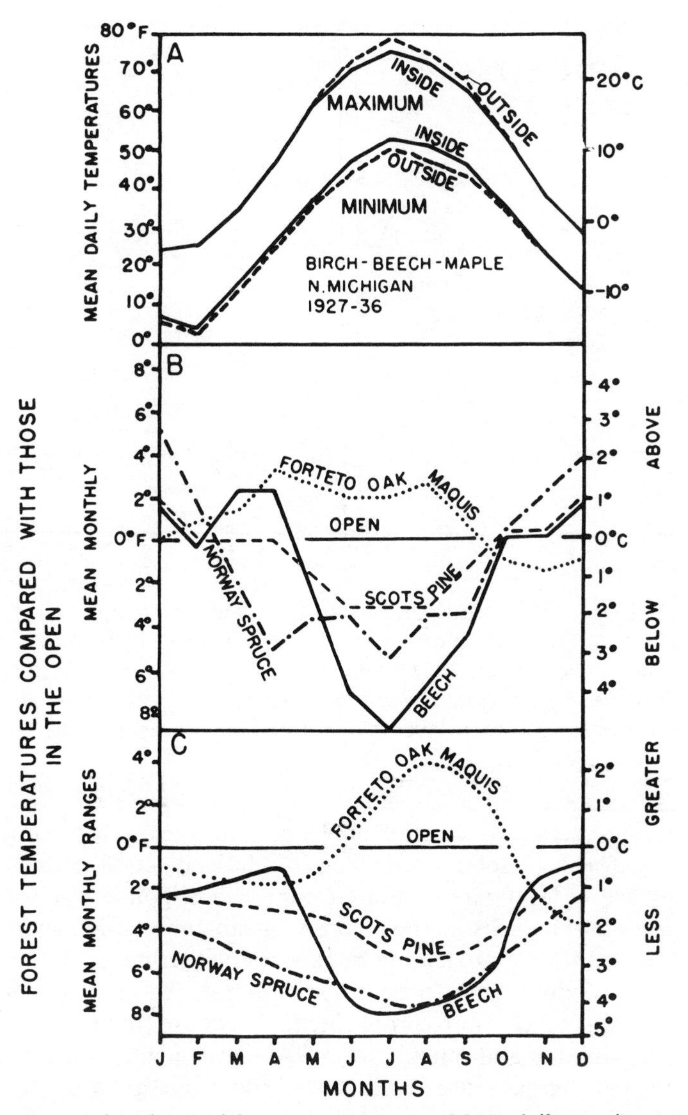

Fig. 7.17. Seasonal regimes of forest temperatures. Mean daily maximum and minimum temperatures (A) inside and outside a birch–beech–maple forest in Michigan (*after US Dept. Agriculture Yearbook 1941*). Mean monthly temperatures (B) and mean monthly temperature ranges (C), compared with those in the open, for four types of Italian forest (*FAO 1962*). Note the anomalous conditions associated with the *forteto* oak maquis, which transpires little.

do not transpire greatly in the summer (e.g. the *forteto* oak maquis of the Mediterranean) the high day temperatures reached in the sheltered woods may cause mean monthly figures to reverse the trend exhibited by temperate forests. Even within individual climatic regions it is difficult to generalize, however, for at elevations of 1000 m the lowering of temperate forest mean temperatures below those in open country may be double that at sea-level.

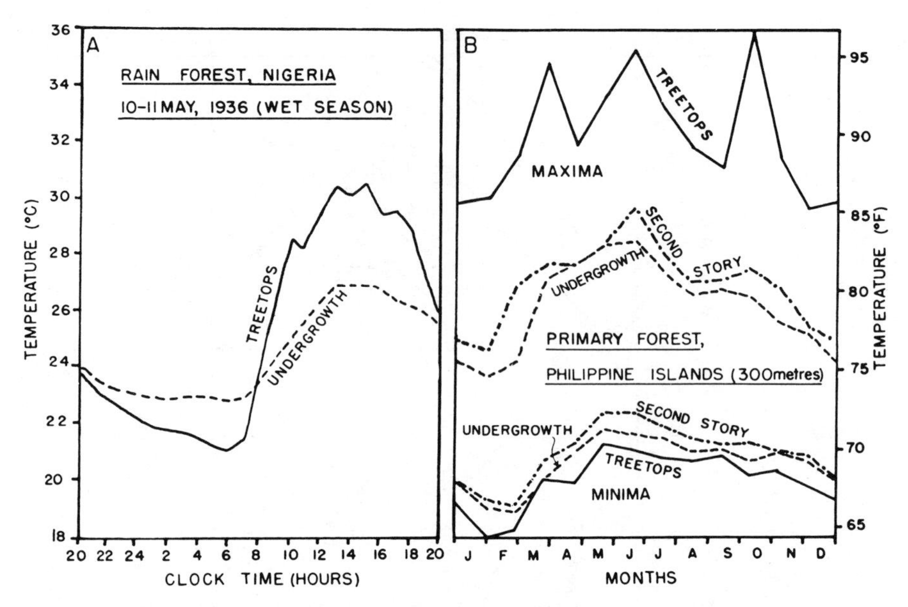

Fig. 7.18. The effect of tropical rain-forest stratification on temperature (*after Richards 1952*). A Daily march of temperature (10–11 May 1936) in the tree tops (24 m) and in the undergrowth (0·7 m) during the wet season in primary rain forest at Shasha Reserve, Nigeria (*after Evans*). B Average weekly maximum and minimum temperatures in three layers of primary (Dipterocarp) forest, Mt Maquiling, Philippine Islands (*after Brown*).

The complex vertical structure of forest stands is a further factor in complicating forest temperatures. Even in relatively simple stands vertical differences are very apparent. For example, in a ponderosa pine forest (*Pinus ponderosa*) in Arizona the recorded mean June–July maximum was increased by 0·8°C (1·5°F) simply by raising the thermometer from 1·5 to 2·4 m (5 to 8 ft) above the forest floor. In stratified tropical forests the thermal picture is much more complicated. The dense canopy heats up very much during the day and quickly loses its heat at night, showing a much greater diurnal temperature range than the undergrowth (fig. 7.18A). Whereas daily maximum temperatures of the second storey are intermediate between those of the tree tops and the undergrowth, the nocturnal minima are higher than either tree tops or undergrowth because the second storey is insulated by trapped air both above and below (fig. 7.18B).

D Urban surfaces

The construction of every house, road or factory destroys existing microclimates and creates new ones of great complexity depending on the design, density and function of the building. Despite the great internal variation of urban climatic influences, it is possible to make certain generalizations regarding the effects of urban structures under three main headings:

(1) Modification of atmospheric composition.
(2) Modification of the heat budget.
(3) Other effects of modifications of surface roughness and composition.

1 Modification of atmospheric composition

The city atmosphere is notoriously liable to pollution, particularly as the result of combustion, and this has effects that involve changes in the thermal properties of the atmosphere, cutting down the passage of sunlight, and providing abundant condensation nuclei. Although pollution poses a general problem both for city dwellers and planners, it is convenient to examine its sources under the following headings:

(a) *Aerosols*. The production of suspended particulate matter (measured in $mg\,m^{-3}$ or $\mu g\,m^{-3}$) chiefly of carbon, lead and aluminium compounds and silica.
(b) *Gases*. The production of gases (measured in parts per million: ppm) can be viewed either from the traditional standpoint with its concentration on industrial and domestic coal burning and the production of such gases as sulphur dioxide (SO_2), or from the newer standpoint of petrol and oil combustion and the production of carbon monoxide (CO), hydrocarbons (Hc), nitrogen oxides (NO_x), ozone (O_3) and the like.

In dealing with atmospheric pollution it must be remembered, firstly, that the diffusion or concentration of pollutants is a function both of atmospheric stability (especially the presence of inversions) and of the features of horizontal air motion, secondly, that aerosols are removed from the atmosphere by settling out and by washing out, and, thirdly, that certain gases are susceptible to complex chains of photochemical changes which may destroy some gases but produce others.

a Aerosols As has already been pointed out (see ch. 1, A.2 and 4), the thermal economy of the globe is significantly affected by the natural production of aerosols which are deflated from deserts, erupted from volcanoes, produced by fires and so on. Their overall thermal effect is probably one of warming, due to increased absorption, and this augments the temperature rise associated with increasing amounts of carbon dioxide and certain trace gases (see fig. 1.3). Over the last century the average dust concentration has increased, particularly in Eurasia, due only in part to such eruptions as those of Mt Agung in Bali (1963) and Kamchatka (1966). The proportion of atmospheric dust directly or indirectly attributable to human activity has been estimated at 30% (see ch. 1, A.4). As an example of the latter, it is interesting that the north African tank battles of the Second World War disturbed the desert surface to such an extent that the material subsequently deflated was visible in clouds over the Caribbean.

The concentration of small nuclei (0·01–0·1 μm diameter) averages some 9500 cm^{-3} in the British countryside, but is typically 150,000 cm^{-3} and can

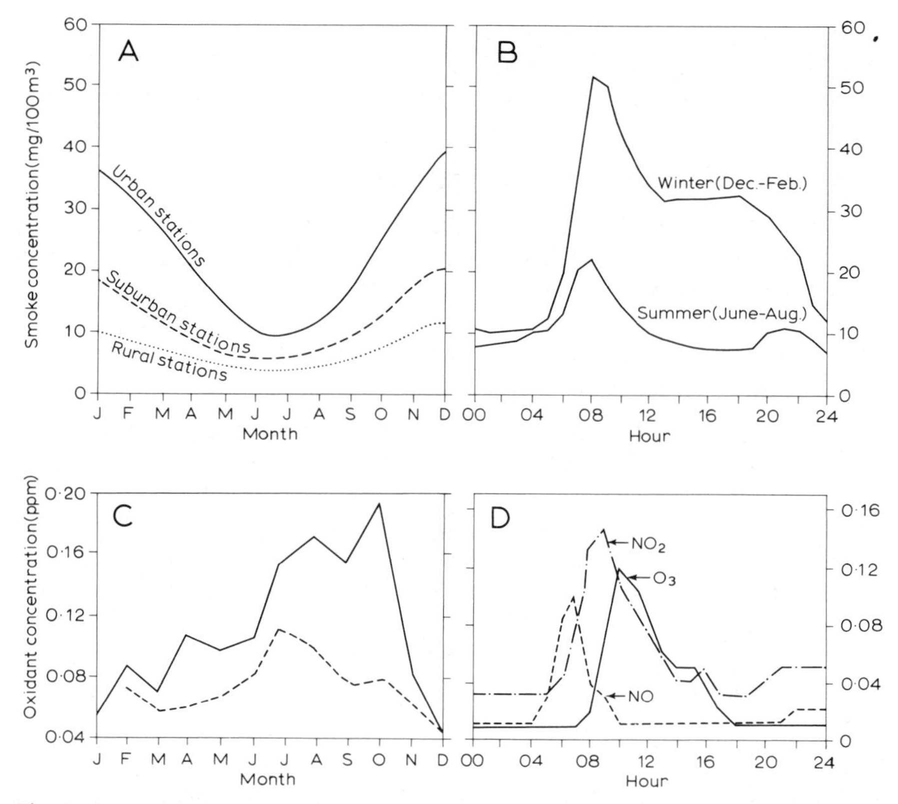

Fig. 7.19. Annual and daily pollution cycles. A Annual cycle of smoke pollution in and around Leicester, England, during the period 1937–9, before smoke abatement legislation was introduced (*after Meetham 1952*). B Diurnal cycle of smoke pollution at Leicester during summer and winter, 1937–9 (*after Meetham 1952*). C Annual cycle of mean daily maximum 1-hour average oxidant concentrations for Los Angeles (1964–5) and Denver (1965) (dashed) (*after US DHEW 1970 and Oke 1978*). D Diurnal cycles of nitric oxide (NO), nitrogen dioxide (NO_2) and ozone (O_3) concentrations in Los Angeles on 19 July 1965 (*after US DHEW 1970 and Oke 1978*).

reach 4,000,000 cm^{-3} in cities, as was measured near ground level in the industrial section of Vienna in 1946. Similarly, the concentration of larger particles (0·5–10 μm diameter) has been measured at 25–30 cm^{-3} in the city of Leipzig, as against 1–2 cm^{-3} in the rural environs. The greatest concentrations of smoke generally occur with low wind speeds, low vertical turbulence, temperature inversions, high relative humidities and air moving from the pollution sources of factory districts or areas of high density housing (pl. 38). The character of domestic heating and power demands causes city smoke pollution to take on striking seasonal and diurnal cycles, with the greatest concentrations occuring at about 0800 in early winter (fig. 7.19). The sudden morning increase is also partly a result of natural processes. Pollution trapped during the night beneath a stable layer a few

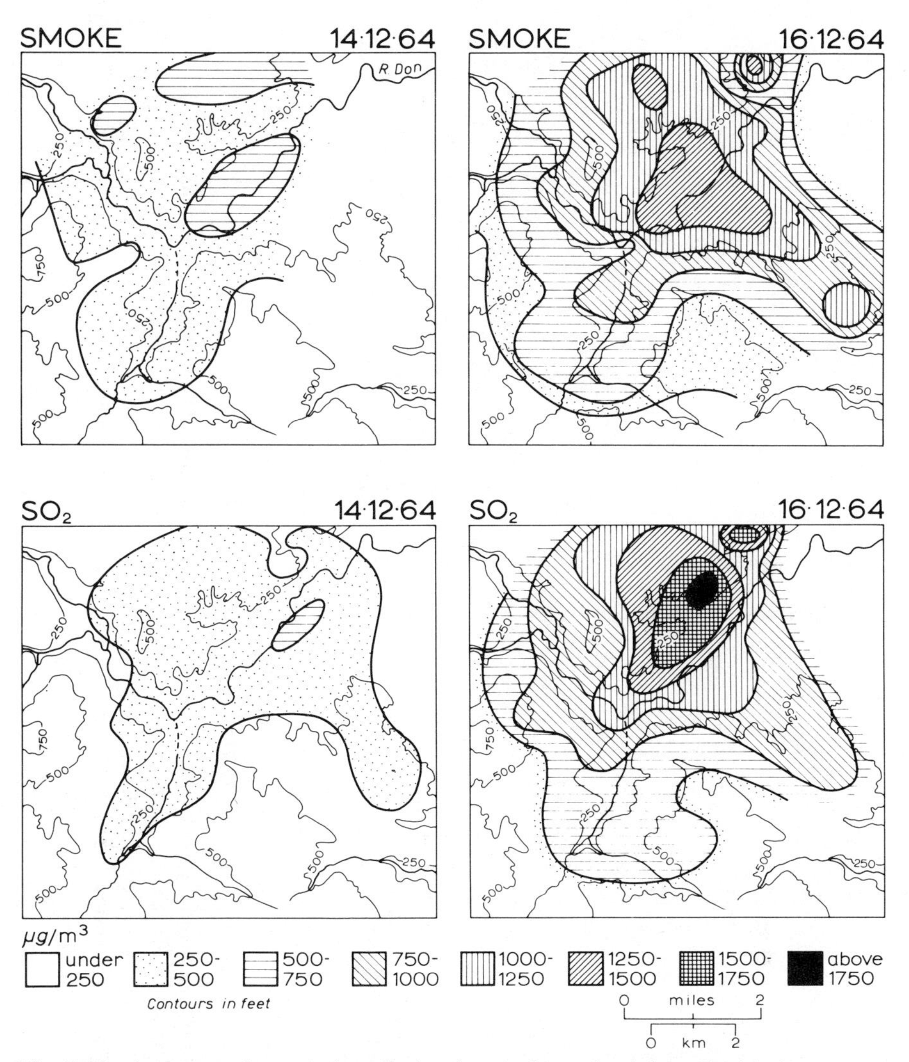

Fig. 7.20. Average values of air pollution by smoke and sulphur dioxide for Sheffield, England, on 14 and 16 December 1964 (*from Garnett 1967*).

hundred metres above the surface may be brought back to ground level when thermal convection sets off vertical mixing. Figure 7.20 shows the striking results of the accumulation of air pollution that occurred over the British industrial city of Sheffield in mid-December 1964 during a period of cloudless skies, weak air flow, maximum long-wave radiation and the development of near-surface temperature inversions and radiation fog. These conditions were associated with a smoke concentration of 10% above the monthly average on 14 December, which increased to 100% above the average on 16 December.

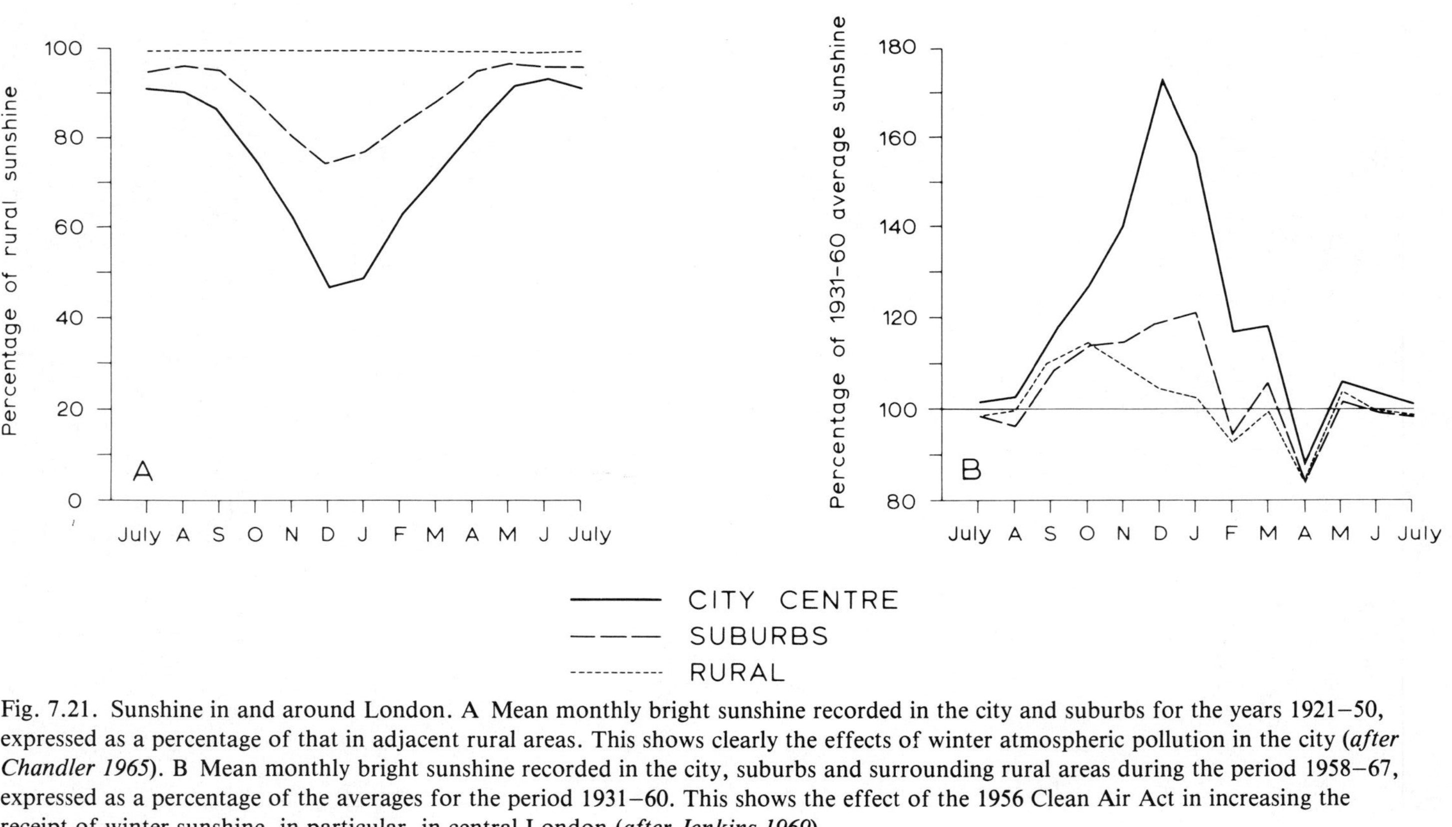

Fig. 7.21. Sunshine in and around London. A Mean monthly bright sunshine recorded in the city and suburbs for the years 1921–50, expressed as a percentage of that in adjacent rural areas. This shows clearly the effects of winter atmospheric pollution in the city (*after Chandler 1965*). B Mean monthly bright sunshine recorded in the city, suburbs and surrounding rural areas during the period 1958–67, expressed as a percentage of the averages for the period 1931–60. This shows the effect of the 1956 Clean Air Act in increasing the receipt of winter sunshine, in particular, in central London (*after Jenkins 1969*).

The most direct effect of particulate pollution is to reduce incoming radiation and sunshine. Pollution, plus the associated fogs (termed *smog*), used to cause some British cities to lose 25–55% of the incoming solar radiation during the period November to March (pl. 38). In 1945 it was estimated that the city of Leicester lost 30% of incoming radiation in winter, as against 6% in summer. These losses are naturally greatest when the sun's rays strike the smog layer at low angle. Compared with the radiation received in the surrounding countryside, Vienna loses 15–21% of radiation when the sun's altitude is 30%, but the loss rises to 29–36% with an altitude of 10°. The effect of smoke pollution is dramatically illustrated by fig. 7.21 by comparing conditions before and after the enforcing of the Clean Air Act of 1956 in London. Before 1950 there was a striking difference of sunshine between the surrounding rural areas and the city centre (fig. 7.21A) which could mean a loss of mean daily sunshine of 16 minutes in the outer suburbs, 25 minutes in the inner suburbs and 44 minutes in the city centre. It must be remembered, however, that smog layers also impeded the re-radiation of surface heat at night and that this blanketing effect contributed to higher night-time city temperatures. The use of smokeless fuels and other pollution controls cut London's total smoke emission from $1 \cdot 4 \times 10^8$ kg (141,000 tons) in 1952 to $0 \cdot 9 \times 10^8$ kg (89,000 tons) in 1960, and fig. 7.21B shows the increase in average monthly sunshine figures for the period 1958–67 as compared with those of 1931–60.

The abundance of condensation nuclei in city atmospheres, particularly those situated on low-lying land adjacent to large rivers, explains the former abundance of city fogs. Between August 1944 and December 1946, for example, suburban Greenwich had a monthly average of more than 20 days with good visibility of 0900 hours, whereas central London had less than 15. Occasionally very stable atmospheric conditions combined with excessive pollution production to give dense smog of a lethal character. During the period 5–9 December 1952 a temperature inversion over London caused a dense fog with visibility less than 10 m for 48 consecutive hours, resulting in 12,000 more deaths (mainly from chest complaints) during the period December 1952 to February 1953 compared with the same period the previous year. The close association of the incidence of fog with increasing industrialization and urbanization was well shown by the city of Prague, where the mean annual number of days with fog rose from 79 during the period 1860–80 to 217 during 1900–20.

b Gases Along with the pollution by smoke and other particulate matter produced by the traditional urban and industrial activities involving the combustion of coal and coke, has been associated the generation of pollutant gases. Before the Clean Air Act it was estimated that, whereas 80–90% of London's smoke was produced by domestic fires, these were responsible for only 30% of the sulphur dioxide released in the atmosphere – the remainder being contributed by electricity power stations (41%) and factories (29%). Figure 7.20 illustrates the association between pollution by smoke and that by sulphur dioxide in Sheffield some 20 years ago; it is significant that on 16 December 1964 the concentration of sulphur dioxide in the city air had risen to three times the monthly average.

Urban complexes are being affected by a newer less obvious, but nevertheless equally serious, form of pollution resulting from the combustion of petrol and oil by cars, lorries and aircraft, as well as from petro-chemical industries. Los Angeles, lying in a topographically constricted basin and often subject to temperature inversions, is the prime example of such pollution, although this affects all modern cities to some extent. In Los Angeles seven million people use some four million private cars, consuming 30 million litres of gasoline per day and producing more than 12,000 tones of pollutants. To this is added the results of the consumption of 0·5 million litres per day of diesel fuel by 13,500 lorries and buses and 2·5 million litres of aviation fuel consumed in the vicinity of the city. Even with controls, 7% of the gasoline from private cars is emitted in an unburned or poorly oxidized form, another 3·5% as photochemical smog and 33–40% as carbon monoxide. The production of the Los Angeles smog which, unlike traditional city smogs, occurs characteristically during the daytime in summer and autumn (see fig. 7.19C and D) is the result of a very complex chain of chemical reactions termed the disrupted photolytic cycle (fig. 7.22). Ultraviolet radiation (0·37–0·42 μm) dissociates natural NO_2 into NO and O. Monatomic oxygen (O) may then combine with natural oxygen (O_2) to produce ozone (O_3). The ozone in turn reacts with the artificial NO to produce NO_2 (which goes back into the photochemical cycle forming a dangerous positive feedback loop) and oxygen. The hydrocarbons produced

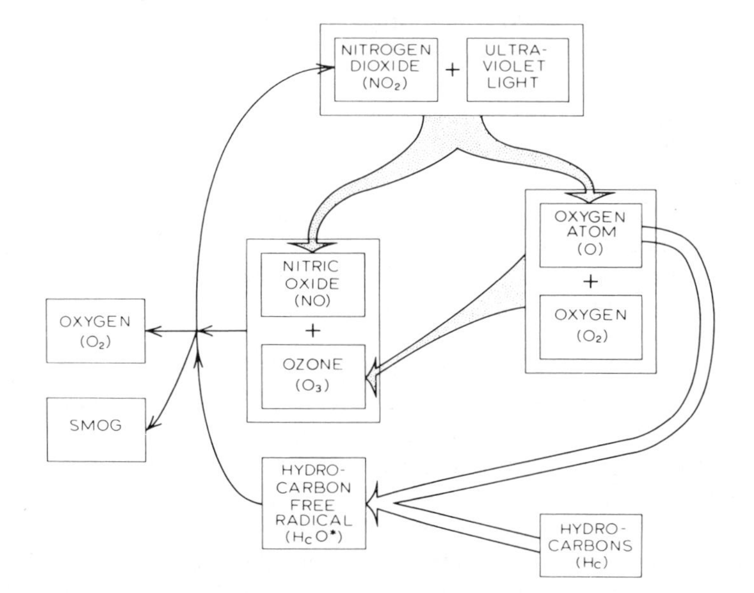

Fig. 7.22. The NO_2 photolytic cycle disrupted by hydrocarbons to produce photochemical smog (*US DHEW 1970 and Oke 1978*).

by the combustion of petrol combine with oxygen atoms to produce the hydrocarbon free radical H_cO^*, and these react with the products of the O_3–NO reaction to generate oxygen and photochemical smog. This smog exhibits well developed annual and diurnal cycles in the Los Angeles basin (fig. 7.19C and D). Annual levels of photochemical smog pollution in Los Angeles (derived from averages of the daily highest hourly figures) are greatest in late summer and autumn when clear skies, light winds and temperature inversions combine with amounts of high solar radiation. The diurnal figures of variations in individual components of the disrupted photolytic cycle reflect the complex reactions with, for example, an early morning concentration of NO_2 due to the build-up of traffic, and a peak of O_3 when receipts of incoming radiation are high. The effect of this smog is not only to modify the radiation budget of cities but to produce a human health hazard: in Tokyo, for instance, citizens sometimes wear respiratory masks in the street in self-defence!

c Pollution distribution Polluted atmospheres commonly assume well marked physical configurations around urban areas, which are very dependent upon environmental lapse rates, particularly the presence of temperature inversions and on wind speed. A pollution dome forms as a result of the collection of pollution below an inversion forming the urban boundary layer (fig. 7.23A). A wind speed as low as 2 m s^{-1} (5 mph) is sufficient to displace the Cincinnati pollution dome downwind, and a wind speed of $3{\cdot}5 \text{ m s}^{-1}$ (8 mph) will disperse it into a plume. Figure 7.23B shows a section of an urban plume with the volume above the urban canopy of the building tops filled by buoyant mixing circulations. *Fumigation* is the term used when an inversion lid prevents upward dispersion, but lapse conditions due to morning heating of the surface air allow convective plumes and associated downdraughts to bring pollution back to the surface. Downwind, *lofting* occurs above the temperature inversion at the top of the rural boundary layer dispersing the pollution upwards. Figure 7.23C illustrates some of the features of a pollution plume up to 160 km (100 miles) downwind of St Louis on 18 July 1975. In view of the complexity of photochemical reactions, it is of note that ozone increases downwind due to photochemical reactions within the plume but decreases over power plants as the result of other reactions with the emissions. This plume was observed to stretch for a total distance of 240 km (150 miles), but under conditions of an intense pollution source, steady large-scale surface airflow and vertical atmospheric stability pollution plumes may extend downwind for hundreds of kilometres. Plumes originating in the Chicago–Gary conurbation have been observed from high-flying aircraft to extend almost to Washington DC, 950 km (600 miles) away.

2 Modification of the heat budget

The energy balance of the built surface is similar to those described earlier in this chapter except for the heat production resulting from human energy

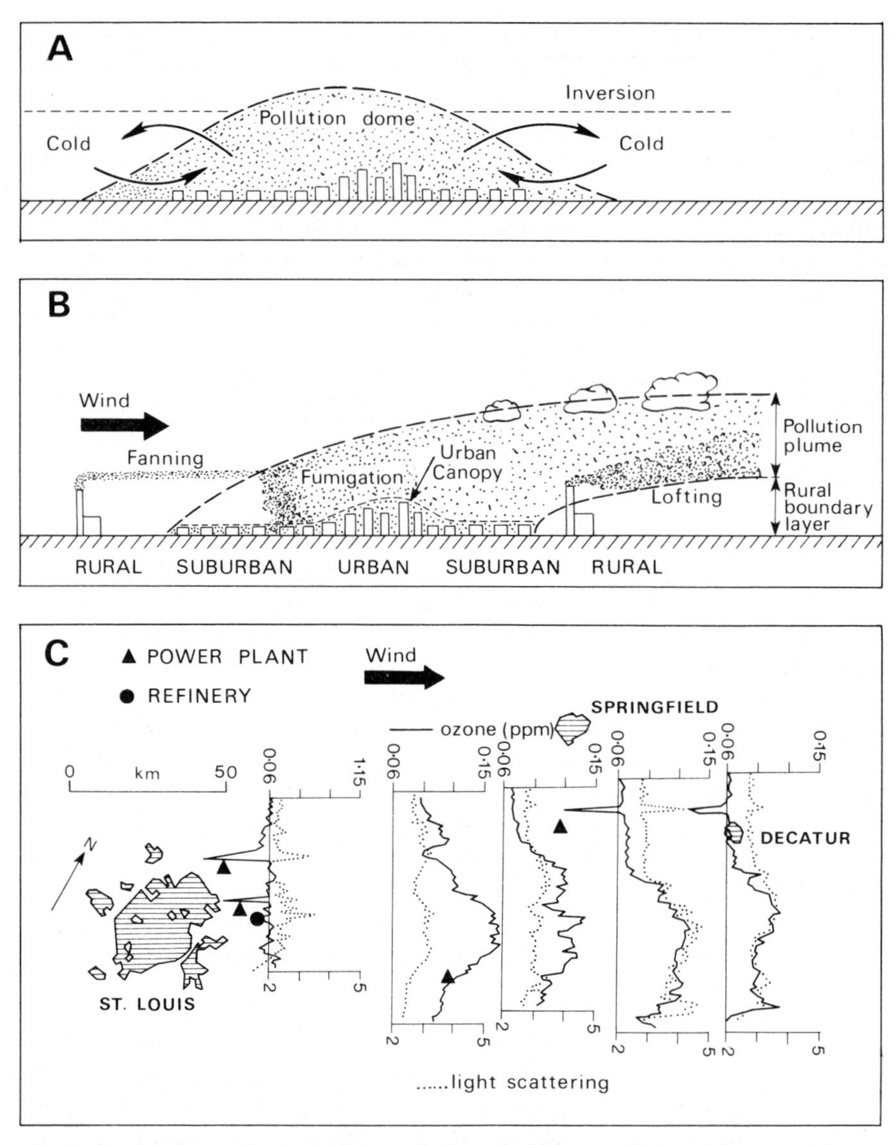

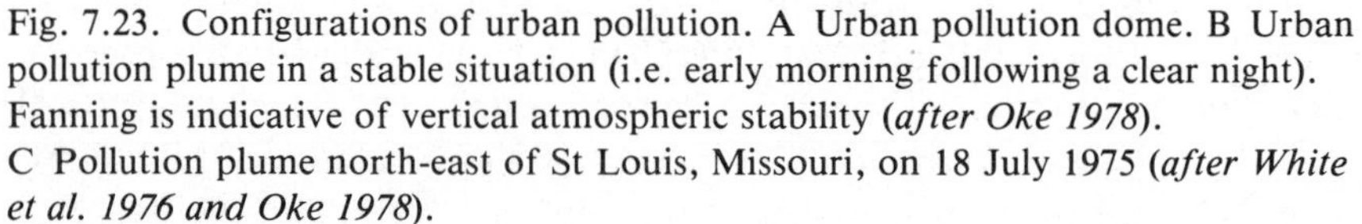

Fig. 7.23. Configurations of urban pollution. A Urban pollution dome. B Urban pollution plume in a stable situation (i.e. early morning following a clear night). Fanning is indicative of vertical atmospheric stability (*after Oke 1978*).
C Pollution plume north-east of St Louis, Missouri, on 18 July 1975 (*after White et al. 1976 and Oke 1978*).

consumption by combustion, which may even exceed R_n during the winter in some cities. Although R_n may not be greatly different from that in nearby rural areas (except during times of significant pollution) heat storage by surfaces is greater (20–30% of R_n by day) leading to greater nocturnal values of H and, in particular, LE is much less in city centres. After long dry periods evapotranspiration may be zero in city centres, except for certain industrial operations, and in the case of irrigated parks and gardens where LE may exceed R_n. This lack of LE means that by day 70–80% of R_n may be transferred to the atmosphere as sensible heat (H). Beneath the urban canopy, the microclimates of the streets and 'urban canyons' are dominated by the effects of elevation and aspect on the energy balance, which may vary strikingly within even one street.

The thermal characteristics of urban areas are in marked contrast to those of the surrounding countryside, and the generally higher urban temperatures are the result of the interaction of the following factors:

(a) Changes in the radiation balance due to atmospheric composition.
(b) Changes in the radiation balance due to albedo, heat conductivity and thermal capacity or urban surface materials.
(c) The production of heat by human activities.
(d) The reduction of heat diffusion due to changes in airflow patterns as the result of urban surface roughness.
(e) The reduction in thermal energy required for evaporation and evapotranspiration due to the surface character, rapid drainage and generally lower wind speeds of urban areas.

Consideration of the last two factors will be left to ch. 7, D.3.

a Atmospheric composition Air pollution makes the transmissivity of urban atmospheres significantly less than that of nearby rural areas. For example, during the period 1960–9 the atmospheric transmissivity over Detroit averaged 9% less than that for nearby areas, and reached 25% less under calm conditions. Table 7.1 gives comparable urban and rural energy budget figures for the Cincinnati region during the summer of 1968 under anticyclonic conditions with <3/10 cloud and a wind speed of $<2\,\mathrm{m\,s^{-1}}$ (5 mph).

b Urban surfaces Primary controls over a city's thermal climate are the character and density of urban surfaces, that is, the total *surface* area of buildings and roads, as well as the building geometry. Table 7.1 shows the relatively high heat absorption of the city surface. A problem of measurement is that, the stronger the urban thermal influence, the weaker is the heat absorption *at street level*, and, consequently, observations made only in streets may lead to erroneous results.

c Human heat production Numerous studies show that large urban conurbations now produce energy through combustion at rates comparable with incoming solar radiation in winter. Winter receipts from solar radiation average around $25\,\mathrm{W\,m^{-2}}$ in Europe, compared with similar heat production from large cities. Figure 7.24 illustrates the magnitude and spatial scale of artificial and natural energy fluxes. In Cincinnati, a significant proportion of the energy budget is generated by human activity, even in summer (table 7.1). This heat production averaged $26\,\mathrm{W\,m^{-2}}$ or more, of which two-thirds

Table 7.1 Energy budget figures for the Cincinnati region during the summer of 1968 (W m^{-2}) (*from Bach and Patterson 1969*).

	Central business district			*Surrounding country*		
	0800	*1300*	*2000*	*0800*	*1300*	*2000*
Short wave, incoming $(Q + q)$	288*	763	—	306	813	—
Short wave, reflected $[Q + q)a]$	42†	120†	—	80	159	—
Net long wave radiation (L_n)	−61	−100	−98	−61	−67	−67
Net radiation (R_n)	184	543	−98	165	587	−67
Heat produced by human activity	36	29	26‡	0	0	0

* Pollution peak.
† An urban surface reflects less than agricultural land, and a rough skyscraper complex can absorb up to 6 times more incoming radiation.
‡ Replaces more than $\frac{1}{4}$ of the long wave radiation loss in the evening.

was produced by industrial, commercial and domestic sources and one third by cars. In terms of the future, it has been estimated that by AD 2000 the Boston–Washington megapolis may accommodate up to 56 million people in a continuous urban area of 30,000 km^2 (11,600 sq miles), and that this concentration of human activity would produce heat equivalent to 50% of the total winter solar radiation recorded on a horizontal surface and 15% of the total summer solar radiation.

d Heat islands The net effect of the urban thermal processes is to make city temperatures generally higher than those in the surrounding rural areas, mainly due to the turbulent diffusion of sensible heat from warm buildings and the absorption of long wave radiation emitted by the city surface and its pollution blanket. This *heat island* effect may result in minimum urban temperatures being 5°–6°C (9°–11°F) greater than those of the surrounding countryside, and these differences may be as much as 6°–8°C (11°–14°F) in the early hours of calm, clear nights in large cities, when the heat stored by urban surfaces during the day (augmented by combustion heating) is released. It should be noted that, because this is a *relative* phenomena, the heat island effect also depends on the rate of rural cooling, which is influenced by the magnitude of the regional environmental lapse rate.

For the period 1931–60, the centre of London had a mean annual temperature of 11·0°C (51·8°F), comparing with 10·3°C (50·5°F) for the suburbs, and 9·6°C (49·2F) for the surrounding countryside. Calculations for London in the 1950s indicated that domestic fuel consumption gave rise to a 0·6°C warming in the city in winter and that accounted for one-third to one-half of the average city heat excess compared with adjacent rural areas. Differences are even more marked during still air conditions, especially at night under a regional inversion (fig. 7.25). For this heat island effect to operate effectively there must be wind speeds of less than 5–6 m s^{-1} (12–14 mph), and it is especially in evidence on calm nights during summer and early autumn when it has steep cliff-like margins and the highest

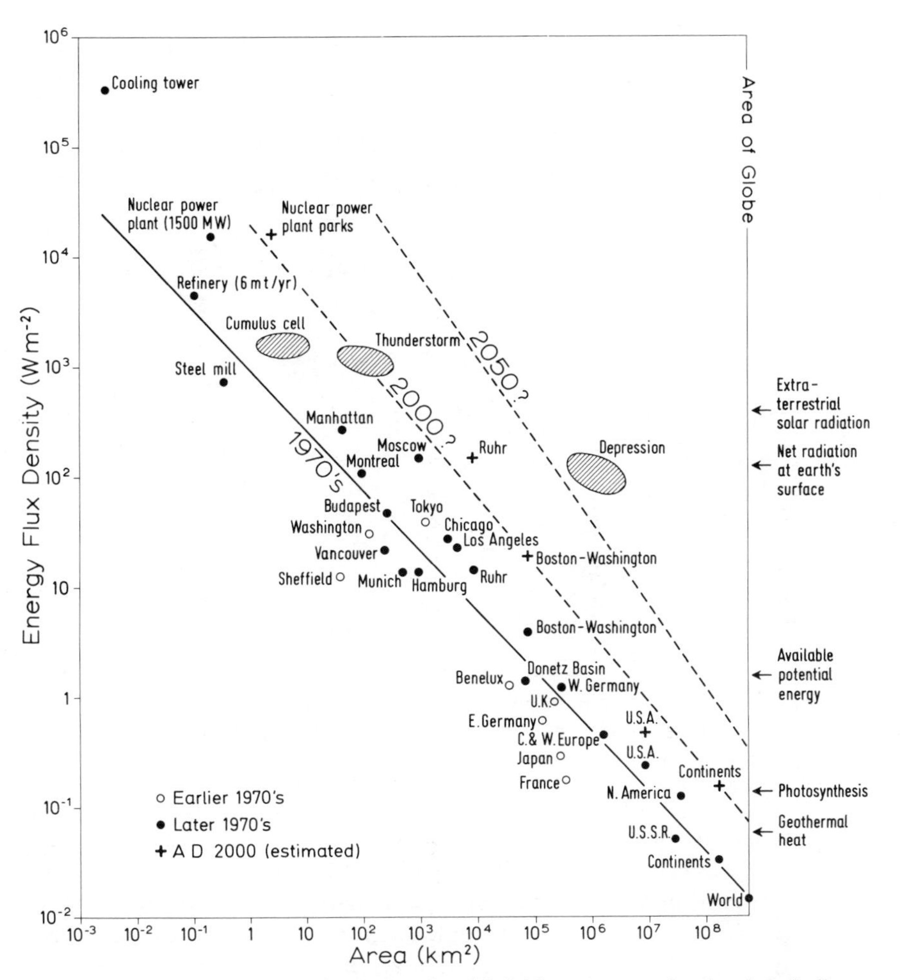

Fig. 7.24. A comparison of natural and artificial heat sources in the global climate system on small, meso- and synoptic scales. Generalized regressions are given for artificial heat releases in the 1970s (early 1970s circles, late 1970s dots), together with predictions for the years 2000 (crosses) and 2050 (*after Pankrath 1980 and Bach 1979*).

temperatures associated with the highest density of urban dwellings. In the absence of regional winds, a well-developed heat island may even generate its own inward local wind circulation at the surface. Thus the thermal contrasts of a city, like those of many other of its climatic features, depend on its topographic situation, and are greatest for sheltered sites with light winds. The fact that for London urban/rural temperature differences are greatest in summer, when direct heat combustion and atmospheric pollution are at a minimum, indicates that heat loss from buildings by radiation is the most important single factor contributing to the heat island effect. Seasonal differences are not necessarily the same, however, in other macroclimatic zones.

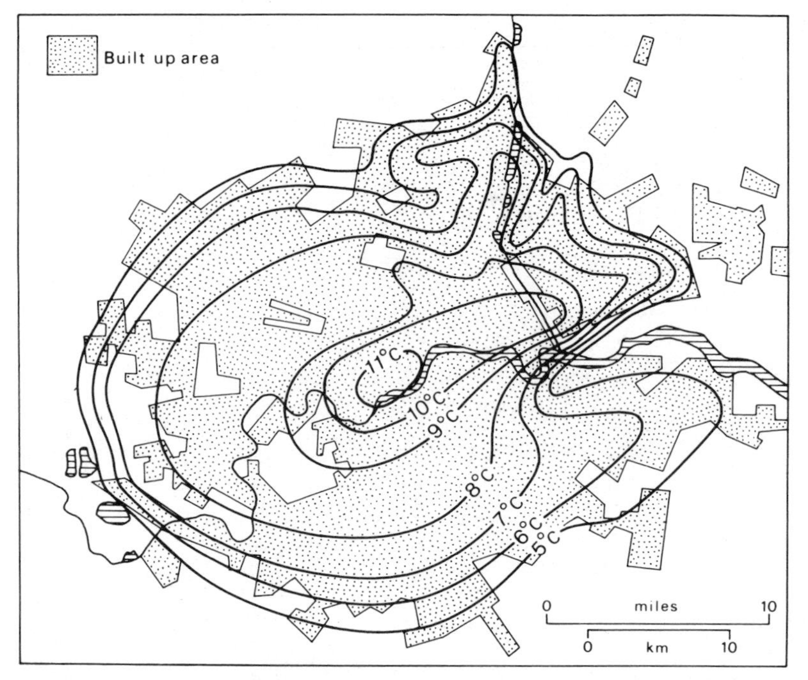

Fig. 7.25. Distribution of minimum temperatures (°C) in London on 14 May 1959, showing the relationship between the 'urban heat island' and the build-up area (*after Chandler 1965*).

The effects on minimum temperatures are especially significant. Cologne, for example, has an average of 34% less days with minima below 0°C (32°F) than its surrounding area, the corresponding figure for Basle being 25% less. In London, Kew has an average of some 72 more days with frost-free screen temperatures than rural Wisley. Precipitation characteristics are also affected; in pre-1917 Berlin 21% of the incidences of rural snowfall were associated with either sleet or rain in the city centre.

Although it is difficult to isolate changes in temperatures that are due to urban controls from those due to other influences (see ch. 8), it has been suggested that city growth is often accompanied by an increase in mean annual temperature, that of Osaka, Japan, rising by 2·6°C (4·5°F) in the last 100 years and that of Tokyo by 1·5°C (3°F). These results may be coincidental, however, since there appears to be no simple relationship between city size and intensity of the heat island. Leicester, when it had a population of 270,000, exhibited warming comparable in intensity with that of central London over smaller sectors. Thus suggests that the thermal influence of city size is not as important as that of urban density. The vertical extent of the heat island is little known, but is thought to exceed 100–300 m, especially early in the night. In the case of cities with skyscrapers, the vertical and horizontal patterns of wind and temperature are very complex (see fig. 7.26) for wind conditions).

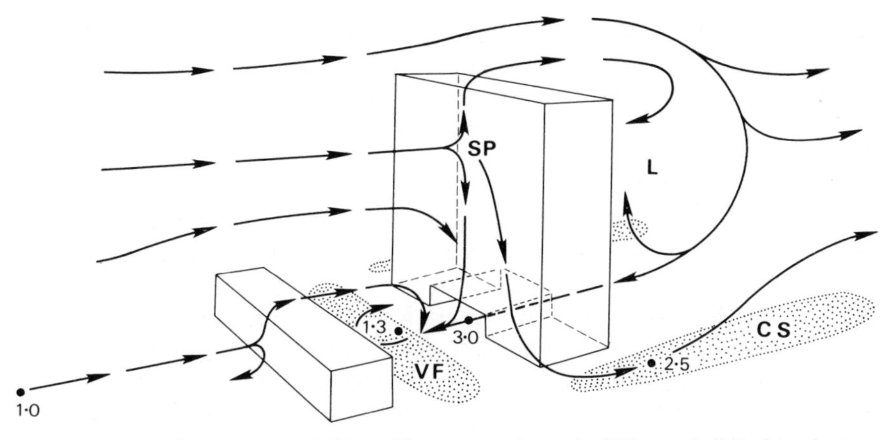

Fig. 7.26. Details of urban airflow. Flow around two buildings of differing size and shape. Numbers give relative wind speeds; stippled areas are those of high wind velocity and turbulence at street level; SP = stagnation point; CS = corner-stream; VF = vortex flow; L = lee eddy (*after Plate 1972 and Oke 1978*).

3 Modification of surface characteristics

a Airflow On the average, city wind speeds are lower than those recorded in the surrounding open country owing to the sheltering effect of the buildings, and central average wind speeds are usually at least 5% less than those of the suburbs. In 1935, for example, winds exceeding 10·5 m s^{-1} (24 mph) were recorded on the relatively open Croydon Airport (London suburbs) for a total of 371 hours, whereas the corresponding figure was only 13 hours for the closely built-up South Kensington. However, the urban effect on air motion varies greatly depending on the time of day and season. During the day, city wind speeds are considerably less than those surrounding rural areas, but during the night the greater mechanical turbulence over the city means that the higher wind speeds aloft are transferred to the air at lower levels by turbulent mixing. During the day (1300 hours) the mean annual wind speed for London Airport (open country within the suburbs) was 2·9 m s^{-1} (6·4 mph), compared with a 2·1 m s^{-1} (4·7 mph) in central London for the period 1961–2. The comparative figures for night (0100 hours) were 2·2 m s^{-1} (4·9 mph) and 2·5 m s^{-1} (5·6 mph). Rural–urban wind speed differences are most marked with strong winds and the effects are therefore more evident during winter than during summer, when a higher proportion of low speeds is recorded in temperate latitudes.

Urban structures have considerable effects on the movement of air both by producing turbulence as a result of their roughening the surface and by the channelling effects of the urban canyons. Some idea of the complexity of airflow round urban structures is given by fig. 7.26, illustrating the great differences in ground-level wind velocity and direction, the development of vortices and lee eddies, and the reverse flows which may occur. Structures play a major role in the diffusion of pollution within the urban canopy; for example, narrow streets often cannot be flushed by vortices. The formation of high-velocity streams and eddies in the usually dry and dusty urban atmosphere, where there is ample debris supply, leads to general urban airflows of only 5 m s^{-1} (11 mph) being annoying, and those of more than 20 m s^{-1} (45 mph) being dangerous.

b Moisture The effect of urbanization on surface-moisture relationships is also important. The absence of large bodies of standing water and the rapid removal of surface run-off through drains decreases local evaporation. In addition, the lack of an extensive vegetational cover eliminates much evapotranspiration and this is an important source of augmenting urban heat. For these reasons, the air of mid-latitude cities has a tendency towards lower absolute humidities than that of their surroundings, especially under conditions of light wind and cloudy skies. On other occasions of calm, clear weather the streets trap warm air, which retains its moisture because less dew is deposited on the warm surfaces of the city. Humidity contrasts between urban and rural areas are most noticeable in the case of relative humidity, which can be as much as 30% less in the city by night as a result of the higher temperatures.

Urban influences on precipitation (excluding fog) are much more difficult to determine with any precision, partly because there are few rain gauges in

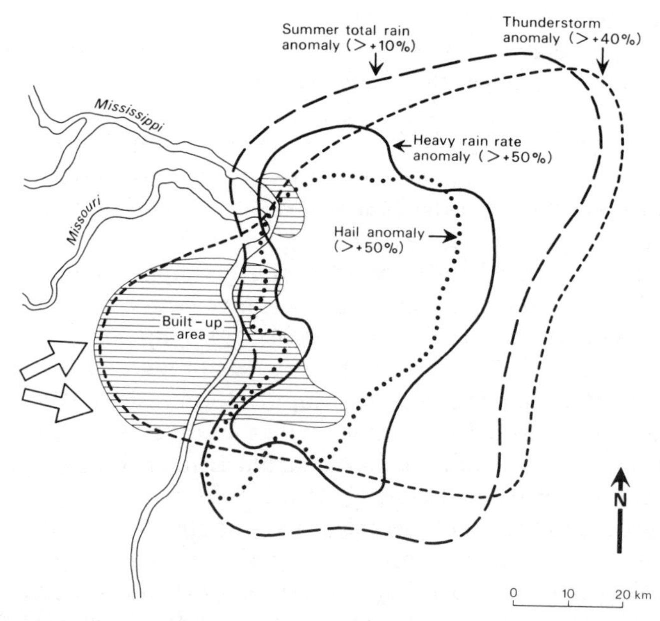

Fig. 7.27. Anomalies of summer rainfall, rate of heavy rains, hail frequency and thunderstorm frequency downwind of the St Louis metropolitan area. Large arrows indicate the prevailing direction of motion of summer rain systems (*after Changnon 1979*).

cities and because air turbulence makes their 'catch' unreliable. It is now fairly certain, however, that urban areas in Europe and North America are responsible for local conditions which, in summer especially, can trigger off excesses of precipitation under marginal conditions. These triggers involve the orographic and turbulence effects of the buildings, the increased density of condensation nuclei, and thermal convection. Recordings for Munich showed 11% more days of light rain (0·1–0·5 mm or 0·004–0·12 in) than in the surrounding countryside, and Nürnberg has recorded 14% more thunderstorms than its rural environs. European and North American cities apparently record 6–7% more days with rain per year than their surrounding regions, giving 5–10% increase in urban precipitation. The effect is generally more marked in the cold season in North America, although urban areas in the Midwest of the United States significantly increase summer convective activity with more frequent thunderstorms and hail downwind of industrial areas of St Louis (for a distance of 30–40 km) compared with rural areas (fig. 7.27). The anomalies shown in fig. 7.27 are among the best documented of urban effects. Over south-east England during 1951–60, summer thunderstorm rain (which comprised 5–15% of the total precipitation) was especially concentrated in west, central and southern London (fig. 7.28) and contrasted strikingly with the distribution of mean annual total rainfall. During this period London's thunderstorm rain was of the order of 20–25 cm (8–10 in) greater than that in rural south-east England.

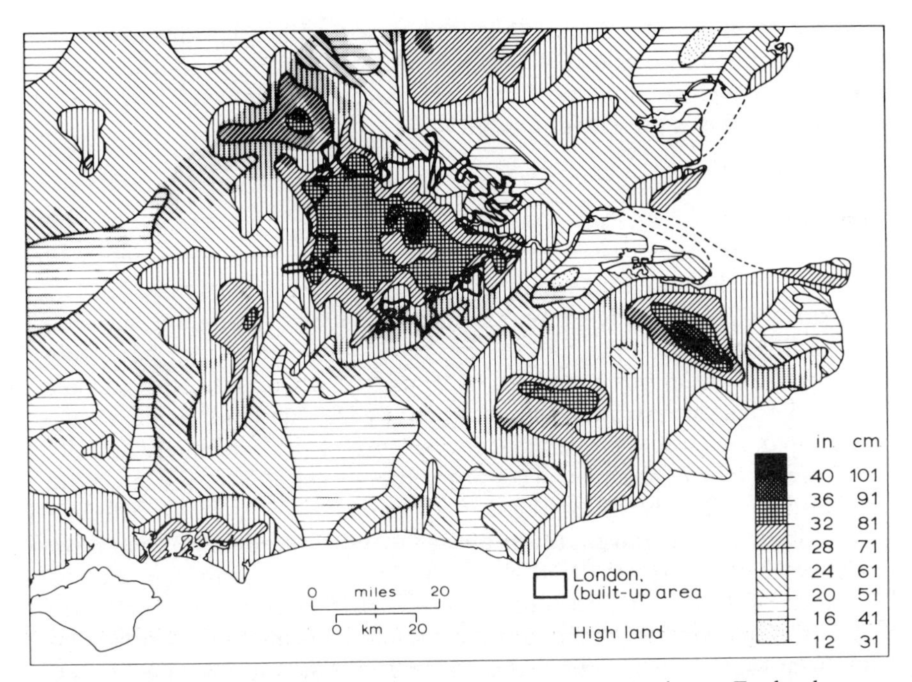

Fig. 7.28. The distribution of total thunderstorm rain in south-east England during the period 1951–60 (*after Atkinson 1968*).

Table 7.2 Average urban climatic conditions compared with those of surrounding rural areas (*partly after WMO 1970*).

atmospheric composition	carbon dioxide	× 2
	sulphur dioxide	× 200
	nitrogen oxides	× 10
	carbon monoxide	× 200(+)
	total hydrocarbons	× 20
	particulate matter	× 3 to 7
radiation	global solar	−15 to 20%
	ultra-violet (winter)	−30%
	sunshine duration	−5 to 15%
temperature	winter minimum (average)	+1° to 2°C
	heating degree days	−10%
wind speed	annual mean	−20 to 30%
	number of calms	+5 to 20%
fog	winter	+100%
	summer	+30%
cloud		+5 to 10%
precipitation	total	+5 to 10%
	days with <5 mm (<0·2 in)	+10%

Many of the results discussed in connection with urban influences are based on limited case studies. It is appropriate, therefore, to conclude with a summary of average climatic differences between cities and their surrounding rural areas. This is presented in table 7.2.

Summary

Small-scale climates are determined largely by the relative importance of the surface-energy-budget components, which vary in amount and sign depending on time of day and season. Bare land surfaces may have wide temperature variations controlled by H and G, whereas those of surface water bodies are strongly conditioned by LE and advective flows. Vegetated surfaces have more complex exchanges which are usually dominated by LE; this may account for more than 50% of the incoming radiation, especially where there is an ample water supply (including irrigation). Forests have a lower albedo (<0·10 for conifers) than most other vegetated surfaces (0·20–0·25). Their vertical structure produces a number of distinct microclimatic layers, particularly in tropical rain forests. Wind speeds are characteristically low in forests and trees form important shelter belts. Unlike short vegetation, the various types of trees

continued

exhibit a variety of rates of evapotranspiration and thereby differentially affect local temperatures and forest humidity. The effect of forests on precipitation has not yet been resolved but they may have a marginal topographic effect under convective conditions in temperate regions. The disposition of forest moisture is very much affected by canopy interception and evaporation, but forested catchments appear to have greater evapotranspiration losses than ones with a grass cover. Another major feature of forest microclimates is their lower temperatures and smaller diurnal ranges compared with surrounding areas.

Urban climates are dominated by the geometry and composition of built-up surfaces and by the effects of human urban activities. The composition of the urban atmosphere is modified by the addition of aerosols, producing smoke pollution and fogs, by industrial gases such as sulphur dioxide, and by a chain of chemical reactions, initiated by automobile exhaust fumes, which causes smog and inhibits both incoming and outgoing radiation. Pollution domes and plumes are produced around cities under appropriate conditions of vertical temperature structure and wind velocity. The urban heat budget is dominated by H and G, except in city parks, and as much as 70–80% of incoming radiation may become sensible heat which is very variably distributed between the complex urban-built forms. Urban influences combine to give generally higher temperatures than in the surrounding countryside, not least because of the growing importance of heat production by human activities. These factors give rise to the urban heat island which may be 6°–8°C warmer than surrounding areas in the early hours of calm, clear nights when heat stored by urban surfaces is being released. The heat island may be a few hundred metres deep, depending on the building configuration. Urban wind speeds are generally less than for rural areas by day, but the wind flow is extremely complex, depending on the geometry of city-built forms. Cities naturally tend to be less humid than rural areas but their topography, roughness and thermal qualities tend to intensify the effects of summer convective activity over and downwind of the urban areas, giving more thunderstorms and heavier storm rainfall.

8

Climatic variability, trends and fluctuations

Probably the aspect of climate which most interests the layman is speculation regarding its possible trends. Unfortunately, as well as being the most interesting, it is also the most uncertain aspect of meteorological research. Realization that climate has changed radically with time came only during the 1840s when indisputable evidence of former ice ages was obtained, yet in many parts of the world the climate has altered sufficiently, even within the last few thousand years, to affect the possibilities for agriculture and settlement. Reliable weather records have only been kept during the last hundred years or so and therefore it is only the recent climatic fluctuations which can be investigated adequately. The discussion in this chapter is mainly limited to these events, but first it is worthwhile considering the methods of handling the meteorological records which are available.

A Climatic data

1 Averages

The climate of a place is often regarded simply as its 'average weather', but vital climatic information is overlooked if the range and frequency of extremes are neglected. Averages can be markedly affected by extreme values and this is particularly true of the arithmetic mean (which is obtained by totalling the individual values and dividing by the number of occurrences). For this reason a 30- or 35-year period is normally required for the determination of climatic averages. Even so, certain types of data are very inadequately summarized by the arithmetic mean, especially when small values are frequent but very large ones occur occasionally. This situation is illustrated by fig. 8.1, where the *histograms* or frequency-distribution graphs of annual rainfall at Helwan (Egypt) and Aden are clearly dissimilar to those at Greenwich (England) and Padua (Italy). The profile for Padua approximates a symmetrical 'normal' distribution, where half the values lie above and half below the mean, and where the most frequent (or *modal*) category is equal to the mean (see note 1). For 1725–1924, the annual rainfall at Padua has a mean of 859 mm, and a mode of about 884 mm. The *median* for the

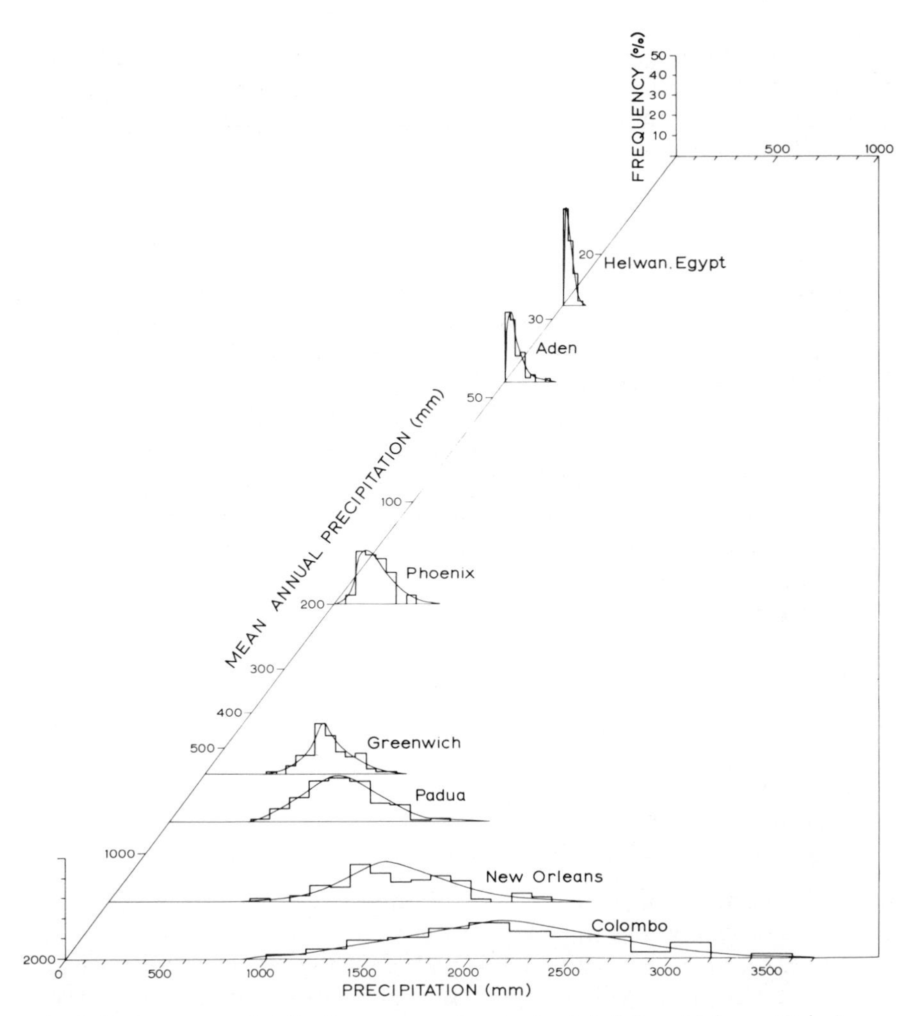

Fig. 8.1. Frequency distribution curves of annual rainfall for: Helwan (Cairo), Egypt (averaged over 37 years); Aden (55 years); Phoenix, Arizona (45 years); Greenwich (London), England (100 years); Padua, Italy (200 years); New Orleans (88 years); and Colombo, Ceylon (70 years).

same period at Padua is 847 mm. This is another useful measure of central tendency, which has exactly half the number of items in the data series above it and half below it. Consequently it is not biased by the occurrence of a few extreme figures. Modal or median values would be much more meaningful indicators of annual precipitation at Aden and Helwan where the graphs exhibit strong 'positive' skewness, i.e. 'tail' of the distribution is towards values greater than the mean. Positive skewness is especially marked when the mean of the distribution approaches zero, when a large number of records includes some infrequent events of high magnitude, and when the length of time to which the event is related is short (for example, rainfall figures referring to any one month are usually more skewed than those relating to the whole year). Another problem of frequency distributions is that they may have more than one frequency peak, as shown in fig. 8.1 for New Orleans, which is bimodal.

2 Variability

Variability about the average can be expressed in several ways. When the median is used it is common also to determine the upper and lower quartiles (Q_1 and Q_2), which are the central values between the median and the upper and lower extremes, respectively, i.e. the 25 and 75% points in the frequency distribution. The average deviation from the median is given by $(Q_1 - Q_2)/2$. A more widely used measure of variability is the standard deviation (σ, pronounced *sigma*) which is calculated by summing the square of the deviation of each value from the mean, dividing by the number of cases and then taking the square root.

$$\sigma = \sqrt{\frac{\Sigma (x_1 - \bar{x})^2}{n}}$$

where:

x = an individual value

$\bar{x} = \frac{x_1}{n}$ (the mean)

n = number of cases

Σ = sum of all values for $i = 1$ to n.

It is therefore a measure of average deviation, where the difficulty created by positive and negative departures (i.e. values greater and less than the mean) is removed by squaring each deviation and rectifying this by finally calculating the square root. Variability of rainfall may be compared between stations if the standard deviation is expressed as a percentage of the mean (the *coefficient of variation, CV*):

$$CV = \frac{\sigma}{\bar{x}} \times 100(\%)$$

Where the standard deviation is not available, use is sometimes made of the *mean deviation, MD.*

$$MD = \frac{\Sigma |x_1 - \bar{x}|}{n}$$

where |—| denotes the absolute difference, disregarding sign.

This measure of variability, standardized against the mean, ranges for annual precipitation from about 10–20% in western Europe and parts of monsoon India to over 50% in arid areas of the world (fig. 8.2). It is in such areas that a small change in the frequency of rainstorms can markedly affect

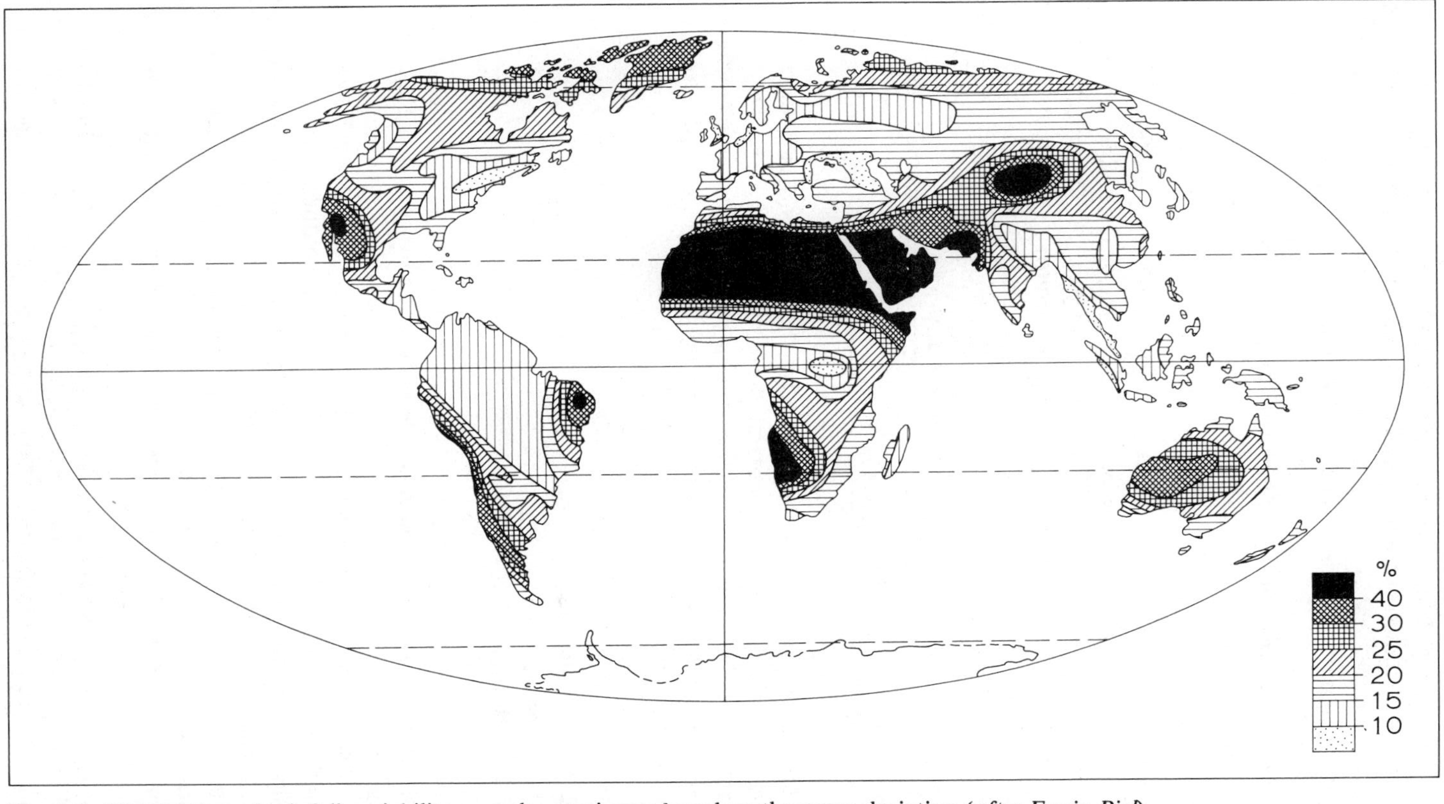

Fig. 8.2. Distribution of rainfall variability over the continents based on the mean deviation (*after Erwin Biel*).

the 'average' rainfall over a given period of years. It should be noted, however, that detailed examination of the precipitation in many diverse climatic regions shows that the apparent inverse relationship between annual total and variability is only very approximate. Moreover, a coefficient of variation $\geqslant 50\%$ in fact violates the statistical assumption of a normal frequency distribution on which this statistic is based.

3 Trends

It is obvious that the great year-to-year variability of climatic conditions may conceal gradual trends from one type of regime towards another. The effect of short-term irregularities can be removed by various statistical techniques, of which the simplest is the *running mean* (or *moving average*). The method is to calculate mean values for successive, overlapping periods of perhaps 5, 10 or 30 years, e.g.

$$\frac{\text{Year 1} + \text{Year 2} + \text{Year 3} + \text{Year 4} + \text{Year 5}}{5} = \text{Mean for Year 3}$$

$$\frac{\text{Year 2} + \text{Year 3} + \text{Year 4} + \text{Year 5} + \text{Year 6}}{5} = \text{Mean for Year 4, etc.}$$

This device smooths out the short-term fluctuations if periods of 20 or 30 years are used, thereby emphasizing the long-term trends. However, running means can also generate apparent regular periodic fluctuations where none exist. This can be shown by calculating running means for a series of random numbers.

B The climatic record

1 Evidence of climatic change

To understand the significance of climatic trends over the last hundred years they need to be viewed against the background of our general knowledge about past conditions.

Land and ocean sediments record numerous alternations between glacial and interglacial conditions during the last several million years. At least eight such cycles have occurred within the last million years, each averaging 125,000 years, with only about 10% of each cycle as warm as the present day. Only four or five glaciations are identified in the land evidence due to the absence of continuous sedimentary records, but it is likely that in each of these glacial intervals large ice sheets covered northern North America and northern Europe. Sea-levels were also lowered by about 100–150 m due to the large volume of water locked up in the ice. Records from tropical lake basins show that these regions were generally arid at these times. The last such glacial maximum climaxed about 18,000 years ago, but 'modern' climatic conditions only became established during post-glacial time – conventionally dated to

10,000 years Before Present (BP). This time scale is used whenever dates are based on radiocarbon (carbon-14, ^{14}C) dating or other radiometric methods involving isotopic decay processes, such as potassium–argon (K–Ar).

For the last glacial cycle and post-glacial time, information on climatic conditions is obtained indirectly from *proxy* records. For example, the advance and retreat of glaciers represents a response to winter snowfall and summer melt. The history of vegetation, which indicates temperature and moisture conditions, can be traced from pollen types preserved in lake sediments and peat bogs. Former lake shorelines indicate changes in moisture balance. Estimates of seasonal climatic elements can be made from studies of annual snow/ice layers in cores taken from polar ice sheets where no melt occurs. These layers also record past volcanic events through the inclusion of microparticles and chemical compounds in the ice. In forested areas where trees form an annual growth layer, the ring-width can be interpreted through dendroclimatological studies in terms of moisture availability (in semi-arid regions) and summer warmth (near the polar and alpine tree-lines). Pollen sequences, lake-level records and ice cores usually span the last 10,000–20,000 years (and exceptionally back to 125,000 BP), while tree rings rarely extend more than 1500 years ago. For more recent time, historical documents often record crop harvests or extreme weather events (droughts, river and lake freezing, etc.).

2 *Post-glacial conditions*

Following the final retreat of the continental ice sheets from Europe and North America between 10,000 and 7000 years ago the climate rapidly ameliorated in middle and higher latitudes. In the subtropics this interval was also generally wetter, with high lake levels in Africa and the Middle East. A *thermal maximum* was reached in the mid-latitudes about 5000 years ago, when summer temperatures are known to have been 1–2°C higher than today and the arctic tree-line was several hundred kilometres further north in Eurasia and North America. By this time, subtropical desert regions were again very dry and were largely abandoned by primitive man. A temperature decline set in around 2000 years ago with colder, wetter conditions in Europe and North America. Although temperatures have not since equalled those of the thermal maximum there was certainly a warmer period in many parts of the world between about AD 1000 and 1250, and this phase was marked by the Viking colonization of Greenland and the occupation of Ellesmere Island in the Canadian Arctic by Eskimos. A further deterioration followed and severe winters between AD 1550 and 1700 gave a 'Little Ice Age' with extensive arctic pack-ice and glacier advances in some areas to maximum positions since the end of the Ice Age. These advances occurred at dates ranging from the mid-seventeenth to the late nineteenth century in Europe, as a result of the lag in glacier response and minor climatic fluctuations. Figure 8.3 attempts to summarize these trends, but it must be stressed that at present only the gross features are represented, as we know little or nothing

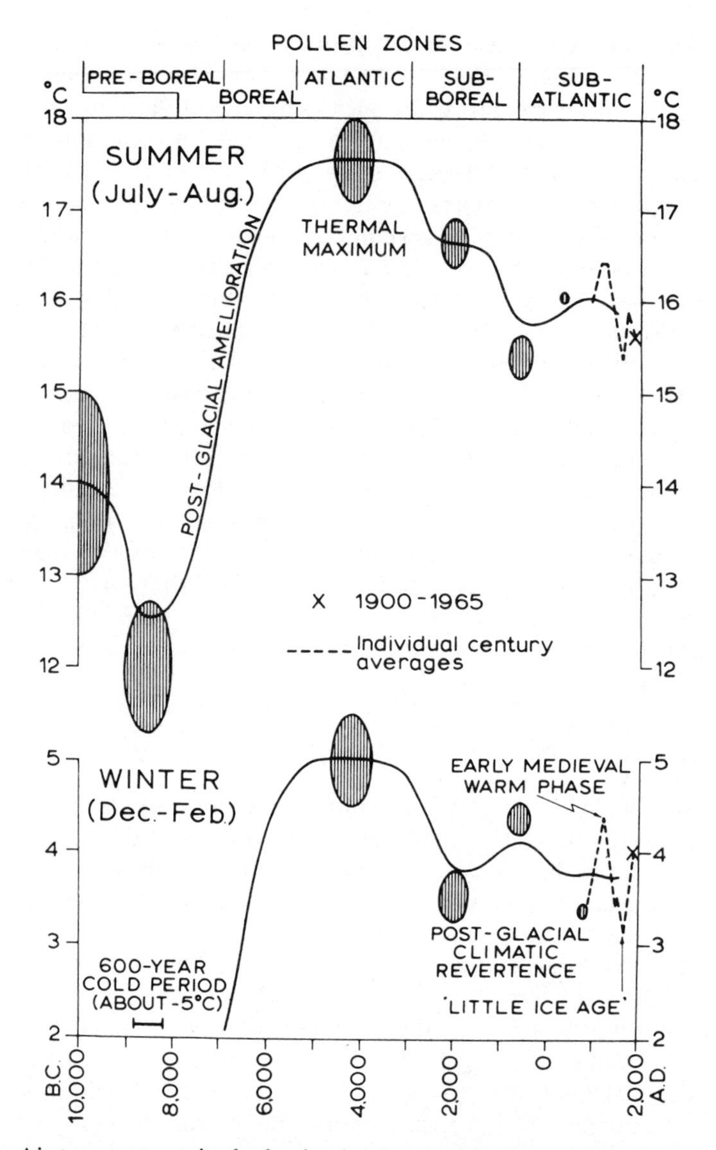

Fig. 8.3. Air temperatures in the lowlands of central England. Trends of the supposed 1000-year and 100-year averages since 10,000 BC (the latter calculated for the last millennium) (*after Lamb 1966*). Shaded ovals indicate the approximate ranges within which the temperature estimates lie and error margins of the radiocarbon dates.

about short-term fluctuations before the Medieval period, for example, and even the relative magnitudes of the changes before about AD 1700 can only be indicated in a very general way.

3 The last 100 years

Long instrumental records for stations in Europe and the eastern United

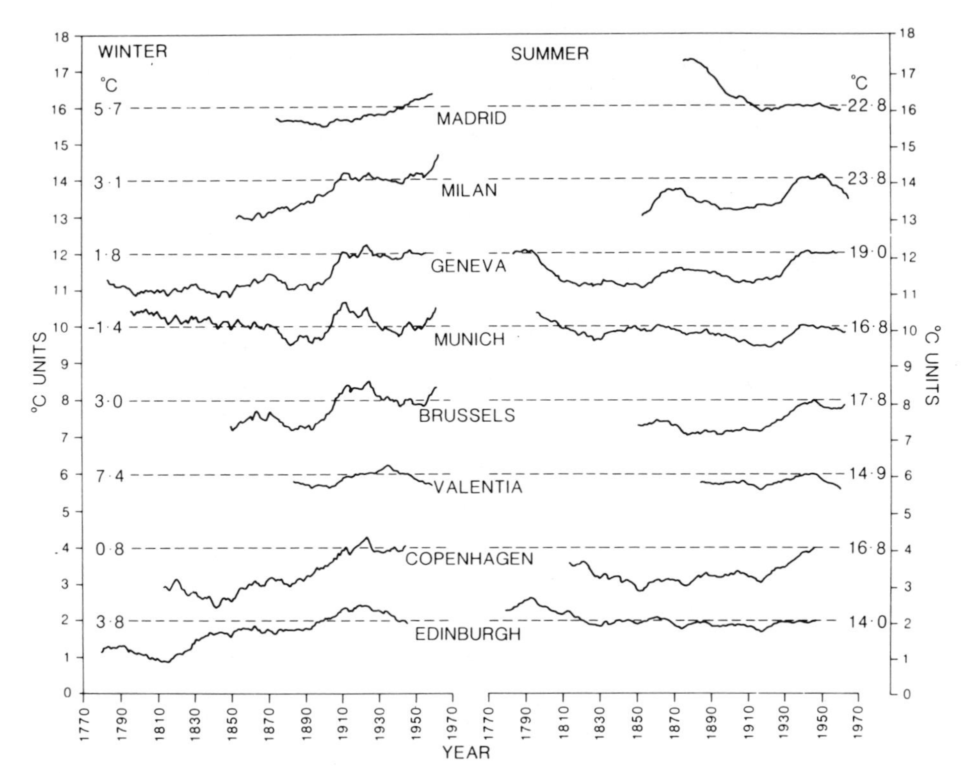

Fig. 8.4. Long-term records of winter and summer 30-year running mean temperatures at European stations (*from Schuurmans 1984*). Station mean values are shown at each side. (*Copyright © 1980/1982 by D. Reidel Publishing Company. Reprinted by permission*).

States indicate that the warming trend which ended the 'Little Ice Age' began early in the nineteenth century, although it was interrupted in some areas about 1880–90 (fig. 8.4). This trend was least in the tropics and greatest in cloudy, maritime regions of high latitudes (fig. 8.5). Winter temperatures were most affected, and on Svalbard the 20-year change in January mean temperature from 1900–19 to 1920–39 amounted to +7·8°C (14·0°F). This pattern is strong evidence of a heat-transporting mechanism. However, increased storminess might cause more frequent movement of warm air masses towards high latitudes, although increases in winter temperatures also occur in areas that are predominantly affected by northerly winds and therefore cannot receive heat directly from the source regions of warm air.

The effects of the temperature rise are apparent in many ways. There has been, for example, a rapid retreat of most of the world's glaciers. Glaciers in the North Atlantic area retreated extensively between about 1920 and the mid-1960s, due largely to temperature increases that have the effect of lengthening the ablation season with a corresponding raising of the snowline. Another tendency illustrating world warming was the retreat of arctic sea ice. Ports in the Arctic remained free of ice for longer periods

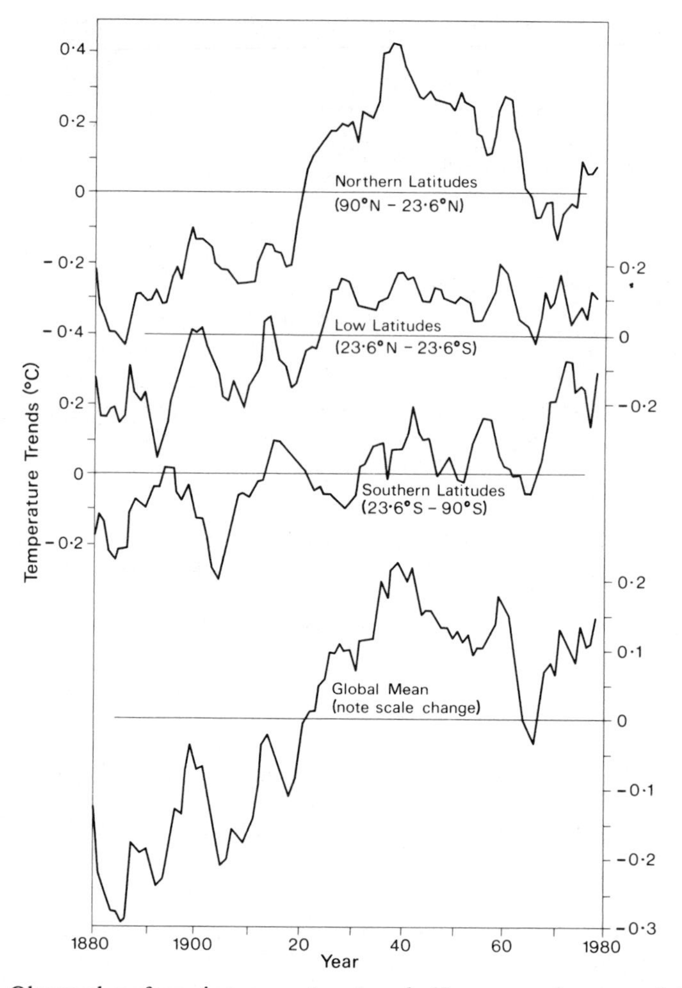

Fig. 8.5. Observed surface air temperature trends (5-year running means) for three latitudinal bands and the entire globe. Scales for low latitudes and the global mean are on the right (*after Hansen et al. 1981*).

during the 1920s–50s and cod extended their feeding northwards off west Greenland by 9° of latitude between 1919 and 1948.

Recent evidence (see fig. 8.5) suggests that the twentieth-century warming has been interrupted. Cooling has taken place especially over northern Siberia, the eastern Canadian Arctic and Alaska, with changes of the order of –2° to –3°C (–4° to –6°F) in winter mean temperature from 1940–9 to 1950–9. Perhaps in compensation slight winter warming has at the same time affected the western United States, eastern Europe and Japan. Whether or not this *downturn* of temperature represents a minor fluctuation or a longer-term trend remains to be seen, but it is clear that the latter would have important economic implications in many parts of the northern hemisphere. In the southern hemisphere the twentieth-century warming was delayed and

Table 8.1 Seasonal rainfall averages at Freetown (*after Kraus 1955*).

	May–October	*November–April*
1875–96	452·3 cm (178·05 in)	41·7 cm (16·41 in)
1907–31	312·1 cm (122·85 in)	23·1 cm (9·10 in)

may still be in progress (fig. 8.5). Note that on a global scale, the changes are within ±0·2° to 0·3°C of the 100-year mean.

While temperatures in the tropics seem to have undergone little change since the 1880s, the same is not true of rainfall totals. Over extensive parts of the tropics, but excluding monsoon Asia, there was a general decrease of annual rainfall about 1900, of the order of 30%. For example, at Freetown, Sierra Leone, in West Africa (see table 8.1), the early rains (May–June) and the late rains (September–October) showed a greater *relative* decrease than the very wet months of July and August, which points to a lengthening of the dry season as the main factor. The greater aridity in the tropics was apparently not accompanied by any compensating increase in temperate latitudes. Indeed, rainfall also decreased in south-east Australia and eastern North America to about latitude 40°, with a recovery of about 10% after 1940.

The west African records for this century (fig.8.6) show a tendency for both wet and dry years to occur in runs of up to 10–18 years. Precipitation minima were experienced in the 1910s, 1940s and post-1968, with intervening wet years, in all of sub-Saharan west Africa. Throughout the two northern zones outlined in fig. 8.6, means for 1970–84 were generally <50% of those for 1950–9 with deficits during 1981–4 equal to or exceeding those of the disastrous early 1970s' drought. It has been suggested that this is related to weakening of the tropical easterly jet stream and limited northward penetration of the west Africa south-westerly monsoon flow. However, S. E. Nicholson attributes the precipitation fluctuations to contraction/expansion of the Saharan arid core, rather than to north–south shifts of the desert margin. In Australia rainfall changes have been related to changes in location and intensity of the subtropical anticyclones. The arid area appears to have increased in extent from the turn of the century to the 1930s and in central Australia annual amounts appear to have declined between 1910 and 1970.

In the middle latitudes, precipitation changes are usually less pronounced. Figure 8.7 illustrates long-term fluctuations for England and Wales and for individual stations. For the country as a whole, decadal departures are only about ±10%. The individual station graphs show that even over relatively short distances there may be considerable differences in the magnitude of anomalies (e.g. Manchester and Oxford). This is further illustrated by the pattern of fluctuations in central Europe (fig. 8.8). The rising trend in the Bauer series, representing primarily West Germany (the Federal Republic), is distinctly different from the other series. Poland in the early part of its

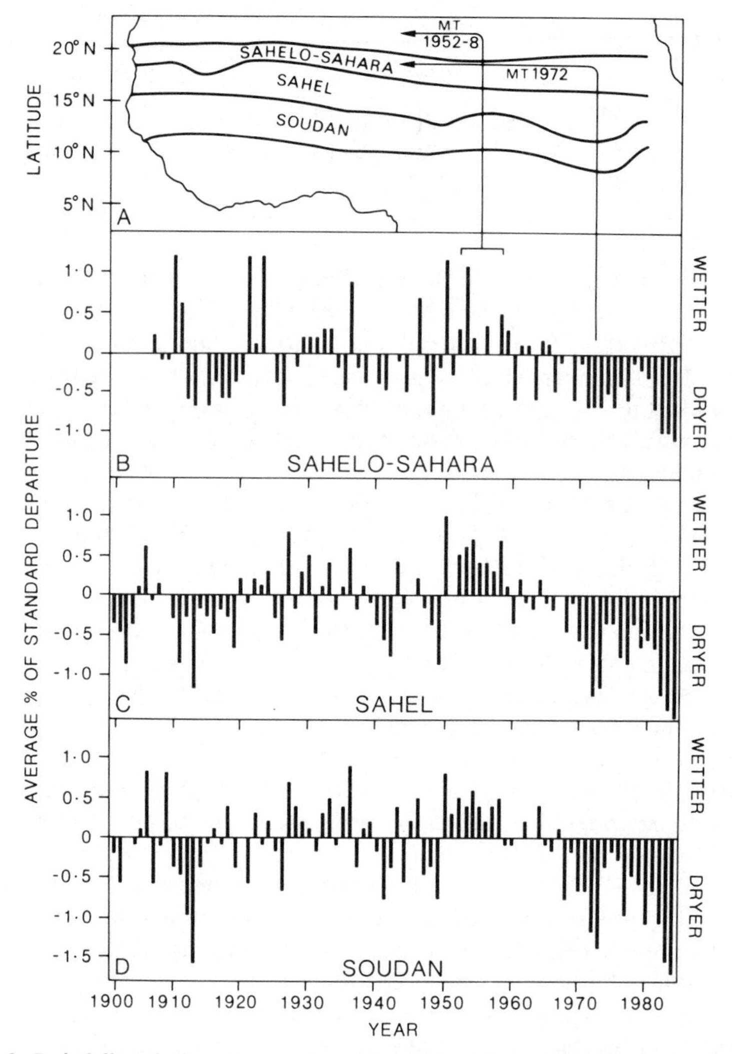

Fig. 8.6. Rainfall variations (percentage of standard departure) since 1900 for the Sahelo-Sahara, Sahel and Soudan zones of west Africa. The average positions of the Monsoon Trough (MT) in northern Nigeria during 1952–8 and in 1972 are shown (*from Nicholson 1985*).

record also differs from the regions to the south. For these countries the decadal range is approximately ±7% about the longer-term means.

C Possible causes of climatic change

Two categories of causal factors affecting the earth's climate system can be distinguished (fig. 8.9). The first involves processes external to the atmosphere–ocean domain such as changes in extraterrestrial solar radiation or

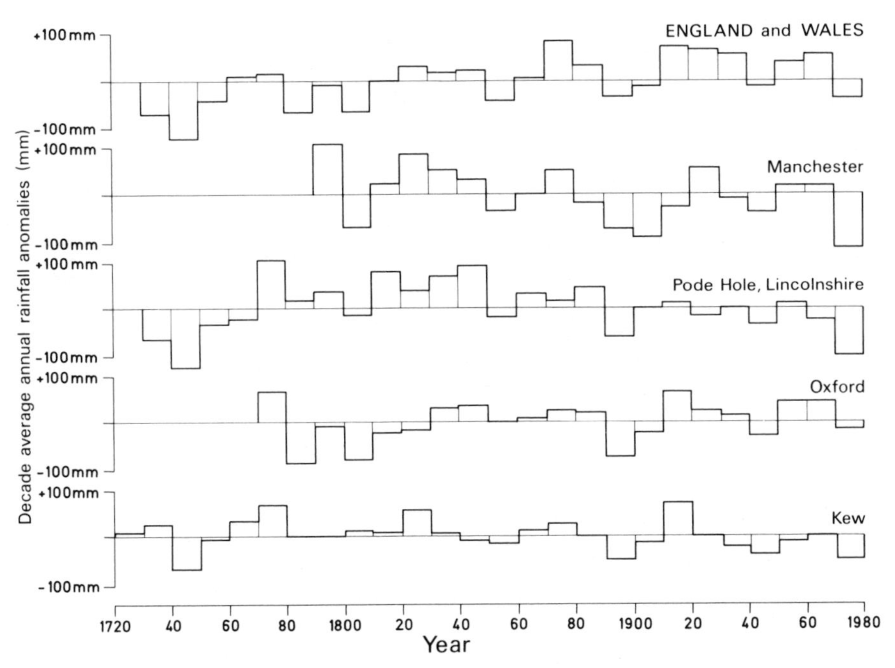

Fig. 8.7. Decadal anomalies of average annual rainfall (mm) for England and Wales and for four individual stations in England (the 'last decade' includes data up to September 1978) (*after Kelly 1980*).

continental drift. The second concerns internal factors such as changes in atmospheric composition, cloudiness or sea-surface temperature. This second category, especially, involves complex *feedback effects* between atmosphere, ocean and cryosphere. Thus, a more extensive snow/ice cover creates higher albedo and lower temperature, which in turn will further extend the snow/ice limit, producing additional cooling; this is an example of *positive* feedback. In other cases, an initial anomaly may be self-cancelling (negative feedback).

1 Long-term changes

Several ice ages have been identified in the geologic past at intervals of 250 million years or more. These seem to coincide with the occurrence of continents in high northern or southern latitudes through continental drift or, in its modern interpretation, 'plate tectonics'. Nevertheless, it appears that although this locational effect is a necessary factor, it may not be a sufficient cause for ice ages or very warm climates of the geological past. The Cretaceous period apparently experienced an anomalously warm global climate and, even allowing for the effects of a clustering of the continents in low latitudes, it seems that additional factors must be sought. The possibility that atmospheric CO_2 concentrations were several times greater than now – giving an amplified 'greenhouse' effect – is one suggested explanation.

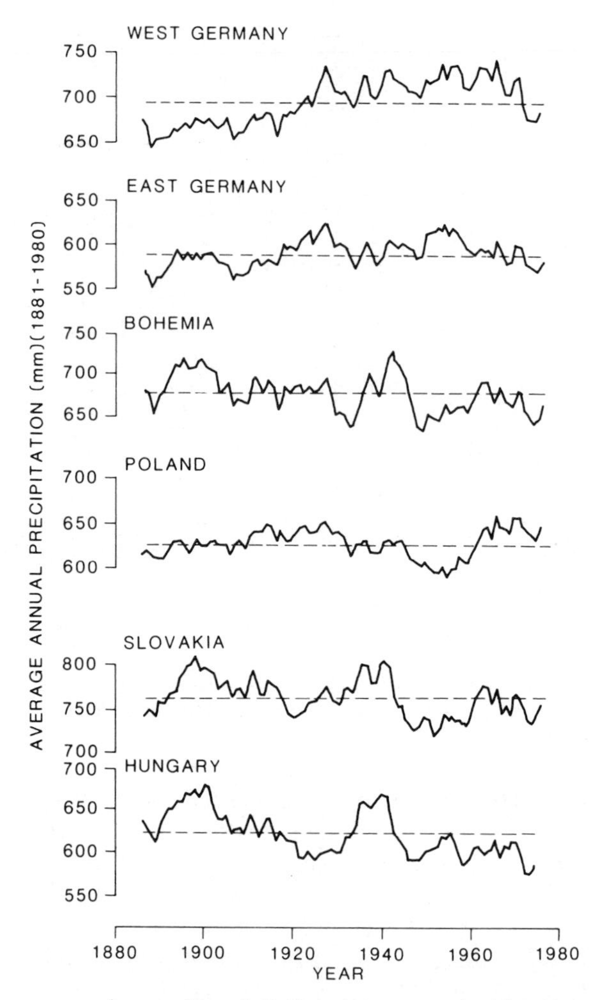

Fig. 8.8. Area-averaged annual precipitation data, smoothed by 11-year running means, for central Europe, 1881–1980 (*from Brádzil et al., 1985*). The series for West Germany is an arithmetic mean of 12 stations in the Federal Republic and 2 in the German Democratic Republic (GDR); the GDR series is a mean of 18 stations, the Hungarian series a mean of 10 stations. The series for Poland and Slovakia made double use of weighted arithmetic means of station data, for station heights and secondly for the geographical area of height intervals. The Bohemian series was derived by planimetry of monthly isohyetal maps. (*By permission of the Royal Meteorological Society*).

Another possible variable over long time scales is solar output. Sunspot cycles show no undisputed evidence of a link with weather conditions, although certain other short-term indicators of solar activity do suggest some possible relationships. Despite the absence of any observational evidence for changes of even 1–2% in total output (see p. 12), however, astronomical theories of star birth, evolution and eventual 'death', spanning many billion years, suggest the likelihood of a variable sun.

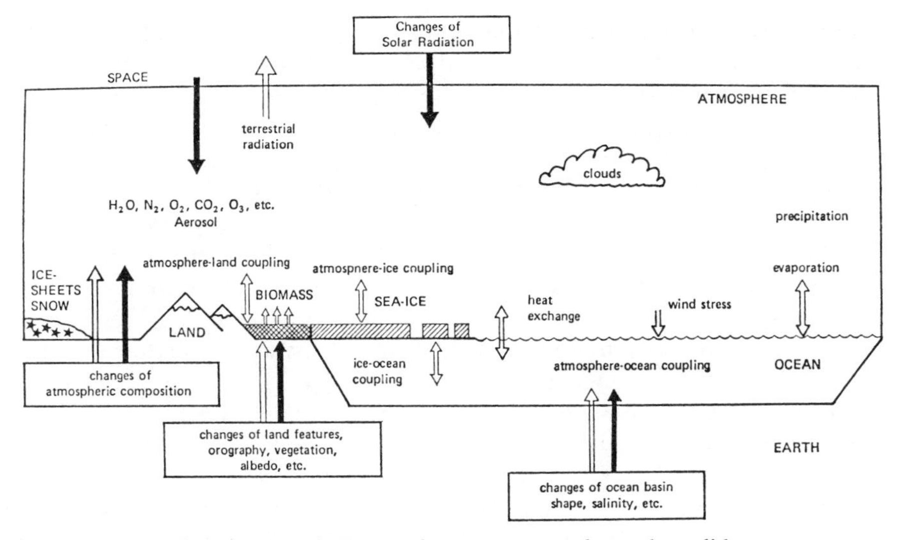

Fig. 8.9. The 'climate system'. External processes are shown by solid arrows; internal processes by open arrows (*after US GARP Committee 1974*).

A more certain link has been established between the astronomical variations in the earth's orbit around the sun and the earth's climate, building on the theory of M. Milankovitch. Three interacting variations occur that involve regular changes in (1) the shape of the elliptical orbit (with a time scale of about 95,000 years); (2) the tilt of the earth's axis of rotation (approximately 42,000 years); and (3) the time of year when the earth is closest to the sun, or perihelion (about 21,000 years). The first affects the total annual radiation received by the earth, whereas the amount of axial tilt (21·8°–24·4°) modifies the summer–winter contrast. The time of perihelion – at present on 3 January (see fig. 1.10) – determines the relative warmth of summer/winter in the respective hemisphere. It has been shown that the timing of glacial conditions is favoured by a diminished seasonal contrast with mild winters and cool summers, i.e. small axial tilt and perihelion in the northern winter. Compared with the beginning of post-glacial time, when perihelion occurred in June, astronomical conditions today are closer to those of the last glaciation. However, a future (minor) glaciation is not expected for at least 3000–4000 years. These orbital effects have also been suggested as the basic cause of global fluctuations in the levels of tropical lakes (9°–27°N). These were high around 9000 years ago, when incoming solar radiation was about 25 W m^{-2} greater (less) than now in July (January) as a result of increased axial tilt. These lakes were very low around 15,000 years ago and again after 5000 years ago.

2 Short-term fluctuations

Remarkably, the causes of the observed climatic changes during the last few centuries are less well understood than those of the last glaciation. A multiplicity of possible explanations exists and, indeed, more than one factor is likely to be involved. The immediate cause of the recent climatic fluctuations appears to be the strength of the global wind circulation. The first 30 years of this century saw a pronounced increase in the vigour of the westerlies over the North Atlantic, the north-east trades, the summer monsoon of southern Asia and the southern hemisphere westerlies (in summer). Over the North Atlantic, these changes consisted of an increased pressure gradient between the Azores high and the Icelandic low, as the latter deepened, and also between the Icelandic low and the Siberian high which spread westward. These changes were accompanied by more northerly depression tracks, and this resulted in reduced continentality in Russia (see fig. 5.2) and a significant increase in the frequency of mild south-westerly airflow over the British Isles between about 1900 and 1930, as reflected by the average annual frequency of Lamb's Westerly airflow type (see ch. 5, A.3). For 1873–97, 1898–1937, 1938–61 and 1962–85 the figures are 27, 38, 30 and 20% respectively.

The recent decrease of westerly airflow, especially in winter, is linked with greater continentality in Europe and Russia (see fig. 5.2) and an increase of northerly airflow, giving more frequent snowfalls in Britain. The southward shift of the main depression tracks produced a number of cool, wet summers in Britain (notably 1954, 1956, 1958 and 1960). These regional indicators reflect a general decline in the overall strength of the mid-latitude circumpolar westerlies, accompanying an apparent expansion of the polar vortex.

As has already been pointed out (ch. 3, D.4), global climate is closely related to the position and strength of the subtropical high-pressure cells. It has been estimated that a warming of the Arctic tropopause (winter +10°C; summer +3°C; annual +7°C), without changing equatorial or antarctic temperatures, would cause an annual shift of the subtropical high-pressure belt from its present average position of 37°N to 41°–43°N (i.e. some 100–200 km in summer but as much as 800 km in winter). This would bring drought to the Mediterranean, California, Middle East, Turkestan and the Punjab; as well as displacing the thermal equator from 6°N to 9°–10°N, increasing the desertification in the belt 0°–20°S.

The key to these atmospheric variations must be linked to the heat balance of the earth-atmosphere system and this forces us to return to the fundamental energy considerations with which we began this book. The evidence for fluctuations in the 'solar constant' is inconclusive, although variations apparently do occur in the emission of high-energy particles and ultraviolet radiation during brief solar flares. All solar activity follows the well-known cycle of approximately 11 years, which is usually measured with reference to the period between sunspot maximum and minimum (see fig. 1.9) but numerous attempts to establish secure correlations between sunspot activity and terrestial climates have produced mostly negative results. A statistical relationship has, nevertheless, been found between the occurrence of drought in the western United States over the last 300 years and the approximately 22-year double (Hale) cycle of the reversal of the solar magnetic polarity. Drought areas are most extensive in the 2–5 years following a Hale sunspot minimum (i.e. alternate 11-year sunspot minima).

Changes in atmospheric composition may also have modified the atmospheric heat budget. The presence of increased amounts of volcanic dust and sulphate aerosols in the stratosphere is one suggested cause of the 'Little Ice Age'. Major eruptions can result in a surface cooling of perhaps 0·2°C for a few years after the event. Hence, frequent volcanic activity would be required for persistently cooler conditions. Conversely, it is suggested that reduced volcanic activity after 1914 may have contributed in part to the early twentieth-century warming. New interest in this question has been aroused by eruptions of Mount St Helens (May 1981) and El Chichón (March 1982), although their climatic effects were minor.

The role of tropospheric aerosols is complex. Aerosols originate naturally, from wind-blown soil and silt for example, as well as from atmospheric pollution due to human activities (industry, domestic heating and modern transportation). Indirect anthropogenic factors, such as increasing population pressures leading to overgrazing and forest clearance, may increase desertification which also contributes to the increase of wind-blown soil. The 'dust-bowl' years of the 1930s in the United States and the African Sahel drought since 1972 illustrates this. Evidence from the Soviet Union shows a sharp rise in dust-fall on mountain snowfields since 1930, and atmospheric turbidity has increased by 57% over Washington DC, over the period 1905–64, and by 85% over Davos, Switzerland (1920–58). The presence of particles in the atmosphere increases the backscatter of short-wave radiation, thereby increasing the planetary albedo and causing cooling, but the effect on infrared radiation is one of surface warming. The net result is complicated by the surface albedo. Man-made aerosols cause net warming over snow and ice and most land surfaces, but cooling over the oceans which have a low albedo. Natural aerosols probably cause general cooling. The overall effect on global surface temperature remains uncertain.

Another change in atmospheric composition involves the steady increase in carbon dioxide as a result of the burning of fossil fuels (see fig. 1.3). Amounts have increased from pre-industrial levels of about 290 ppm to around 345 ppm in 1986. Based on experiments with atmospheric models, a doubling of atmospheric carbon dioxide is expected to raise the average surface-air temperature by about 2°–4°C, with much larger increases in polar regions, due to the stable atmosphere and snow/ice retreat. Increases in other atmospheric trace gases (especially methane and chlorofluorcarbons) are expected to augment the CO_2 effect. The temperature rise is caused by the increased atmospheric retention of the outgoing terrestrial infrared radiation due to the CO_2 and H_2O absorption bands (see fig. 1.8). Doubling of CO_2 levels is anticipated by the middle of the next century, given world levels of energy consumption, the growth of population and economics of 'third-world' countries, and the long lead-times required for development of alternative energy sources (solar, wind or nuclear). Based on these calculations, the observed increase of CO_2 since the late nineteenth century should have produced a warming of about 0·3°C, or close to half of the change observed in the northern hemisphere between 1890 and the 1940s. However, this leaves the subsequent cooling unexplained! It may be a result of increased atmospheric particles, mainly those of volcanic origin in the stratosphere, or other, as yet unidentified, causal effects.

A further anthropogenic cause of climatic change results from the clearing and burning of tropical rain forests, which amounts to an annual average of 110,000 km^2 (i.e. 0.08% of the non-glacial land surface of the globe). This annual figure is more than half the total land surface at present under irrigation, and twice the annual loss of marginal land to desertification. Deforestation affects world climate in two main ways – by altering the atmospheric composition and by affecting the hydrological cycle:

1 Forests store great amounts of carbon dioxide, so buffering the carbon dioxide cycle in the atmosphere. The carbon retained in the vegetation of the Amazon basin is equivalent to at least 20% of the entire atmospheric CO_2. Destruction of the vegetation would release about four-fifths of this to the atmosphere, of which about one-half would dissolve in the oceans but the other half would be added to the 16% increase of atmospheric CO_2 already observed this century. The effect of this would be to accelerate the increase of world temperatures. A further effect of tropical forest destruction would be to reduce the natural production of nitrous oxide. Tropical forests and their soils produce up to one-half of the world's nitrous oxide which helps to destroy stratospheric ozone. Any increase in ozone would warm the stratosphere, but lower global surface temperatures.

2 Dense tropical forests have a great effect on the hydrological cycle through their high evapotranspiration and their reduction of surface runoff (about one-third of the rain never reaches the ground, being intercepted and evaporating off the leaves). Forest destruction is resulting in a lowering of atmospheric humidity, an increase in runoff, a decrease in local rainfall and an increase in its seasonality, a fall of the water table and a tendency to degrade existing forest to savanna.

The interactions between ocean and atmosphere also introduce a variety of complications. The sea can store vast quantities of heat and may therefore greatly modify the heat and moisture exchanges with the air above. However, detailed investigations by J. Bjerknes show that variations in Atlantic sea-surface temperatures are due to *initial* changes in the wind regime. Similar conclusions have been reached regarding changes over the North Pacific.

The natural range of global temperature variability on the 100-year time scale is about ±0·5°C. Clearly, man-induced climatic change may quite soon exceed this. However, as a result of the considerable geographical variations in short-term climatic regimes – greater sensitivity in high latitudes and

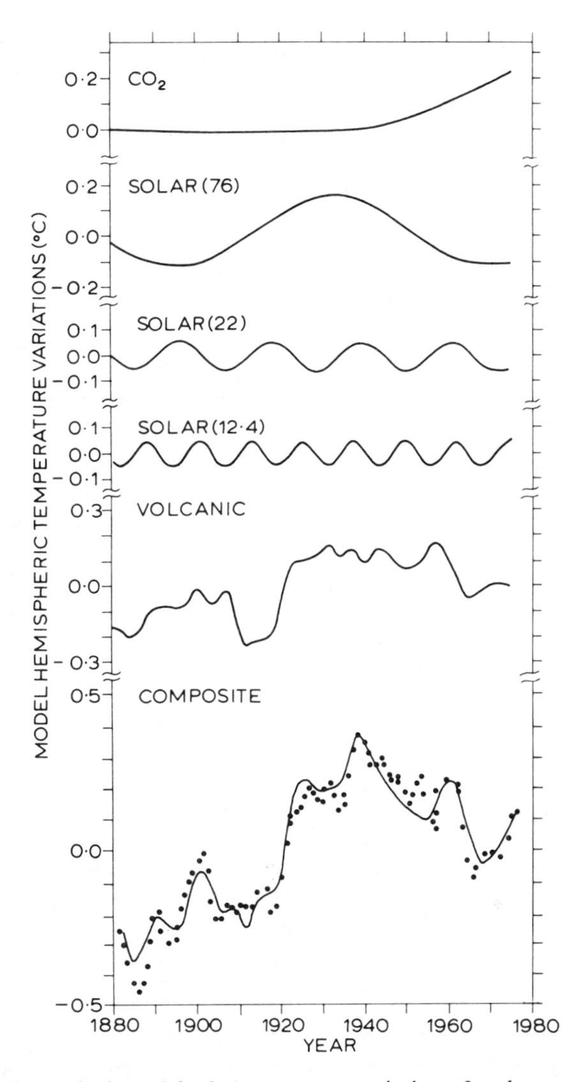

Fig. 8.10. Mathematical model of temperature variations for the northern hemisphere associated with observed CO_2 and volcanic dust measurements and with three solar energy cycles over the past century. The composite curve for these five causes compares well with the observed temperature record (dots) which has been smoothed with a 1·7-year filter (*from Gilliland 1982*). (*Copyright* © *1980/1982 by D. Reidel Publishing Company. Reprinted by permission*).

strong spatial differences on a sub-continental scale – some of the projected trends may be well underway before they can be reliably detected with routine instrumental observations. This concern is leading to a growing interest in obtaining better understanding of how our climate system works because, increasingly in the modern world, even minor climatic shifts can have major environmental and economic implications. For this reason mathematical models are being employed to simulate and forecast world climate. One such model for the northern hemisphere has employed estimates of temperature variations associated with the observed increase of CO_2, with the 76, 22 and 12·4-year oscillations of solar output, together with the observed 100-year effects of volcanic aerosols to produce a composite temperature curve (see fig. 8.10) which compares well with the observed temperature record. However, not all workers would agree with the particular temperature response functions adopted for this simulation.

Climate modification by man is both inadvertent and, potentially at least, intentional. Although ideas in the latter category are still highly speculative, they merit at least brief note here. Suggestions for modifying surface energy budgets and temperature include the production of widespread cirrus cloud by seeding from aircraft or rockets, and injecting large quantities of dust particles or metallic needles into orbit in the upper atmosphere. Other ideas are to increase cloudiness and precipitation in arid areas by creating *thermal mountains* (painting desert surfaces black) in order to decrease the albedo and stimulate convection, or by the creation of major inland seas in arid basins with interior drainage, such as Lake Eyre, Australia, and thereby modify the moisture balance. Another *geographical enginering* project is the Soviet plan to divert northward-flowing Siberian rivers southward into central Asia. The reduced freshwater runoff into the Arctic Basin could have important consequences for the sea ice cover in the Kara–Laptev seas and adjacent ocean, and consequently for the climate of the northern hemisphere. Initial analyses point to an increased summer ice cover, but more detailed studies are required (note 2). Hopefully, the unknown attendant dangers in such permanent large-scale tampering with the earth's surface and atmosphere will postpone such schemes – perhaps permanently!

Summary

Climatic data are described in terms of an average value (mean, median, mode) and a measure of the variability about the average (standard deviation, range, and so on. For skewed distributions, such as daily rainfall amounts, it is essential to know the frequency distribution.

Changes in climate involve factors both external to and within the climate system. External ones include solar variability, astronomical effects of the earth's orbit and volcanic activity. Internal factors include variability within the atmosphere and ocean, and their feedbacks. During the last century, man-induced climatic change on local and global scales has become a reality primarily through changes in atmospheric composition and surface properties.

Climatic changes on geological time-scales involve continental drift, volcanic activity and possible changes in solar output. Within the last few million years glacial–interglacial cycles appear to be strongly controlled by astronomical variations in the earth's orbit, although atmosphere–ocean–cryosphere feedbacks must also be involved. Shorter-term fluctuations appear to result from changes in atmospheric circulation regimes, but it is uncertain whether specific factors such as solar variability and atmospheric composition are the primary causes of such changing regimes. A warming trend in the northern hemisphere up to the 1940s and slight subsequent cooling is well documented. Although not yet fully understood, changes in the atmospheric composition are a prime candidate to account for these trends.

Notes

1 Fuller details of elementary statistical procedures may be found in S. Gregory (1973) *Statistical Methods and the Geographer* (Longman), R. Hammond and P. McCullogh (1974) *Quantitative Techniques in Geography* (Oxford University Press) or J. Silk (1979) *Statistical Concepts in Geography* (George Allen and Unwin).

2 The USSR recently announced cancellation of these plans, primarily on economic grounds.

Bibliography

General

Anthes, R. A., Panofsky, H. A., Cahir, J. J. and Rango, A. (1981) *The Atmosphere*, 3rd edn. C. E. Merrill, Columbus, Ohio. 531 pp.

Atkinson, B. W. (1981) *Meso-Scale Atmospheric Circulations*. Academic Press, London. 496 pp.

*Atkinson, B. W. (ed.) (1981) *Dynamical Meteorology*. Methuen, London. 250 pp.

Barrett, E. C. (1974) *Climatology from Satellites*. Methuen, London. 418 pp.

Barrett, E. C. and Martin, D. W. (1981) *The Use of Satellite Data in Rainfall Monitoring*. Academic Press, London. 340 pp.

Barry, R. G. and Perry, A. H. (1973) *Synoptic Climatology: Methods and Applications*. Methuen, London. 555 pp.

Barry, R. G. (1981) *Mountain Weather and Climate*. Methuen, London. 313 pp.

Battan, L. J. (1984) *Fundamentals of Meteorology*, 2nd edn. Prentice-Hall, Englewood Cliffs, New Jersey. 304 pp.

*Berry, F. A., Bollay, E. and Beers, N. R. (eds.) (1945) *Handbook of Meteorology*. McGraw-Hill, New York. 1068 pp.

Blüthgen, J. (1966) *Allgemeine Klimageographie*, 2nd edn. W. de Gruyter, Berlin. 720 pp.

Bruce, J. P. and Clark, R. H. (1966) *Introduction to Hydrometeorology*. Pergamon, Oxford. 319 pp.

*Byers, H. R. (1974) *General Meteorology*, 4th edn. McGraw-Hill, New York. 480 pp.

Chang, J. H. (1972) *Atmospheric Circulation Systems and Climates*. Oriental Publishing Co., Honolulu. 326 pp.

Cole, F. W. (1980) *Introduction to Meteorology*, 3rd edn. Wiley, New York. 505 pp.

Craig, R. A. (1968) *The Edge of Space*. Doubleday, New York. 150 pp.

Crowe, P. R. (1971) *Concepts in Climatology*. Longman, London. 589 pp.

Fairbridge, R. W. (ed.) (1967) *The Encyclopaedia of Atmospheric Sciences and Astrogeology*. Rheinold, New York. 1200 pp.

Fleagle, R. G. and Businger, J. A. (1980) *Introduction to Atmospheric Physics*, 2nd edn. Academic Press, New York. 432 pp.

*Flohn, H. (ed.) (1969) *General Climatology* 2. World Survey of Climatology, 2. Elsevier, Amsterdam. 266 pp.

* Indicates more advanced texts with reference particularly to the physical bases of meteorology, often including mathematical material.

*Flohn, H. (ed.) (1981) *General Climatology* 3. World Survey of Climatology, 3. Elsevier, Amsterdam. 408 pp.

Gedzelman, S. D. (1980) *The Science and Wonders of the Atmosphere*. Wiley, New York. 535 pp.

Geiger, R. (1965) *The Climate near the Ground*, 2nd edn. Harvard University Press, Cambridge, Mass. 611 pp.

*Haltiner, G. J. and Martin, F. L. (1957) *Dynamical and Physical Meteorology*. McGraw-Hill, New York. 470 pp.

Hecht, A. D. (ed.) (1985) *Paleoclimate Analysis and Modeling*. Wiley, New York. 445 pp.

Herman, J. R. and Goldberg, R. A. (1985) *Sun, Weather and Climate*. Dover, New York. 360 pp.

*Hess, S. L. (1959) *Introduction to Theoretical Meteorology*. Henry Holt, New York. 362 pp.

Hewson, E. W. and Longley, R. W. (1944) *Meteorology, Theoretical and Applied*. Wiley, New York. 468 pp.

Hobbs, J. (1980) *Applied Climatology*. Westview Press, Boulder, Colorado. 250 pp.

Houghton, D. D. (1985) *Handbook of Applied Meteorology*. Wiley, New York. 1461 pp.

Houghton, H. G. (1985) *Physical Meteorology*. MIT Press, Cambridge, Mass. 442 pp.

Houghton, J. T. (ed.) (1984) *The Global Climate*. Cambridge University Press, Cambridge. 233 pp.

*Humphreys, W. J. (1929) *Physics of the Air*. McGraw-Hill, New York. 654 pp.

Huschke, R. E. (ed.) (1959) *Glossary of Meteorology*. American Meteorological Society, Boston. 638 pp.

Hutchinson, G. E. (1957) *A Treatise on Limnology* 1. Wiley, New York. 1015 pp.

Kendrew, W. G. (1961) *The Climates of the Continents*, 5th edn. OUP, London. 608 pp.

Lamb, H. H. (1972) *Climate: Present, Past and Future* 1: *Fundamentals and Climate Now*. Methuen, London. 613 pp.

List, R. J. (1951) *Smithsonian Meteorological Tables*, 6th edn. Smithsonian Institution, Washington. 527 pp.

Lockwood, J. G. (1974) *World Climatology: An environmental approach*. Arnold, London. 330 pp.

Lockwood, J. G. (1979) *Causes of Climate*. Arnold, London. 260 pp.

Lutgens, F. K. and Tarbuck, E. J. (1986) *The Atmosphere: An introduction to meteorology*. 2nd edn. Prentice-Hall, Englewood Cliffs, New Jersey. 492 pp.

McBoyle, G. (ed.) (1973) *Climate in Review*. Houghton-Mifflin, Boston. 314 pp.

McIntosh, D. H. and Thom, A. S. (1972) *Essentials of Meteorology*. Wykeham Publications, London. 239 pp.

*Malone, T. F. (ed.) (1951) *Compendium of Meteorology*. American Meteorological Society, Boston. 1334 pp.

Manley, G. (1952) *Climate and the British Scene*. Collins, London. 314 pp.

Miller, A. and Anthes, R. A. (1980) *Meteorology*, 4th edn. C. E. Merrill, Columbus, Ohio. 170 pp.

Monteith, J. L. (1973) *Principles of Environmental Physics*. Arnold, London. 241 pp.

Monteith, J. L. (ed.) (1975) *Vegetation and the Atmosphere*. Vol. 1, *Principles*. Academic Press, London. 278 pp.

Munn, R. E. (1966) *Descriptive Micrometeorology*. Academic Press, New York. 245 pp.
Neiburger, M., Edinger, J. G. and Bonner, W. D. (1982) *Understanding Our Atmospheric Environment*, 2nd edn. W. H. Freeman, San Francisco. 453 pp.
Oliver, J. E. and Fairbridge, R. W. (eds.) (1987) *The Encyclopedia of Climatology*. Van Nostrand Reinhold, New York. 986 pp.
*Palmén, E. and Newton, C. W. (1969) *Atmosphere Circulation Systems: Their structure and physical interpretation*. Academic Press, New York. 603 pp.
Pedgley, D. E. (1962) *A Course of Elementary Meteorology*. HMSO, London. 189 pp.
Perry, A. H. and Walker, J. M. (1977) *The Ocean–Atmosphere System*. Longman, London. 160 pp.
*Petterssen, S. (1956) *Weather Analysis and Forecasting*, 2 vols. McGraw-Hill, New York. 428 pp and 266 pp.
Petterssen, S. (1969) *Introduction to Meteorology*, 3rd edn. McGraw-Hill, New York. 333 pp.
Pickard, G. L (1975) *Descriptive Physical Oceanography*, 2nd edn. Pergamon Press, Oxford. 214 pp.
*Reiter, E. R. (1963) *Jet Stream Meteorology*. University of Chicago Press, Chicago, Ill. 515 pp.
Rex, D. F. (ed.) (1969) *Climate of the Free Atmosphere*. World Survey of Climatology, 4. Elsevier, Amsterdam. 450 pp.
Sabins, F. F. (1978) *Remote Sensing*. W. H. Freeman, Reading. 426 pp.
Schaefer, V. J. and Day, J. A. (1981) *A Field Guide to the Atmosphere*. Houghton Mifflin, Boston, Mass. 359 pp.
Schwerdtfeger, W. (1984) *Weather and Climate of the Antarctic*. Elsevier, Amsterdam. 261 pp.
*Sellers, W. D. (1965) *Physical Climatology*. University of Chicago Press, Chicago, Ill. 272 pp.
Smith, W. L. *et al.* (1986) The meteorological satellite: overview of 25 years of operation. *Science* **231**, 455–62.
Strahler, A. N. (1965) *Introduction to Physical Geography*. Wiley, New York. 455 pp.
Stringer, E. T. (1972a) *Foundations of Climatology*. W. H. Freeman, San Francisco. 586 pp.
Stringer, E. T. (1972b) *Techniques of Climatology*. W. H. Freeman, San Francisco. 539 pp.
Sverdrup, H. V. (1945) *Oceanography for Meteorologists*. Allen and Unwin, London. 235 pp.
Sverdrup, H. V., Johnson, M. W. and Fleming, R. H. (1942) *The Oceans; Their physics, chemistry and general biology*. Prentice-Hall, New York. 1087 pp.
Taljaard, J. J., van Loon, H., Crutcher, H. L. and Jenne, R. L. (1969) *Climate of the upper Air, Part 1. Southern Hemisphere* 1. Naval Weather Service Command, Washington, DC. NAVAIR 50-1C-55.
Taylor, J. A. and Yates, R. A. (1967) *British Weather in Maps*, 2nd edn. Macmillan, London. 315 pp.
Trewartha, G. T. (1981) *The Earth's Problem Climates*, 2nd edn. University of Wisconsin Press, Madison. 371 pp.
Trewartha, G. T. and Horne, L. H. (1980) *An Introduction to Climate*, 5th edn. McGraw-Hill, New York. 416 pp.
van Loon, H. (ed.) (1984) *Climates of the Oceans: World Survey of Climatology*, Vol. 15. Elsevier, Amsterdam. 716 pp.

Wallace, J. M. and Hobbs, P. V. (1977) *Atmospheric Science: An introductory survey*. Academic Press, New York. 467 pp.
Willett, H. C. and Sanders, F. (1959) *Descriptive Meteorology*, 2nd edn. Academic Press, New York. 355 pp.
World Meteorological Organization (1962) *Climatological Normals (CLINO) for CLIMAT and CLIMAT SHIP stations for the period 1931–60*. World Meteorological Organization, Geneva.
World Meteorological Organization (1973) The use of satellite pictures in weather analysis and forecasting. *WMO Technical Note* 124, Geneva. 275 pp.

Chapter 1 Atmospheric composition and energy

Ahmad, S. A. and Lockwood, J. G. (1979) Albedo. *Prog. Phys. Geog.* **3**, 520–43.
Barry, R. G. (1985) The cryosphere and climatic change; In MacCracken, M. C. and Luther, F. M. (eds.) *Detecting the Climatic Effects of Increasing Carbon Dioxide*. DOE/ER-0235. US Department of Energy, Washington, DC. pp. 109–48.
Barry, R. G. and Chambers, R. E. (1966) A preliminary map of summer albedo over England and Wales. *Quart. J. Roy. Met. Soc.* **92**, 543–8.
Beckinsale, R. P. (1945) The altitude of the zenithal sun: A geographical approach. *Geog. Rev.* **35**, 596–600.
Budyko, M. I. *et al.* (1962) The heat balance of the surface of the earth. *Soviet Geography* **3**(5), 3–16.
Craig, R. A. (1965) *The Upper Atmosphere; meteorology and physics*. Academic Press, New York. 509 pp.
Defant, F. R. and Taba, H. (1957) The threefold structure of the atmosphere and the characteristics of the tropopause. *Tellus* **9**, 259–74.
Fröhlich, C. and London, J. (1985) *Radiation Manual*. World Meteorological Organization, Geneva.
Garnett, A. (1937) Insolation and relief. *Trans. Inst. Brit. Geog.* **5**. 71 pp.
Hare, F. K. (1962) The stratosphere. *Geog. Rev.* **52**, 525–47.
Hastenrath, S. L. (1968) Der regionale und jahrzeithliche Wandel des vertikalen Temperaturgradienten und seine Behandlung als Wärmhaushaltsproblem. *Meteorologische Rundschau* **1**, 46–51.
Henderson-Sellers, A. and Wilson, M. F. (1983) Surface albedo data for climate modeling. *Rev. Geophys. Space Phys.* **21**, 1743–78.
Keeling, C. D. (1973) The carbon dioxide cycle: reservoir models to depict the exchange of atmospheric carbon dioxide with the oceans, and land plants; In Rasool, S. I. (ed.) *Chemistry of the Lower Atmosphere*. Plenum Press, New York, pp. 251–329.
Kellogg, W. W. and Schware, R. (1981) *Climate Change and Society*. Westview Press, Boulder, Colorado. 178 pp.
Kondratyev, K. Ya. and Moskalenko, N. I. (1984) The role of carbon dioxide and other minor gaseous components and aerosols in the radiation budget; In Houghton, J. T. (ed.) *The Global Climate*. Cambridge University Press, Cambridge, pp. 225–33.
Kung, E. C., Bryson, R. A. and Lenschow, D. H. (1964) Study of a continental surface albedo on the basis of flight measurements and structure of the earth's surface cover over North America. *Monthly Weather Review* **92**, 543–64.
Lautensach, H. and Bogel, R. (1956) Der Jahrsgang des mittleren geographischem

Hohengradienten der Lufttemperatur in den verschiedenen Klimagebieten der Erde. *Erdkunde* **10**, 270–82.

London, J. and Angell, J. K. (1982) The observed distribution of ozone and its variations; In Bower, F. A. and Ward, R. B. (eds.) *Stratospheric Ozone and Man*. Chemical Rubber Company, CRC Press, Boca Raton, Fla. pp. 7–12.

London, J. and Sasamori, T. (1971) Radiative energy budget of the atmosphere. *Space Research* **11**, 639–49.

London, J. (1985) The observed distribution of atmospheric ozone and its variations; In Whitten, R. C. and Prasad, T. S. (eds.) *Ozone in the Free Atmosphere*. Van Nostrand Reinhold, New York. pp. 11–80.

Lumb, F. E. (1961) *Seasonal variation of the sea surface temperature in coastal waters of the British Isles*, Sci. Paper No. 6. Meteorological Office, HMSO, London. 21 pp.

McFadden, J. D. and Ragotzkie, R. A. (1967) Climatological significance of albedo in central Canada. *Jour. Geophys. Res.* **72**, 1135–43.

Machta, L. (1972) The role of the oceans and biosphere in the carbon dioxide cycle; In Dyrssen, D. and Jagner, D. (eds.) *The Changing Chemistry of the Oceans*, Nobel Symposium 20. Wiley, New York. pp. 121–45.

Miller, D. H. (1968) A Survey Course: the energy and mass budget at the surface of the earth. *Pub. No. 7, Assn. Amer. Geog.* Washington, DC. 142 pp.

Newell, R. E. (1964) The circulation of the upper atmosphere. *Sci. American* **210**, 62–74.

Newton, H. W. (1958) *The Face of the Sun*. Pelican, London. 208 pp.

Paffen, K. (1967) Das Verhältniss der Tages-zur Jahreszeitlichen Temperaturschwankung, *Erdkunde* **21**, 94–111.

Plass, G. M. (1959) Carbon dioxide and climate. *Sci. American* **201**, 41–7.

Ramanathan, V., Cicerone, R. J., Singh, H. B. and Kiehl, J. T. (1985) Trace gas trends and their potential role in climatic change. *J. Geophys. Res.* **90**(D3), 5547–66.

Rampino, M. R. and Self, S. (1984) The atmospheric effect of El Chichón. *Scientific American* **250**(1), 34–43.

Ransom, W. H. (1963) Solar radiation and temperature. *Weather* **8**, 18–23.

Ratcliffe, J. A. (ed.) (1960) *Physics of the Upper Atmosphere*. Academic Press, New York and London. 586 pp.

Sellers, W. D. (1980) A comment on the cause of the diurnal and annual temperature cycles. *Bull. Amer. Met. Soc.* **61**, 741–55.

Stephens, G. L., Campbell, G. G. and Vonder Haar, T. H. (1981) Earth radiation budgets. *J. Geophys. Res.* **86**(C10), 9739–60.

Stone, R. (1955) Solar heating of land and sea. *Geography* **40**, 288.

Sutton, G. (1963) Scales of temperature. *Weather* **18**, 130–4.

Valley, S. L. (ed.) (1965) *Handbook of Geophysics and Space Environments*. McGraw-Hill, New York.

World Meteorological Organization (1964) Regional basic networks. *WMO Bulletin* **13**, 146–7.

Chapter 2 Atmospheric moisture

Armstrong, C. F. and Stidd, C. K. (1967) A moisture-balance profile in the Sierra Nevada. *J. of Hydrology* **5**, 258–68.

Atlas, D., Chou, S.-H. and Byerly, W. P. (1983) The influence of coastal shape on winter mesoscale air–sea interactions. *Monthly Weather Review* **111**, 245–52.

Ayoade, J. A. (1976) A preliminary study of the magnitude, frequency and distribution of intense rainfall in Nigeria. *Hydro. Sci. Bull.* **21**(3), 419–29.

Bannon, J. K. and Steele, L. P. (1960) Average water-vapour content of the air. *Geophysical Memoirs Meteorological Office* **102**, 38 pp.

Baumgartner, A. and Reichel, E. (1975) *The World Water Balance: Mean annual global, continental and maritime precipitation, evaporation and runoff.* Elsevier, Amsterdam. 179 pp.

Bennetts, D. A., McCallum, E. and Grant, J. R. (1986) Cumulonimbus clouds: An introductory review. *Met. Mag.* **115**, 242–56.

Bergeron, T. (1960) Problems and methods of rainfall investigation; In *The Physics of Precipitation*, Geophysical Monograph 5. Amer. Geophys. Union, Washington, DC, 5–30.

Biswas, M. R. and Biswas, A. K. (eds.) (1980) *Desertification*. Pergamon, Oxford. 523 pp.

Borchert, J. R. (1971) The dust bowl in the 1970s. *Ann. Assn. Amer. Geogr.* **61**, 1–22.

Braham, R. R. (1959) How does a raindrop grow? *Science* **129**, 123–9.

Browning, K. A. (1980) Local weather forecasting. *Pro. Roy. Soc. Lond. Sect. A* **371**, 179–211.

Browning, K. A. (1985) Conceptual models of precipitation systems. *Met. Mag.* **114**, 293–319.

Browning, K. A. and Hill, F. F. (1981) Orographic rain. *Weather* **36**, 326–29.

Bryson, R. A. (1973) Drought in the Sahel: Who or what is to blame? *The Ecologist* **3**(10), 366–71.

Byers, H. R. and Braham, R. R. (1949) *The Thunderstorm*. US Weather Bureau.

Chacon, R. E. and Fernandez, W. (1985) Temporal and spatial rainfall variability in the mountainous region of the Reventazon River basin, Costa Rica. *J. Climatology* **5**, 175–88.

Dalby, D. and Harrison Church, R. J. (eds.) (1973) *Drought in Africa*. Centre for African Studies, School of Oriental and African Studies, University of London. 124 pp.

Deacon, E. L. (1969) Physical processes near the surface of the earth; In Flohn, H. (ed.), *General Climatology*, World Survey of Climatology 2. Elsevier, Amsterdam, pp. 39–104.

Dobbie, C. H. and Wolf, P. O. (1953) The Lynmouth flood of August 1952. *Pro. Inst. Civ. Eng.* Part III, 522–88.

Doornkamp, J. C. and Gregory, K. J. (eds.) (1980) *Atlas of Drought in Britain 1975–6*. Institute of British Geographers, London. 82 pp.

Dorman, C. E. and Bourke, R. H. (1981) Precipitation over the Atlantic Ocean, 30°S to 70°N. *Monthly Weather Review* **109**, 554–63.

Durbin, W. G. (1961) An introduction to cloud physics. *Weather* **16**, 71–82 and 113–25.

East, T. W. R. and Marshall, J. S. (1954) Turbulence in clouds as a factor in precipitation. *Quart. J. Roy. Met. Soc.* **80**, 26–47.

Garcia-Prieto, P. R., Ludlam, F. H. and Saunders, P. M. (1960) The possibility of artificially increasing rainfall on Tenerife in the Canary Islands. *Weather* **15**, 39–51.

Gilman, C. S. (1964) Rainfall; In Chow, V. T. (ed.), *Handbook of Applied Hydrology*. McGraw-Hill, New York, Section 9.

Harrold, T. W. (1966) The measurement of rainfall using radar. *Weather* **21**, 247–9 and 256–8.

Hastenrath, S. L. (1967) Rainfall distribution and regime in Central America. *Archiv. Met. Geophys. Biokl.* B, **15**(3), 201–41.

Hershfield, D. M. (1961) Rainfall frequency atlas of the United States for durations from 30 minutes to 24 hours and return periods of 1 to 100 years. *US Weather Bureau, Tech. Rept.* 40.

Hopkins, M. M. Jr (1967) An approach to the classification of meteorological satellite data. *J. Appl. Met.* **6**, 164–78.

Houze, R. A., Jr and Hobbs, P. V. (1982) Organization and structure of precipitating cloud systems. *Adv. Geophys.* **24**, 225–315.

Howe, G. M. (1956). The moisture balance in England and Wales. *Weather* **11**, 74–82.

Ilesanmi, O. O. (1971) An empirical formulation of an ITD rainfall model for the tropics: A case study for Nigeria. *J. App. Met.* **10**(5), 882–91.

Jaeger, L. (1976) Monatskarten des Niederschlags für die ganze Erde. *Berichte des Deutsches Wetterdienstes* **18**(139). Offenbach am Main, 38 pp. + plates.

Jiusto, J. E. and Weickmann, H. K. (1973) Types of snowfall. *Bull. Amer. Met. Soc.* **54**, 148–62.

Kelly, P. M. and Wright, P. B. (1978) The European drought of 1975–6 and its climatic context. *Prog. Phys. Geog.* **2**, 237–63.

Landsberg, H. E. (1974) Drought, a recurring element of climate. *Graduate Program in Meteorology, University of Maryland*, Contribution No. 100. 47 pp.

Latham, J. (1966) Some electrical processes in the atmosphere. *Weather* **21**, 120–7.

Ligda, M. G. H. (1951) Radar storm observation; In Malone, T. F. (ed.), *Compendium of Meteorology*. American Meteorological Society, Boston, Mass., pp. 1265–82.

Linsley, R. K. and Franzini, J. B. (1964) *Water-Resources Engineering*. McGraw-Hill, New York. 654 pp.

Ludlam, F. H. (1980) *Clouds and Storms. The behavior and effect of water in the atmosphere.* Pennsylvania State University, University Park and London. 405 pp.

MacDonald, J. E. (1962) The evaporation-precipitation fallacy. *Weather* **17**, 168–77.

Markham, C. G. and McLain, D. R. (1977) Sea-surface temperature related to rain in Ceará, north-eastern Brazil. *Nature* **265**, 320–3.

Mason, B. J. (1959) Recent developments in the physics of rain and rainmaking. *Weather* **14**, 81–97.

Mason, B. J. (1962) Charge generation in thunderstorms. *Endeavour* **21**, 156–63.

Mason, B. J. (1975) *Clouds, Rain and Rainmaking*, 2nd edn. Cambridge University Press, Cambridge and New York. 189 pp.

Miller, D. H. (1977) *Water at the Surface of the Earth*. Academic Press, New York. 557 pp.

Möller, F. (1951) Vierteljahrkarten des Niederschlags für die ganze Erde. *Petermanns Geographische Mitteilungen*, 95 Jahrgang, 1–7.

More, R. J. (1967) Hydrological models and geography; In Chorley, R. J. and Haggett, P. (eds.) *Models in Geography*. Methuen, London, pp. 145–85.

Palmer, W. C. (1965) Meteorological drought. *Research Paper No. 45*, US Weather Bureau, Washington, DC.

Paulhus, J. L. H. (1965) Indian Ocean and Taiwan rainfall set new records. *Monthly Weather Review* **93**, 331–5.

Pearl, R. T. *et al.* (1954) *The calculation of irrigation need*. Tech. Bull. No. 4, Min. Agric., Fish and Food. HMSO, London. 35 pp.
Penman, H. L. (1963) *Vegetation and Hydrology*. Tech. Comm. No. 53, Commonwealth Bureau of Soils, Harpenden. 124 pp.
Ratcliffe, R. A. S. (1978) Meteorological aspects of the 1975–6 drought. *Pro. Roy. Soc. Lond. Sect. A* **363**, 3–20.
Reitan, C. H. (1960) Mean monthly values of precipitable water over the United States, 1946–56. *Monthly Weather Review* **88**, 25–35.
Rodda, J. C. (1970) Rainfall excesses in the United Kingdom. *Trans. Inst. Brit. Geog.* **49**, 49–60.
Sawyer, J. S. (1956) The physical and dynamical problems of orographic rain. *Weather* **11**, 375–81.
Schermerhorn, V. P. (1967) Relations between topography and annual precipitation in western Oregon and Washington. *Water Resources Research* **3**, 707–11.
Simpson, C. G. (1941) On the formation of clouds and rain. *Quart. J. Roy. Met. Soc.* **67**, 99–133.
So, C. L. (1971) Mass movements associated with the rainstorm of June 1966 in Hong Kong. *Trans. Inst. Brit. Geog.* **53**, 55–65.
Sutcliffe, R. C. (1956) Water balance and the general circulation of the atmosphere. *Quart. J. Roy. Met. Soc.* **82**, 385–95.
Thompson, B. W. (1986) Small-scale katabatics and cold hollows. *Weather* **41**, 146–53.
Ward, R. C. (1963) Measuring potential evapotranspiration. *Geography* **47**, 49–55.
Weischet, W. (1965) Der tropische-konvective und der ausser tropischeadvektive Typ der vertikalen Niederschlagsverteilung. *Erdkunde* **19**, 6–14.
Weston, K. J. (1977) Cellular cloud patterns. *Weather* **32**, 446–50.
Widger, W. K. (1961) Satellite meteorology – fancy and fact. *Weather* **16**, 47–55.
Wilhite, D. A. and Glantz, M. H. (1982) Understanding the drought phenomenon: The role of definitions. *Water Internat.* **10**, 111–30.
World Meteorological Organization (1956) *International Cloud Atlas*. Geneva.
World Meteorological Organization (1972) *Distribution of precipitation in mountainous areas*, 2 vols. WMO No. 326, Geneva. 228 and 587 pp.
Yarnell, D. L. (1935) Rainfall intensity–frequency data. *US Dept. Agr., Misc. Pub.* 204.

Chapter 3 Atmospheric motion

Barry, R. G. (1967) Models in meteorology and climatology; In Chorley, R. J. and Haggett, P. (eds.) *Models in Geography*. Methuen, London, pp. 97–144.
Barry, R. G. (1979) Recent advances in climate theory based on simple climate models. *Prog. Phys. Geog.* **3**, 259–86.
Beran, W. D. (1967) Large amplitude lee waves and chinook winds. *J. Appl. Met.* **6**, 865–77.
Borchert, J. R. (1953) Regional differences in world atmospheric circulation. *Ann. Assn. Amer. Geog.* **43**, 14–26.
Brinkmann, W. A. R. (1971) What is a foehn? *Weather* **26**, 230–9.
Brinkmann, W. A. R. (1974) Strong downslope winds at Boulder, Colorado. *Monthly Weather Review* **102**, 592–602.
Buettner, K. J. and Thyer, N. (1965) Valley winds in the Mount Rainer area. *Archiv. Met. Geophys. Biokl.* B **14**, 125–47.

Corby, G. A. (ed.) (1970) *The global circulation of the atmosphere*. Roy. Met. Soc., London. 257 pp.
Crowe, P. R. (1949) The trade wind circulation of the world. *Trans. Inst. Brit. Geog.* **15**, 38–56.
Crowe, P. R. (1950) The seasonal variation in the strength of the trades. *Trans. Inst. Brit. Geog.* **16**, 23–47.
Defant, F. and Taba, H. (1957) The threefold structure of the atmosphere and the characteristics of the tropopause. *Tellus* **9**, 259–74.
Dietrich, G. (1963) *General Oceanography: An introduction*. Wiley, New York. 588 pp.
Druyan, L. M., Somerville, R. J. C. and Quirk, W. J. (1975) Extended-range forecasts with the GISS model of the global atmosphere. *Monthly Weather Review* **103**, 779–95.
Eddy, A. (1966) The Texas coast sea-breeze: A pilot study. *Weather* **21**, 162–50.
Ernst, J. A. (1976) SMS-1 Night-time infrared imagery of low-level mountain waves. *Monthly Weather Review* **104**, 207–9.
Flohn, H. (1969) Local wind systems; In Flohn, H. (ed.) *General Climatology*, World Survey of Climatology 2. Elsevier, Amsterdam, pp. 139–71.
Flohn, H. and Fantechi, R. (eds.) (1984) *The Climate of Europe: Past, present and future*. D. Reidel, Dordrecht. 356 pp.
Garbell, M. A. (1947) *Tropical and Equatorial Meteorology*. Pitman, London. 237 pp.
Geiger, R. (1969) Topoclimates; In Flohn, H. (ed.) *General Climatology*, World Survey of Climatology 2. Elsevier, Amsterdam, pp. 105–38.
Glenn, C. L. (1961) The chinook. *Weatherwise* **14**, 175–82.
Hare, F. K. (1965) Energy exchanges and the general circulation. *Geography* 50, 229–41.
Johnson, A. and O'Brien, J. J. (1973) A study of an Oregon sea breeze event. *J. Appl. Met.* **12**, 1267–83.
Kuenen, Ph. H. (1955) *Realms of Water*. Cleaver-Hulme Press, London. 327 pp.
Lamb, H. H. (1960) Representation of the general atmospheric circulation. *Met. Mag.* **89**, 319–30.
LeMarshall, J. F., Kelly, G. A. M. and Karoly, D. J. (1985) An atmospheric climatology of the southern hemisphere based on 10 years of daily numerical analyses (1972–1982). I, Overview. *Austral. Met. Mag.* **33**, 65–86.
Levitus, S. (1982) Climatological atlas of the world ocean. *NOAA Professional Paper No. 13*, Rockville, Md. 173 pp.
Lockwood, J. G. (1962) Occurrence of föhn winds in the British Isles. *Met. Mag.* **91**, 57–65.
Lorenz, E. N. (1967) *The nature and theory of the general circulation of the atmosphere*. World Meteorological Organization, Geneva. 161 pp.
McDonald, J. E. (1952) The Coriolis effect. *Sci. American* **186**, 72–8.
Meehl, G. A. (1984) Modeling the earth's climate. *Climatic Change* **6**, 259–86.
Namias, J. (1972) Large-scale and long-term fluctuations in some atmospheric and ocean variables; In Dyrssen, D. and Jagner, D. (eds.) *The Changing Chemistry of the Oceans*, Nobel Symposium 20. Wiley, New York, pp. 27–48.
O'Connor, J. F. (1961) Mean circulation patterns based on 12 years of recent northern hemispheric data. *Monthly Weather Review* **89**, 211–28.
Oke, T. R. (1978) *Boundary Layer Climates*. Methuen, London. 372 pp.
Palmén, E. (1951) The role of atmospheric disturbances in the general circulation. *Quart. J. Roy. Met. Soc.* **77**, 337–54.

Pfeffer, R. L. (1964) The global atmospheric circulation. *Trans. New York Acad. Sci.*, ser. 11, **26**, 984–97.
Riehl, H. (1962a) General atmospheric circulation of the tropics. *Science* **135**, 13–22.
Riehl, H. (1962b) *Jet streams of the atmosphere*. Tech. Paper No. 32. Colorado State University. 117 pp.
Riehl, H. (1969) On the role of the tropics in the general circulation of the atmosphere. *Weather* **24**, 288–308.
Riehl, H. *et al.* (1954) The jet stream. *Met. Monogr.* **2**(7). American Meteorological Society, Boston, Mass. 100 pp.
Riley, D. and Spalton, L. (1981) *World Weather and Climate*, 2nd edn. Cambridge University Press, London, 128 pp.
Rossby, C.-G. (1941) The scientific basis of modern meteorology. US Dept. of Agriculture Yearbook *Climate and Man*, pp. 599–655.
Rossby, C.-G. (1949) On the nature of the general circulation of the lower atmosphere; In Kuiper, G. P. (ed.) *The Atmosphere of the Earth and Planets*. University of Chicago Press, Chicago, Ill., pp. 16–48.
Sawyer, J. S. (1957) Jet stream features of the earth's atmosphere. *Weather* **12**, 333–4.
Scorer, R. S. (1958) *Natural Aerodynamics*. Pergamon Press, Oxford. 312 pp.
Scorer, R. S. (1961) Lee waves in the atmosphere. *Sci. American* **204**, 124–34.
Starr, V. P. (1956) The general circulation of the atmosphere. *Sci. American* **195**, 40–5
Steinacker, R. (1984) Area–height distribution of a valley and its relation to the valley wind. *Contrib. Atmos. Phys.* **57**, 64–74.
Streten, N. A. (1980) Some synoptic indices of the southern hemisphere mean sea level circulation 1972–77. *Monthly Weather Review* **108**, 18–36.
Taljaard, J. J., van Loon, H., Crutcher, H. L. and Jenne, R. L. (1969) *Climate of the upper Air, Part 1. Southern Hemisphere* 1. Naval Weather Service Command, Washington, DC. NAVAIR 50-1C-55.
Tucker, G. B. (1962) The general circulation of the atmosphere. *Weather* **17**, 320–40.
van Arx, W. S. (1962) *Introduction to Physical Oceanography*. Addison-Wesley, Reading, Mass. 422 pp.
van Loon, H. (1964) Mid-season average zonal winds at sea level and at 500 mb south of 25°S and a brief comparison with the northern hemisphere. *J. Appl. Met.* **3**, 554–63.
van Loon, H. (ed.) (1984) *Climates of the Oceans;* In Landsberg, H. E. (ed.) *World Survey of Climatology*, Vol. 15. Elsevier, Amsterdam. 716 pp.
Waco, D. E. (1968) Frost pockets in the Santa Monica Mountains of southern California. *Weather* **23**, 456–61.
Walker, J. M. (1972) Monsoons and the global circulation. *Met. Mag.* 101, 349–55.
Wallington, C. E. (1960) An introduction to lee waves in the atmosphere. *Weather* **15**, 269–76.
Wallington, C. E. (1969) Depressions as moving vortices. *Weather* **24**, 42–51.
Wickham, P. G. (1966) Weather for gliding over Britain. *Weather* **21**, 154–61.

Chapter 4 Air masses, fronts and depressions

Barrett, E. C. (1964) Satellite meteorology and the geographer. *Geography* **49**, 377–86.

Bates, F. C. (1962) Tornadoes in the central United States. *Trans. Kansas Acad. Sci.* **65**, 215–46.

Belasco, J. E. (1952) Characteristics of air masses over the British Isles. Meteorological Office, *Geophysical Memoirs* **11**(87). 34 pp.

Bosart, L. (1985) Weather forecasting; In Houghton, D. D. (ed.) *Handbook of Applied Meteorology*. Wiley, New York, pp. 205–79.

Boucher, R. J. and Newcomb, R. J. (1962) Synoptic interpretation of some TIROS vortex patterns: A preliminary cyclone model. *J. Appl. Met.* **1**, 122–36.

Boyden, C. J. (1960) The use of upper air charts in forecasting. *The Marine Observer* **30**, 27–31.

Boyden, C. J. (1963) Development of the jet stream and cut-off circulations. *Met. Mag.* **92**, 287–99.

Browning, K. A. (1968) The organization of severe local storms. *Weather* **23**, 429–34.

Browning, K. A. (1980) Local weather forecasting. *Pro. Roy. Soc. Lond. Sect. A* **371**, 179–211.

Browning, K. A. (ed.) (1983) *Nowcasting*. Academic Press, New York. 256 pp.

Browning, K. A. (1985) Conceptual models of precipitation systems. *Met. Mag.* **114**, 293–319.

Browning, K. A. (1986) Weather radar and FRONTIERS. *Weather* **41**, 9–16.

Browning, K. A. and Hill, F. F. (1981) Orographic rain. *Weather* **36**, 326–9.

Businger, S. (1985) The synoptic climatology of polar low outbreaks. *Tellus* **37A**, 419–32.

Carleton, A. M. (1985) Satellite climatological aspects of the 'polar low' and 'instant occlusion'. *Tellus* **37A**, 433–50.

Crowe, P. R. (1949) The trade wind circulation of the world. *Trans. Inst. Brit. Geog.* **15**, 38–56.

Crowe, P. R. (1965) The geographer and the atmosphere. *Trans. Inst. Brit. Geog.* **36**, 1–19.

Freeman, M. H. (1961) Fronts investigated by the Meteorological Research Flight. *Met. Mag.* **90**, 189–203.

Fulks, J. R. (1951) The instability line; In Malone, T. F. (ed.) *Compendium of Meteorology*. American Meteorological Society, Boston, Mass., pp. 647–52.

Gadd, A. J. (1985) The 15-level weather prediction model. *Met. Mag.* **114**, 222–6.

Galloway, J. L. (1958a) The three-front model: Its philosophy, nature, construction and use. *Weather* **13**, 3–10.

Galloway, J. L. (1958b) The three-front model, the tropopause and the jet stream. *Weather* **13**, 395–403.

Galloway, J. L. (1960) The three-front model, the developing depression and the occluding process. *Weather* **15**, 293–309.

Godson, W. L. (1950) The structure of North American weather systems. *Cent. Proc. Roy. Met. Soc.* London, pp. 89–106.

Hare, F. K. (1960) The westerlies. *Geog. Rev.* **50**, 345–67.

Harman, J. R. (1971) Tropical waves, jet streams, and the Unites States weather patterns. Association of American Geographers, Commission on College Geography, *Resource Paper No. 11*. 37 pp.

Harrold, T. W. (1973) Mechanisms influencing the distribution of precipitation within baroclinic disturbances. *Quart. J. Roy. Met. Soc.* **99**, 232–51.

Hindley, K. (1977) Learning to live with twisters. *New Scientist* **70**, 280–2.

Hobbs, P. V. (1978) Organization and structure of clouds and precipitation on the meso-scale and micro-scale of cyclonic storms. *Rev. Geophys. and Space Phys.* **16**, 1410–11.

Houghton, D. M. (1965) Current forecasting practice. *Quart. J. Roy. Met. Soc.* **91**, 524–6.

Houze, R. A. and Hobbs, P. V. (1982) Organization and structure of precipitating cloud systems. *Adv. Geophys.* **24**, 225–315.

Hunt, R. D. (1985) The models in action. *Met. Mag.* **114**, 261–72.

Jackson, M. C. (1977) Meso-scale and small-scale motions as revealed by hourly rainfall maps of an outstanding rainfall event: 14–16 September 1968. *Weather* **32**, 2–16.

Kessler, E. (ed.) (1983) *The Thunderstorm in Human Affairs*, 2nd edn. University of Oklahoma Press, Norman, Okla. 250 pp.

Klein, W. H. (1948) Winter precipitation as related to the 700-mb circulation. *Bull. Amer. Met. Soc.* **29**, 439–53.

Klein, W. H. (1957) Principal tracks and mean frequencies of cyclones and anti-cyclones in the Northern Hemisphere. *Research Paper No. 40*, Weather Bureau, Washington DC. 60 pp.

Klein, W. H. (1982) Statistical weather forecasting on different time scales. *Bull. Amer. Met. Soc.* **63**, 170–7.

Lamb, H. H. (1951) Essay on frontogenesis and frontolysis. *Met. Mag.* **80**, 35–6, 65–71 and 97–106.

Ludlam, F. H. (1961) The hailstorm. *Weather* **16**, 152–62.

Lyall, I. T. (1972) The polar low over Britain. *Weather* **27**, 378–90.

Maddox, R. A. (1980) Mesoscale convective complexes. *Bull. Amer. Met. Soc.* **61**, 1374–87.

Mason, B. J. (1974) The contribution of satellites to the exploration of the global atmosphere and to the improvement of weather forecasting. *Met. Mag.* **103**, 181–201.

Miles, M. K. (1961) The basis of present-day weather forecasting. *Weather* **16**, 349–63.

Miles, M. K. (1962) Wind, temperature and humidity distribution at some cold front over SE England. *Quart. J. Roy. Met. Soc.* **88**, 286–300.

Miller, R. C. (1959) Tornado-producing synoptic patterns. *Bull. Amer. Met. Soc.* **40**, 465–72.

Miller, R. C. and Starrett, L. G. (1962) Thunderstorms in Great Britain. *Met. Mag.* **91**, 247–55.

Newton, C. W. (1966) Severe convective storms. *Adv. Geophys.* **12**, 257–308.

Newton, C. W. (ed.) (1972) Meteorology of the Southern Hemisphere. *Met. Monogr. No. 13* (35). American Meteorological Society, Boston, Mass. 263 pp.

Pedgley, D. E. (1962) A meso-synoptic analysis of the thunderstorms on 28 August 1958. *Geophysical Memoirs Meteorological Office* **14**(1). 30 pp.

Penner, C. M. (1955) A three-front model for synoptic analyses. *Quart. J. Roy. Met. Soc.* **81**, 89–91.

Petterssen, S. (1950) Some aspects of the general circulation of the atmosphere. *Cent. Proc. Roy. Met. Soc.* London, pp. 120–55.

Pothecary, I. J. W. (1956) Recent research on fronts. *Weather* 12, 147–50.

Reed, R. J. (1960) Principal frontal zones of the northern hemisphere in winter and summer. *Bull. Am. Met. Soc.* **41**, 591–8.

Richter, D. A. and Dahl, R. A. (1958) Relationship of heavy precipitation to the jet maximum in the eastern United States. *Monthly Weather Review* **86**, 368–76.

Riley, D. and Spalton, L. (1974) *World Weather and Climate*. Cambridge University Press, 120 pp.

Showalter, A. K. (1939) Further studies of American air mass properties. *Monthly*

Weather Review **67**, 204–18.
Slater, P. M. and Richards, C. J. (1974) A memorable rainfall event over southern England. *Met. Mag.* **103**, 255–68 and 288–300.
Smagorinsky, J. (1974) Global atmospheric modeling and the numerical simulation of climate; In Hess, W. N. (ed.) *Weather and Climate Modification*. Wiley, New York, pp. 633–86.
Smith, W. L. (1985) Satellites; In Houghton, D. D. (ed.) *Handbook of Applied Meteorology*. Wiley, New York, pp. 380–472.
Snow, J. T. (1984) The tornado. *Scientific American* **250**(4), 56–66.
Sutcliffe, R. C. and Forsdyke, A. G. (1950) The theory and use of upper air thickness patterns in forecasting. *Quart. J. Roy. Met. Soc.* **76**, 189–217.
United States Department of Commerce (1981) *Operations of the National Weather Service*. National Oceanic and Atmospheric Administration, Silver Springs, Md. 249 pp.
Vederman, J. (1954) The life cycles of jet streams and extratropical cyclones. *Bull. Amer. Met. Soc.* **35**, 239–44.
Wallington, C. E. (1963) Meso-scale patterns of frontal rainfall and cloud. *Weather* **18**, 171–81.
Weatherwise (1983) A conversation with Donald Gilman. Predicting the weather for the long term. *Weatherwise* **36**, 290–7.
Wendland, W. M. and Bryson, R. A. (1981) Northern hemisphere airstream regions. *Monthly Weather Review* **109**, 255–70.
Wendland, W. M. and McDonald, N. S. (1986) Southern hemisphere airstream climatology. *Monthly Weather Review* **114**, 88–94.
Wick, G. (1973) Where Poseidon courts Aeolus. *New Scientist*, 18 January, pp. 123–6.
Yoshino, M. M. (1967) Maps of the occurrence frequencies of fronts in the rainy season in early summer over east Asia. *Science Reports of the Tokyo University of Education* **89**, 211–45.

Chapter 5 Weather and climate in temperate latitudes

Bailey, H. P. (1964) Toward a unified concept of the temperate climate. *Geog. Rev.* **54**(4), 516–45.
Balling, R. C., Jr (1985) Warm seasonal nocturnal precipitation in the Great Plains of the United States. *J. Climate Applied Met.* **24**, 1383–7.
Barry, R. G. (1963) Aspects of the synoptic climatology of central south England. *Met. Mag.* **92**, 300–8.
Barry, R. G. (1967a) Seasonal location of the arctic front over North America. *Geog. Bull.* **9**, 79–95.
Barry, R. G. (1967b) The prospect for synoptic climatology: A case study; In Steel, R. W. and Lawton, R. (eds.) *Liverpool Essays in Geography*. Longman, London, pp. 85–106.
Barry, R. G. (1973) A climatological transect on the east slope of the Front Range, Colorado. *Arct. Alp. Res.* **5**, 89–110.
Barry, R. G. and Hare, F. K. (1974) Arctic climate; In Ives, J. D. and Barry, R. G. (eds.) *Arctic and Alpine Environments*. Methuen, London, pp. 17–54.
Belasco, J. E. (1948) The incidence of anticyclonic days and spells over the British Isles. *Weather* **3**, 233–42.

Belasco, J. E. (1952) Characteristics of air masses over the British Isles. Meteorological Office, *Geophysical Memoirs* **11**(87). 34 pp.

Boast, R. and McQuinigle, J. B. (1972) Extreme weather conditions over Cyprus during April 1971. *Met. Mag.* **101**, 137–53.

Borchert, J. (1950) The climate of the central North American grassland. *Ann. Assn. Amer. Geog.* **40**, 1–39.

Browning, K. A. and Hill, F. F. (1981) Orographic rain. *Weather* **36**, 326–9.

Bryson, R. A. (1966) Air masses, streamlines and the boreal forest. *Geog. Bull.* **8**, 228–69.

Bryson, R. A. and Hare, F. K. (eds.) (1974) *Climates of North America.* World Survey of Climatology, 11. Elsevier, Amsterdam. 420 pp.

Bryson, R. A. and Lahey, J. F. (1958) *The March of the Seasons.* Meteorological Department, University of Wisconsin. 41 pp.

Bryson, R. A. and Lowry, W. P. (1955) Synoptic climatology of the Arizona summer precipitation singularity. *Bull. Amer. Met. Soc.* **36**, 329–39.

Burbidge, F. E. (1951) The modification of continental polar air over Hudson Bay. *Quart. J. Met. Soc.* **77**, 365–74.

Butzer, K. W. (1960) Dynamic climatology of large-scale circulation patterns in the Mediterranean area. *Meteorologische Rundschau* **13**, 97–105.

Carleton, A. M. (1986) Synoptic-dynamic character of 'bursts' and 'breaks' in the southwest US summer precipitation singularity. *J. Climatol.* **6**, 605–23.

Chandler, T. J. and Gregory, S. (eds.) (1976) *The Climate of the British Isles.* Longman, London. 390 pp.

Derecki, J. A. (1976) Heat storage and advection in Lake Erie. *Water Resources Research* **12**(6), 1144–50.

Durrenberger, R. W. and Ingram, R. S. (1978) Major storms and floods in Arizona 1862–1977. *State of Arizona, Office of the State Climatologist, Climatological Publications, Precipitation Series* No. 4. 44 pp.

Easterling, D. R. and Robinson, P. J. (1985) The diurnal variation of thunderstorm activity in the United States. *J. Climate Appl. Met.* **24**, 1048–58.

Elsom, D. M. and Meaden, G. T. (1984) Spatial and temporal distribution of tornadoes in the United Kingdom 1960–1982. *Weather* **39**, 317–23.

Environmental Science Services Administration (1965) *APT Users' Guide.* US Department of Commerce, Washington DC. 80 pp.

Environmental Science Services Administration (1968) *Climatic Atlas of the United States.* US Department of Commerce, Washington DC. 80 pp.

Evenari, M., Shanan, L. and Tadmor, N. (1971) *The Negev.* Harvard University Press, Cambridge, Mass. 345 pp.

Ferguson, E. W., Ostby, F. P., Leftwich, P. W., Jr and Hales, J. E., Jr (1986) The tornado season of 1984. *Monthly Weather Review* **114**, 624–35.

Flohn, H. (1954) *Witterung und Klima in Mitteleuropa.* Zurich. 218 pp.

Gentilli, J. (ed.) (1971) *Climates of Australia and New Zealand.* World Survey of Climatology, 13. Elsevier, Amsterdam. 405 pp.

Gorczynski, W. (1920) Sur le calcul du degré du continentalisme et son application dans la climatologie. *Geografiska Annaler* **2**, 324–31.

Green, C. R. and Sellers, W. D. (1964) *Arizona Climate.* University of Arizona Press, Tucson. 503 pp.

Hales, J. E., Jr (1974) South-western United States summer monsoon source – Gulf of Mexico or Pacific Ocean. *J. Appl. Met.* **13**, 331–42.

Hare, F. K. (1968) The Arctic. *Quart. J. Roy. Met. Soc.* **74**, 439–59.

Hare, F. K. and Thomas, M. K. (1979) *Climate Canada*, 2nd edn. Wiley, Canada. 230 pp.

Hawke, E. L. (1933) Extreme diurnal range of air temperature in the British Isles. *Quart. J. Roy. Met. Soc.* **59**, 261–5.

Hill, F. F., Browning, K. A. and Bader, M. J. (1981) Radar and rain gauge observations of orographic rain over South Wales. *Quart. J. Roy. Met. Soc.* **107**, 643–70.

Horn, L. H. and Bryson, R. A. (1960) Harmonic analysis of the annual march of precipitation over the United States. *Ann. Assn. Am. Geog.* **50**, 157–71.

Huttary, J. (1950) Die Verteilung der Niederschläge auf die Jahreszeiten im Mittelmeergebeit. *Meteorologische Rundschau* **3**, 111–19.

Klein, W. H. (1963) Specification of precipitation from the 700-mb circulation. *Monthly Weather Review* **91**, 527–36.

Knox, J. L. and Hay, J. E. (1985) Blocking signatures in the northern hemisphere: Frequency distribution and interpretation. *J. Climatology* **5**, 1–16.

Lamb, H. H. (1950) Types and spells of weather around the year in the British Isles: Annual trends, seasonal structure of the year, singularities. *Quart. J. Roy. Met. Soc.* **76**, 393–438.

Linacre, W. and Hobbs, J. (1977) *The Australian Climatic Environment*. Wiley, Brisbane. 354 pp.

Longley, R. W. (1967) The frequency of Chinooks in Alberta. *The Albertan Geographer* **3**, 20–2.

Lumb, F. E. (1961) Seasonal variations of the sea surface temperature in coastal waters of the British Isles. *Met. Office Sci. Paper No. 6*, MO 685. 21 pp.

Manley, G. (1944) Topographical features and the climate of Britain. *Geog. Jour.* **103**, 241–58.

Manley, G. (1945) The effective rate of altitude change in temperate Atlantic climates. *Geog. Rev.* **35**, 408–17.

Meteorological Office (1952) *Climatological Atlas of the British Isles*, MO 488. HMSO, London. 139 pp.

Meteorological Office (1962) *Weather in the Mediterranean I, General Meteorology*, 2nd edn. MO 391, HMSO, London. 362 pp.

Meteorological Office (1964a) *Weather in the Mediterranean II,* 2nd edn. MO 391b. HMSO, London. 372 pp.

Meteorological Office (1964b) *Weather in Home Fleet Waters I, The Northern Seas,* Part 1, MO 732a. HMSO, London. 265 pp.

Namias, J. (1964) Seasonal persistence and recurrence of European blocking during 1958–60. *Tellus* **16**, 394–407.

Nickling, W. G. and Brazel (1984) Temporal and spatial characteristics of Arizona dust storms (1965–1980). *Climatology* **4**, 645–60.

Poltaraus, B. V. and Staviskiy, D. B. (1986) The changing continentality of climate in central Russia. *Soviet Geography* **27**, 51–8.

Rayner, J. N. (1961) *Atlas of Surface Temperature Frequencies for North America and Greenland*. Arctic Meteorological Research Group, McGill University, Montreal.

Rex, D. F. (1950–51) The effect of Atlantic blocking action upon European climate. *Tellus* **2**, 196–211 and 275–301; **3**, 100–11.

Schick, A. P. (1971) A desert flood. *Jerusalem Studies in Geography* **2**, 91–155.

Shaw, E. M. (1962) An analysis of the origins of precipitation in Northern England, 1956–60. *Quart. J. Roy. Met. Soc.* **88**, 539–47.

Sivall, T. (1957) Sirocco in the Levant. *Geografiska Annaler* **39**, 114–42.

Stone, J. (1983) Circulation type and the spatial distribution of precipitation over central, eastern and southern England. *Weather* **38**, 173–7, 200–5.

Storey, A. M. (1982) A study of the relationship between isobaric patterns over the UK and central England temperature and England–Wales rainfall. *Weather* **37**, 2–11, 46, 88–9, 122, 151, 170, 208, 244, 260, 294, 327, 360.

Sumner, E. J. (1959) Blocking anticyclones in the Atlantic-European sector of the northern hemisphere. *Met. Mag.* **88**, 300–11.

Thomas, M. K. (1964) *A survey of Great Lakes snowfall.* Great Lakes Research Division, University of Michigan, Publication No. 11, pp. 294–310.

Thornthwaite, C. W. and Mather, J. R. (1955) *The Moisture Balance*, Publications in Climatology, **8**(1). Laboratory of Climatology, Centerton, NJ. 104 pp.

Tout, D. G. and Kemp, V. (1985) The named winds of Spain. *Weather* **40**, 322–9.

United States Weather Bureau (1947) *Thunderstorm Rainfall.* Vicksburg, Mississippi. 331 pp.

Villmow, J. R. (1956) The nature and origin of the Canadian dry belt. *Ann. Assn. Amer. Geog.* **46**, 221–32.

Visher, S. S. (1954) *Climatic Atlas of the United States.* Harrd. 403 pp.

Wallace, J. M. (1975) Diurnal variations in precipitation and thunderstorm frequency over the conterminous United States. *Monthly Weather Review* **103**, 406–19.

Wallén, C. C. (1960) Climate; In Somme, A. (ed.) *The Geography of Norden.* Cappelens Forlag, Oslo, pp. 41–53.

Wallén, C. C. (ed.) (1970) *Climates of Northern and Western Europe.* World Survey of Climatology, 5. Elsevier, Amsterdam. 253 pp.

Chapter 6 Tropical weather and climate

Anthes, R. A. (1982) Tropical cyclones: Their evolution, structure, and effects. *Met. Monogr.* Amer. Met. Soc., Boston, Mass. **19**(41), 208 pp.

Academica Sinica (1957–58) On the general circulation over eastern Asia. *Tellus* **9**, 432–46; **10**, 58–75 and 299–312.

Arakawa, H. (ed.) (1969) *Climates of Northern and Eastern Asia.* World Survey of Climatology, 8. Elsevier, Amsterdam. 248 pp.

Atkinson, G. D. (1971) *Forecasters Guide to Tropical Meteorology.* Headquarters Air Weather Service, US Air Force. Tech. Rep. 240. 360 pp.

Barry, R. G. (1978) Aspects of the precipitation characteristics of the New Guinea mountains. *J. Trop. Geog.* **47**, 13–30.

Beckinsale, R. P. (1957) The nature of tropical rainfall. *Tropical Agriculture*, **34**, 76–98.

Blumenstock, D. I. (1958) Distribution and characteristics of tropical climates. *Proc. 9th Pacific Sci. Congr.* **20**, 3–23.

Breed, C. S. *et al.* (1979) Regional studies of sand seas, using Landsat (ERTS) imagery. *US Geological Survey Professional Paper No. 1052*, 305–97.

Chang, J.-H. (1962) Comparative climatology of the tropical western margins of the northern oceans. *Ann. Assn. Amer. Geog.* **52**, 221–7.

Chang, J.-H. (1967) The Indian summer monsoon. *Geog. Rev.* **57**, 373–96.

Chang, J.-H. (1971) The Chinese monsoon. *Geog. Rev.* **61**, 370–95.

Chopra, K. P. (1973) Atmospheric and oceanic flow problems introduced by islands. *Adv. Geophys.* **16**, 297–421.

Crowe, P. R. (1949) The Trade Wind circulation of the world. *Trans. Inst. Brit. Geog.* **15**, 37–56.

Crowe, P. R. (1951) Wind and weather in the equatorial zone. *Trans. Inst. Brit. Geog.* **17**, 23–76.

Cry, G. W. (1965) Tropical cyclones of the North Atlantic Ocean. *Tech. Paper No. 55*, Weather Bureau, Washington DC. 148 pp.

Curry, L. and Armstrong, R. W. (1959) Atmospheric circulation of the tropical Pacific ocean. *Geografiska Annaler* **41**, 245–55.

Dubief, J. (1963) Le climat du Sahara. *Memoire de L'Institut de Recherches Sahariennes, Université d'Alger, Algiers*. 275 pp.

Dunn, G. E. and Miller, B. I. (1960) *Atlantic Hurricanes*. Louisiana State University Press. 326 pp.

Eldridge, R. H. (1957) A synoptic study of West African disturbance lines. *Quart. J. Roy. Met. Soc.* **83**, 303–14.

Fett, R. W. (1964) Aspects of hurricane structure: New model considerations suggested by TIROS and Project Mercury observations. *Monthly Weather Review* **92**, 43–59.

Findlater, J. (1974) An extreme wind speed in the low-level jet-stream system of the western Indian Ocean. *Met. Mag.* **103**, 201–5.

Flohn, H. (1968) *Contributions to a meteorology of the Tibetan Highlands*. Atmos. Sci. Paper No. 130. Colorado State University, Fort Collins. 120 pp.

Flohn, J. (1971) Tropical circulation patterns. *Bonn. Geogr. Abhandl.* **15**. 55 pp.

Fosberg, F. R., Garnier, B. J. and Küchler, A. W. (1961) Delimitation of the humid tropics. *Geog. Rev.* **51**, 333–47.

Frank, N. L. and Hubert, P. J. (1974) Atlantic tropical systems of 1973. *Monthly Weather Review* **102**, 290–5.

Frost, R. and Stephenson, P. H. (1965) Mean streamlines and isotachs at standard pressure levels over the Indian and west Pacific Oceans and adjacent land areas. *Geophys. Mem.* **14**(109). HMSO, London. 24 pp.

Gao, Y.-X. and Li, C. (1981) Influences of Qinghai-Xizang plateau on seasonal variation of general atmospheric circulation; In *Geoecological and Ecological Studies of Qinghai-Xizang Plateau*, Vol. 2. Science Press, Beijing, pp. 1477–84.

Garnier, B. J. (1967) Weather conditions in Nigeria. *Climatological Research Series No. 2*. McGill University Press, Montreal. 163 pp.

Gentilli, J. (ed.) (1971) *Climates of Australia and New Zealand*. World Survey of Climatology, 13. Elsevier, Amsterdam. 405 pp.

Goudie, A. and Wilkinson, J. (1977) *The Warm Desert Environment*. Cambridge University Press. 88 pp.

Gray, W. M. (1968) Global view of the origin of tropical disturbances and hurricanes. *Monthly Weather Review* **96**, 669–700.

Gray, W. M. (1979) Hurricanes: Their formation, structure and likely role in the tropical circulation; In Shaw, D. B. (ed.) *Meteorology over the Tropical Oceans*. Royal Meteorological Society, Bracknell, pp. 155–218.

Gray, W. M. and Jacobson, R. W. (1977) Diurnal variation of deep cumulus convection. *Monthly Weather Review* **105**, 1171–88.

Gregory, S. (1965) *Rainfall over Sierra Leone*. Geography Department, University of Liverpool, Research Paper No. 2. 58 pp.

Griffiths, J. F. (ed.) (1972) *Climates of Africa*. World Survey of Climatology, 10. Elsevier, Amsterdam. 604 pp.

Hamilton, M. G. (1979) *The South Asian Summer Monsoon*. Arnold, Australia. 72 pp.

Hastenrath, S. (1985) *Climate and Circulation of the Tropics*. D. Reidel, Dordrecht. 455 pp.

Houze, R. A., Goetis, S. G., Marks, F. D. and West, A. K. (1981) Winter monsoon convection in the vicinity of North Borneo. *Monthly Weather Review* **109**, 1595–614.

Houze, R. A. and Hobbs, P. V. (1982) Organization and structure of precipitating cloud systems. *Adv. Geophys.* **24**, 225–315.

Hutchings, J. W. (ed.) (1964) *Proceedings of the Symposium on Tropical Meteorology*. New Zealand Meteorological Service, Wellington. 737 pp.

Indian Meteorological Department (1960) *Monsoons of the World*. Delhi. 270 pp.

Jackson, I. J. (1977) *Climate, Water and Agriculture in the Tropics*. Longman, London. 248 pp.

Jalu, R. (1960) Étude de la situation météorologique au Sahara en Janvier 1958. *Ann. de Géog.* **69**(371), 288–96.

Jordan, C. L. (1955) Some features of the rainfall at Guam. *Bull. Amer. Met. Soc.* **36**, 446–55.

Kamara, S. I. (1986) The origins and types of rainfall in West Africa. *Weather* **41**, 48–56.

Koteswaram, P. (1958) The easterly jet stream in the tropics. *Tellus* **10**, 43–57.

Krishnamurti, T. N. (ed.) (1977) Monsoon meteorology. *Pure Appl. Geophys.* **115**, 1087–529.

Kurashima, A. (1968) Studies on the winter and summer monsoons in east Asia based on dynamic concept. *Geophys. Mag.* (Tokyo) **34**, 145–236.

Kurihara, Y. (1985) Numerical modeling of tropical cyclones; In Manabe, S. (ed.) *Issues in Atmospheric and Oceanic Modeling. Part B, Weather Dynamics. Advances in Geophysics*. Academic Press, New York, pp. 255–87.

Lau, K.- M. and Li, M.- T. (1984) The monsoon of East Asia and its global associations – a survey. *Bull Amer. Met. Soc.* **65**, 114–25.

Le Borgue, J. (1979) Polar invasion into Mauretania and Senegal. *Ann. de Géog.* **88**(485), 521–48.

Lighthill, J. and Pearce, R. P. (eds.) (1979) *Monsoon Dynamics*. Cambridge University Press. 735 pp.

Lockwood, J. G. (1965) The Indian monsoon – a review. *Weather* **20**, 2–8.

Logan, R. F. (1960) The Central Namib Desert, South-west Africa. *National Academy of Sciences, National Research Council, Publication 758*. Washington DC. 162 pp.

Lowell, W. E. (1954) Local weather of the Chicama Valley, Peru. *Archiv Met. Geophys. Biokl.* B **5**, 41–51.

Lydolph, P. E. (1957) A comparative analysis of the dry western littorals. *Ann. Assn. Amer. Geog.* **47**, 213–30.

Maejima, I. (1967) Natural seasons and weather singularities in Japan. *Geog. Report No. 2*. Tokyo Metropolitan University, pp. 77–103.

Maley, J. (1982) Dust, clouds, rain types, and climatic variations in tropical North Africa. *Quaternary Res.* **18**, 1–16.

Malkus, J. S. (1955–56) The effects of a large island upon the trade-wind air stream. *Quart. J. Roy. Met. Soc.* **81**, 538–50; **82**, 235–38.

Malkus, J. S. (1958) Tropical weather disturbances: Why do so few become hurricanes? *Weather* **13**, 75–89.

Malkus, J. S. and Riehl, H. (1964) *Cloud Structure and Distributions over the Tropical*

Pacific Ocean. University of California Press, Berkeley and Los Angeles. 229 pp.
Mason, B. J. (1970) Future developments in meteorology: An outlook to the year 2000. *Quart. J. Roy. Met. Soc.* **96**, 349–68.
Mink, J. F. (1960) Distribution pattern of rainfall in the leeward Koolau Mountains, Oahu, Hawaii. *J. Geophys. Res.* **65**, 2869–76.
Musk, L. (1983) Outlook – changeable. *Geog. Mag.* **55**, 532–3.
Neal, A. B., Butterworth, L. J. and Murphy, K. M. (1977) The Morning Glory. *Weather* **32**, 176–83.
Nicholson, S. E. and Flohn, H. (1980) African environmental and climatic changes and the general atmospheric circulation in late Pleistocene and Holocene. *Climatic Change* **2**, 313–48.
Nieuwolt, S. (1977) *Tropical Climatology*. Wiley, London. 207 pp.
Omotosho, J. B. (1985) The separate contributions of line squalls, thunderstorms and the monsoon to the total rainfall in Nigeria. *J. Climatology* **5**, 543–52.
Palmén, E. (1948) On the formation and structure of tropical hurricanes. *Geophysica* **3**, 26–38.
Palmer, C. E. (1951) Tropical meteorology; In Malone, T. F. (ed.) *Compendium of Meteorology*. American Meteorological Society, Boston, Mass., pp. 859–80.
Physik, W. L. and Smith, R. K. (1985) Observations and dynamics of sea-breezes in northern Australia. *Austral. Met. Mag.* **33**, 51–63.
Raghavan, K. (1967) Influence of tropical storms on monsoon rainfall in India. *Weather* **22**, 250–5.
Ramage, C. S. (1952) Relationships of general circulation to normal weather over southern Asia and the western Pacific during the cool season. *J. Met.* **9**, 403–8.
Ramage, C. S. (1964) Diurnal variation of summer rainfall in Malaya. *J. Trop. Geog.* **19**, 62–8.
Ramage, C. S. (1968) Problems of a monsoon ocean. *Weather* **23**, 28–36.
Ramage, C. S. (1971) *Monsoon Meteorology*. Academic Press, New York and London. 296 pp.
Ramage, C. S. (1986) El Niño. *Scientific American* **254**, 76–83.
Ramage, C. S., Khalsa, S. J. S. and Meisner, B. N. (1980) The central Pacific near-equatorial convergence zone. *J. Geophys. Res.* **86**(7), 6580–98.
Ramaswamy, C. (1956) On the sub-tropical jet stream and its role in the development of large-scale convection. *Tellus* **8**, 26–60.
Ramaswamy, C. (1962) Breaks in the Indian summer monsoon as a phenomenon of interaction between the easterly and the sub-tropical westerly jet streams. *Tellus* **14**, 337–49.
Reiter, E. R. and Heuberger, H. (1960) A synoptic example of the retreat of the Indian summer monsoon. *Geografiska Annaler* **42**, 17–35.
Reynolds, R. (1985) Tropical meteorology. *Prog. Phys. Geog.* **9**, 157–86.
Riehl, H. (1954) *Tropical Meteorology*. McGraw-Hill, New York. 392 pp.
Riehl, H. (1963) On the origin and possible modification of hurricanes. *Science* **141**, 1001–10.
Riehl, H. (1979) *Climate and Weather in the Tropics*. Academic Press, New York. 611 pp.
Rossignol-Strick, M. (1985) Mediterranean Quaternary sapropels, an immediate response of the African monsoon to variation in insolation. *Palaeogeography, Palaeoclimatology, Palaeoecology* **49**, 237–63.
Sadler, J. C. (1975a) The monsoon circulation and cloudiness over the GATE area. *Monthly Weather Review* **103**, 369–87.

Sadler, J. C. (1975b) *The Upper Tropospheric Circulation over the Global Tropics*. UHMET-75-05. Department of Meteorology, University of Hawaii. 35 pp.

Saha, R. R. (1973) Global distribution of double cloud bands over the tropical oceans. *Quart. J. Roy. Met. Soc.* **99**, 551–5.

Saito, R. (1959) *The climate of Japan and her meteorological disasters*. Proceedings of the International Geophysical Union, Regional Conference in Japan, Tokyo, pp. 173–83.

Sawyer, J. S. (1970) Large-scale disturbance of the equatorial atmosphere. *Met. Mag.* **99**, 1–9.

Shaw, D. B. (ed.) (1978) *Meteorology over the Tropical Oceans*. Royal Meteorological Society, Bracknell. 278 pp.

Shukla, J. and Wallace, J. (1983) Numerical simulation of the atmospheric response to the equatorial Pacific sea surface temperature anomalies. *J. Atmos. Sci.* **40**, 1613–40.

Sikka, D. R. (1977) Some aspects of the life history, structure and movement of monsoon depressions. *Pure and Applied Geophysics* **115**, 1501–29.

Thompson, B. W. (1951) An essay on the general circulation over South-East Asia and the West Pacific. *Quart. J. Roy. Met. Soc.* 569–97.

Trenberth, K. E. (1976) Spatial and temporal oscillations of the Southern Oscillation. *Quart. J. Roy. Met. Soc.* **102**, 639–53.

Trewartha, G. T. (1958) Climate as related to the jet stream in the Orient. *Erdkunde* **12**, 205–14.

Trewartha, G. T. (1981) *The Earth's Problem Climates*, 2nd edn. University of Wisconsin Press, Madison. 371 pp.

Watts, I. E. M. (1955) *Equatorial Weather, with particular reference to South-east Asia*. OUP, London. 186 pp.

World Meteorological Organization (1972) Synoptic analysis and forecasting in the tropics of Asia and the south-west Pacific. *WMO No. 321*, Geneva. 524 pp.

World Meteorological Organization (n.d.) *The Global Climate System. A critical review of the climate system during 1982–1984*. World Climate Data Programme, WMO, Geneva. 52 pp.

Wyrtki, K. (1982) The Southern Oscillation, ocean–atmosphere interaction and El Niño. *Marine Tech. Soc. J.* **16**, 3–10.

Yarnal, B. (1985) Extratropical teleconnections with El Niño/Southern Oscillation (ENSO) events. *Prog. Phys. Geog.* **9**, 315–52.

Ye, D. (1981) Some characteristics of the summer circulation over the Qinghai-Xizang (Tibet) Plateau and its neighbourhood. *Bull. Amer. Met. Soc.* **62**, 14–19.

Ye, D. and Gao, Y.- X. (1981) The seasonal variation of the heat source and sink over Qinghai-Xizang plateau and its role in the general circulation; In *Geoecological and Ecological Studies of Qinghai-Xizang Plateau*, Vol. 2. Science Press, Beijing, pp. 1453–61.

Yoshino, M. M. (1969) Climatological studies on the polar frontal zones and the intertropical convergence zones over South, South-east and East Asia. *Climatol. Notes*, Hosei University, **1.** 71 pp.

Yoshino, M. M. (ed.) (1971) *Water Balance of Monsoon Asia*. University of Tokyo Press. 308 pp.

Young, J. A. (coordinator) (1972) *Dynamics of the Tropical Atmosphere* (Notes from a Colloquium). National Center for Atmospheric Research, Boulder, Colorado. 587 pp.

Chapter 7 Small-scale climates

Anderson, G. E. (1971) Mesoscale influences on wind fields. *J. App. Met.* **10**, 377–86.

Atkinson, B. W. (1968) A preliminary examination of the possible effect of London's urban area on the distribution of thunder rainfall 1951–60. *Trans. Inst. Brit. Geog.* **44**, 97–118.

Atkinson, B. W. (1977) Urban effects on precipitation: An investigation of London's influence on the severe storm of August 1975. *Dept. Geog. Queen Mary Coll. London, Occ. Paper* **8**. 31 pp.

Bach, W. (1971) Atmospheric turbidity and air pollution in Greater Cincinnati. *Geog. Rev.* **61**, 573–94.

Bach, W. (1979) Short-term climatic alterations caused by human activities. *Prog. Phys. Geog.* **3**(1), 55–83.

Bach, W. and Patterson, W. (1969) Heat budget studies in Greater Cincinnati. *Proc. Assn. Amer. Geog.* **1**, 7–16.

Bryson, R. A. and Kutzbach, J. E. (1968) Air pollution. *Association of American Geographers, Commission on College Geography, Resource Paper 2.* 42 pp.

Caborn, J. M. (1955) The influence of shelter-belts on microclimate. *Quart. J. Roy. Met. Soc.* **81**, 112–15.

Chandler, T. J. (1965) *The Climate of London*. Hutchinson, London. 292 pp.

Chandler, T. J. (1967) Absolute and relative humidities in towns. *Bull. Amer. Met. Soc.* **48**, 394–9.

Changnon, S. A. (1969) Recent studies of urban effects on precipitation in the United States. *Bull. Amer. Met. Soc.* **50**, 411–21.

Changnon, S. A. (1979) What to do about urban-generated weather and climate changes. *J. Amer. Plan. Assn.* **45**(1), 36–48.

Committee on Air Pollution (1955) *Report*, Cmnd 9322. HMSO, London.

Coutts, J. R. H. (1955) Soil temperatures in an afforested area in Aberdeenshire. *Quart. J. Roy. Met. Soc.* **81**, 72–9.

Duckworth, F. S. and Sandberg, J. S. (1954) The effect of cities upon horizontal and vertical temperature gradients. *Bull. Amer. Met. Soc.* **35**, 198–207.

Food and Agriculture Organization of the United Nations (1962) *Forest Influences*, Forestry and Forest Products Studies No. 15, Rome. 307 pp.

Garnett, A. (1967) Some climatological problems in urban geography with special reference to air pollution. *Trans. Inst. Brit. Geog.* **42**, 21–43.

Gay, L. W. and Stewart, J. B. (1974) Energy balance studies in coniferous forests. *Report No. 23*, Inst. Hydrol., Nat. Env. Res. Coun., Wallingford.

Goldreich, Y. (1984) Urban topo-climatology. *Prog. Phys. Geog.* **8**, 336–64.

Goldsmith, J. R. (1969) Los Angeles smog. *Science Jour.* **5**, 44–9.

Hewson, E. W. (1951) Atmospheric pollution; In Malone, T. F. (ed.) *Compendium of Meteorology*. American Meteorological Society, Boston, Mass., pp. 1139–57.

Jäger, J. (1983) *Climate and Energy Systems. A review of their interactions*. Wiley, New York. 231 pp.

Jenkins, I. (1969) Increases in averages of sunshine in Greater London. *Weather* **24**, 52–4.

Kessler, A. (1985) Heat balance climatology; In Essenwanger, B. M. (ed.) *World Survey of Climatology. Vol. 1A, General climatology*. Elsevier, Amsterdam. 224 pp.

Kittredge, J. (1948) *Forest Influences*. McGraw-Hill, New York. 394 pp.

Koppány, Gy. (1975) Estimation of the life span of atmospheric motion systems by means of atmospheric energetics. *Met. Mag.* **104**, 302–6.

Landsberg, H. E. (1981) City climate; In Landsberg, H. E. (ed.) *General Climatology* 3. World Survey of Climatology, 3. Elsevier, Amsterdam, pp. 299–334.

Long, I. F., Monteith, J. L., Penman, H. L. and Szeicz, G. (1964) The plant and its environment. *Meteorologische Rundschau* **17**(4), 97–101.

Lowry, W. P. (1969) *Weather and Life*. Academic Press, New York. 305 pp.

McNaughton, K. and Black, T. A. (1973) A study of evapotranspiration from a Douglas fir forest using the energy balance approach. *Water Resources Research* **9**, 1579–90.

Marshall, W. A. L. (1952) *A Century of London Weather*. Met. Office, Air Ministry, Rept MO 508. HMSO, London. 103 pp.

Meetham, A. R. (1955) Know your fog. *Weather* **10**, 103–5.

Meetham, A. R. *et al.* (1980) *Atmospheric Pollution*, 4th edn (1st edn 1952). Pergamon Press, Oxford and London.

Miess, M. (1979) The climate of cities; In Laurie, I. C. (ed.) *Nature in Cities*. Wiley, Chichester, pp. 91–104.

Miller, D. H. (1965) The heat and water budget of the earth's surface. *Adv. Geophys.* **11**, 175–302.

Nicholas, F. W. and Lewis, J. E. (1980) Relationships between aerodynamic roughness and land use and land cover in Baltimore, Maryland. *US Geol. Surv. Prof. Pap. 1099-C*. 36 pp.

Oke, T. R. (1978) *Boundary Layer Climates*. Methuen, London. 372 pp.

Oke, T. R. (1980) Climatic impacts of urbanization; In Bach. W., Pankrath, J. and Williams, J. (eds.) *Interactions of Energy and Climate*. D. Reidel, Dordrecht, Holland, pp. 339–56.

Oke, T. R. and East, C. (1971) The urban boundary layer in Montreal. *Boundary-Layer Met.* **1**, 411–37.

Pankrath, J. (1980) Impact of heat emissions in the Upper-Rhine region; In Bach, W., Pankrath, J. and Williams, J. (eds.) *Interactions of Energy and Climate*. D. Reidel, Dordrecht, Holland, pp. 363–81.

Parry, M. (1966) The urban 'heat island'; In Tromp, S. W. and Weite, W. H. (eds.) *Biometeorology*, 2. Pergamon, Oxford and London, pp. 616–24.

Pease, R. W., Jenner, C. B. and Lewis, J. E. (1980) The influences of land use and land cover on climate analysis: An analysis of the Washington–Baltimore area. *US Geol. Surv. Prof. Pap. 1099-A*. 39 pp.

Peel, R. F. (1974) Insolation and weathering: Some measures of diurnal temperature changes in exposed rocks in the Tibesti region, central Sahara. *Zeit.f.Geomorph. Supp.* **21**, 19–28.

Peterson, J. T. (1971) Climate of the city; In Detwyler, T. R. (ed.) *Man's Impact on Environment*. McGraw-Hill, New York, pp. 131–54.

Plate, E. (1972) Berücksichtigung von Windströmungen in der Bauleitplanung; In *Seminarberichte Rahmenthema Unweltschutz*. Institut für Stätebau und Landesplanung, Selbstverlag, Karlsruhe, pp. 201–29.

Reynolds, E. R. C. and Leyton, L. (1963) Measurement and significance of throughfall in forest stands; In Whitehead, F. M. and Rutter, A. J. (eds.) *The Water Relations of Plants*. Blackwell Scientific Publications, Oxford, pp. 127–41.

Richards, P. W. (1952) *The Tropical Rain Forest*. Cambridge University Press. 450 pp.

Rutter, A. J. (1967) Evaporation in forests. *Endeavour* **97**, 39–43.
Scorer, R. (1968) *Air Pollution*. Pergamon, Oxford and London. 151 pp.
Sellers, W. D. (1965) *Physical Climatology*. University of Chicago Press. 272 pp.
Smagorinsky, J. (1974) Global atmospheric modelling and the numerical simulation of climate; In Hess, W. N. (ed.) *Weather and Climate Modification*. Wiley, New York, pp. 633–86.
Sopper, W. E. and Lull, H. W. (eds.) (1967) *International Symposium on Forest Hydrology*. Pergamon, Oxford and London. 813 pp.
Stearn, A. C. (ed.) (1968) *Atmospheric Pollution*, 3 vols. Academic, New York.
Sukachev, V. and Dylis, N. (1968) *Fundamentals of Forest Biogeocoenology*. Oliver and Boyd, Edinburgh. 672 pp.
Terjung, W. H. (1970) Urban energy balance climatology. *Geog. Rev.* **60**, 31–53.
Terjung, W. H. and Louis, S. S.-F. (1973) Solar radiation and urban heat islands. *Ann. Assn. Amer. Geog.* **63**, 181–207.
Terjung, W. H. and O'Rourke, P. A. (1980) Simulating the causal elements of urban heat islands. *Boundary-Layer Met.* **19**, 93–118.
Terjung, W. H. and O'Rourke, P. A. (1981) Energy input and resultant surface temperatures for individual urban interfaces. *Archiv. Met. Geophys. Biokl.* B, **29**, 1–22.
Turner, W. C. (1955) Atmospheric pollution. *Weather* **10**, 110–19.
Tyson, P. D., Garstang, M. and Emmitt, G. D. (1973) The structure of heat islands. *Occasional Paper No. 12*. Dept. of Geography and Environmental Studies, University of the Witwatersrand, Johannesburg. 71 pp.
US DHEW (1970) *Air Quality Criteria for Photochemical Oxidants*. National Air Pollution Control Administration, US Public Health Service, Publication No. AP-63, Washington DC.
Vehrencamp, J. E. (1953) Experimental investigation of heat transfer at an air-earth interface. *Trans. Amer. Geophys. Union* **34**, 22–30.
White, W. H., Anderson, J. A., Blumenthal, D. L., Husar, R. B., Gillani, N. V. Husar, J. D. and Wilson, W. E. (1976) Formation and transport of secondary air pollutants: Ozone and aerosols in the St Louis urban plume. *Science* **194**, 187–9.
World Meteorological Organization (1970) Urban climates. *WMO Technical Note No. 108*. 390 pp.
Zon, R. (1941) Climate and the nation's forests. US Dept. of Agriculture Yearbook, *Climate and Man*, pp. 477–98.

Chapter 8 Climatic variability, trends and fluctuations

Bayce, A. (ed.) (1979) *Man's Influence on Climate*. D. Reidel, Dordrecht, Holland. 113 pp.
Beckinsale, R. P. (1965) Climatic change: A critique of modern theories; In Whittow, J. B. and Wood, P. D. (eds.) *Essays in Geography for Austin Miller*. University of Reading Press, pp. 1–38.
Bradley, R. S. (1985) *Quaternary Paleoclimatology*. Allen and Unwin, Boston. 472 pp.
Brádzil, R., Šamaj, F. and Valovič, Š. (1985) Variation of spatial annual precipitation sums in central Europe in the period 1881–1980. *J. Climatology* **5**, 617–31.
Bryson, R. A. (1968) All other factors being constant. . . . *Weatherwise* **21**, 51–61.
Bryson, R. A. (1974) A perspective on climatic change. *Science* **184**, 753–60.

Callendar, G. S. (1961) Temperature fluctuations and trends over the earth. *Quart. J. Roy. Met. Soc.* **87**, 1–12.

Conrad, V. and Pollak, L. W. (1950) *Methods in Climatology*. Harvard University Press, Cambridge, Mass. See ch. 2, Statistical analysis of climatic elements, pp. 17–60.

Frakes, L. A. (1979) *Climates throughout Geologic Time*. Elsevier, Amsterdam. 310 pp.

Gilliland, R. L. (1982) Solar, volcanic and CO_2 forcing of recent climatic change. *Climatic Change* **4**, 111–31.

Goudie, A. (1981) *The Human Impact: Man's role in environmental change*. Blackwell, Oxford. 316 pp.

Gregory, S. (1962 and subsequent editions) *Statistical Methods and the Geographer*. Longman, London. 240 pp.

Gregory, S. (1969) Rainfall reliability; In Thomas, M. F. and Whittington, G. W. (eds.) *Environment and Land Use in Africa*. Methuen, London, pp. 57–82.

Gribbin, J. (ed.) (1978) *Climatic Change*. Cambridge University Press. 280 pp.

Hansen, J., Johnson, D., Lacis, A., Lebedeff, S., Lee, R., Rind, D. and Russell, G. (1981) Climate impact of increasing carbon dioxide. *Science* **213**, 957–66.

Hughes, M. K., Kelly, P. M., Pilcher, J. R. and La Marche, V. (eds.) (1981) *Climate from Tree Rings*. Cambridge University Press. 400 pp.

Imbrie, J. and Imbrie, K. P. (1979) *Ice Ages: Solving the mystery*. Macmillan, London. 224 pp.

Kelly, P. M. (1980) Climate: Historical perspective and climatic trends; In Doornkamp, J. C. and Gregory, K. J. (eds.) *Atlas of Drought in Britain, 1975–6*. Institute of British Geographers, London, pp. 9–11.

Kraus, E. B. (1955) Secular changes of tropical rainfall regimes. *Quart. J. Roy. Met. Soc.* **81**, 198–210.

Kraus, E. B. (1963) Recent changes of east-coast rainfall regimes. *Quart. J. Roy. Met. Soc.* **89**, 145–6.

Kutzbach, J. E. and Street-Perrott, A. (1985) Milankovitch forcings of fluctuations in the level of tropical lakes from 18 to 0 kyr BP. *Nature* **317**, 130–9.

Lamb, H. H. (1965) Frequency of weather types. *Weather* **20**, 9–12.

Lamb, H. H. (1966) *The Changing Climate: Selected Papers*. Methuen, London. 236 pp.

Lamb, H. H. (1970) Volcanic dust in the atmosphere; with a chronology and an assessment of its meteorological significance. *Phil. Trans. Roy. Soc.*, A **266**, 425–533.

Lamb, H. H. (1977) *Climate: Present, Past and Future, 2: Climatic History and the Future*. Methuen, London. 835 pp.

Landsberg, H. E. (1970) Man-made climatic changes. *Science* **170**, 1265–74.

Leopold, L. B. (1951) Rainfall frequency: An aspect of climatic variation. *Trans. Amer. Geophys. Union*. **32**(3), 347–57.

Lewis, P. (1960) The use of moving averages in the analysis of time-series. *Weather* **15**, 121–6.

McCormac, B. M. and Seliga, T. A. (1979) *Solar–Terrestrial Influences on Weather and Climate*. D. Reidel, Dordrecht, Holland. 340 pp.

Macdonald, G. J. F. (1966) Weather and climate modification: Problems and prospects. *Bull. Amer. Met. Soc.* **47**, 4–19.

Manley, G. (1958) Temperature trends in England, 1698–1957. *Archiv Met. Geophy. Biokl.* (Vienna), B, **9**, 413–33.

Mitchell, J. M., Jr (ed.) (1968) Causes of climatic change. *Amer. Met. Soc. Monogr.* **8**(30). 159 pp.

Mitchell, J. M., Jr (1972) The natural breakdown of the present interglacial and its possible intervention by human activities. *Quat. Res.* **2**, 436–45.

Mitchell, J. M., Jr, Stockton, C. W. and Meko, D. M. (1979) Evidence of a 22-year rhythm of drought in the western United States related to the Hale solar cycle since the seventeenth century; In McCormac, B. M. and Seliga, T. A. (eds.) *Solar–Terrestrial Influences on Weather and Climate*. Reidel, Dordrecht, Holland, pp. 125–43.

Nicholson, S. E. (1980) The nature of rainfall fluctuations in subtropical West Africa. *Mon. Wea. Rev.* **108**, 473–87.

Nicholson, S. E. (1985) Sub-Saharan rainfall 1981–84. *J. Climate Applied Met.* **24**, 1388–91.

Pfister, C. (1985) Snow cover, snow lines and glaciers in central Europe since the 16th century; In Tooley, M. J. and Sheail, G. M. (eds.) *The Climatic Scene*. Allen and Unwin, London, pp. 154–74.

Pittock, A. B., Frakes, L. A., Jenssen, D., Peterson, J. A. and Zillman, J. W. (eds.) (1978) *Climatic Change and Variability: A southern perspective*. Cambridge University Press. 455 pp.

Schneider, S. H. and Kellogg, W. W. (1973) The chemical basis for climate change; In Rasool, S. I. (ed.) *Chemistry of the Lower Atmosphere*. Plenum, New York, pp. 203–49.

Schuurmans, C. J. E. (1984) Climate variability and its time changes in European countries, based on instrumental observations; In Flohn, H. and Fantechi, R. (eds.) *The Climate of Europe: Past, present and future*. D. Reidel, Dordrecht, pp. 65–101.

Sewell, W. R. D. (ed.) (1966) *Human dimensions of weather modification*. University of Chicago, Dept. of Geography, Research Paper No. 105. 423 pp.

Sioli, H. (1985) The effects of deforestation in Amazonia. *Geog. J.* **151**, 197–203.

Street, F. A. (1981) Tropical palaeoenvironments. *Progr. Phys. Geog.* **5**, 157–85.

Study of Man's Impact on Climate (SMIC) (1971) *Inadvertent Climate Modification*. Massachusetts Institute of Technology Press, Cambridge, Mass. 308 pp.

Toon, O. B. and Pollack, J. B. (1980) Atmospheric aerosols and climate. *Amer. Scientist* **68**, 268–78.

Tucker, G. B. (1964) Solar influences on the weather. *Weather* **19**, 302–11.

Wigley, T. M. L., Ingram, M. J. and Farmer, G. (eds.) (1981) *Climate and History*. Cambridge University Press, Cambridge. 530 pp.

Williams, J. (ed.) (1978) *Carbon Dioxide, Climate and Society*. Pergamon, Oxford. 332 pp.

APPENDIX 1

Climatic classification

The purpose of any classification system is to obtain an efficient arrangement of information in a simplified and generalized form. Thus, climatic statistics can be organized in order to describe and delimit the major types of climate in quantitative terms. Obviously no single classification can serve more than a limited number of purposes satisfactorily and many different schemes have therefore been developed. Some schemes merely provide a convenient nomenclature system, whereas others are an essential preliminary to further study. Many climatic classifications, for instance, are concerned with the relationships between climate and vegetation or soils, but surprisingly few attempts have been made to base a classification on the direct effects of climate on man.

Only the basic principles of the four groups of the most widely known classification systems are summarized here. Further information may be found in the listed references.

A Generic classifications related to plant growth or vegetation

The numerous schemes that have been suggested for relating climatic limits to plant growth or vegetation groups rely on two basic criteria – the degree of aridity and of warmth.

Aridity is not simply a matter of low precipitation, but of the 'effective precipitation' (i.e. precipitation minus evaporation). The ratio of rainfall/temperature has been used as such an index of precipitation effectiveness, on the grounds that higher temperatures increase evaporation. The ratio r/t was proposed by R. Lang in 1915 (where r = mean annual rainfall in mm, and t = mean annual temperature in °C), such that $r/t < 40$ is considered arid and $r/t > 160$ perhumid.

The work of W. Köppen is the prime example of this type of classification. Between 1900 and 1936 he published several classification schemes involving considerable complexity in their full detail. Nevertheless, the system has been used extensively in geographical teaching. The key features of Köppen's final classification are temperature criteria and aridity criteria.

Temperature criteria: five of the six major climatic types are recognized on the basis of monthly mean temperature.

A Tropical rainy climate: coldest month > 18°C (64·4°F).
B Dry climates.
C Warm temperate rainy climates: coldest month between −3° and 18°C, warmest month > 10°C (50°F).
D Cold boreal forest climates: coldest month < −3°C (26·6°F), warmest month > 10°C (see note 1).
E Tundra climate: warmest month 0°–10°C.
F Perpetual frost climate: warmest month < 0°C.

The arbitrary temperature limits stem from a variety of criteria, the supposed significance of the selected values being as follows: the 10°C summer isotherm correlates with the poleward limit of tree growth; the 18°C winter isotherm is critical for certain tropical plants; and the −3°C isotherm indicates a few weeks of snow cover. However, these correlations are far from precise! The criteria were determined from the study of vegetation groups defined on a physiological basis (i.e. according to the internal functions of plant organs) by De Candolle in 1874.

Aridity criteria:

	steppe (BS)/desert (BW) boundary	*forest/steppe boundary*
winter precipitation maximum	$r/t = 1$	$r/t = 2$
precipitation evenly distributed	$r/(t+7) = 1$	$r/(t+7) = 2$
summer precipitation maximum	$r/(t+14) = 1$	$r/(t+14) = 2$

Where: r = annual precipitation (cm)
t = mean annual temperature (°C)

The criteria imply that, with winter precipitation, arid (desert) conditions occur where $r/t < 1$, semi-arid conditions where $1 < r/t < 2$. If the rain falls in summer, a larger amount is required to offset evaporation and maintain an equivalent total of effective precipitation.

Subdivisions of each major category are made with reference, firstly, to the seasonal distribution of precipitation (the most common of which are: f = no dry season; m = monsoonal, with a short dry season and heavy rains during the rest of the year; s = summer dry season; w = winter dry season) and, secondly, to additional temperature characteristics. (For the B climates: h = mean annual temperature > 18°C; k = mean annual temperature < 18°C (warmest month > 18°C); k' = mean annual temperature (and warmest month) < 18°C.) Figure App. 1.1A illustrates the distribution of the major Köppen climatic types on a hypothetical continent of low and uniform elevation.

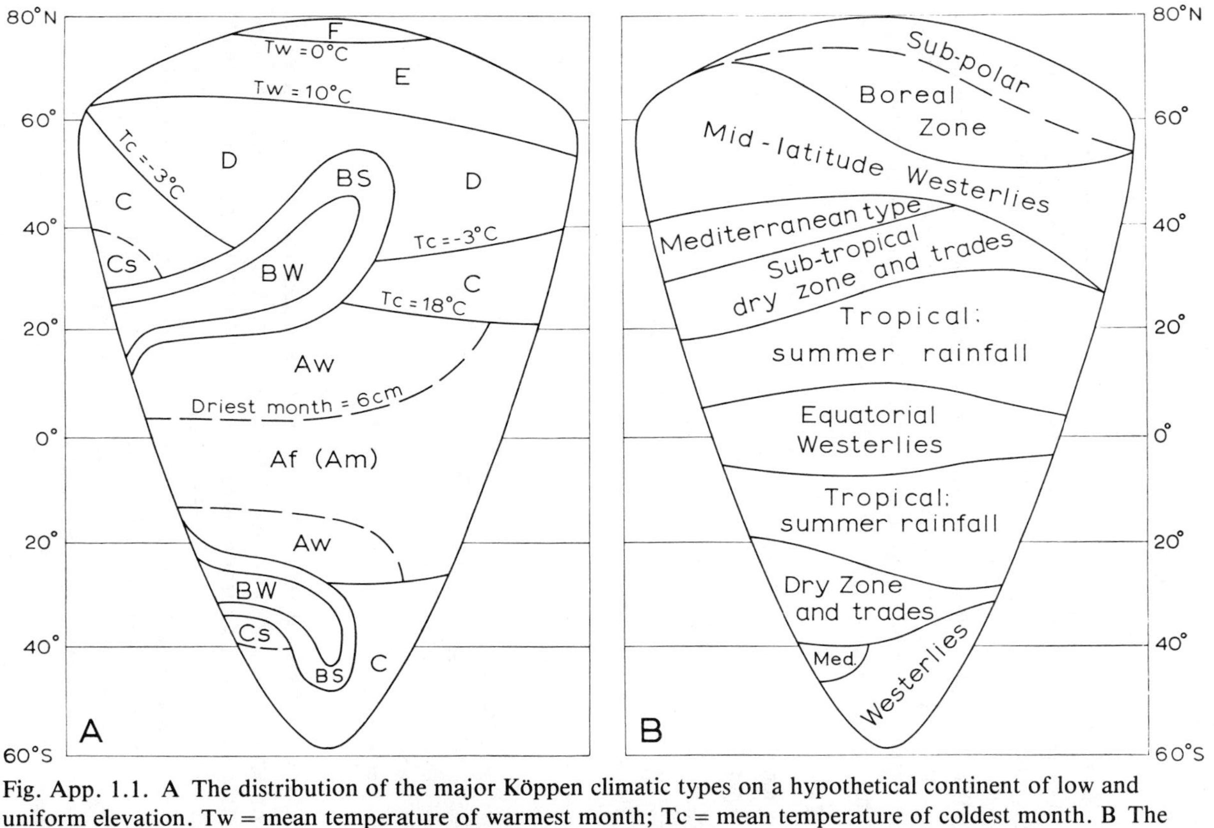

Fig. App. 1.1. A The distribution of the major Köppen climatic types on a hypothetical continent of low and uniform elevation. Tw = mean temperature of warmest month; Tc = mean temperature of coldest month. B The distribution of Flohn's climatic types on a hypothetical continent of low and uniform elevation (*from Flohn 1950*).

A somewhat similar scheme has been proposed by A. A. Miller (1951), using the following criteria:

Boundary of arid conditions: $r/t = 1/5$.
Boundary of semi-arid conditions: $r/t = 1/3$.
Where: r = mean annual rainfall (in)
t = mean annual temperature (°F).

The thermal units relate to the *accumulated temperature*, which Miller estimated approximately by using 'month-degrees' – the excess of mean monthly temperatures above 43°F (6°C) – rather than the usual day-degrees based on daily mean temperatures above this limit.

C. W. Thornthwaite introduced a complex, empirical classification in 1931. An expression for *precipitation efficiency* was obtained by relating measurements of pan evaporation to temperature and precipitation. For each months the ratio

$$115(r/t - 10)^{10/9}$$

where: r = mean monthly rainfall (in)
t = mean monthly temperature (°F)

is calculated. The sum of the twelve monthly ratios gives the *precipitation efficiency* (*P-E*) index. By determining boundary values for the major vegetation regions the following humidity provinces were defined:

	P-E index
A Rain forest	>127
B Forest	64–127
C Grasslands	32–63
D Steppe	16–31
E Desert	<16

The second element of the classification is an index of *thermal efficiency* (*T-E*), expressed by the positive departure of monthly mean temperatures from freezing point. The index is thus the annual sum of $(t - 32)/4$ for each month. On this scale zero is 'frost climate' and over 127 is 'tropical'. Unlike Köppen, Thornthwaite makes moisture the primary classificatory factor for a *T-E* index of over 31 (the taiga/cool temperate boundary). Maps of the distribution of these climatic provinces in North America and over the world have been published, but the classification is now largely of historical interest.

B Energy and moisture budget classifications

Thornthwaite's most important contribution was his second (1948) classification. It is based on the concept of potential evapotranspiration and the moisture budget (see chs. 2, A and 5, B.3.c). The potential evapotranspiration (*PE*) is calculated from the mean monthly temperature (in °C), with corrections for day length. For a 30-day month (12-hour days):

PE (in cm) $= 1\cdot6(10t/I)^a$
where: $I =$ the sum for 12 months of $(t/5)^{1\cdot514}$
$a =$ a further complex function of I.

Tables have been prepared for the easy computation of these factors.

The monthly water surplus (*S*) or deficit (*D*) is determined from a moisture budget assessment, taking into account stored soil moisture. A moisture index (*Im*) is given by:

$$Im = (100S - 60D)/PE$$

The weighting of a deficit by 0·6 is supposed to allow for the beneficial action of a surplus in one season when moisture is stored in the subsoil, to be drawn on during subsequent droughts by deep-rooted perennials. In 1955 this weighting factor was omitted since it was recognized that a deficit can begin as soon as any moisture is removed from the soil by evaporation. The later revision also allows for a variable soil moisture storage according to vegetation cover and soil type, and permits the evaporation rate to vary with the actual soil moisture content.

A novel feature of the system is that the thermal efficiency is derived from the *PE* value because this itself is a function of temperature. The climatic types defined by these two factors are:

*Im (1955 system)**		*PE*		
		cm	*in*	
>100	Perhumid (*A*)	>114	>44·9	Megathermal (A')
20 to 100	Humid (B_1 to B_4)	57 to 114	22·4 to 44·9	Mesothermal (B_1' to B_4')
0 to 20	Moist Subhumid (C_2)	28·5 to 57	11·2 to 22·4	Microthermal (C_1' to C_2')
−33 to 0	Dry Subhumid (C_1)	14·2 to 28·5	5·6 to 11·2	Tundra (D')
−67 to −33	Semi-arid (*D*)	<14·2	<5·6	Frost (E')
−100 to −67	Arid (*E*)			

* $Im = 100(S - D)/PE$ is equivalent to $100\ (r/PE - 1)$, where $r =$ annual precipitation.

Both elements are subdivided according to the season of moisture deficit or surplus and the seasonal concentration of thermal efficiency.

The system has been applied to many regions, although no world map has yet been published. In tropical and semi-arid areas the method is not very satisfactory, but in eastern North America, for example, vegetation boundaries have been shown to coincide reasonably closely with particular *PE* values. This classification, unlike that of 1931, Köppen's and many others, does not use vegetation boundaries to determine climatic ones.

M. I. Budyko in the Soviet Union developed a similar, but more fundamental approach using net radiation rather than temperature (see ch. 2, A). He related the net radiation available for evaporation from a wet surface (R_o) to the heat required to evaporate the mean annual precipitation (Lr). This ratio R_o/Lr (where L = latent heat of evaporation) is called the *radiational index of dryness*. It has a value less than unity in humid areas and greater than unity in dry areas. Boundary values are:

R_o/Lr	
>3·0	Desert
2·0–3·0	Semi-desert
1·0–2·0	Steppe
0·33–1·0	Forest
<0·33	Tundra

By way of comparison with the revised Thornthwaite index ($Im = 100(r/PE - 1)$) it may be noted that $Im = 100(L\text{r}/R_o - 1)$ if all the net radiation is used for evaporation from the wet surface (i.e. none is transferred into the ground by conduction or into the air as sensible heat). A general world map of R_o/Lr has appeared but over large parts of the earth there are as yet no measurements of net radiation.

Energy fluxes have also been used by Terjung and Louie (1972) to categorize the magnitude of energy input (net radiation and advection) and outputs (sensible heat and latent heat), and their seasonal range. On this basis, sixty-two climatic types are distinguished (in six broad groups), and a world map is presented.

C Genetic classifications

The genetic basis of large-scale (or macro-) climates is the atmospheric circulation, and this can be related to regional climatology in terms of wind regimes or air masses. One attempt, made by A. Hettner in 1931, incorporated the wind system, continually, rainfall amount and duration, position relative to the sea, and elevation. A very generalized scheme using air masses, according to their seasonal dominance, was put forward by B. P. Alissov in 1936.

A more satisfactory system, however, was proposed in 1950 by H. Flohn.

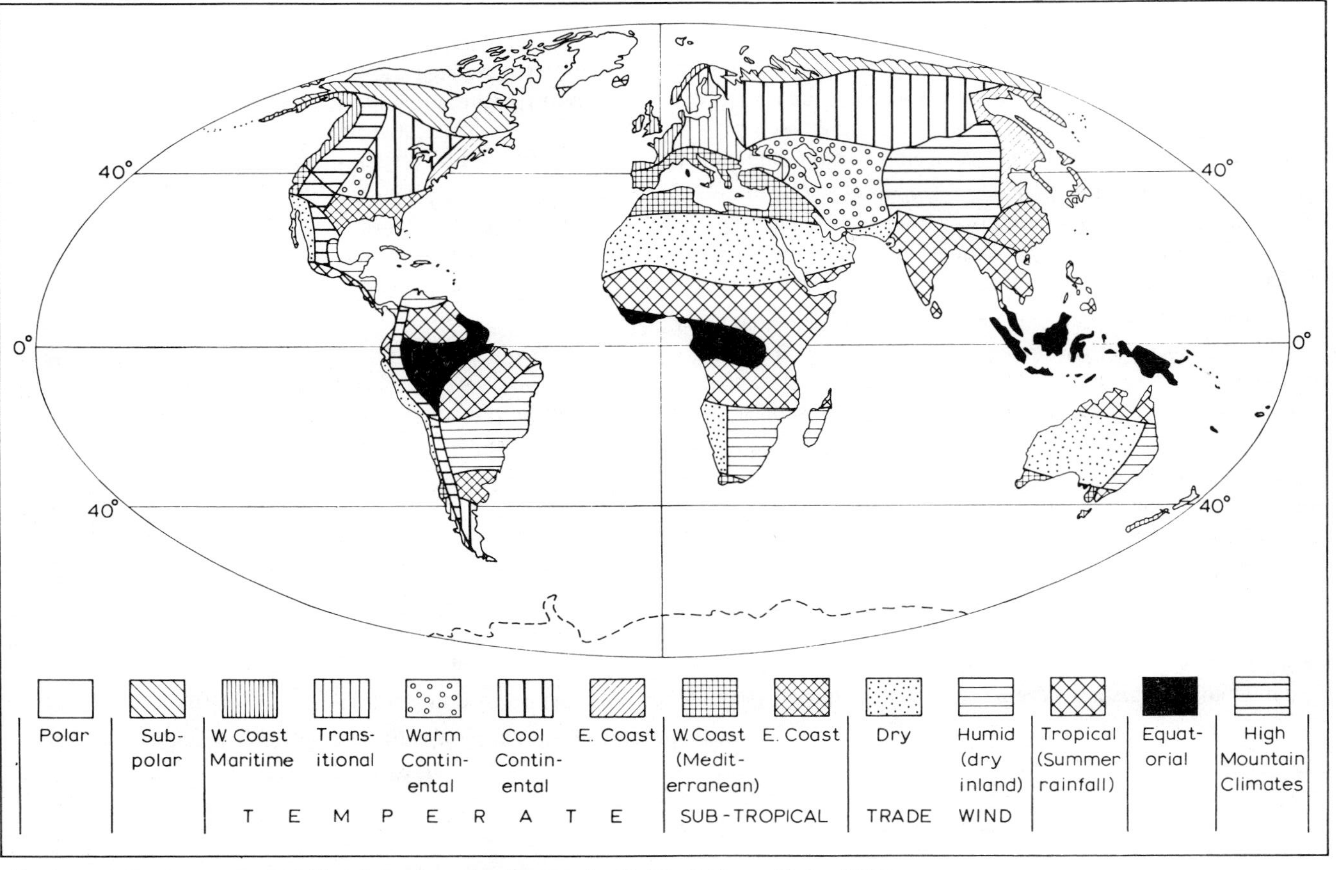

Fig. App. 1.2. A genetic classification of world climates by E. Neef (*from Flohn 1957*).

Table App. 1.1. Strahler's climatic classification (*from A. N. Strahler 1969*)

Climate name	*Köppen symbol*		*Air mass source regions and frontal zones, general climate characteristics*
Group 1: Low-latitude climates (controlled by equatorial and tropical air masses)			
1 Wet equatorial climate 10°N–10°S lat. (Asia 10°–20°N)	*Af* *Am*	Tropical rain forest climate, and Tropical rain forest climate, monsoon type	Equatorial trough (convergence zone) climates are dominated by warm, moist tropical maritime (mT) and equatorial (mE) air masses yielding heavy rainfall through convectional storms. Remarkably uniform temperatures prevail throughout the year.
2 Trade wind littoral climate 10°–25°N and S lat.	*Af-Am*	Included in climate 1	Tropical easterlies (trades) bring maritime tropical (mT) air masses from moist western sides of oceanic subtropical high-pressure cells to give narrow east-coast zones of heavy rainfall and uniformly high temperatures. Rainfall shows strong seasonal variation.
3 Tropical desert and steppe climates 15°–35°N and S lat.	*BWh* *BSh*	Desert climate, hot, and Steppe climate, hot	Source regions of continental-tropical (cT_s) air masses in high-pressure cells at high level over lands astride the Tropics of Cancer and Capricorn give arid to semi-arid climate with very high maximum temperatures and moderate annual range.
4 West-coast desert climate 15°–30°N and S lat.	*BWk* *BWh*	Desert climate, cool, and Desert climate, hot (*BWn* in earlier versions, *n* meaning frequent fog)	On west coasts bordering the oceanic subtropical high-pressure cells, subsiding maritime tropical (mT_s) air masses are stable and dry. Extremely dry, but relatively cool, foggy desert climates prevail in narrow coastal belts. Annual temperature range is small.
5 Tropical wet-dry climate 5°–25°N and S lat.	*Aw* *Cwa*	Tropical rainy climate, savanna; also Temperate rainy (Humid mesothermal) climate, dry winter, hot summer	Seasonal alternation of moist mT or mE air masses with dry cT air masses gives climate with wet season at time of high sun, dry season at time of low sun.

Group 2: Middle-latitude climates (controlled by both tropical and polar air masses)

Climate	Code	Description	Characteristics
6 Humid subtropical climate 20°–35°N and S lat.	*Cfa*	Temperate rainy (Humid mesothermal) climate, hot summers	Subtropical, eastern continental margins dominated by moist maritime (mT) air masses flowing from the western sides of oceanic high-pressure cells. In high-sun season, rainfall is copious and temperatures high. Winters are cool with frequent continental polar (cP) air mass invasions. Frequent cyclonic storms.
7 Marine west-coast climate 40°–60°N and S lat.	*Cfb*	Temperate rainy (Humid mesothermal) climate, warm summers, and	Windward, middle-latitude west coasts receive frequent cyclonic storms with cool, moist maritime polar (mP) air masses. These bring much cloudiness and well-distributed precipitation, but with winter maximum. Annual temperature range is small for middle latitudes.
	Cfc	same but cool, short summers	
8 Mediterranean climate 30°–45°N and S lat.	*Csa*	Temperate rainy (Humid mesothermal) climate, dry, hot summer, and	This wet-winter, dry-summer climate results from seasonal alternation of conditions causing climates 4 and 7; mP air masses dominate in winter with cyclonic storms and ample rainfall, mT air masses dominate in summer and extreme drought. Moderate annual temperature range.
	Csb	same, but dry, warm summer	
9 Middle-latitude desert and steppe-climates 35°–50°N and S lat.	*BWk*	Desert climate, cool	Interior, middle-latitude deserts and steppes of regions shut off by mountains from invasions of maritime air masses (mT or mP), but dominated by continental tropical (cT) air masses in summer and continental polar (cP) air masses in winter. Great annual temperature range; hot summers, cold winters.
	BWk'	same, but cold; and	
	BSk	Steppe climate, cool	
	BSk'	same, but cold	
10 Humid continental climate 35°–60°N lat.	*Dfa*	Cold, snowy forest (Humid microthermal) climate, moist all year, hot summers; and	Located in central and eastern parts of continents of middle latitudes, these climates are in the polar front zone, the 'battle ground' of polar and tropical air masses. Seasonal contrasts are strong and weather highly variable. Ample precipitation throughout the year is increased in summer by invading maritime tropical (mT) air masses. Cold winters are dominated by continental polar (cP) air masses invading frequently from northern source regions.
	Dfb	same, but warm summers; also	
	Dwa	Cold, snowy forest (Humid microthermal) climate, dry winters, hot summers; and	
	Dwb	same, but warm summers	

Table App. 1.1.—*continued*

Climate name	*Köppen symbol*		*Air mass source regions and frontal zones, general climate characteristics*
Group 3: High-latitude climates (controlled by polar and arctic air masses)			
11 Continental sub-artic climate 50°–70°N lat.	*Dfc*	Cold, snowy forest (Humid microthermal) climate, moist all year, cool summers; and	This climate lies in source region of continental polar (cP) air masses, which in winter are stable and very cold. Summers are short and cool. Annual temperature range is enormous. Cyclonic storms, into which maritime polar (mP) air is drawn, supply light precipitation, but evaporation is small and the climate is therefore effectively moist.
	Dfd	same, but very cold winters; also	
	Dwc	Cold, snowy forest (Humid microthermal) climate, dry winter, cool summer; and	
	Dwd	same, but very cold winter	
12 Marine subarctic climate 50°–60°N and 45°–60°S	*ET*	Polar, tundra climate	Located in the arctic frontal zones of the winter season, these windward coasts and islands of subarctic latitudes are dominated by cool mP air masses. Precipitation is relatively large and annual temperature range small for so high a latitude.
13 Tundra climate north of 55°N south of 50°S		Polar, tundra climate	The arctic coastal fringes lie along a frontal zone, in which polar (mP, cP) air masses interact with arctic (*A*) air masses in cyclonic storms. Climate is humid and severely cold with no warm season or summer. Moderating influence of ocean water prevents extreme winter severity as in climate 11.
14 Icecap climate (Greenland, Antarctica)	*EF*	Polar climate, perpetual frost	Source regions of arctic (*A*) and antarctic (*AA*) air masses situated upon the great continental icecaps have climate with annual temperature average far below all other climates and no above-freezing monthly average. High altitudes of ice plateaus intensify air mass cold.
Highland climates			Cool to cold moist climates, occupying high-altitude zones of the world's mountain ranges, are localized in extent and not included in classification system.

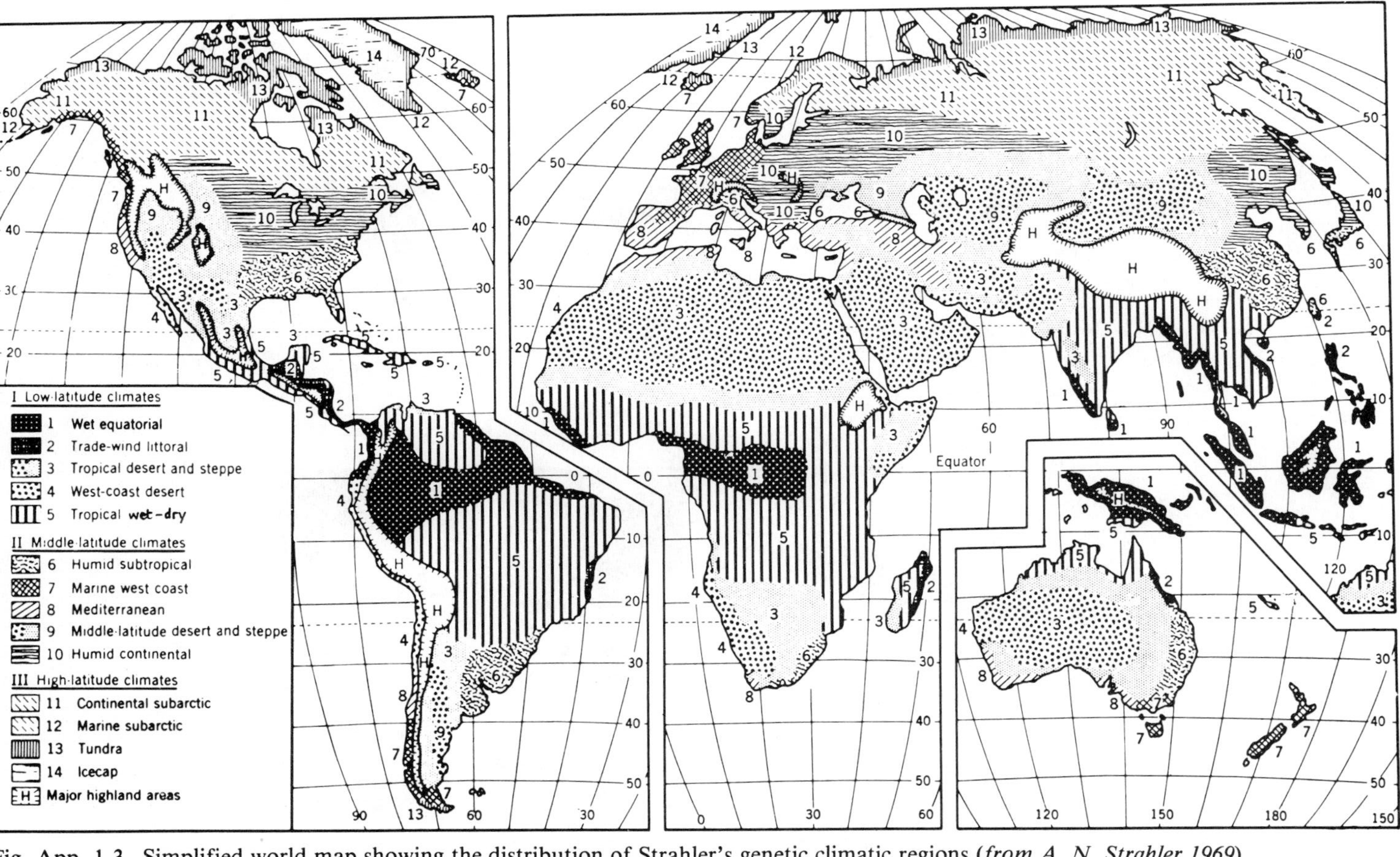

Fig. App. 1.3. Simplified world map showing the distribution of Strahler's genetic climatic regions (*from A. N. Strahler 1969*).

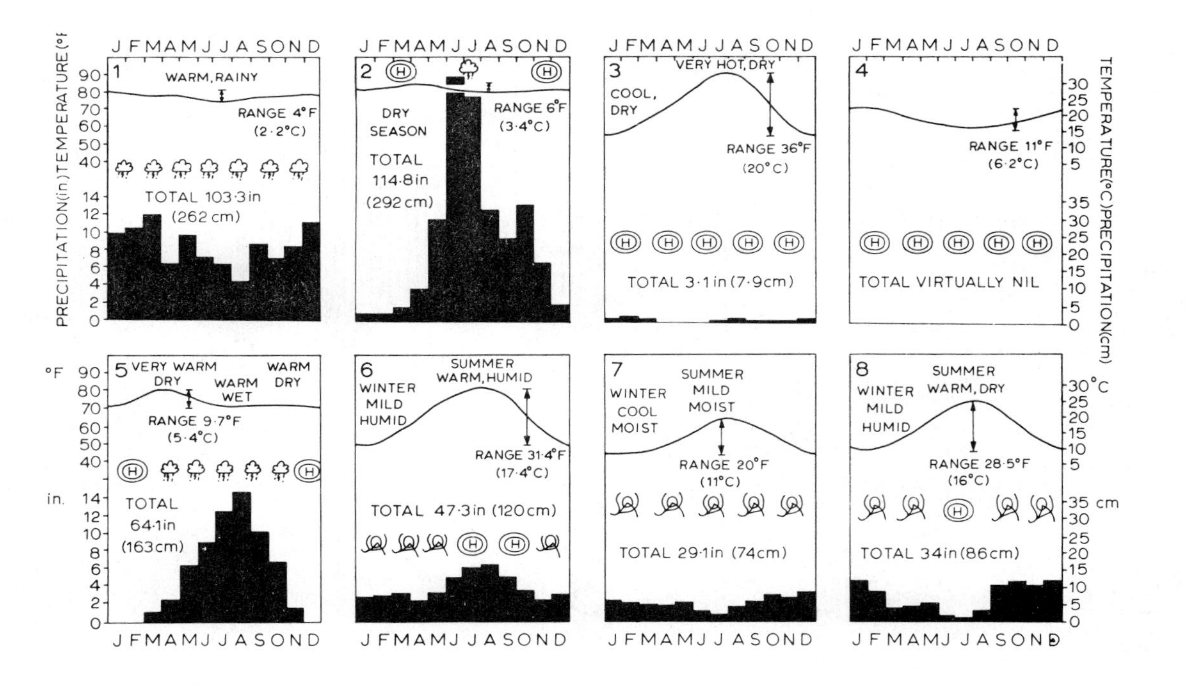
PRECIPITATION(in) TEMPERATURE(°F)
TEMPERATURE(°C) PRECIPITATION(cm)
J F M A M J J A S O N D
1
WARM, RAINY
RANGE 4°F (2·2°C)
TOTAL 103·3 in (262 cm)
2
DRY SEASON
RANGE 6°F (3·4°C)
TOTAL 114·8 in (292 cm)
3
COOL, DRY
VERY HOT, DRY
RANGE 36°F (20°C)
TOTAL 3·1 in (7·9 cm)
4
RANGE 11°F (6·2°C)
TOTAL VIRTUALLY NIL
5
VERY WARM DRY
WARM WET
WARM DRY
RANGE 9·7°F (5·4°C)
TOTAL 64·1 in (163 cm)
6
WINTER MILD HUMID
SUMMER WARM, HUMID
RANGE 31·4°F (17·4°C)
TOTAL 47·3 in (120 cm)
7
WINTER COOL MOIST
SUMMER MILD MOIST
RANGE 20°F (11°C)
TOTAL 29·1 in (74 cm)
8
WINTER MILD HUMID
SUMMER WARM, DRY
RANGE 28·5°F (16°C)
TOTAL 34 in (86 cm)
°F
in.
°C
cm

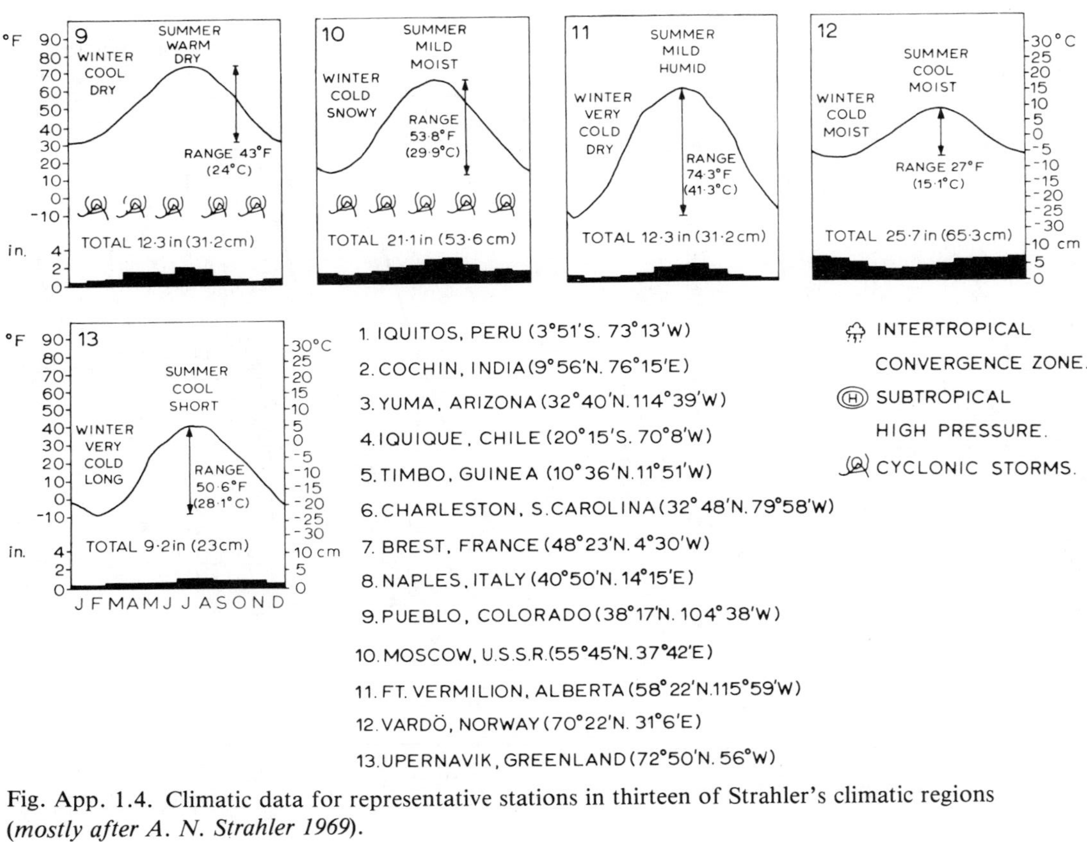

Fig. App. 1.4. Climatic data for representative stations in thirteen of Strahler's climatic regions (*mostly after A. N. Strahler 1969*).

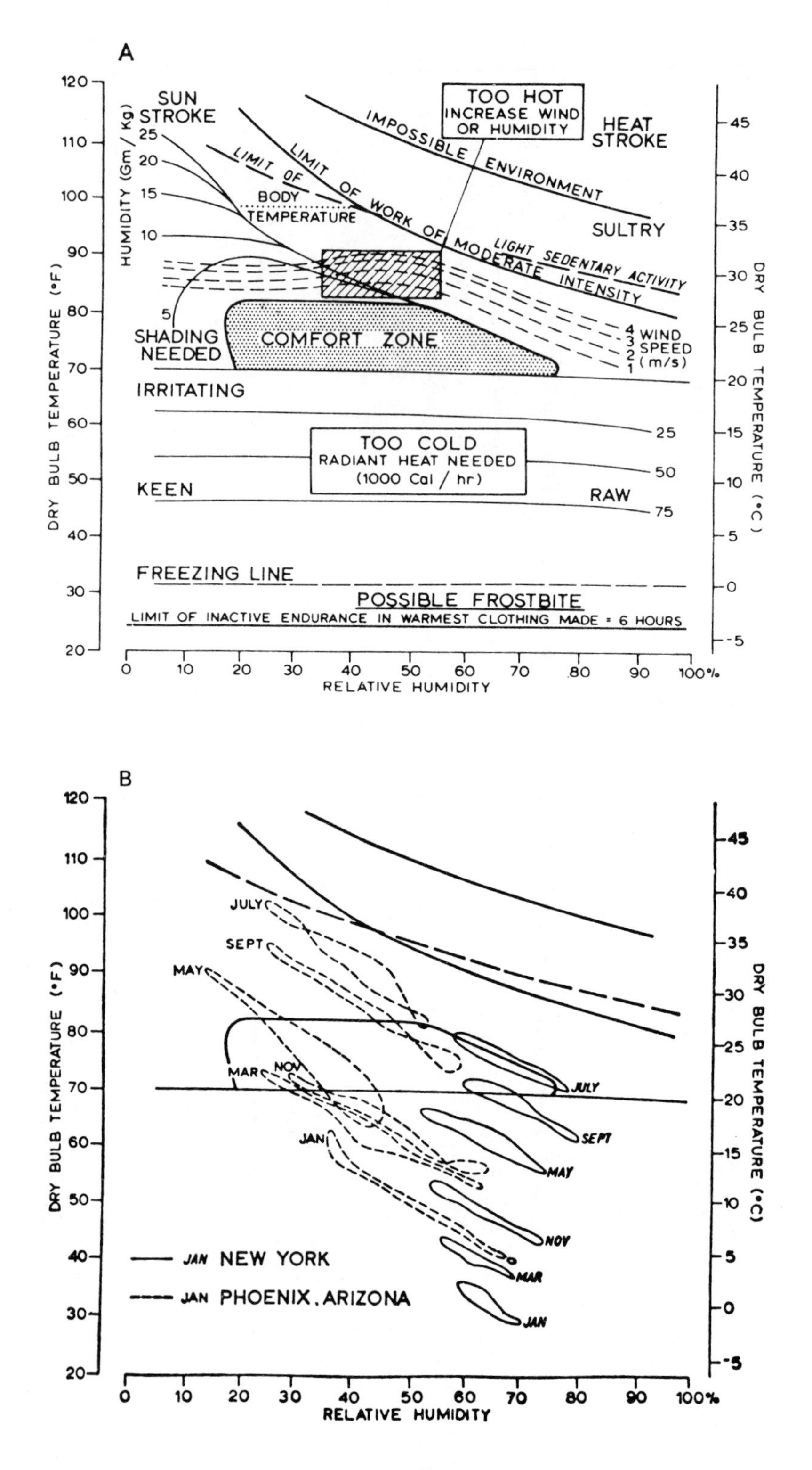
A
SUN STROKE
HUMIDITY (Gm / Kg)
25
20
15
10
5
TOO HOT
INCREASE WIND OR HUMIDITY
HEAT STROKE
IMPOSSIBLE ENVIRONMENT
LIMIT OF WORK OF MODERATE INTENSITY
LIMIT OF
BODY TEMPERATURE
SULTRY
LIGHT SEDENTARY ACTIVITY
SHADING NEEDED
COMFORT ZONE
4 3 2 1 WIND SPEED (m/s)
IRRITATING
TOO COLD
RADIANT HEAT NEEDED
(1000 Cal / hr)
25
50
75
KEEN
RAW
FREEZING LINE
POSSIBLE FROSTBITE
LIMIT OF INACTIVE ENDURANCE IN WARMEST CLOTHING MADE = 6 HOURS
DRY BULB TEMPERATURE (°F)
DRY BULB TEMPERATURE (°C)
RELATIVE HUMIDITY
B
JULY
SEPT
MAY
MAR
NOV
JAN
JAN NEW YORK
JAN PHOENIX, ARIZONA
DRY BULB TEMPERATURE (°F)
DRY BULB TEMPERATURE (°C)
RELATIVE HUMIDITY

His major categories, which are based on the global wind belts and the precipitation characteristics, are as follows:

1 Equatorial westerly zone: constantly wet.
2 Tropical zone, winter trades: summer rainfall.
3 Subtropical dry zone (trades or subtropical high pressure): dry conditions prevail.
4 Subtropical winter-rain zone (Mediterranean type): winter rainfall.
5 Extratropical westerly zone: precipitation throughout the year.
6 Subpolar zone: limited precipitation throughout the year.
6a Boreal, continental subtype: summer rainfall; limited winter snowfall.
7 High polar zone: meagre precipitation; summer rainfall, early winter snowfall.

It will be noted that temperature does not appear explicitly in the scheme. Figure App. 1.1B shows the distribution of these types on a hypothetical continent. Rough general agreement between these types and those of Köppen's scheme is apparent. Note that the boreal sub-type is restricted to the northern hemisphere and that the subtropical zones do not occur on the east side of a land mass. Flohn's approach has much to commend it as an introductory teaching outline. Although no world map of the distribution of these zones has been published, two maps prepared along similar lines by E. Neef and E. Kupfer were presented and discussed by Flohn in 1957. Neef's map is reproduced for reference in fig. App. 1.2.

Another simple, but extremely effective, genetic classification of world climates has been proposed by Strahler. He makes a major tripartite diversion into:

1 Low-latitude climates, controlled by equatorial and tropical air masses.
2 Middle-latitude climates, controlled by both tropical and polar air masses.
3 High-latitude climates, controlled by polar and arctic air masses.

Fig. App. 1.5. The human relevance of climatic ranges (*after Olgyay 1963*).
A Atmospheric comfort, discomfort and danger for inhabitants of temperate climatic zones. Within the stippled limits of dry-bulb temperature and relative humidity the body feels comfortable at elevations of less than 300 m with customary indoor clothing, and performing sedentary or light work. Outside these limits corrective measures are necessary to restore the feeling of comfort. Below the 'comfort zone' radiant heat is needed, above it the skin temperature can be lowered by increasing the evaporation and heat convection either by raising the wind speed (for humid air) or by an increase of atmospheric moisture (for dry air).
B Mean daily dry-bulb temperatures and relative humidities during alternate months for New York City and Phoenix, Arizona, plotted on the 'comfort chart', indicating the need for both central heating and air conditioning.

These are subdivided into fourteen climatic regions, to which is added that of Highland Climates (table App. 1.1). Figure App. 1.3 shows the world distribution of these fifteen regions, and fig. App. 1.4 gives mean monthly climatic data for representatives of thirteen of them.

D Classifications of climatic comfort

Climatic comfort indices have been established mainly by physiological experiments in test chambers. The most significant bioclimatic elements are air temperature, relative humidity, and wind speed (K. J. Buettner). Figure App. 1.5A illustrates the ranges of human comfort, discomfort and danger (heat stroke, frostbite) for one type of bioclimatic chart developed in the United States by V. Olgyay. The frequency of mean climatic conditions outside the *comfort range* in New York and Phoenix is shown in fig. App. 1.5B, indicating the need for supplementary heat or clothing and air conditioning or evaporative cooling systems. Body heat is lost primarily by net radiation (60%) and evaporation from the lungs and skin (25%) in indoor conditions whereas outdoors additional heat is lost convectively due to the wind. At subfreezing temperatures the effective *wind-chill* temperature may be substantially below the air temperature. Thus, a 15 $m\,s^{-1}$ wind with an air temperature of −10°C gives a wind-chill equivalent temperature of −25°C.

A bioclimatic classification which incorporates estimates of comfort using temperature, relative humidity, sunshine and wind-speed data has been proposed for the United States by W. H. Terjung.

Note

1 Note that many American workers use a modified version with 0°C as the C/D boundary.

Bibliography

Bailey, H. P. (1960) A method for determining the temperateness of climate. *Geografiska Annaler* **42**, 1–16.

Basile, R. M. and Corbin, S. W. (1969) A graphical method for determining Thornthwaite climatic classification. *Ann. Assn. Amer. Geog.* **59**, 561–72.

Budyko, M. I. (1956) *The Heat Balance of the Earth's Surface* (trans. by N. I. Stepanova). US Weather Bureau, Washington, DC.

Budyko, M. I. (1974) *Climate and Life* (trans. by D. H. Miller). Academic Press, New York. 508 pp.

Buettner, K. J. (1962) Human aspects of bioclimatological classification; In Tromp, S. W. and Weihe, W. H. (eds.) *Biometeorology*. Pergamon, Oxford and London, pp. 128–40.

Carter, D. B. (1954) Climates of Africa and India according to Thornthwaite's 1948 classification. *Publications in Climatology* **7**(4). Laboratory of Climatology, Centerton, NJ.

Chang, J.-H. (1959) An evaluation of the 1948 Thornthwaite classification. *Ann. Assn. Amer. Geog.* **49**, 24–30.

Crowe, P. R. (1957) Some further thoughts on evapotranspiration: A new estimate. *Geographical Studies* **4**, 56–75.

Flohn, H. (1950) Neue Anschauungen über die allgemeine Zirkulation der Atmosphäre und ihre klimatische Bedeutung. *Erdkunde* **4**, 141–62.

Flohn, H. (1957) Zur Frage der Einteilung der Klimazonen. *Erdkunde* **11**, 161–75.

Gentilli, J. (1958) *A Geography of Climate*. University of Western Australia Press, pp. 120–66.

Gregory, S. (1954) Climatic classification and climatic change. *Erdkunde* **8**, 246–52.

Hare, F. K. (1951) Climatic classification; In Stamp, L. D. and Wooldridge, S. W. (eds.) *London Essays in Geography*. Longmans Green, London, pp. 111–34.

Miller, A. A. (1951) Three new climatic maps. *Trans. Inst. Brit. Geog.* **17**, 13–20.

Olgyay, V. (1963) *Design with Climate: Bioclimatic approach to architectural regionalism*. Princeton University Press, Princeton, NJ. 190 pp.

Oliver, J. E. (1970) A genetic approach to climatic classification. *Ann. Assn. Amer. Geog.* **60**, 615–37. (Commentary, see **61**, 815–20.)

Papadakis, J. (1975) *Climates of the World and their Potentialities*. Buenos Aires. 200 pp.

Shear, J. A. (1966) A set-theoretic view of the Köppen dry climates. *Ann. Assn. Amer. Geog.* **56**, 508–15.

Sibbons, J. L. H. (1962) A contribution to the study of potential evapotranspiration. *Geografiska Annaler* **44**, 279–92.

Strahler, A. N. (1965) *Introduction to Physical Geography*. Wiley, New York. 455 pp.

Strahler, A. N. (1969) *Physical Geography*, 3rd edn. Wiley, New York. 733 pp.

Terjung, W. H. (1966) Physiologic climates of the conterminous United States: A bioclimatological classification based on man. *Ann. Assn. Amer. Geog.* **56**, 141–79.

Terjung, W. H. and Louie, S. S.-F. (1972) Energy input–output climates of the world. *Arch. Met. Geophys. Biokl.* B, **20**, 129–66.

Thornthwaite, C. W. (1933) The climates of the earth. *Geog. Rev.* **23**, 433–40.

Thornthwaite, C. W. (1943) Problems in the classification of climates. *Geog. Rev.* **33**, 233–55.

Thornthwaite, C. W. (1948) An approach towards a rational classification of climate. *Geog. Rev.* **38**, 55–94.

Thornthwaite, C. W. and Hare, F. K. (1955) Climatic classification in forestry. *Unasylva* **9**, 50–9.

Thornthwaite, C. W. and Mather, J. R. (1955) The water balance. *Publications in Climatology* **8**(1). Laboratory of Climatology, Centerton, NJ. 104 pp.

Thornthwaite, C. W. and Mather, J. R. (1957) Instructions and tables for computing potential evapotranspiration and the water balance. *Publications in Climatology* **10**(3). Laboratory of Climatology, Centerton, NJ. 127 pp.

Troll, C. (1958) Climatic seasons and climatic classification. *Oriental Geographer* **2**, 141–65.

APPENDIX 2

Système International (SI) units

Quantity	*Dimensions*	*SI*	*cgs metric*	*British*
length	L	m	10^2 cm	3·2808 ft
area	L^2	m^2	10^4 cm^2	10·7640 ft^2
volume	L^3	m^3	10^6 cm^3	35·3140 ft^3
mass	M	kg	10^3 g	2·2050 lb
density	ML^{-3}	kg m^{-3}	10^{-3} g cm^{-3}	
time	T	s	s	
velocity	LT^{-1}	m s^{-1}	10^2 cm s^{-1}	2·24 mi hr^{-1}
acceleration	LT^{-2}	m s^{-2}	10^2 cm s^{-2}	
force	MLT^{-2}	Newton (kg m s^{-2})	10^5 dynes (10^5g cm^{-1} s^{-2})	
pressure	$ML^{-1}T^{-2}$	N m^{-2} (Pascal)	10^{-2} mb	
energy, work	ML^2T^{-2}	Joule (kg m^2 s^{-2})	10^7 ergs (10^7g cm^2 s^{-2})	
power	ML^2T^{-3}	Watt (kg m^2 s^{-3})	10^7 ergs s^{-1}	$1·340 \times 10^{-3}$ hp
temperature	θ	Kelvin (K)	°C	1·8°F
heat energy	ML^2T^{-2} (or H)	Joule (J)	0·2388 cal	$9·470 \times 10^{-4}$ BTU
heat/radiation flux	HT^{-1}	Watt (W) or J s^{-1}	0·2388 cal s^{-1}	3·412 BTU hr^{-1}
heat flux density	$HL^{-2}T^{-1}$	W m^{-2}	$2·388 \times 10^{-5}$ cal cm^{-2} s^{-1}	

The *basic SI units* are metre, kilogram, second (m, kg, s):

1 m	= 3·2808 feet	1 ft	= 0·3048 m
1 km	= 0·6214 miles	1 mi	= 1·6090 km
1 kg	= 2·2046 lb	1 lb	= 0·4536 kg
1 m s^{-1}	= 2·2400 mi hr^{-1}	1 mi hr^{-1}	= 0·4460 m s^{-1}
1 m^2	= 10·7640 ft^2	1 ft^2	= 0·0929 m^2
1 km^2	= 0·3861 mi^2	1 mi^2	= 2·5900 km^2
1°C	= 1·8°F	1°F	= 0·555°C

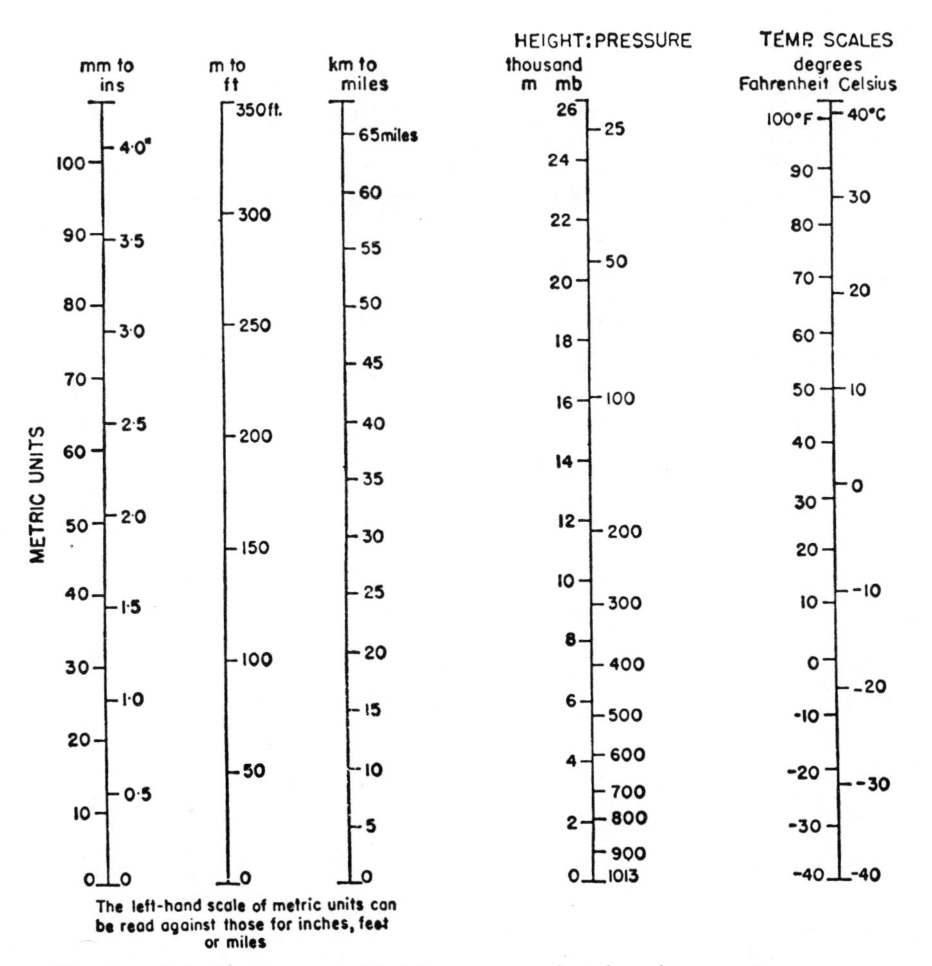

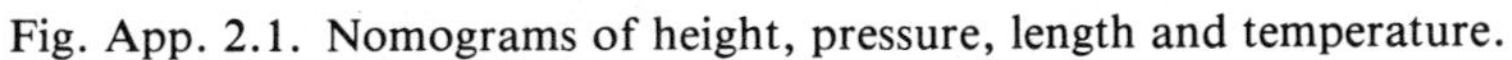

Fig. App. 2.1. Nomograms of height, pressure, length and temperature.

Temperature conversions can be determined by noting that:

$$\frac{T\,(^\circ C)}{5} = \frac{T\,(^\circ F) - 32}{9}$$

Energy conversion factors:

4·1868 J = 1 calorie

J cm^{-2} = 0·2388 cal cm^{-2}

Watt = J s^{-1}

W m^{-2} = $1{\cdot}433 \times 10^{-3}$ cal^{-2} min^{-1}

697·8 W m^{-2} = 1 cal cm^{-2} min^{-1}

For time sums:

Day: $1\ \mathrm{W\ m^{-2}} = 8{\cdot}64\ \mathrm{J\ cm^{-2}\ dy^{-1}} = 2{\cdot}064\ \mathrm{cal\ cm^{-2}\ dy^{-1}}$
Day: $1\ \mathrm{W\ m^{-2}} = 8{\cdot}64 \times 10^{4}\ \mathrm{J\ m^{-2}\ dy^{-1}}$
Month: $1\ \mathrm{W\ m^{-2}} = 2{\cdot}592\ \mathrm{M\ J\ m^{-2}\ (30\ dy)^{-1}} = 61{\cdot}91\ \mathrm{cal\ cm^{-2}\ (30\ dy)^{-1}}$
Year: $1\ \mathrm{W\ m^{-2}} = 31{\cdot}536\ \mathrm{M\ J\ m^{-2}\ yr^{-1}} = 753{\cdot}4\ \mathrm{cal\ cm^{-2}\ yr^{-1}}$

Gravitational acceleration (g) = $9{\cdot}81\ \mathrm{m\ s^{-2}}$

Latent heat of vaporization (288 K) = $2{\cdot}47 \times 10^{6}\ \mathrm{J\ kg^{-1}}$

Latent heat of fusion (273 K) = $3{\cdot}33 \times 10^{5}\ \mathrm{J\ kg^{-1}}$

APPENDIX 3

Synoptic weather maps

The synoptic weather map provides a generalized view of weather conditions over a large area at a given time. The map analysis smooths out local pressure and wind departures from the broad pattern. Such maps are usually prepared at 6- or 12-hourly intervals. Maps are generally prepared for mean sea-level pressure (or of height contours for the 1000 mb pressure surface) and at standard isobaric surfaces – 850, 700, 500, 300 mb, etc. The MSL pressure map typically shows isobars at 4 or 5 mb intervals, surface fronts and weather information.

Weather phenomena shown on the map are as follows:

temperature	type and height of cloud base
dew point	present weather
wind direction	past weather (last 6 hours)
wind speed	pressure tendency
pressure	pressure change (last 3 hours)
cloud amount	visibility

These data are presented in coded or symbolic form for each weather station. The plotting convention ('station model') is illustrated in fig. App. 3.1. The basic weather symbols are illustrated in fig. App. 3.2, and the synoptic code is given in table App. 3.1.

Note: Meteorological Office Leaflet No. 12. HMSO, London, gives titles and prices of meteorological reports, maps, reporting forms and diagrams.

Reference

Stubbs, M. W. (1981) New code for reporting surface observations – an introduction. *Weather* **36**, 357–66.

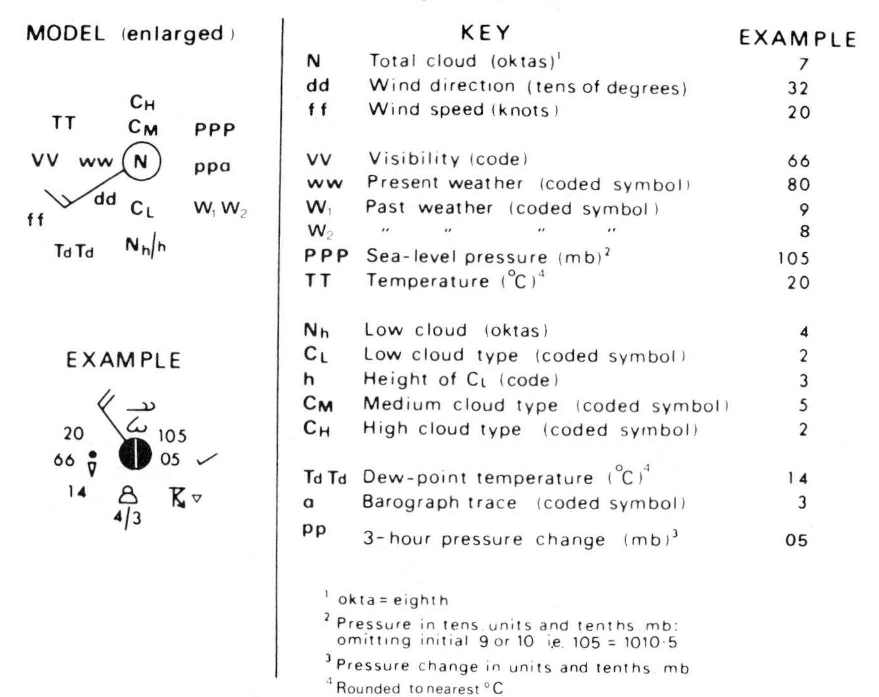

Fig. App. 3.1. Basic station model for plotting weather data. The key and example are tabulated in the internationally agreed sequence for teletype messages. These data would be preceded by an identifying station number, date and time.

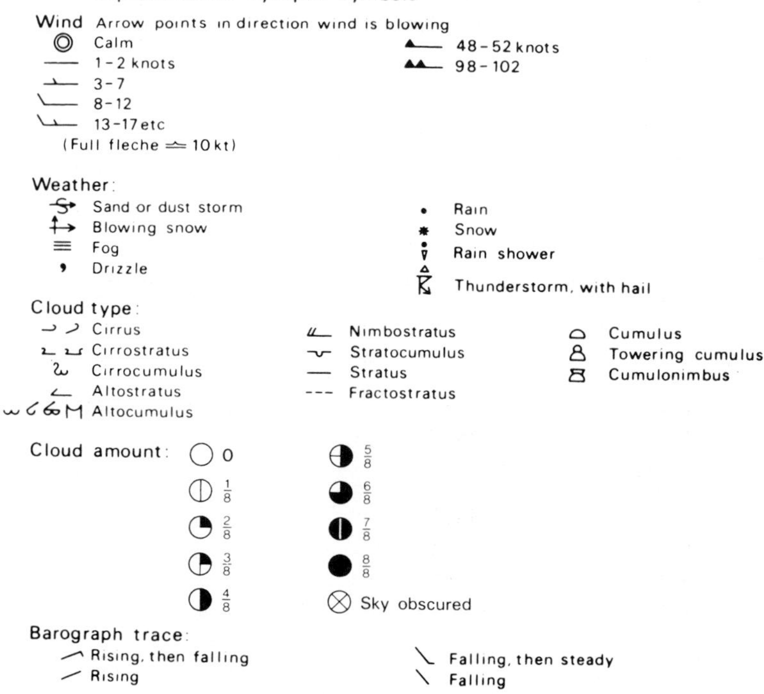

Fig. App. 3.2. Representative synoptic symbols.

Table App. 3.1 Synoptic code (World Meteorological Organization, January 1982)

Symbol	*Key*	*Example*	*Comments*
yy	Day of the month (GMT)	05	All group are in blocks of 5 digits
GG	Time (GMT) to nearest hour	06	
i_w	Indicator for type of wind speed observation and units	4	Measured by anemometer (knots)
IIiii	International index number of station		
i_R	Indicator: precipitation data included/omitted (code)	3	Data omitted
i_X	Indicator: station type + ww W_1W_2 included/omitted (code)	1	Manned station with ww W_1W_2 included
h	Height of lowest cloud (code)	3	
vv	Visibility (code)	66	
N	Total cloud amount (oktas)	7	
dd	Wind direction (tens of degrees)	32	
ff	Wind speed (knots, or m s^{-1})	20	Knots
1	Header	0	
s_n	Sign of temperature (code)	0	Positive value
TTT	Temperature (0·1°C), plotted rounded to nearest 1°C	203	(1 = negative value)
2	Header	2	
s_n	Sign of temperature (code)	0	
$T_dT_dT_d$	Dewpoint temperature (as TTT)	138	
4	Header	4	
PPPP	Mean sea-level pressure (tenths of mb, omitting thousands)	0105	
5	Header	5	
a	Characteristic of pressure tendency (coded symbol)	3	
ppp	3-hour pressure tendency (tenths of mb)	005	
7	Header	7	
ww	Present weather (coded symbol)	80	
W_1	Past weather (coded symbol)	9	(W_1 must be greater than W_2)
W_2	Past weather (coded symbol)	8	
8	Header	8	
N_h	Amount of low cloud (oktas)	4	
C_L	Low cloud type (coded symbol)	2	
C_M	Medium cloud type (coded symbol)	5	
C_H	High cloud type (coded symbol)	2	

Note: Group 3 is for a report of surface pressure and group 6 for precipitation data.

APPENDIX 4

Data Sources

A Daily weather maps and data

Western Europe/North Atlantic: *Daily Weather Summary* (synoptic chart, data for the UK). London Weather Centre, 284 High Holborn, London WC1V 7HX, England.

Western Europe/North Atlantic: *Monthly Weather Report* (published about 15 months in arrears; tables for approximately 600 stations in the UK). London Weather Centre.

Europe – eastern/North Atlantic: *European Daily Weather Report* (synoptic chart). Deutsche Wetterdienst, Zentralamt D6050 Offenbach, Federal Republic of Germany.

Europe – eastern/North Atlantic: *Weather Log* (daily synoptic chart, supplement to *Weather* magazine). Royal Meteorological Society, Bracknell, Berkshire, England.

North America: *Daily Weather Reports* (weekly publication). National Environmental Satellite Data and Information Service, NOAA, US Government Printing Office, Washington, DC 20402, USA.

B Satellite data

NOAA operational satellites (imagery, digital data). Satellite Data Services Division, NOAA/NESDIS, World Weather Building, Washington, DC 20023, USA.

Defense Meteorological Satellite Program (imagery). National Snow and Ice Data Center, University of Colorado, Boulder, Co. 80309-0449, USA.

Metsat (imagery, digital data). ESOC, Robert-Bosil Str. 5, D-6100 Darmstadt, Federal Republic of Germany.

NASA research satellites (digital data). National Space Science Data Center, Goddard Space Flight Center, Greenbelt, Md. 20771, USA.

C Climatic data

Canadian Climate Center: *Climatic Perspectives* (197 weekly and monthly summary charts). Atmospheric Environment Service, 4905 Dufferin Street, Downsview, Ontario, Canada M3H 5T4.

National Climatic Center, NOAA/NESDIS: *Local Climatological Data* (1948–) (monthly tabulations, charts); *Monthly Climatic Data for the*

World (May 1948–). National Climatic Data Center, Federal Building, Asheville, NC 28801, USA.

Climate Analysis Center, NOAA/NESDIS: *Climatic Diagnostics Bulletin* (1983–) (monthly summaries of selected diagnostic product from NMC analyses). Climate Analysis Center, NMC NOAA/NWS, World Weather Building, Washington, DC 20233, USA.

Climatic Research Unit, University of East Anglia: *Climate Monitor* (1976–) (monthly summaries, global and UK). Climatic Research Unit, University of East Anglia, Norwich N34 7TJ, England.

World Climate Data Programme: *Climate System Monitoring Bulletin* (1984–) (monthly). World Climate Data Programme, WMO Secretariat, CP5, Geneva 20 CH-1211, Switzerland.

World Meteorological Centre, Melbourne: *Climate Monitoring Bulletin for the Southern Hemisphere* (1986–). Bureau of Meteorology, GPO Box 1289 K, Melbourne, Victoria 3001, Australia.

Bibliography

European Space Agency (1978) *Introduction to the Med-sat System*. European Space Operations Center, Darmstadt. 54 pp.

European Space Agency (1981) *Atlas of Meteosat Imagery. Atlas Metosat*. ESA-SP-1030. ESTEC, Nordwijk, Netherlands. 494 pp.

Finger, F. G., Laver, J. D., Bergman, K. H. and Patterson, V. L. (1985) The Climate Analysis Center's user information service. *Bull. Amer. Met. Soc.* **66**, 413–20.

Jenne, R. L. and McKee, T. B. (1985) Data; In Houghton, D. D. (ed.) *Handbook of Applied Meteorology*. Wiley, New York, pp. 1175–281.

Meteorological Office (1958) *Tables of Temperature, Relative Humidity and Precipitation for the World*. HMSO, London.

Singleton, F. (1985) Weather data for schools. *Weather* **40**, 310–13.

US Department of Commerce (1983) *NOAA Satellite Programs Briefing*. National Oceanic and Atmospheric Administration, Washington, DC. 203 pp.

US Department of Commerce (1984) *North American Climatic Data Catalog. Part 1.* National Environmental Data Referral Service, Publication NEDRES-1, National Oceanic and Atmospheric Administration, Washington, DC. 614 pp.

World Meteorological Organization (1965) *Catalogue of Meteorological Data for Research*. WMO No. 174. TP-86. World Meteorological Organization, Geneva.

Problems

Chapter 1

1 The solar energy received at the top of the atmosphere (S) is proportional to $1/D^2$ where D is the solar distance. For mean solar distance ($149 \cdot 5 \times 10^6$ km), $S = 1 \cdot 35$ kW m^{-2}. What are the amounts for maximum distance (152×10^6 km) and minimum distance (147×10^6 km)?

2 Calculate the noontime solar radiation received on a horizontal surface at the top of the atmosphere (ignoring variations in solar distance) at latitudes 0°, $23\frac{1}{2}$°, 45°, $66\frac{1}{2}$°, 90°N for the following dates: 22 December, 20 March and 21 June, and graph your results. The noontime solar elevation angles are:

	22 December	*20 March*	*21 June*
0°	$66\frac{1}{2}$°	90°	$66\frac{1}{2}$°
$23\frac{1}{2}$°	43°	$66\frac{1}{2}$°	90°
45°	$21\frac{1}{2}$°	45°	$68\frac{1}{2}$°
$66\frac{1}{2}$°	0°	$23\frac{1}{2}$°	47°
90°N	—	0°	$23\frac{1}{2}$°

3 How does terrestrial radiation differ from solar radiation? Explain the physical basis of this difference.

4 Determine the radiation emitted from black bodies with temperatures of 6000 K and 300 K, respectively. The Stefan–Boltzmann constant is, $\sigma = 5 \cdot 67 \times 10^{-8}$ W m^{-2} K^{-4}.

5 Show that the effective planetary temperature is approximately 255 K using Stefan's law and the estimated emitted infrared radiation (from fig. 1.27).

6 What is the role of (*a*) ozone, (*b*) carbon dioxide, (*c*) dust particles, (*d*) water vapour, in the earth's radiation budget?

7 Assess the importance of cloud cover as a factor determining climate conditions at the earth's surface. (Note the typical range of cloud albedo, cloud amounts and surface albedos.)

8 What sections of the electromagnetic spectrum would be suitable for determining the following from a satellite?

(a) surface and cloud-top temperature;

(b) night-time cloud cover;
(c) planetary albedo;
(d) stratospheric temperatures.
(Refer to fig. 1.8.)

9 What is the basis for the division of the vertical structure of the atmosphere? Explain the zones of temperature increase in the upper atmosphere.

Chapter 2

1 Why does cooling eventually cause a mass of air to reach its saturation point? Describe the cooling processes which may result in cloud formation.
2 Mean daily evaporation over the globe is about 2·5 mm. What percentage of average incoming solar radiation absorbed at the surface does this represent?
3 Air at 10°C with a relative humidity of 50% is cooled at constant pressure. At approximately what temperature will it reach its dew point? (Use fig. 1.7.)
4 If the temperature of an air parcel at 1000 mb is 20°C, what is its temperature at 700 mb following (*a*) unsaturated ascent, (*b*) saturated ascent? (Use fig. 2.10.)
5 By how much is an air parcel warmed if it ascends, saturated, from the 1000 mb level at 10°C to 800 mb and descends unsaturated to 1000 mb?
6 Discuss the physical conditions that make 'cloud seeding' possible.
7 Explain the different types of cloud pattern visible on satellite imagery. How are the various patterns related to weather systems?
8 What methods would you use to describe the areal variation of magnitude and frequency of rainstorms?
9 How do fogs form and in what geographical regions are they common?
10 Discuss the definition of drought. Which regions of the world are most susceptible to drought?
11 Determine the annual moisture regime for a station in your area by the Penman and Thornthwaite methods (see the references to ch. 2: Pearl *et al.* 1954, and App. 1: Thornthwaite and Mather, 1957).
12 Consider what physical processes serve to limit extreme rainfall amounts? Some new records are described in *Weather* **39** (1984), p. 12; compare these with the graph of expected extremes. (Use fig. 2.24).

Chapter 3

1 Determine the balance of forces for cyclonic and anticyclonic gradient wind flow in the southern hemisphere. Compare these with the geostrophic wind case.

2 Calculate the geostrophic wind speed ($m\,s^{-1}$) at latitudes 20° and 43°N for a pressure gradient of 1·5 mb/100 km. Assume air density $\varrho = 1 \cdot 2\,kg\,m^{-3}$. (Note 1 mb = $100\,kg\,m^{-1}\,s^{-2}$.)
3 Using figs. 1.23 and 3.24 plot a graph for 40°N of latitudinal temperature departure and the sign of the meridional MSL wind component in January. (Take winds of SE-SW as positive, NW-NE as negative; W or E winds = zero.) Note the correlation between the sign of the temperature anomaly and the meridional component of the wind.
4 Explain the locations of the major centres of low and high pressure in fig. 3.24.
5 Using figs. 1.23 and 3.24, determine the direction of the thermal wind component in July over (*a*) 35°N, 20°W, (*b*) 30°N, 100°W.
6 Explain the location of the major desert areas of the world.
7 Under what circumstances may local climatic influences be more important than large-scale controls?
8 Outline the characteristics of jet streams in the upper troposphere. What are their relationships with surface weather and climate?
9 Describe the role of the tropics in the general circulation of the atmosphere.
10 What is the zonal index and how is it related to characteristics of the circulation in middle latitudes?

Chapter 4

1 What are the three primary factors determining air-mass weather?
2 Determine the source regions from which air masses are likely to affect your home area in summer and winter. Outline the weather conditions likely to be associated with them following air-mass modifications en route.
3 What weather conditions are typically associated with a noontime tropical air mass moving northward (*a*) over sea (*b*) over land?
4 Explain the relationship between frontal zones and upper tropospheric jet streams. (Consider a vertical section and a plan view of a frontal cyclone.)
5 Explain how a low pressure system may deepen or fill.
6 Compare figs. 4.18 and 3.19A and comment on the features identified.
7 Describe the types of *non*-frontal low pressure system and explain their occurrence.
8 What are the bases of methods of short-range and long-range weather forecasts?
9 Maintain a daily log of forecast weather conditions for your location (based on newspaper, TV or radio broadcast information) and tabulate the actual temperature, wind and weather conditions that occurred. In what synoptic situations and seasons are the forecasts more/less reliable?

Chapter 5

1 Examine figs. 1.23, 3.24 and 3.38 in relation to winter and summer temperature conditions in north-western Europe, eastern Siberia and north-eastern Canada. What principal factors are operative in each region?
2 What is meant by 'continentality' and what factors determine it?
3 List the major influences of the large-scale orography on climate in western North America.
4 Discuss the circumstances that give rise to long spells of a particular type of weather over Europe.
5 Using daily weather maps and daily precipitation records for your locality, determine the proportion of the precipitation occurring over a winter and summer season with frontal and non-frontal situations and different air masses.
6 Select cases of strong zonal flow and of blocking for Scandinavia or Alaska from daily weather maps and analyse the patterns of temperature and precipitation that result. If upper air charts are available, compare also the jet-stream patterns.

Chapter 6

1 What are the major differences between weather systems in the tropics and in middle latitudes?
2 Why are hurricanes absent from the South Atlantic and the eastern South Pacific?
3 What are the equatorial westerlies and what is their climatological significance?
4 What effects does the trade wind inversion have on tropical weather and climate?
5 Examine the role of synoptic systems in shaping the character of the monsoon regime of southern Asia.
6 In what respects is the monsoon regime of West Africa similar to that of southern Asia?
7 What are the most important local climatic influences in tropical regions?

Chapter 7

1 What are the main determinants of urban heat islands? Which factor is considered to be most important in mid-latitude cities in (*a*) winter; (*b*) summer?
2 What effects to differences in surface type (urban area, forest, lake, etc.) have on climatic parameters?
3 Enumerate meteorological considerations which could mitigate some of the undesirable features of urban climate if taken into account in city planning?

4 Discuss the view that a city has many small-scale climates.
5 What are the principal microclimatic effects of (*a*) shelter belts (*b*) forest clearings?
6 Topics for group topoclimatic investigations:

> Spatial and temporal features of fog occurrence; climatic measurements (wind speed, temperature, light intensity, etc.) inside and outside a forested area; comparison of the duration of sunshine and of the diurnal course of temperature on slopes of northerly and southerly aspect.

7 Analyse the monthly frequency of light winds (less than 2 m s^{-1}) in your area. Use hourly measurements if possible. Compile comparative data on fog frequency and air quality if appropriate. (A source of data on air quality is the *Journal of the Air Pollution Control Association*.)

Chapter 8

1 Plot frequency distributions of annual precipitation totals using data for a station in your locality and a contrasting climatic regime. (Use at least 30 years of data and not more than eight classes.) Compare with fig. 8.1. Determine appropriate averages and measures of variability.
2 For corresponding 30-year series of temperature data (such as mean daily temperature for January and July) determine arithmetic means and standard deviations.
3 Enumerate the various terrestrial and extraterrestrial factors which may be involved in global climatic change and consider the time scales over which each is likely to be significant. Discuss the processes which are involved in their climatic effects.
4 Using some long-term temperature and precipitation records, compare the trends since the late nineteenth century in high, middle and low latitudes. Graph the values by individual years and as 10-year running means.

Solutions to problems

Chapter 1

1. 1·40 and 1·31 kW m^{-2}

2.

Lat.	*22 December*	*20 March*	*21 June*
0°	1·26	1·37	1·26 kW m^{-2}
$23\frac{1}{2}$°	0·93	1·26	1·37
45°	0·50	0·97	1·27
$66\frac{1}{2}$°	0·0	0·55	1·00
90°N	0·0	0·0	0·55

4. 73×10^3 kW m^{-2}; 459·3 kW m^{-2}

Chapter 2

2. 20·6%
3. 0°C
4. (a) −8·5°C, (b) 6°C
5. 8°C

Chapter 3

2. (a) 25 m s^{-1}
   (b) 12·5 m s^{-1}
5. (a) Westerly
   (b) Southerly

Geographical index

Subject index